# INTRAOPERATIVE CLINICAL NEUROPHYSIOLOGY

# INTRAOPERATIVE CLINICAL NEUROPHYSIOLOGY

## A Comprehensive Guide to Monitoring and Mapping

*Edited by*

**Mirela V. Simon, MD**

*Director, Intraoperative Monitoring Unit*
*Massachusetts General Hospital*
*Department of Neurology*
*Harvard Medical School*
*Boston, Massachusetts*

demosMEDICAL

NEW YORK

*Acquisitions Editor:* Beth Barry
*Cover Design:* Steve Pisano
*Compositor:* Publication Services, Inc.
*Printer:* Sheridan Press

Visit our website at www.demosmedpub.com



Medicine is an ever-changing science. Research and clinical experience are continually expanding our knowledge, in particular our understanding of proper treatment and drug therapy. The authors, editors, and publisher have made every effort to ensure that all information in this book is in accordance with the state of knowledge at the time of production of the book. Nevertheless, the authors, editors, and publisher are not responsible for errors or omissions or for any consequences from application of the information in this book and make no warranty, express or implied, with respect to the contents of the publication. Every reader should examine carefully the package inserts accompanying each drug and should carefully check whether the dosage schedules mentioned therein or the contraindications stated by the manufacturer differ from the statements made in this book. Such examination is particularly important with drugs that are either rarely used or have been newly released on the market.

Library of Congress Cataloging-in-Publication Data
Intraoperative clinical neurophysiology / edited by Mirela V. Simon.
p. ; cm.
Includes bibliographical references and index.
ISBN 978-1-933864-46-4
1. Neurophysiologic monitoring. I. Simon, Mirela V.
[DNLM: 1. Monitoring, Intraoperative--methods. 2. Neurosurgical Procedures--adverse effects. 3. Diagnostic Techniques, Neurological. 4. Intraoperative Complications--prevention & control. WL 368 I613 2010]
RD52.N48I555 2010
616.85'2--dc22

2009040425

Special discounts on bulk quantities of Demos Medical Publishing books are available to corporations, professional associations, pharmaceutical companies, health care organizations, and other qualifying groups. For details, please contact:

Special Sales Department
Demos Medical Publishing
11W. 42nd Street, 15th Floor
New York, NY 10036
Phone: 800–532–8663 or 212–683–0072
Fax: 212–941–7842
Email: rsantana@demosmedpub.com

Made in the United States of America
09 10 11 12 13 5 4 3 2 1

*To my son, Siddharth*

# CONTENTS

# PREFACE

Rapid developments in neuroscience, especially within the past two decades, have fueled the growth of intraoperative neurophysiologic monitoring and mapping, also known as IOM or IONM, as a new and exciting field. By offering early detection of reversible neurophysiologic dysfunction during surgery, neuro-monitoring allows prompt intervention and prevents the occurrence of permanent neurological damage. Similarly, neuromapping allows identification with subsequent preservation of functionally viable nervous tissue. Thus, the most important role of intraoperative neurophysiology is the promise of improving neurological outcome after surgery. It also provides prognostic information about clinical outcomes and allows determination of the time when and mechanism through which an injury occurred. This helps the surgeon to identify those surgical maneuvers or instruments that introduce additional risks. However, there is much still to be learned and discovered, making this field not only of immediate clinical interest but also a tremendous resource for research.

Until recently, the neurophysiology fellowship programs throughout the country did not offer special training in intraoperative neurophysiology. Rather, fellows graduating from such programs pursued careers either as epileptologists, neuromuscular neurologists, or sleep specialists. I was no exception: after my two years of EEG/epilepsy fellowship at Massachusetts General Hospital, I was a trained epileptologist with a keen interest in intraoperative neurophysiology, but unfortunately I did not have many theoretical or practical materials at hand to help me build easily the necessary foundation. Years later, I sat down to write the book that I wished I'd had as an aspiring IOM neurophysiologist.

This book is thus geared towards neurologists, particularly those with formal training in clinical neurophysiology, who would like to learn the ins and outs of intraoperative neurophysiology. I am confident that one can master this field if equipped with solid knowledge of anatomy and physiology of the central and peripheral nervous systems, and of each neurophysiological test used in the operating room—while simultaneously understanding their particularities in the operating theater environment—as well as of the key moments of the surgical procedures that require monitoring. This book tries to accomplish the difficult task of delivering these elements. It also offers a comprehensive algorithm of troubleshooting in the process of recognizing clinically significant neurophysiologic changes and provides a step-by-step approach to managing these changes.

On a more personal level, I wanted to share my take, as an epileptologist, on how useful EEG recordings can be in intraoperative neurophysiology. I do believe that this test has been underused in the OR, and thus, over the span of three chapters, we have tried to emphasize and promote its irreplaceable utility.

I hope that this book, written by two neurologists and 11 surgeons, will also convince other surgeons to employ new resources in their practice, so as to further reduce the operative risk of neurological damage. Finally, I believe that IOM technologists will find it useful and inspiring.

Mirela V. Simon

# ACKNOWLEDGMENTS

I would like to acknowledge all those whose help made this book possible.

I would like to begin by thanking Kathy Carlson, for her infectious enthusiasm and support in easing the editing process. The IOM team at Massachusetts General Hospital: Margaret Barlow, Michelle Brennan, Marcia Mulligan, Michelle Rouke, Kelley Salvagno, Jennifer Tower, Aaron Tripp, Adasha Turner and Ling Wang whose hard work and perseverance, whether in early hours of morning or late at night, constantly propelled us forward as a division and as a team. I would especially like to thank Margaret Barlow, for her dedication and leadership. And to Aaron Tripp for his help in selecting the most appropriate case examples that are used throughout the book. I would like to thank my colleague, Ronan Kilbride, for his advice and important contributions to our work as a team.

I would like to thank Lynn Santos for her great illustrations and helping me deliver the manuscript on time.

To all my colleagues epileptologists and neurophysiologists at MGH, from whom I never stopped learning over the years.

I would like to thank all the collaborating authors for their invaluable and prompt contributions. I am especially grateful to Keith Chiappa and Emad Eskandar for their thoughtful insights and without whose guidance this book endeavor may have faltered seriously. I would like to thank my Chairperson, Anne Young, for constantly inspiring me as a scientist, leader and person, for her support in all my professional endeavors.

To Beth Barry, my editor, for her valuable advice and patience throughout this process. I would also like to thank my friend, Irena for being such a graceful listener of my daily detailed reports on the progress of this book.

In the end, I would like to thank my family, who have been and will always be the core of my Universe. To my parents, who inculcated in me since an early age the value of hard work, for their unflinching support and faith in me. To my best friend, and husband, Sandeep for his tremendous help with correcting my English and cooking the most delicious dinners and for his unmatchable and most reassuring smiles. And to my eight year old son, Siddharth, who at this tender age, showed extraordinary patience to "let mommy finish her book" before he could go skiing again. And who was the first to openly acknowledge his great faith in the fruits of this labor.

# CONTRIBUTORS

**Wael F. Asaad, MD, PhD**
Clinical Fellow in Surgery
Department of Neurosurgery
Harvard Medical School
Massachusetts General Hospital
Boston, Massachusetts

**Lawrence F. Borges, MD**
Associate Professor
Director of Neurological Spine Center
Department of Neurosurgery
Harvard Medical School
Massachusetts General Hospital
Boston, Massachusetts

**Keith H. Chiappa, MD**
Associate Professor
Director, EEG/EP Laboratory
Department of Neurology
Harvard Medical School
Massachusetts General Hospital
Boston, Massachusetts

**Mark F. Conrad, MD**
Instructor in Surgery
Department of Vascular and Endovascular Surgery
Harvard Medical School
Massachusetts General Hospital
Boston, Massachusetts

**Jason S. Cooper, MD**
Department of Plastic and Reconstructive Surgery
Harvard Medical School
Massachusetts General Hospital
Boston, Massachusetts

**Jean-Valerie C.E. Coumans, MD**
Assistant Professor of Neurosurgery
Department of Neurosurgery
Harvard Medical School
Massachusetts General Hospital
Boston, Massachusetts

**Emad N. Eskandar, MD**
Associate Professor of Neurosurgery
Director Stereotactic and Functional Neurosurgery
Department of Neurosurgery
Harvard Medical School
Massachusetts General Hospital
Boston, Massachusetts

**Jason L. Gerrard, MD, PhD**
Department of Neurosurgery
Harvard Medical School
Massachusetts General Hospital
Boston, Massachusetts

**Brian E. Grottkau, MD**
Assistant Professor of Orthapaedic Surgery
Chief of Pediatric Orthopaedics
Department of Orthopaedic Surgery
Harvard Medical School
Massachusetts General Hospital
Boston, Massachusetts

**Erin S. Hart, RN, CPNP**
Department of Orthopaedic Surgery
Harvard Medical School
Massachusetts General Hospital
Boston, Massachusetts

**Michael A. Peck, MD**
Clinical Fellow in Surgery
Department of Vascular and Endovascular Surgery
Harvard Medical School
Massachusetts General Hospital
Boston, Massachusetts

**Mirela V. Simon, MD**
Director, Intraoperative Monitoring Unit
Massachusetts General Hospital
Department of Neurology
Harvard Medical School
Boston, Massachusetts

**Donald C. Shields, MD, PhD**
Department of Neurosurgery
Harvard Medical School
Massachusetts General Hospital
Boston, Massachusetts

**Jonathan M. Winograd, MD**
Assistant Professor of Surgery
Department of Plastic and Reconstructive Surgery
Harvard Medical School
Massachusetts General Hospital
Boston, Massachusetts

# INTRAOPERATIVE CLINICAL NEUROPHYSIOLOGY

# 1

# NEUROPHYSIOLOGIC TESTS IN THE OPERATING ROOM

MIRELA V. SIMON



## I. INTRODUCTION

The success of intraoperative neurophysiology relies in a major part on the ability of the neurophysiologist to use appropriately all of the tests available at hand in the outpatient laboratory and adapt them to the specific environment of the operating theater and, as needed, for different surgical procedures. In my view, this goal cannot be fully achieved without a good understanding of the principles and methodology of each neurophysiologic technique.

This chapter provides an overview of the tests that constitutes the tools in intraoperative neurophysiology. It is not meant as an approach to their use for clinical diagnosis in the outpatient lab or for dealing with abnormal results or troubleshooting. The latter is an essential step in the complex algorithms of interpreting diverse intraoperative monitoring and mapping (IOM) procedures that is further detailed in the corresponding chapters. Instead, this chapter gives a practical presentation of each neurophysiologic method regarding the following:

1. The principle on which it is based
2. The methodology for stimulation and recording
3. The normal electrical potentials to be recorded their correlation with the neuroanatomy and their main roles in IOM
4. The applications of each method in different types of surgery

## II. ELECTROENCEPHALOGRAM (EEG)

Despite the exclusive reference to the term EEG, this subchapter applies in a significant way to electrocorticography (ECoG) as well, because the latter is, in fact, an EEG that is recorded directly from the cortical surface. Thus, with the exception of its technical setup, which is extensively described on pages 115–17, the following principles, recording technique, and consideration regarding interpretation apply to both EEG and ECoG. Moreover, detailed examples of such recordings, troubleshooting, and interpretation of findings can be found in Chapters **2**, **3**, and **4**, respectively.

## 1. Principle

The EEG recordings reflect the extracellular or field potentials generated by the intracellular and extracellular electrical activities. The neuron has a transmembrane potential of 60–70 mV, with the negative polarity in the intracellular space. This potential is under the influence of the synaptic connections of its cell body with other neurons. If the synapse is excitatory, the action potentials arriving from a different cell generate an excitatory postsynaptic potential (EPSP) in the neuron. This potential starts the process of depolarizing the cellular membrane. If this process repeats several times within a relatively short period of time, several EPSPs generated in the neuron reach a depolarization threshold, generating an action potential. Similarly, if the synaptic connection is inhibitory, the action potential arriving to the neuron causes an inhibitory postsynaptic potential (IPSP), which leads to hyperpolarization of the neuronal membrane (Hubbard et al., 1969; Shepherd, 1974). The dynamic fluctuations of the postsynaptic potentials are considered by many authors (Creutzfeldt and Houchin, 1974; Speckmann et al., 1984) as being essential for the generation of the extracellular field potentials. The question arises as to why other types of potentials, such as action potentials, do not contribute to the EEG signals. According to Lopes da Silva and Van Rotterdam (1999), there are at least two explanations. First, these action potentials, by depolarizing a very small piece of neuronal membrane at a given instant in time, generate a small field that relies on a single small dipole, perpendicular on the membrane. This is in contrast to the field generated by the postsynaptic potentials, which extend over a larger portion of the membrane, and thus the field generated corresponds to a layer of dipoles, perpendicular on the membrane. This field attenuates with distance less rapidly than the field generated by the action potentials. Second, the duration of an action potential is much shorter (1–2 msec) than the duration of a postsynaptic potential (~10–250 msec). Thus, action potentials overlap much less than postsynaptic potentials. Because EEG on the scalp is considered to be the result of field potentials occurring at distance (from the recording sites) and concomitantly in different populations of neurons, it becomes evident why action potentials do not contribute to EEG signals. Moreover, the latter reflect the dynamics of the electrical activity in such populations characterized by an appropriate spatial organization and directly rely on the capacity of the neurons to work in synchrony through connectivity and networking (Lopes da Silva, 1999). Even though the discussion has revolved around the role of neurons in generation of the field potentials, one has to acknowledge that the electrical activity within and around the glial cells significantly influences these processes in a similar manner. Furthermore, these potentials are recorded via scalp electrodes and compared to each other, amplifying differences and providing a map of electrical activity of the brain.

## 2. Methodology

### *A. Type of Recording Electrodes*

Conductive electrodes are attached to the skin, after its preparation abrasive gel (NUPREP) and application of collodion. In order to record differences in electrical potentials accurately, few conditions have to be met. First, the electrodes used have to be nonpolarizable, reversible, or also called inert (e.g. Ag AgCl electrode). This is essential in order to allow a steady current flow between the skin, where a DC voltage exists (as determined by a distant field potential), through the electrolyte jelly and toward the electrode. If the recording electrode is polarizable, the flow occurs only in the beginning and then, as the electrode polarizes, rapidly ceases. We use 10-mm gold-plated disc reusable electrodes (XLTEK). Second, the resistance to flow, also called impedance and representing the sum of the resistance of the body fluid, skin, and electrode jelly, has to be relatively small. This is important so that the voltage input delivered to the amplifier represents, as accurately as possible, the DC voltage to be measured. An impedance of 5 kOhms or less is the goal. Third, it is important that the impedances at all contacts are relatively equal. Otherwise, false differences in the voltages to be compared can be introduced in the amplifier.

### *B. Recording Sites and Channels*

In order to give a spatial representation of the field potentials over the scalp, the EEG needs to record simultaneously from different electrodes placed in different regions over the scalp, thus giving an adequate coverage of all parts of the head. The positions of the electrodes are standard, as determined by the 10–20 international electrode system (Fig. **1-1**). We strongly recommend use of at least 16 electrodes; insufficient channels significantly decrease the sensitivity of the EEG recordings in picking up abnormalities. Positions of the electrodes should always be determined by exact measurements from standard landmarks on the skull (nasion, inion, and the preauricular regions). Designations of the positions are in terms of

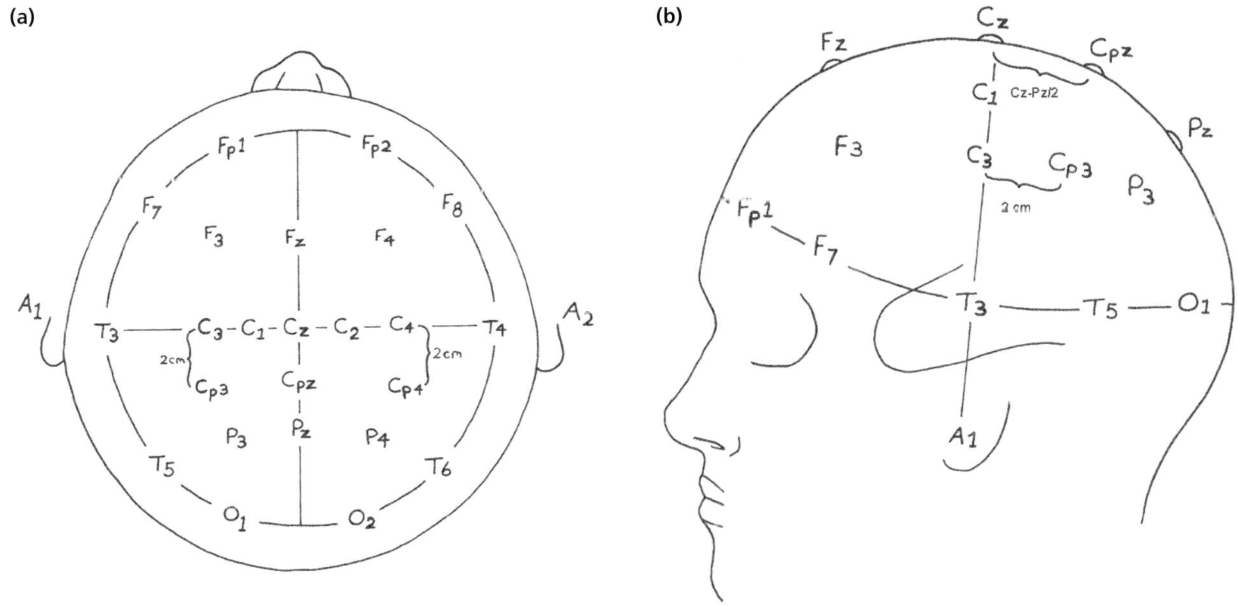


**FIGURE 1-1** The 10–20 International Electrode System: **a:** axial view; **b:** sagittal view. Notice the position of C1 and C2 electrodes, situated half distance between C3-Cz and C4-Cz, respectively. Also, note the position of Cp3 and Cp4, situated 2 cm posterior to C3 and C4, respectively, in the regions corresponding to the somatosensory parietal cortex. Cpz is located half distance between Cz and Pz.

the brain areas. To these, numbers are added, even for the right hemisphere and uneven for the left hemisphere. This results in the following positions:

1. Frontal electrodes: Fp1/Fp2 (frontopolar), F3/F4, Fz and F7/F8 (the latter also regarded as reflecting the activity in the anterior temporal regions)
2. Temporal electrodes: F7/F8 (anterior temporal), T3/T4 (mid temporal), T5/T6 (posterior temporal)
3. Parietal electrodes: P3/P4 and Pz
4. Occipital electrodes (O1/O2)
5. Central electrodes: C3/C4 and Cz. A1 and A2 are the ear electrodes.

Other electrode positions (e.g. T1, T2) can be added to the montage. Moreover, the more current 10–10 system allows an extension of the 10–20 system, defining the 10% electrode positions that are unnamed in the latter. However, all of these additional electrode positions are not commonly used in the operating room (OR) recording so are left aside. For detailed description of the electrodes set up in the 10–20 system, see Tyner et al. (1983).

A variety of recording montages can be used, and digital EEG recording allows rapid changes between these montages. For OR recordings, we use a bipolar montage (preferably longitudinal bipolar montage) as well as a reference montage. We find that a reference montage has a better localizing value than a bipolar montage, which is useful in procedures such as ECoG for seizure foci mapping. On the other hand, we find bipolar recordings more helpful in detecting interhemispheric asymmetries, and thus we rely on them during carotid endarterectomies. We always add electrocardiogram (EKG) monitoring to the montage. This helps in detecting cardiac arrhythmias and also in differentiating between real brain activity (e.g. rhythmic spikes or rhythmic delta slowing) and EKG/pulsation artifact (see Figs. **3-18**, **4-12**, and **4-17**).

For more examples on recording channels (including EKG), reference electrodes, and the utility of EKG channel, see also pages 66–72; page 124; and pages 166–67. Jasper (1974) described the position of the central electrodes, which were considered to overlie the sensorimotor regions. These positions become important for somatosensory evoked potentials (SSEPs) recording as well as for motor evoked potentials (MEPs) monitoring (see pages 10 and 16–17).

## *C. Recording Technique*

EEG recording consists of measurement, amplification, and registration of differences between fluctuating electrical field potentials as a function of time. Because these extracellular fields fluctuate in time, it means that EEG reflects those fluctuations that fall within its recording bandwidth. The EEG recordings have a bandwidth or frequency response from 0.1 to 1000 Hz. The measurement of the electrical field potentials is done by the input amplifier connected to the recording electrodes. The real signal voltage at an electrode lead is the difference between the input signal for that lead and the voltage at a reference electrode. All of the leads have to be referred to the same reference electrode. Also, a reference electrode needs to be as quiet as possible; a noisy reference introduces "clipping" because of saturation of the amplifier. Amplifiers with high dynamic range that faithfully follow large ranges of input voltage, usually plus or minus several hundred microvolts, are necessary to solve this problem (Wong, 1996). EEG recording uses a differential amplifier; this amplifies only the difference in voltage between two inputs. Thus, it does not amplify equal voltages at the two inputs; these equal voltages instead are rejected. How well this is accomplished is given by the common mode rejection (CMR) expressed in terms of a ratio. For example, a ratio of 5000/1 means that the unwanted signal has its amplitude decreased 5,000 times in the output signal. Frequently encountered in the OR recordings, 60-Hz noise is supposed to be rejected in this way. The remainder of it, which is further transmitted in the output, can be further eliminated from the recording by applying a 60-Hz notch filter. Similarly, the amplification factor shows how many times the difference in the voltage between two input signals is amplified. For example, an amplification factor of 1,000 means that a difference in the input of 0.001 V results in an output signal of 1 V. An A/D converter translates the voltage difference in units that further translate in a certain amplitude of the recorded waveform, according to the desired sensitivity.

An EEG recording channel is provided with adjustable high-frequency filters (HFF) and low-frequency (LFF) filters; these limit the frequencies of a channel to a desired band (between LFF frequency and HFF frequency). The usual set up for our OR recordings is 1 Hz for the LFF and 70 Hz for the HFF. The output amplitudes for activities less than 1 Hz increase logarithmically from 0 to ~70%, with the 1-Hz activities represented at 70% amplitude from their amplitudes at zero attenuation. Similarly, with a cutoff frequency of 70 Hz for the HFF, the output amplitudes of activities at more than 70 Hz drop logarithmically to zero, starting with the activities at 70 Hz, which have an output amplitude of 70% from their amplitudes at zero attenuation. The output signal is then registered and displayed. In digital EEG, the sensitivity is a matter of display magnification and thus can be varied according to the needs. Similarly, the time base can also be varied. As known, a consistent effect of the anesthetics on EEG recordings is amplitude attenuation and slowing. Thus, in order to appreciate accurately asymmetries in slow frequency activities, as well as changes in amplitudes, we may need to increase the sensitivity to 2–3 μV/div and decrease the time base to 15, 10, or 5 mm/sec, as described on pages 69–71.

## 3. Interpretation

### A. EEG Analysis

EEG recordings are notorious for a wide variety of normal patterns that differ from one individual to another but also within the same individual, and they are subject to influence by many factors, from a change in state to effects of drugs. With such variability of the recordings, it becomes challenging to distinguish abnormal patterns. This is the main reason why EEG analysis requires experience and skills at pattern recognition. Thus, this chapter is not supposed to and cannot be a substitute for formal training in analyzing EEGs. Instead, it summarizes the significance of presence of certain activities, of their dynamic changes, and of their topography for EEG interpretation in intraoperative neurophysiology. These important findings are listed below.

*A1. Slowing.* Slow-frequency activities, either within delta (1–4 Hz) or theta range (4–8 Hz), are invariably seen during EEG recordings in anesthetized patients. It is well known that increasing amounts of anesthetics cause diffuse cortical suppression of increasing severity, translating in different degrees of slowing until a burst-suppression pattern is obtained (see pages 75–80 and 155). The slowing induced by anesthetics is expected to be diffuse and without focality. However, in cases of preexisting cerebral insult or dysfunction, the postinduction background can become asymmetric, as if the two hemispheres react differently to the same anesthetic concentration. On pages 73–80, this situation is further detailed, describing the postinduction preclamp baseline. Focal slowing usually points toward an old or new injury. For example, it can signal the presence of a tumor or old infarct (see pages 117–20 and Fig. **3-15**). On the other hand, acute appearance of a new focal, lateralized, or even diffuse increase in slowing can be a sign of acute cerebral dysfunction, usually as a result of ischemia. A good example is acute EEG changes occurring with clamping of the carotid artery in carotid endarterectomies or with sudden and significant drop in the systemic blood pressure (see pages 80 and 84; Figs. **2-1**, **2-23**, and **2-31**).

On the other hand, recording from different structures can show different activities. For example, recording from the subcortical white matter is characterized by significant low-frequency delta slowing, which is, at times, semirhythmic. One is likely to see this type of recorded activity, for example, at the deepest contacts of a depth electrode.

*A2. Fast-frequency activities.* EEG recordings in the OR are characterized not only by significant slowing but also by a significant amount of fast-frequency activities, which are generally seen with lower levels of anesthetics (see Fig. **2-25a,b**, page 76; and Chapter **8**, pages 327–36). Asymmetries in these fast frequencies usually signal underlying dysfunction and are a particularly useful sign in cases of acute ischemic changes, as seen with acute hypoperfusional states that are due to clamping of the carotid in carotid endarterectomies (see pages 84–87). A particularly challenging situation is the distinction between these activities and possible epileptiform discharges in cases of ECoG for mapping the seizure foci (see pages 117–20; and Fig. **3-16c**) or in cases of ECoG monitoring for afterdischarges (see pages 163–64).

*A3. Epileptiform and epileptic activities.* The characteristics of such activities and their significance are extensively detailed on pages 121–24 and 163–65. The entire role and importance of ECoG as a mapping tool for the irritative zones relies on the ability to detect, characterize, and interpret the spikes, sharps, and polyspikes, as well as ictal events. Similarly, monitoring for afterdischarges relies on a certain degree of comfort with interpreting such activities.

*A4. Presence of artifact.* One of the main challenges of using EEG for monitoring and mapping in the OR is to distinguish brain activity from artifact. This is an important factor for any EEG recording, but in the OR environment, in particular, the presence of artifact can be overwhelming. Pages 124 and 166–70 give many such examples and offer troubleshooting suggestions.

*A5. The effects of anesthesia.* For a reliable interpretation of EEG monitoring and mapping procedures, it is very important to be familiar with the effects of different anesthetics on EEG. A more detailed discussion can be found in Chapter **8** (pages 327–36). Chapters **2** (e.g. Fig. **2-25**, page 76) and **3** (e.g. Fig. **3-16**, page 120), and **4** (e.g. Fig. **4-14**, page 165) show many examples of such effects and how they may interfere with EEG interpretation. It is also important to realize that anesthetics cause a general suppression, with an overall attenuation of all activities, and thus, as emphasized on pages 69–70**,** the sensitivity of the scalp recording has to be increased; this effect is less cumbersome for direct cortical recordings via ECoG.

### *B. Applications in Intraoperative Neurophysiology*

EEG is used in a variety of surgical procedures as a monitoring and also as a mapping tool. First, it is used to monitor the cerebral activity. This allows detection of early dysfunction caused, for example, by cerebral ischemia. Thus, any type of neurosurgical or vascular procedure that could result in a hypoperfusional state of the cerebral hemispheres could be monitored with EEG. Because different levels of anesthetics influence the cerebral activities differently, EEG can also be used to assess the depth of anesthesia. Moreover, EEG recorded directly from the cortical surface (ECoG) is very useful in detecting changes in cortical excitability and thus in the appearance of afterdischarges during electrical stimulation in functional mapping. Second, EEG can be used for mapping the irritative zones in epilepsy surgery. Table **1-1** shows the main roles of EEG in intraoperative neurophysiology.

## III. SHORT-LATENCY SOMATOSENSORY EVOKED POTENTIALS (SSEPS)

### 1. Principle

SSEPs recording is a neurophysiologic test that assesses the integrity of the large fiber sensory system. The latter

**TABLE 1-1** EEG: Applications in intraoperative neurophysiology

| Monitoring | Mapping |
|---|---|
| 1. Carotid endarterectomy<br>2. Other cases of vascular neck surgery (e.g. reconstruction of neck vessels, bypass surgery)<br>3. Aortic arch aneurysm repair<br>4. Monitoring for afterdischarges in functional mapping<br>5. Assessing the depth of anesthesia | Mapping of seizure foci in epilepsy surgery |

is responsible for accurate perception of vibration and joint position sense. SSEPs are obtained by direct electrical stimulation of peripheral nerves (e.g. posterior tibialis at ankle, median, or ulnar nerves at the wrist) and recording at different levels within the neuraxis (e.g. brachial or lumbar plexus, cervical spine, parietal somatosensory cortex) of the far- and near-field potentials generated by the transmitted electrical volley (Fig. **1-2**). Thus, SSEPs have been extensively used in research and clinical practice for diagnosis, anatomical localization, and management of neurologic conditions affecting the large fiber sensory system. Patients with deficit in pain or temperature modalities have normal SSEPs, because the corresponding pathways are of much smaller diameter and not sufficiently stimulated at the intensities used in clinical practice. By assessing the electrical transmission through the large fiber sensory system, SSEPs have also an important role in monitoring its integrity or mapping its location, during a variety of surgeries that could result in its damage.

## 2. Methodology

This section contains a description of SSEPs methodology used in IOM procedures. The SSEPs methodology used for mapping techniques employed in the localization of central sulcus and in dorsal column mapping are described in detail on pages 134–36 and 195–201, respectively.

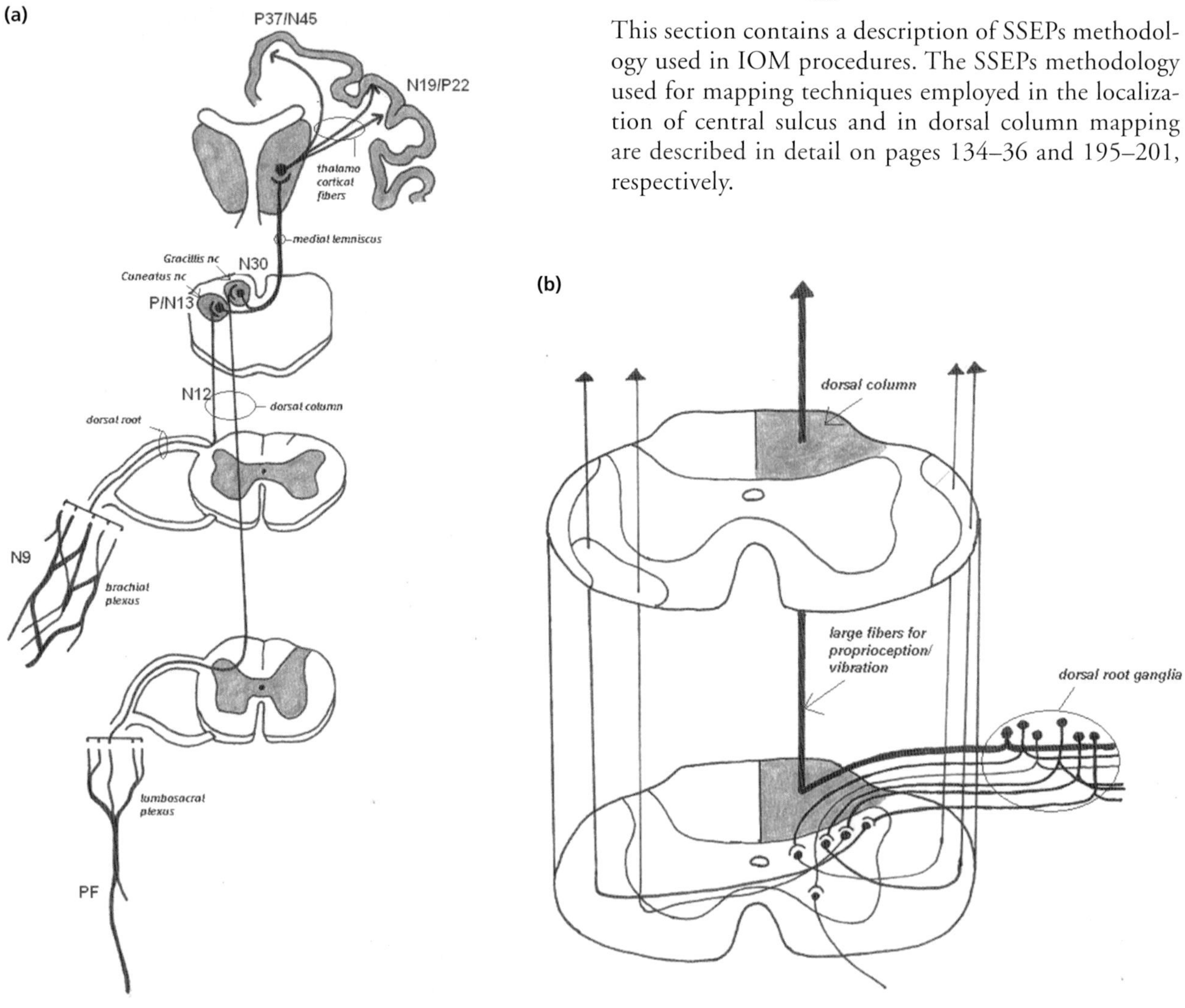


**FIGURE 1-2** The large fiber sensory system, responsible for proprioception and vibration senses. **a:** Notice the neurophysiologic-neuroanatomic correlation between SSEPs and different levels within the sensory pathway. Upper limbs SSEPs: N9: brachial plexus; N12: ascending volley through the dorsal column; P/N13: cervicomedullary junction (Cuneate nucleus); N19/ P22: thalamocortical structures. Lower limbs SSEPs: PF: popliteal fossa; N30: cervicomedullary junction (Gracillis nucleus); P37/N45: thalamocortical structures. **b:** Diagram representing dorsal root ganglia, the dorsal root entry, and the distribution of sensory fibers. The thick line shows the trajectory of the large fiber sensory system within the spinal cord, ascending through the dorsal column.

## *A. Stimulation*

In order to obtain reliable SSEPs, appropriate stimulation of the peripheral nerves has to occur. This depends on using appropriate setup, stimulation technique, and stimulus parameters.

*A1. Types of stimulating electrodes.* We use two disposable conductive solid-gel disc electrodes (ROCHESTER or VIASYS) per nerve, oriented parallel to the nerve, preferably with the cathode situated proximally. Their polarity, however, unlike in the nerve conduction studies (NCS) in the EMG laboratory, does not matter as much, as long as the baselines and the entire following recording are done with similar orientation of these electrodes. However, in presence of certain conditions, such as significant limb edema, excessive adipose tissue, skin breaks, or excessive sweating, stimulation via surface electrodes may not be optimal. This can be easily assessed by presence or absence of a good thumb twitch (see the following). We are sometimes able to improve our stimulation by employing two subdermal needle electrodes (12 mm long/27 G diameter; VIASYS). However, in such cases, an appropriate stimulation can be achieved only when the needles are in close proximity to the nerve. Additional caution needs to be taken when using needle electrodes, to minimize the risk of burns (see the following section) or other injuries (e.g., infections). It is also important to keep track of the numbers of applied needle electrodes, by performing a needle count at the setup and after removal.

*A2. Stimulation sites.* The stimulation is usually done at the ankle for the posterior tibial nerve (2–3 cm posterior to the medial malleolus) and at the wrist for the median or ulnar nerve (Fig. **1-3**). Occasionally, we might stimulate the peroneal nerve at the knee. However, in cases of severe peripheral neuropathy or other circumstances when stimulation of the posterior tibial nerve at the ankle is not possible, stimulation of the tibial nerve

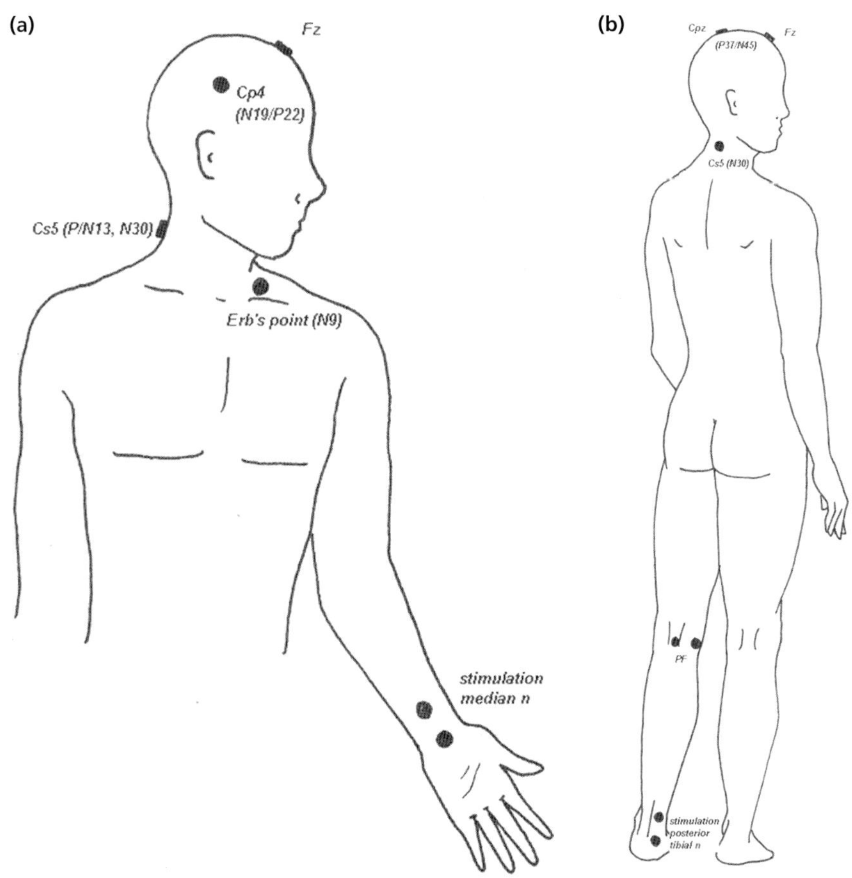


**FIGURE 1-3** Diagrams showing the positions of the stimulating and recording electrodes: **a:** Median nerve SSEPs. **b:** Posterior tibial nerve SSEPs.

proximally, via the popliteal electrodes, can be done. One could attempt proximal stimulation at the elbow for the nerves of the upper limbs in similar situations. The most proximal location for the stimulation of the nerves of the upper limbs is the brachial plexus. In fact, we are using this location quite frequently in brachial plexus repair surgeries; when the surgeon stimulates different nerve roots/plexus or trunks, the evoked potentials are recorded via electrodes placed at the cervical and scalp level (for details, see pages 286–89). When stimulating at more proximal locations, one should be aware of shortened absolute latencies for the following SSEPs.

*A3. Stimulation technique and stimulus parameters.* For stimulating the peripheral nerves, we use constant current stimulation: the level of current flow is kept constant by varying the delivered voltage in direct proportion with the variation of the impedances (the latter significantly affected by the interface between the stimulating electrode and the skin). We prefer the constant current stimulus because the current intensity (as a direct measure of current flow and expressed in milliamperes) is closely related to the number of nerve fibers stimulated. The stimulus is delivered via an FDA-approved stimulator incorporated in our EP recording XLTEK machines. The stimulator is equipped with a stimulus isolation device that helps minimize the stimulus artifact by restricting the leakage of current via the amplifiers. We stimulate the upper or lower limbs one at a time, using interleaved stimulation for upper and lower limbs, respectively. We believe that bilateral simultaneous stimulation, even though it offers higher amplitudes for SSEPs, loses specificity for anatomical localization of potential damage and can, in fact, be misleading because unilateral damage may be overlooked because of a minimal decrease in the amplitudes of SSEPs.

We apply a square wave, with a pulse duration of 0.2–0.3 msec (higher in cases of preexisting peripheral neuropathy). The intensity is gradually increased until a good thumb or toe twitch is obtained. This indicates appropriate stimulation of the nerve. The intensity at which a good thumb twitch is obtained has interindividual variability. For example, a person with significant peripheral neuropathy may need higher current amplitudes than a healthy person in order to obtain an appropriate stimulation of the peripheral nerve. The motor twitches result from antidromic stimulation of the motor fibers in the mixed nerves (e.g. median or posterior tibial nerves). Others consider a stimulation intensity three times higher than the sensory threshold as adequate. Because recording SSEPs intraoperatively (including baselines) is done with the patient asleep, the motor twitch is used to judge appropriate stimulation. It is our experience that once a good motor twitch is obtained, increasing the stimulus intensity further does not improve the SSEPs. The usual intensity range used is 15–25 mA, for the upper limbs SSEPs and usually 30–60 mA, for the lower limbs SSEPs; rarely we go up to 70 mA for the lower limbs stimulation, in attempt to obtain a good toe twitch and only when using surface electrodes (e.g. cases of peripheral neuropathy). If the stimulation is done via a subdermal needle electrode, the current density applied to the underlying soft tissue is higher; thus the maximum current applied is 35–40 mA. Extra caution has to be taken in case of poor vascularization of the respective area of the skin. For further details on safety issues regarding the current density, see pages 42–43.

SSEPs monitoring procedures require long hours of reliable setup. Moreover, in the beginning of the surgery or during the patient's positioning, displacement of the electrodes can occur easily. Securing the electrode with adhesive tape over the stimulated nerve can be helpful.

We use repetitive pulses at 3.17 Hz and average 500–1,000 responses to obtain one waveform. In case signal-to-noise ratio is low, more averaging is required. If the signal is good and noise is minimal, a few hundred averages are enough. This also permits faster results. In order to minimize and, hopefully, eliminate 60-Hz noise, we prefer to stimulate at a frequency that doesn't divide 60 (e.g. because 3 exactly divides 60, we may choose 3.17). Usually a range of 2–5 Hz is used for stimulation. A higher frequency decreases the time for feedback to the surgeon and can work well for stimulating the healthy nervous tissue (Pratt et al., 1980). However, in cases of damaged sensory pathways at baseline (e.g. neuropathy), increasing the frequency results in a decrease in amplitude and poorer morphology of these potentials. This is mainly true for the cortical SSEPs, because a high-frequency impulse is gated by the multi-synapse pathway; however, subcortical potentials (i.e. P/N13 and N30) may maintain unchanged amplitudes and latencies, as long as the interstimulus interval is longer than the normal recovery time. Thus, we are postulating that using high-frequency stimulation, with an interstimulus interval just above the recovery time of the healthy nerve, and monitoring the cervicomedullary junction potentials might help in earlier detection of otherwise latent neuronal dysfunction during spine surgery. We believe that this is particularly important for monitoring larger fibers, such as type I a muscle afferents, whose integrity is essential for sustaining normal proprioception and vibration sense. Although these senses can still be affected, low-frequency stimulation SSEPs would not detect the neurophysiologic abnormality, because the assessment of the corresponding pathways is only partially done at such low rate of stimulation. Additionally, selective assessment with higher stimulation frequencies of the electrical transmission in the muscle afferents from the lower limbs is important because these fibers likely travel through the spinocerebellar pathways, rather than the posterior columns as the cutaneous afferents do (Jones et al., 1982). This would also explain how muscle afferents can suffer selective damage. Many authors looked into improving the sensitivity and specificity of SSEPs by using stimuli in closer approximation to the natural stimuli. Mechanical stimulation (Onofrj et al., 1990), high-frequency vibration (Hamano et al., 1993), magnetic (Kunesch et al., 1993), and even chemical stimuli (Kobal and Hummell, 1988) have been used. However, none of these stimulus modalities have yet proven reliable enough for clinical applications. Although some authors found that stimulation of the mixed nerves generates higher amplitude SSEPs (Kritchevsky and Wiederholt, 1978), others have argued that simultaneous stimulation of muscle afferents antidromically or passively, by muscle contraction, result in a drop in frontal cortical SSEPs because of possible mixture of different potentials. For example, Rossini and colleagues (1990) found that muscle contraction decreased the SSEPs, whereas muscle relaxation increased their amplitudes. If the muscle relaxation is achieved iatrogenically, then the increase in the amplitudes of SSEPs may be due, at least partially, to a decrease in muscle artifact, followed by a subsequent increase in the signal-to-noise ratio. Finally, it is our experience that mixed nerve stimulation, involving both cutaneous and muscle afferent fibers, increases the amplitudes of subcortical potentials (e.g. P/N13), making them more robust for monitoring. These observations are in concordance with our practice of considering adequate and, subsequently, most efficient stimulation that produces a good thumb or toe twitch, resulting in the highest SSEPs amplitudes.

## B. Recording

*B1. Types of recording electrodes.* For recording, we are using collodion-attached reusable EEG disc electrodes (10-mm gold-plated discs, XLTEK) for the scalp positions and disposable conductive solid-gel electrodes elsewhere. For the surface electrode setup, in order to obtain low impedances (below 5k ), the underlying skin is first rubbed with an abrasive skin prepping gel. As with the stimulation electrodes, in case of skin breaks or other situations (e.g. profuse sweating), preventing good contact of the electrode to the skin, using disposable EEG subdermal needle electrode (12 mm long/27 G diameter; VIASYS) is useful. In addition to ensuring a better adhesion to the skin, subdermal needle recording minimizes the ambient noise that may interfere with the recording.

*B2. Recording sites and channels; recorded potentials*

***a. Upper limb SSEPs***

1. *Brachial plexus (peripheral) potential.* The first electrode is placed at the level of the Erb's point (EP). This is found about 2 cm above the midpoint of the clavicle, in the supraclavicular fossa. Its location can be determined by stimulating with this electrode and looking for the best obtained thumb twitch, to give us the most accurate location. The recording channel is Erbi-Erbc (ipsilateral to the stimulation site to the contralateral Erb point) or Erbi-Fz. It displays the first recorded potential as the EP or brachial plexus potential or N9. This consists of a negative upward deflection, occurring at about 9 msec after stimulation of the median nerve at the wrist, and represents the electronegativity caused by the electrical volley reaching the region of the EP.
2. *Cervicomedullary (subcortical) potential.* The second recording electrode is placed at the level of cervical vertebra 5 or 2 (Cs5 or Cs2); this electrode is usually referenced to Fz or a noncephalic reference (such as hand). Cs5-Fz channel displays the second recorded potential, the cervicomedullary junction potential, represented by P/N13 complex. This potential is represented by a negative upward deflection, occurring at about 13 msec after the stimulation of the median nerve at the wrist. However, unlike the hand, Fz is not electrical neutral and contributes to the cervicomedullary junction complex with positivities at 12 through 15 msec, whereas the neck electrode contributes with negativities as slightly earlier latencies (N11, N12, and N13). These latter negativities can be easily seen in the Cs2-hand reference (Chiappa, 1997a). Placement of the cervical electrode can be problematic in cases of cervical spine surgeries with posterior approach. In

such cases, we sometimes try placement of a subdermal needle as recording electrode in the sterile field, inserted in one of the exposed neck muscles. With some luck, it does not get in the way of the surgeon and allows recording of the cervicomedullary junction potentials. Other neurophysiologists prefer to use the anterior neck or the chin for the active electrode in such cases (Restuccia and Mauguiere, 1991). Moreover, when referenced to a posterior electrode, this derivation shows dorsal horn activity, by avoiding the electropositive cortical potentials P13–P14 (Emerson and Pedley, 1984, 1986).

3. *Thalamocortical potential(s).* The next sites for recording are on the scalp, corresponding to the location of the parietal cortex. Thus, surface electrodes are placed on the scalp at Fz, Cp3, Cp4, A1, and A2 (ears). Cp3 and Cp4 locations are 2 cm behind C3 and C4, respectively, in areas thought to correspond to the somatosensory parietal cortex (see Fig. **1-1**). The recording channels for the thalamocortical potentials N19/P22 are Cpc-Fz, Cpc-Cpi, and Cpc-A1 (A2). Cpc is located on the contralateral (to the site of stimulation) parietal area, whereas Cpi is located on the ipsilateral parietal area. The contralateral parietal cortex (Cpc) becomes more electronegative than the ipsilateral parietal (Cpi) and frontal (Fz) cortex, at about 19 msec after the stimulation of the median nerve. Thus, the waveform obtained in the Cpc-Cpi and, respectively, in Cpc-Fz channels has a negative upward deflection at about 19 msec (N19), followed by a positive downward deflection at about 22 msec (P22).

### *b. Lower limbs SSEPs*

1. *Popliteal fossa (peripheral) potential.* Popliteal fossa potential (also called knee potential or PF) has an important role in intraoperative monitoring. The electrodes are placed at the crease and 5 cm proximal to it. The recording channel is Pop1-Pop2. These electrodes, however, could also be used for stimulation of tibial nerve at the popliteal fossa in cases of severe peripheral neuropathy or whenever stimulation at the ankle is not possible (see Fig. **5A-1**, pages 181–82). In the outpatient lab, a second peripheral potential is used: the lumbar potential (LP), which helps estimate the activity in cauda equina. However, this potential is not routinely used in intraoperative neurophysiology.
2. *Cervicomedullary (subcortical) potential.* The second recording channel, as in the case of upper limbs SSEPs, is Cs5-Fz, recorded via a cervical electrode placed at the level of 2nd or 5th cervical vertebra and referenced to Fz. The subcortical potential recorded in this channel is N30, which occurs at about 30 msec after stimulation of the posterior tibial nerve at the ankle and is represented by a negative upward deflection. This potential is not routinely recorded in the outpatient EP lab, yet it becomes important and relatively easy to obtain in the OR. Some authors agree that N30 represents the equivalent of the P/N13. Others think it is the equivalent of N18. Regardless of its specific origin, N30 is a subcortical potential.
3. *Thalamocortical potential(s).* The scalp recording channels for the lower limbs SSEPs are Cpz-Fz, Cpi-Fz, Cpi-Cpc; Cpi, Cpc, Fz electrodes retain the same position as described for the N19/P22 recordings. Cpz is located at mid distance between Cz and Pz. The potentials recorded at this level are P37/N45 and represent the electrical volleys in the thalamocortical pathways. However, the electrical fields of these potentials differ from those of N19/P22, in that the ipsilateral (to the side of stimulation) somatosensory parietal cortex corresponding to the lower limb becomes first more electropositive than the frontal regions (Fz) at about 37 msec after stimulation of the posterior tibial nerve at the ankle. Similarly, Cpz area becomes more electropositive (than Fz) at about the same latency. Thus, in Cpi-Fz channel, P37 is represented by a positive downward deflection at about 37 msec, followed by a negative, upward deflection at about 45 msec (P37/N45). Of note, P37 is sometimes referred at as N/P37 because of its mixed origin of more electropositivity in the parietal region (Cpi) and some electronegativity from the frontal region (Fz), occurring at about the same time, ~37 msec, after stimulation of the nerve.

Fig. **1-3** shows the position of the recording electrodes for both upper and lower limbs SSEPs.

*B3. Recording technique.* The recording of SSEPs is done on the same XLTEK machine that delivers the stimulus. The amplifier headbox is isolated from the stimulator headbox. The recording headbox presents with two inputs for ground (common) and two for reference. We use a common or ground electrode for each neurophysiologic recording. This is an isolated ground which is used to clean the recording from superimposed artifact. For this purpose, we prefer a surface electrode with a larger contact surface area (e.g. disposable gelled

Ag AgCl 40/50 mm or 100/60 mm VIASYS). However, regular pregelled disc electrodes can also be used. We usually set up two of these electrodes, one on the shoulder and the other on the lateral aspect of the knee. The additional ground becomes handy in long recordings, when the initial electrode gets displaced and thus can be easily unplugged and replaced with the spare electrode. We position the common ground electrode between the stimulator and recording channels. Some suggest that its placement should be close to the recording electrodes in order to be more efficient in eliminating the artifact. However, electrical artifact is also picked by wires and at the amplifier itself; thus the proximity of the ground to the patient's recording site (e.g. scalp) may not necessarily help. In case of stimulus artifact, there are indications to place the ground close to the stimulating electrodes. However, similarly, this may not be necessarily more efficient, because the electrical artifact can be the cause of spread of radio frequencies through an antenna mechanism, from the electrical source—in this case, directly from the stimulator. An electrical ground electrode is different than an isolated ground, and its role in safety is discussed on page 44.

For practical handling, the main amplifier headbox is split into 4 separate extensions (breakout boxes), each containing 8 inputs; each extension is attached by the bedside (e.g. toward the head, legs, sideways), allowing an easier access to the needed leads during troubleshooting.

The recording software allows for a setup of up to 16 channels. The time base is set at 10 msec/div for the lower extremity SSEPs and to the 5 msec/div for the upper limbs SSEPs. The sensitivity is varied according to the amplitude of the recorded waveforms, usually 0.3–1 μV/div. Each limb SSEPs (usually 3–4 channels per limb) are presented within a window display. The window display allows presentation of each channel of SSEPs in a chronological order (vertical stack mode) or grouping the live or most recent SSEPs trials versus baseline (replace stack mode). We find the vertical stack mode very helpful in appreciating the dynamic or the evolution of waveforms with time, such as abrupt versus progressive changes. This information is important in establishing the cause for such changes (see pages 187–88). The display monitor shows the following: a stimulator window, enabling direct operation of the stimulators; a raw data window for the recording channels, enabling detection and identification of different types of artifact, including EMG artifact (see Fig. **1-13**, pages 31–32); a group window, allowing assessment of the averaging process and the actual recording windows, one for each limb. The usual recording bandpass is 30 Hz for the LFF and 500 Hz for the HFF. The software allows varying the recording band-pass frequencies, as well as the rejection thresholds, which becomes important in troubleshooting a noisy recording.

Figure **1-4** shows a setup and recording of normal median and posterior tibial SSEPs.

## 3. Interpretation

### A. Anatomy

For a better understanding of how SSEPs could reflect damage at different levels within neuraxis, it is important to first review the anatomy of the system whose function is tested via SSEPs, i.e. the large fiber sensory system.

SSEPs pathways start with large peripheral myelinated fibers of the cutaneous and muscle Ia afferents, serving the vibration and proprioceptive senses. The first neurons are situated in the dorsal root ganglia. The axons of these neurons travel through the ipsilateral dorsal column, up to the cervicomedullary junction. Here, they synapse with the second neurons, situated in nucleus Gracillis (medial) for the legs and in nucleus Cuneatus (lateral) for the arms. The axons of the second neurons cross to the opposite site and subsequently travel proximally in the medial lemniscus, reaching VPL (ventroposterolateral) nucleus of the thalamus, where they synapse with the third neurons. The axons of the third neurons then reach the parietal sensory cortex (Fig. **1-2**).

However, several authors postulate that faster impulses initially transmitted by fibers Ia muscle afferents travel through the dorsal spinocerebellar tract rather than through the posterior columns (Jones et al., 1982). The faster transmission might be due to the fact that the average diameter in the posterior column axons (Ohnishi et al., 1976) is less than the diameter of axons in the spinocerebellar tracts. Thus the conduction of the electrical volleys in the muscle afferents from mixed nerve (the posterior tibial nerve) is faster than that of the electrical volleys from the cutaneous paths (sural nerve). Also, the contribution of the type I muscle afferents translates into higher-amplitude and shorter-latency SSEPs. Subsequently, the posterior tibial SSEPs have higher amplitudes and shorter latencies than the sural nerve SSEPs (Chiappa, 1983; Burke et al., 1981, 1982; Vogel et al., 1986). This is somehow discordant with data reported by other authors, who found that contribution of muscle afferents (stimulated during muscle contraction) actually decrease the amplitudes of SSEPs. For the upper limbs SSEPs, it is generally accepted that both cutaneous and muscle afferents travel in the posterior columns. Other authors also described uncertainty about localization of the vibration and joint position sense in the dorsal columns (Greenberg et al., 1987; Wall and Noordenbos, 1977). However, other studies showed that injury of the dorsal columns results in loss of SSEPs and disturbance of the previously mentioned senses. Cusick and colleagues (1978) showed that segmental resection of the dorsal column at thoracic and cervical levels results in total loss of SSEPs responses proximal to the lesion. The results reported by Macon et al. (1982) support the theory that the pathways responsible for SSEPs generation are located in the posterior part of the spine, because lesions of the anterior spinal artery did not affect them.

Even with this variability in precisely locating the proprioception and vibration sensation, the generally accepted belief is that the tracts responsible for these senses, also known as the large fibers sensory system, travel in the dorsal columns of the spinal cord. For monitoring and mapping purposes, the somatotopic distribution of the legs and arms fibers within the dorsal columns, with the arms sensation traveling laterally and the leg sensation traveling medially, is also important. This topography becomes especially important in cases of myelotomies during intramedullary spinal cord tumor resections (see pages 195–201).

### B. Neuroanatomic-Neurophysiologic Correlation

In the following discussion, I have further detailed the main electrical volleys within the SSEPs pathways, used in IOM procedures, their neurophysiologic-neuroanatomic correlation, and briefly summarized their role in IOM.

(a)

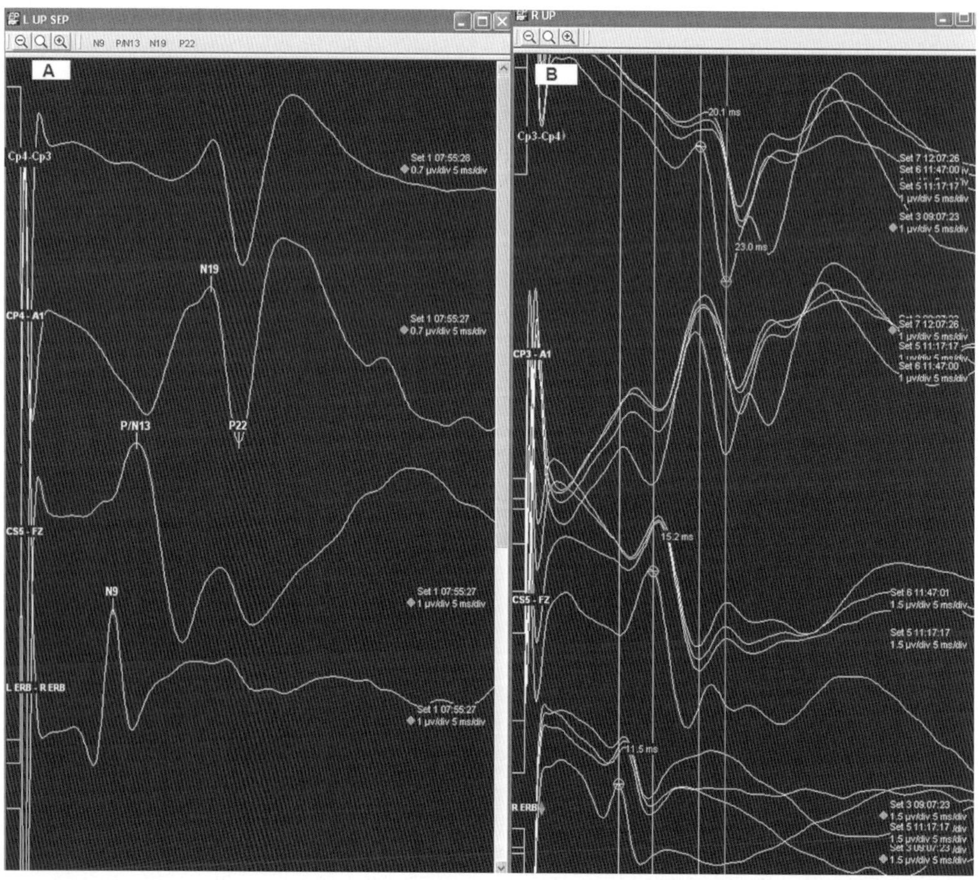


**FIGURE 1-4** Recordings of SSEPs under total intravenous anesthesia (TIVA). **a:** Median nerve SSEPs. Panel **A** shows the baseline left median SSEPs. Panel **B** shows several superimposed trials of the right median SSEPs, demonstrating the reproducibility of the waveforms; notice the absolute latencies of all potentials: N9: 11.5 msec, P/N13: 15.2 msec, N19: 20.1 msec, P22: 23 msec.

(b)

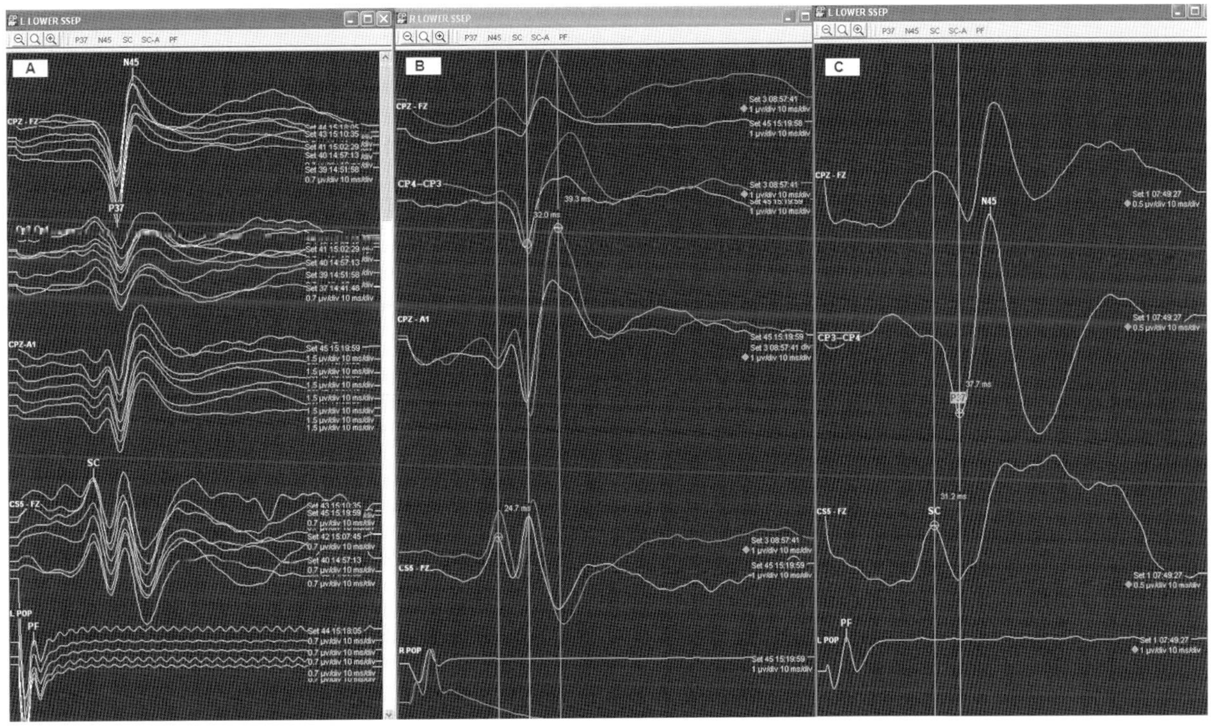


(c)

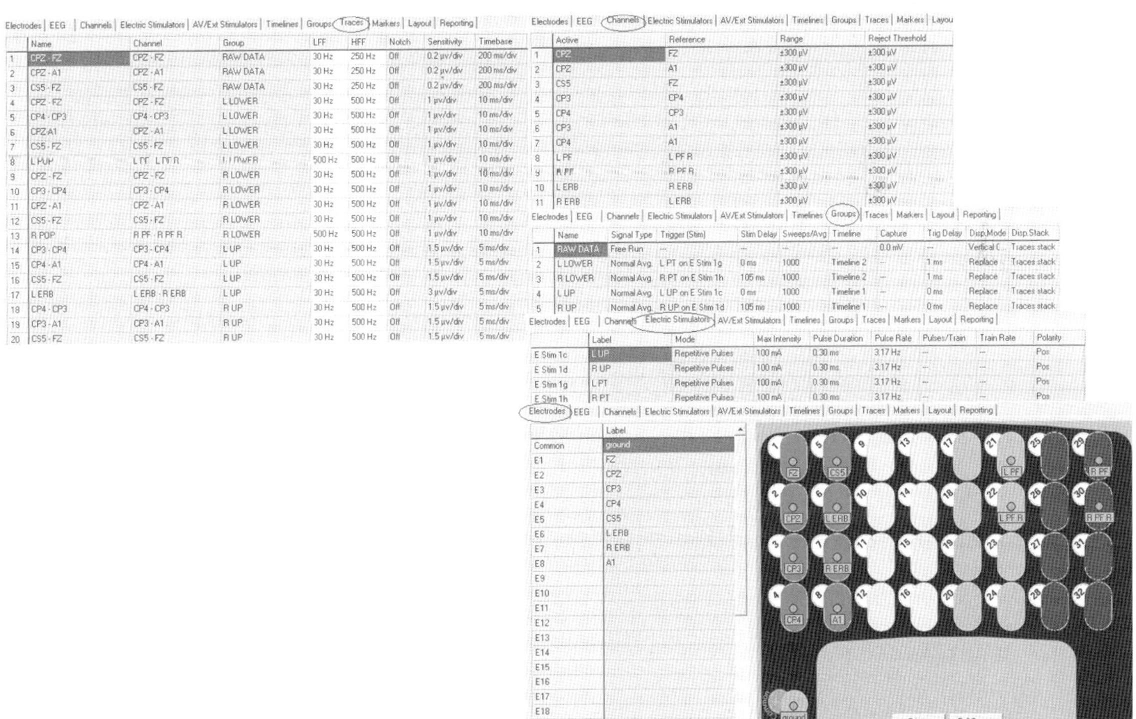


**FIGURE 1-4** *(Continued)* **b:** Posterior tibial nerve SSEPs. Panel **A** shows the last several trials of the left posterior tibial SSEPs, with the most recent trial on top, in vertical stack mode. Panel **B** shows the last trial (set 45) of the right posterior tibial SSEPs, against the baseline (set 3), in replace mode. In both panels, the subcortical N30 is seen at 24.7 msec and P37 at 32 msec (as shown in panel **B**). In panel **C,** the N30 is at 31.2 msec, whereas the absolute latency of P37 is 37.7 msec. Notice the change in the absolute latencies of the posterior tibial SSEPs with the patient's height: the first two panels, **A** and **B,** are recorded from an 8-year-old boy, whereas the left posterior tibial SSEPs shown in panel **C** are from a 30-year-old man. **c:** Setup for SSEPs technique: recording electrodes, channels, traces, groups, stimulators.

*B1. Upper Limbs SSEPs*

***a. Brachial plexus (peripheral) potential.*** The first recorded potential after stimulation of the median or ulnar nerves is the brachial plexus potential, N9 or EP potential (see page 9). This potential represents electrical activity at the level of the brachial plexus, in the nerve trunks and roots, close to the cervical spinal cord. In the outpatient laboratory, this is also called the reference point (Chiappa, 1997a) for reliable SSEPs interpretation, when absolute latencies are abnormal. In the OR, its utility is evident in several circumstances. First, it allows appreciation of proper stimulation. This becomes essential in troubleshooting of SSEP changes, when the loss of subcortical or cortical potentials could be a reflection of displaced stimulating electrodes; the absence of N9 would concur to this "problem localization" (see Fig. **5A-2**, page 184). Second, it helps in detecting improper arm positioning during certain long spine cases that could result in compression brachial plexopathies.

***b. Cervicomedullary (subcortical) potential(s).*** The second recorded potential is P/N13, the cervicomedullary junction potential. In reality, there is more than one waveform participating to it: N11 at the level of the root entry zone in the dorsal horns; N12/P12 at the level of the dorsal columns; N13/P13 at the level of the dorsal column nucleus (Cuneate). However, depending on the channels used, the recorded potential might be either a combination of a far-field and near-field potential, recorded in C5-Fz channel, or the actual near-field potential, recorded in a Cs5-noncephalic reference channel. As already mentioned, in Cs5-Fz channel, Fz also contributes with electropositive charges, which start at 12 msec and continue toward 15 msec (see pages 9–10). Its first role in the OR relates to the fact that the negativity recorded in the back of the neck represents mainly a robust subcortical response, resistant to the effects of anesthetics and reliable during SSEPs spinal cord monitoring in anesthetized patients. Second, this subcortical potential is also used in brachial plexus surgeries; its presence after stimulating parts of the brachial plexus signals intact transmission at the level of the dorsal root (see pages 286–89). Third, for monitoring the large fiber system, we do not particularly pay attention to other potentials, such as N12, representing the ascending electrical volley in the dorsal columns. However, mapping of the Cuneate fascicles in dorsal column mapping relies on the recording of this potential (see pages 196–99).

***c. Thalamocortical potential(s).*** The third recorded waveform is N19/P22, which represents the electrical volleys in the thalamocortical fibers. Different authors have reported relatively different origin for this potential; however, it is generally accepted that N19 is a negative potential, representing the activity in the thalamocortical fibers, whereas P22 is a pure cortical potential, originating in the primary somatosensory parietal cortex (see pages 9–10). First, these potentials are routinely recorded as part of the SSEPs monitoring, and they can be used not only to monitor the integrity of the spinal dorsal columns but also the integrity of the most proximal segments of the large fiber sensory system, from the brainstem up to the cerebral cortex (see pages 66; 177–78; 209–10; 225–26; 246–47; 320). Second, because the parietal postcentral regions become more electronegative than the frontal precentral regions, the dipole created at 19 msec after stimulation of the peripheral nerve is used in the central sulcus localization in functional cortical mapping (see pages 133–40). Because of the fact that this potential relies on a polysynaptic pathway and is sensitive to anesthetics, particularly to inhalational agents (such as nitrous oxide and halogenated agents), it is less robust than P/N13. However, with an adequate anesthetic regimen (e.g. total intravenous anesthesia [TIVA]), the N19/22 thalamocortical potentials can be reliably used in IOM procedures.

*B2. Lower limbs SSEPs*

***a. Popliteal fossa (peripheral) potential.*** Similarly, for the lower limbs, the first recorded potential is PF, the popliteal fossa potential. This potential represents the registration of the sensory nerve action potential (SNAP) of the posterior tibial nerve from the popliteal fossa, thus representing the activity in the large fiber sensory system distal to the knee level. As with the brachial plexus potential, this potential has few important roles in the OR. First, it allows assessment of the adequacy of the stimulation of the peripheral posterior tibial nerve (see page 181). Second, it allows detection of limb ischemia during repair of thoracoabdominal aneurysms, especially of the left leg, because the cannulation of the left femoral artery for placement of the left atrio-femoral bypass and inadequate retrograde blood flow could result in left leg ischemia with loss of knee potential (see Chapter **5**, Fig. **5D-8**, page 249). Third, stimulation of the posterior tibial nerve, rather than recording, can be done through the popliteal electrodes. This can prove helpful in cases of severe peripheral neuropathy, when its distal stimulation at the ankle is unsatisfactory (see pages 181–83 and Fig. **5A-1**).

***b. Cervicomedullary (subcortical) potential.*** The second recorded potential is N30, an electronegative potential, the equivalent of P/N13 and roughly representing the activity at the level of the cervicomedullary junction, or nucleus Gracillis (see pages 9–10). First, similar to the roles of P/N13, N30 is a relatively robust potential, minimally affected by the anesthetics. Its preservation during SSEPs monitoring signals unaffected transmission through the dorsal columns and up to the cervicomedullary junction. Second, the earlier transmitted volley through the dorsal column is not used for monitoring but rather for mapping the dorsal columns, by identifying the median raphae for myelotomy procedures during intramedullary tumor resection (see pages 196–99).

***c. Thalamocortical potential(s).*** The third potential is P37/N45, representing the activity in the thalamocortical fibers projecting at the level of the parietal somatosensory cortex (see pages 9–10). It is generally smaller than its upper limb equivalent, because it lacks the muscle afferent input (when the posterior tibial nerve is stimulated at the ankle). As with N19/P22 potential, it can be used to monitor the electrical transmission through the entire length of the large fiber sensory pathways, from the distal peripheral nerve, this time posterior tibial nerve, to the parietal cortex. Also like N19/P22, despite its increased sensitivity to certain anesthetics such as inhalational agents, it is an important part of the SSEPs IOM procedures, remaining reliable during TIVA.

## C. Applications in Intraoperative Neurophysiology

SSEPs IOM techniques have become an important adjunct in a wide range of surgical procedures. Thus, they are used for monitoring the function of the large fiber sensory system during a variety of surgical procedures that could result in its damage. These include neurosurgical or orthopaedic spine surgeries, neurointerventional procedures, such as embolization of spinal AVMs or highly vascular tumors, supratentorial tumor or vascular malformation resections, brainstem surgeries and thoracoabdominal aneurysm repairs, peripheral nervous system surgeries, such as brachial plexopathies for ruling out/in dorsal root avulsion, thus helping the surgeon in choosing a certain surgical approach. Additionally, SSEPs are used as a mapping tool, to identify the dorsal columns during intramedullary spinal tumor removals or the central sulcus during functional cortical mapping. Table **1-2** summarizes the use of SSEPs during surgical procedures. The specifics of different IOM techniques using SSEPs, including detailed troubleshooting, are further discussed in the respective chapters.

**TABLE 1-2** SSEPs: Applications in intraoperative neurophysiology

| Region | Monitoring | Mapping |
|---|---|---|
| Cortical/subcortical sensory pathways | 1. Supratentorial primary or metastatic tumor resection<br>2. Supratentorial AVM resection<br>3. Brainstem surgery (e.g. tumors, Chiari malformation)<br>4. Vascular surgeries involving the carotid system (e.g. carotid endarterectomy [CEA], reconstructive surgery of the neck, aneurysms of the aortic arch) | Central sulcus localization via median SSEPs phase-reversal technique in supratentorial tumor and AVM removal as well as epilepsy surgery |
| Spinal cord | 1. Intramedullary and extramedullary intradural spinal tumor removal<br>2. Spine deformity surgeries (e.g. scoliosis)<br>3. Trauma of the spine<br>4. Other decompressive surgeries of the spine (e.g. spondylosis, extradural tumors of the spine)<br>5. Embolization of vascular tumors of the spinal cord<br>6. Embolization of spinal AVM<br>7. Thoracoabdominal aneurysm (TAA) repair | Dorsal column mapping before myelotomy, particularly using posterior tibial SSEPs (e.g. intramedullary spinal tumor removal) |
| Peripheral nervous system | 1. TAA repair (for limb ischemia)<br>2. Cervical and lumbar spine surgeries involving nerve roots (yet not very sensitive in detecting damage of single roots) | Establishing the integrity of the dorsal roots in brachial plexus surgery |

## IV. MOTOR EVOKED POTENTIALS (MEPs)

### 1. Principle

In the outpatient lab, the MEPs triggered by transcranial magnetic stimulation (TMS) constitute a method used for clinical diagnosis of certain neurologic conditions as well as research. However, the TMS-triggered MEPs are very susceptible to the effect of anesthetics and thus are not reliable for intraoperative use. On the other hand, use of MEPs triggered by electrical cortical stimulation, quite unpleasant in awake patients, is a technique done successfully under anesthesia, for neurophysiologic IOM of the motor pathways. It involves applying an electrical current, either transcranially (tc) or directly on the cortical surface (dc), with the purpose of depolarizing the corticospinal system proximal to the level of the surgery, above its threshold. The appropriate transmission of the created electrical volley, through the part of the motor system at risk for injury, is then assessed by observing the triggered MEPs responses, recorded distal to the level of surgery. Thus, this method assesses the integrity of the motor pathways, from the cortex to the muscles (Fig. **1-5**). Additionally, this method allows identification of the cortical and subcortical regions that, when stimulated, trigger motor responses (motor mapping). Transcranially triggered MEPs (tcMEPs) are mainly used for monitoring the spinal cord, whereas MEPs triggered by direct cortical stimulation (dcMEPs) are used in monitoring of the corticospinal tract (CST), as well as for mapping of the motor cortex and subcortical motor pathways in supratentorial surgery.

As with the other methods, this chapter addresses neither the MEPs changes related to technical difficulties nor the troubleshooting involved to overcome them. Neither does it address the real MEPs changes that may occur during the surgical procedures as a consequence of neurophysiologic dysfunction of the motor pathways. All of these situations are covered extensively in Chapter **5** (pages 177–266).

### 2. Methodology

#### *A. Stimulation*

During transcranial electrical stimulation, because of relatively high stimulus amplitudes that are necessary to overcome the skull impedance, the stimulation is, on many occasions, not restricted to the cortex and less so to only the primary motor cortex. Thus, the stimulation of the CST can occur at deeper levels, including corona radiata and the internal capsule. The monitoring remains reliable as long as the activation of the CST occurs above the level of the surgery, such as in cases of spine surgery. However, during supratentorial surgery, this nonspecific, deep stimulation may not be adequate, because it may occur distal to the potential site of surgical injury. In such situations, it is essential to stimulate at relatively low stimulus amplitude when stimulating transcranially or, better, to rely on direct cortical stimulation, for selective stimulation of the primary motor cortex.

*A1. Types of stimulating electrodes.* For most of our cases, we use corkscrew electrodes (SOGEVA) because they offer more reliable attachment to the patient's scalp. Occasionally, these may cause local bleeding, but, in our experience, this has never been significant. Some centers opt to use subdermal EEG needle electrodes. However, these electrodes can be less well secured. In the case of direct cortical stimulation for motor mapping, we use a monopolar handheld electrode, connected to the anode, and a sterile subdermal needle, placed in the surgical field as a cathode (see pages 140–45). In the case of direct cortical stimulation for monitoring purposes, a subdural strip, like those used for ECoG and functional mapping recordings (see pages 115–17, Fig **3-13**; pages 134–36) is used, placed over the primary motor cortex, and stimulating is done via two of its contacts.

*A2. Stimulation sites.* Because the main goal is to remain as specific as possible and to activate the primary motor cortex corresponding to the arms/hands and legs/feet as represented in the motor homunculus, the stimulating electrodes should be located overlying these regions. Thus their placement on the scalp is at C3, C4, C1, and C2, with C1 and C2 situated half distance between C3-Cz and C4-Cz, respectively (see Fig. **1-1**). C3 and C4, located lower on the hemispheric convexity, are considered to be at or close to the arm/hand motor area. C1 and C2, situated closer to the midline, are approximately overlying the leg motor regions. Thus, these pairs could be preferentially used for improving the stimulation of either of these regions (e.g. use of C1/C2 position to facilitate the stimulation of leg regions and thus obtaining lower limbs MEPs). However, we have come across many cases when better MEPs of lower limbs muscle were triggered when the stimulating electrodes were located in C3/C4 positions and vice versa; similarly, better muscle responses in the upper limbs were obtained with C1/C2 position. Some authors (Deletis, 2002) have described the successful use of a third location, Cz/6 cm

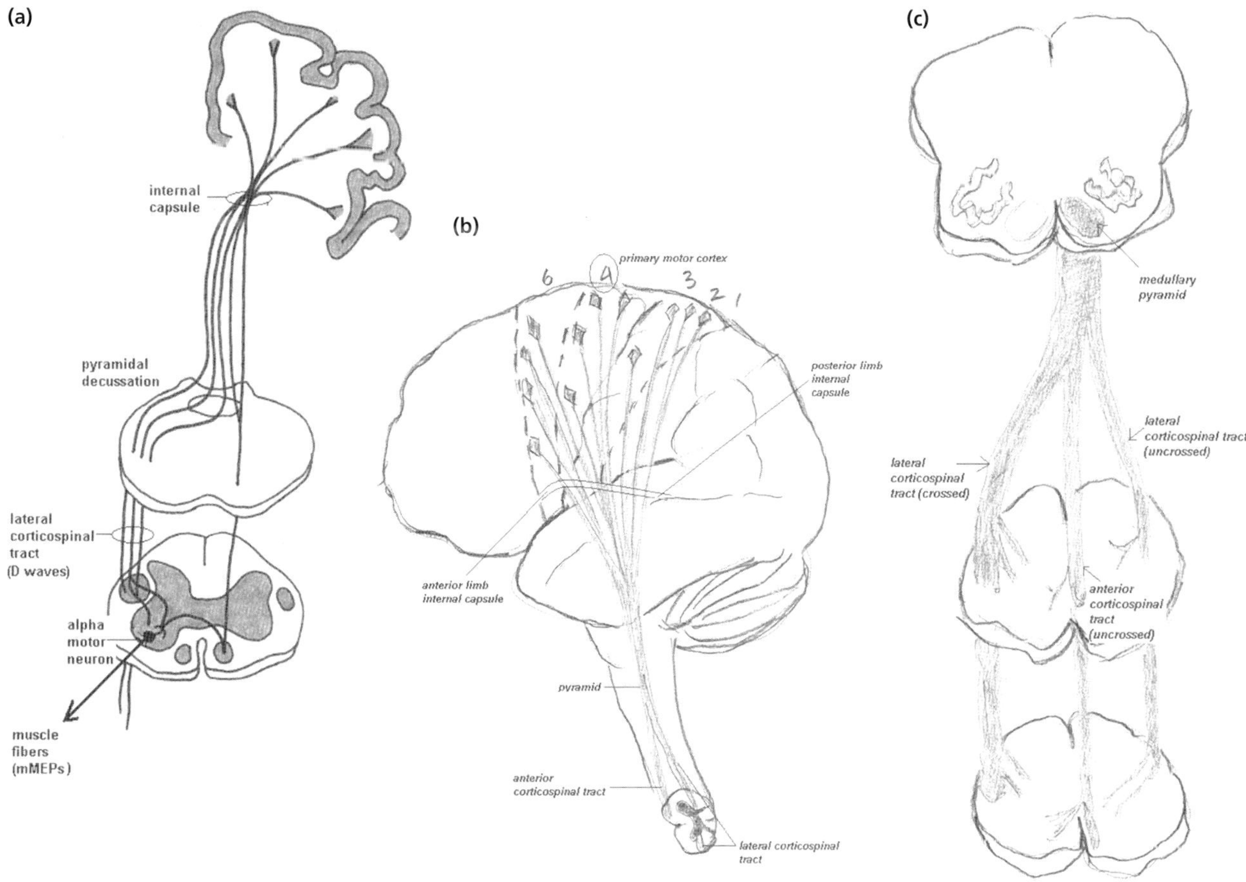


**FIGURE 1-5** Schematic representation of the motor pathways. **a:** The descent of the corticospinal tract. Notice the neurophysiologic-neuroanatomic correlation of motor evoked responses within the motor pathways: D waves are recorded at the level of the corticospinal tract, whereas mMEPs are recorded from limb muscles. **b:** Sagittal view of the motor pathways, showing cortical areas contributing to the generation of the corticospinal system. **c:** Crossed (lateral) and uncrossed (anterior and lateral) corticospinal tracts. (Modeled after Carpenter MB, *Core Text of Neuroanatomy*, 1996.)

anterior to Cz, that seems to stimulate the leg regions selectively, especially when these are located deep within the interhemispheric fissure. Unfortunately, we haven't been able to reproduce their success. Instead, we have come across circumstances in which, probably as a result of differences in the representation of motor homunculus within the two hemispheres, stimulating at C1/C4 or C2/C3 (for one side only or for both sides) improved the quality of some or all of the motor responses.

Because small variations of the locations of the stimulating electrodes can have a significant impact on successful recording of MEPs, we strongly recommend that the position of the electrodes be determined by exact measurements according the to the 10–20 international electrode system. Also, because of the significant interindividual variability, we recommend that stimulating electrodes be placed in all positions. This allows a quick shift between the pairs of the stimulating electrodes during troubleshooting, with the goal of obtaining good quality MEPs in all limbs, while minimizing significant body movements that would interfere with the surgical procedure. It is our experience, that stimulation at C3/C4 position is more likely to cause more movement, because it causes a deeper and broader stimulation of the motor pathways (see also page 183).

*A3. Stimulation technique and stimulus parameters.* The active stimulating electrode is connected to the anode. For example, if the right hemisphere is the one to be stimulated, C4 or C2 electrodes are connected to the anode, and C3 or C1 electrodes are connected to the cathode. In the case when the direct cortical stimulation is done via a strip, the contact situated on the motor strip is the anode. Monopolar stimulation for functional cortical mapping via a handheld stimulator is described on pages 140–45.

We use the multipulse train technique for stimulation. This technique consists of delivering repetitive trains of several pulses. Each train consists of 3 to 8 pulses (we usually use 6 pulses). Each rectangular pulse has a width of 0.5 msec. Rarely, in order to maximize the amplitude (or power) of the stimulus, we increase the stimulus width to 1 msec. The interstimulus interval (ISI) can also vary; in general we use ~3.5 msec ISI. The ISI gives in fact the pulse frequency (within the train). The train frequency can also be varied. We usually use a train frequency of 2 Hz (see Fig. **4-3 b**, page 141). However, several pulses (a multipulse train) are needed to reach the depolarization threshold of the alpha motor neuron, and thus to trigger an action potential distal to it, responsible for depolarization of muscle fibers and generation of muscle MEPs (mMEPs). For obtaining D waves, single pulse trains at 1 or 2 Hz (i.e. repetitive pluses at 1–2 Hz) are applied, while maintaining constant the other parameters (basically 1 pulse per train). This is due to the fact that each pulse of stimulation (or single-pulse train) triggers a D wave. The stimulus intensity can be safely increased up to 200 mA for transcranial stimulation and up to 25 mA for direct cortical stimulation. The stimulation thresholds can be quite variable, ranging from 40 to 200 mA for transcranial stimulation and from 1.5 to 25 mA for direct cortical stimulation. These wide threshold ranges are due to a significant interindividual and intraindividual variability of the thresholds for motor stimulation, to which many factors contribute, including effects of anesthetics, ischemia, blood loss, age, prior pathology, and existing motor deficit (see pages 155–57). Similar stimulus intensities are necessary for obtaining the D waves. For details regarding troubleshooting the transcranial electrical stimulation, see page 183. Direct electrical cortical stimulation for functional motor mapping, although using the same stimulus parameters, poses its own specifics, which are expanded on pages 140–47.

Safety issues regarding electrical cortical stimulation are further detailed on page 44**.**

## B. Recording

*B1. Types of recording electrodes.* For recording the mMEPs, we generally use subdermal needle electrodes (e.g. 23 mm /27 G, XLTEK) placed in the muscles of interest. The advantage of needle electrodes is that they do not require skin preparation and can be placed relatively quickly. Another significant advantage is that they offer cleaner records, by increasing the signal-to-noise ratio. First, they have a lower impedance, and also their impedances tend to be more even, because they do not depend on variable degrees of skin preparation. Second, they reside within or close to the muscle and thus are less likely to pick up ambient noise. Their main disadvantage is a potential risk of needle stick injury and transmission of infection. Thus, we recommend special attention to the setup every time that the needles are involved. Another disadvantage is that, because of their invasive nature, the setup can be done only after induction of anesthesia. The alternative for recording mMEPs remains the surface stick-on (gelled disc) electrodes. In this case, the setup can be done before the patient is brought into the OR. However, it requires more time because it involves appropriate skin prepping. This introduces more variability in the impedances, as well as overall higher impedances. The muscle channels involving surface electrodes are invariably noisier than those involving needle electrodes. If the recorded MEPs are D waves, we use a 3-contact epidural spinal electrode (AD-TECH manufacturer). The size of each contact is 1.2 mm. The distance between the contacts is 15 mm. The electrode's diameter is 1.1 mm and its length is 11.5 cm.

*B2. Recording sites and channels; recorded potentials.* The recording sites depend on the types of MEPs to be recorded. In the case of mMEPs, the chosen muscles depend on the level of the surgery and the location of the preexisting muscle deficit. If the surgery is done at several spine levels, we try to include muscles representing the corresponding myotomes. Table **1-3** summarizes the muscles that we use, with respect to different levels of the spinal segments. However, regardless of the spine level, we consistently include hand muscles, APB-ADM (abductor pollicis brevis-abductor digiti minimi) or APB1-APB2 (two electrodes sampling the APB muscle), as well as AH (abductor hallucis). This is because the motor regions of the hand and foot are relatively well represented on the motor homunculus and thus easier to stimulate. As a consequence, even though the mentioned muscles are small, the triggered mMEPs have high amplitudes and are very reliable for monitoring. For the muscles of the upper limbs, each recording channel represents the activity in one muscle (brachioradialis 1-2, deltoid 1-2, trapezius 1-2, triceps 1-2), with the exception of the hand channel, which usually represents the activity in two muscles (APB and ADM). For the lower limbs recording channels, each muscle is referenced to a common reference electrode, a subdermal needle placed on the lateral aspect of the knee, at the level of the fibular head. Our most commonly used lower limbs muscles are abductor hallucis (AH) and anterior tibialis (AT).

**TABLE 1-3** Muscle sites recommended for mMEPs monitoring in spine surgery

| Spine level | Upper limb | Lower limb |
|---|---|---|
| Cervical | Trapezius (C3, C4)<br>Deltoid (**C5**, C6)<br>Brachioradialis (C5, **C6**)<br>Triceps (C6, **C7**, C8)<br>Hand: APB (C8, **T1**)-ADM (C8, T1) | Anterior tibialis–AT (**L4**, L5)<br>Abductor hallucis–AH (S1, S2) |
| Thoracic | Brachioradialis<br>APB-ADM | AT<br>AH |
| Low thoracic/ Lumbar | APB-ADM | Quadriceps-Quad (L2, **L3**, **L4**)<br>AT<br>AH |

Additional muscles can be added to the setup, depending on the surgical level. For example, if the surgery level is low thoracic or high lumbar, we also include one of the quadriceps muscles. Another special case is if one of the nerve roots is particularly at risk for damage, in which case including a muscle channel representing the corresponding innervated myotome is helpful. For the low cervical, thoracic, and lumbar surgeries, we always have a reference channel represented by a muscle from a myotome situated proximal to the level of surgery. For example, for a surgery at C7 level, deltoid muscle channel can serve as reference. It is important to have a reference channel to enable differentiation between changes related to a systemic factor (e.g. anesthetics, low systemic blood pressure, blood loss) and those related to the spine manipulation, the latter occurring only in the muscle channels representing myotomes distal to the level of surgery. Another reason to have such a reference channel is that, during long procedures, inappropriate positioning of the arms can lead to brachial plexopathies, which, within several hours or sometimes earlier, translates into diffuse SSEPs as well as MEPs changes involving the upper extremities (see page 189).

The setup of the recording electrodes for mMEPs starts with the patient awake. It is important to correctly identify the muscles to be recorded from. This identification sometimes requires the patient's cooperation (i.e. voluntary contraction of the muscle). Additionally, there are topographic landmarks, generally accepted in the EMG lab, by which one can place the recording electrodes in the appropriate locations. Fig. **1-6** shows how to identify the insertional points of the most frequently used muscles for MEPs monitoring, as well as certain maneuvers used to better delineate the corresponding muscles. If identification of the insertional point requires the patient's cooperation, we recommend marking those points for use after the induction, when the actual setup with subdermal needles takes place.

The motor evoked responses recorded in the muscle channels are, in fact, compound muscle action potentials (CMAPs), which are triggered by electrical cortical stimulation. As previously mentioned, the hand and AH mMEPs are in general well represented and thus easy to monitor. By recording changes in the activity in different muscles innervated from different myotomes, one can localize the damage that occurred within a certain level of the neuraxis (for example, loss of hand MEPs, with preservation of the deltoid MEPs, places the injury below C5/C6 levels and above or at C8/T1 levels). MEPs occur as polyphasic potentials with variable latencies, shorter for the cervically innervated muscles and longer for the lumbosacrally innervated muscles (~15 to 60 msec). Their variable amplitudes and variable durations (~10–40 msec) depend on the number of depolarized muscle fibers (Fig **1-7**).

For D waves, the recording electrode is placed by the surgeon, either epidurally but most often subdurally, ideally both proximally (for a reference recording) and distally (for the active recording) of the surgical level. However, the placement of a reference electrode is not always possible, particularly in high cervical cases. On the other hand, if the surgical level is below T11, the recording distal to the surgical level of the small conus medularis D wave becomes challenging. The recording montage for the D wave can include three channels for each electrode: 1-2, 2-3, and 1-3, where contact 1 is situated at the tip of the electrode. Even though not all of them are essential for recording, it is useful to try different channels. Because the spinal electrodes are placed blindly under the dura, some of the contacts may not make good contact with the spine surface. Thus, trying different combinations is essential for an increased signal–to-noise ratio. However, we prefer channel 1-3 (or 3-1), because the distance between the contacts is the longest, about 3 cm. This channel may display a sizable difference between the polarities of the two contacts, because the electrical descending volley travels in proximal electrode from contact 1 to contact 3 and in distal electrode from contact 3 to contact 1, respectively; thus, the polarity difference between contacts 1 and 3 is easier to identify. Fig. **1-8** shows placement of a spinal electrode, as well as a schematic drawing of the orientation of both recording electrodes. The recorded

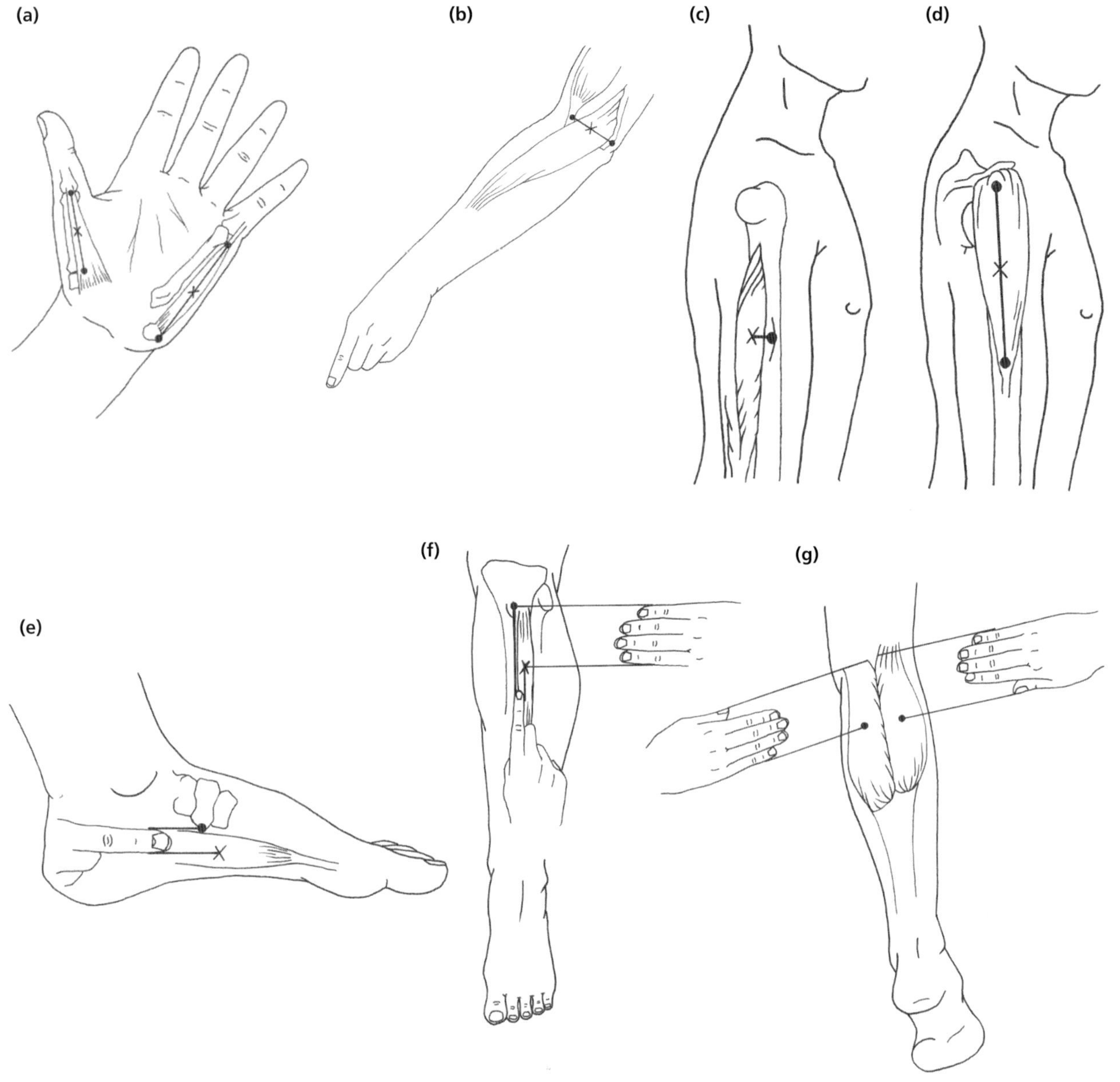


**FIGURE 1-6** Limb muscles most commonly used in mMEPs and CMAPs recordings: insertional points of the recording needle electrodes and enhancing maneuvers. **a:** Abductor pollicis brevis. Insertion: midpoint of a palmar line drawn between the first metacarpophalangeal joint and the carpometacarpal joint. Maneuver: palmar abduction of the thumb. Abductor digiti minimi. Insertion: midpoint of a line drawn between the ulnar aspects of the fifth metacarpophalangeal joint and that of the pisiform. Maneuver: abduction of the small finger. **b:** Brachioradialis. Insertion: midpoint of a line drawn between biceps tendon and lateral epicondyle and parallel to the flexor crease. Maneuver: flexion of the forearm in neutral position. **c:** Triceps (lateral head). Insertion: posterior to the deltoid tubercle. Maneuver: extension at the elbow. **d:** Deltoid (middle head). Insertion: midpoint of a line drawn between the acromion and the deltoid tubercle. Maneuver: abduction of the arm. **e:** Abductor hallucis. Insertion: half distance between calcaneus and base of the proximal phalanx of the great toe, one fingerbreadth below the navicular bone. Maneuver: spreading of the toes. **f:** Tibialis anterior. Insertion: four fingerbreadths below the tibial tuberosity and one fingerbreadth lateral to the tibial crest. Maneuver: dorsiflexion of the foot. **g:** Gastrocnemius. Insertion: one palm below the popliteal crease, on the medial aspect of the calf (medial head), one palm below the popliteal crease, on the lateral aspect of the calf (lateral head). Maneuver: plantar flexion of the foot while the knee is extended.

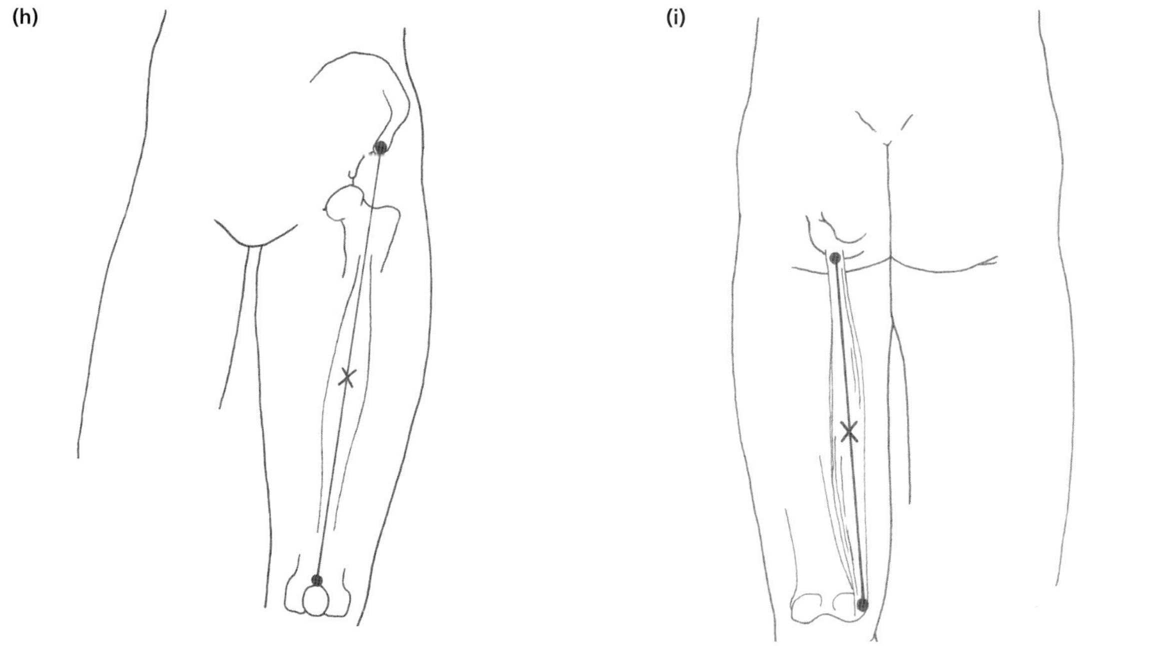


**FIGURE 1-6** *(Continued)* **h:** Rectus femoris (part of quadriceps). Insertion: midpoint of a line drawn between patella and the anterior aspect of the superior iliac spine. Maneuver: flexion of the hip, while the knee is extended. **i:** Semitendinosus (part of hamstring). Insertion: midpoint of a line drawn between the medial femoral epicondyle and the ischial tuberosity. Maneuver: flexion of the knee and internal rotation of the tibia.

D wave (Fig. **1-9**, panel **B**) is the first negative peak, with amplitudes varying mostly from 5–30 μV, although other neurophysiologists obtained D waves as high as 60 μV (Deletis and Sala, 2008). The amplitudes tend to be smaller in the low thoracic regions, and below T11 they become hardly recordable. In cases of pathological involvement of the anterior-lateral spinal cord, dispersion and desynchronization of the descending volley through the CST results in a very small or absent D wave, despite the presence of the distal mMEPs. The latencies of the D waves also vary: they increase as the volley descends in the CST, from 2–3 msec in the cervical region, to 9–10 msec in the thoracic regions.

*B3. Recording technique.* Similar to the recording of the SSEPs, MEPs requires that the recording electrodes are connected to an amplifier, with the aid of 4 extensions, for easier setup and in order to facilitate troubleshooting (see pages 10–11).

The muscle channels for a hemibody, consisting of a group, are displayed in a separate window (e.g. left MEPs). Each group is then connected to a stimulator. For example, the group left MEPs, containing the left hemibody muscles, is connected to a stimulator with the anode at the contralateral hemisphere (see pages 17–18). The recording parameters are time base 10 msec/div for the lower limbs MEPs and 5 msec/div for the upper limbs MEPs. The sensitivity for display varies according to the amplitude of the muscle responses and thus can be anywhere between 15 μV and 500 μV/div.

For the D wave recordings, the same proximal and distal spinal electrodes record both right- and left-sided D waves, depending on the hemisphere that is stimulated. The left- and right-sided D waves recordings are displayed in two separate windows, each connected to the stimulator that stimulates the contralateral hemisphere. Alternatively, four epidural electrodes can be used, two (one proximal and one distal) located on each side of the spinal cord, close to the respective CST; although this recording setup increases the specificity of the CST monitoring (as a result of a more precise positioning), it also causes crowding of the surgical field.

The mMEPs are not averaged. Instead, several trials are obtained at a time (e.g. four trials after four trains, within 2 seconds). This is important because, on many occasions, the first couple of trials may show very small or no muscle responses, followed by higher-amplitude MEPs The reason behind this is not clear. This may be

(a)

(b)

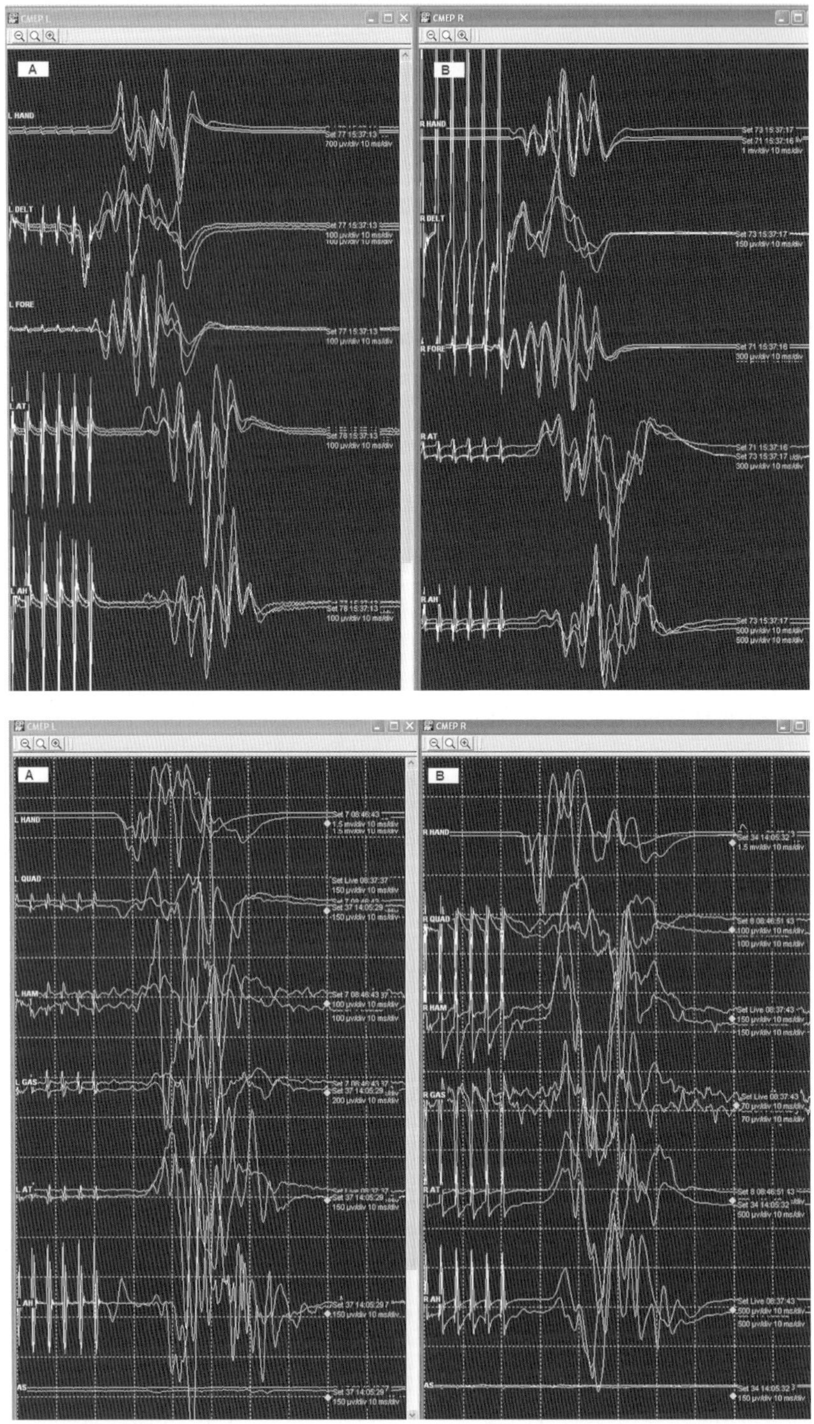


**FIGURE 1-7** Muscle MEPs recordings after transcranial electrical stimulation (tcmMEPs), under TIVA. Notice the high amplitudes, polyphasic morphology, and variability of latencies according to the level of the myotome. **a:** Cervical spine surgery, where recording is done from several arm muscles. Panel **A** shows tcmMEPs obtained in the left hemibody muscles, after stimulation of the right motor cortex. Similarly, panel **B** shows tcmMEPs obtained in the right hemibody muscles, after stimulation of the left motor cortex. The following labels were used for the muscle channels: HAND for APB-ADM; DELT for deltoid1-2; FORE for brachioradialis1-2; AT for AT1-ipsilat knee ref2; AH for AH1-ipsilateral knee ref2. **b:** Low thoracic surgery, where recording is done from several leg muscles. Similar to **a,** panel **A** shows left-sided tcmMEPs, and panel **B** shows right-sided tcmMEPs. The following additional labels were used for muscle channels: QUAD for quadriceps1-ipsi knee ref2; HAM for hamstring1-ipsi knee ref2; GAS for gastrocnemius1-ipsi knee ref2; AS for anal sphincter1-2; HAND, AT, and AH labels as in **a**.

due to hyperpolarization of the stimulated cortical regions in time, requiring more trials to be depolarized. It may also be due to progressive recruitment of several motor systems in addition to the CST. Moreover, the mMEPs amplitudes vary (see pages 18–21). For a D wave, we usually average about four stimulation responses.

All of the details previously mentioned and others regarding the setup for mMEPs and D waves recording are seen in Fig. **1-10**.

## 3. Interpretation

### A. Anatomy

As mentioned previously, the triggered mMEPs result from stimulation of several descending tracts, involved in both the pyramidal and extrapyramidal systems. The fibers descending from the cortex toward the spinal cord constitute the corticospinal system. The cells of origin of this system are located in the lamina V, mainly in the precentral area 4, but also in the premotor 6 area; the postcentral gyrus, areas 3a, 3b,1, 2; and parietal cortex, area 5 (Fig. **1-5 b**; Carpenter, 1996). The largest fibers arise from the giant pyramidal cells of Betz in the precentral gyrus. The integrity of these fibers in particular and of the entire corticospinal system in general is thought to be essential for avoiding permanent motor deficit. Although the mMEPs can be abolished because of damage to other tracts (e.g. propriospinal), the preservation of D waves, as a direct measure of the normal activity in the CST, of at least at 50% from their baseline amplitude, ensures motor strength recovery (see pages 202–6). The origin of the largest fibers in the primary motor cortex may also explain the difference in the stimulation thresholds between the primary and supplementary motor regions (see page 157). Furthermore, these fibers converge in the corona radiata, pass through the internal capsule, and descend to form the crus cerebri. Below, in the medulla, the fibers form the massive pyramids, and, at the cervicomedullary junction, the corticospinal system splits, with most of its fibers (70–90%) crossing to form the large lateral CST, which descends through the lateral funiculus, contralateral to the hemisphere of origin. Two smaller tracts descend uncrossed in the anterior (close to the anterior median fissure) and anterior-lateral parts of the spinal cord (Fig. **1-5 c**). It seems that fibers from the precentral cortex terminate mainly in the dorsolateral parts of lamina IX and in lamina VII, whereas fibers from the somatosensory cortex terminate in parts of the posterior horn.

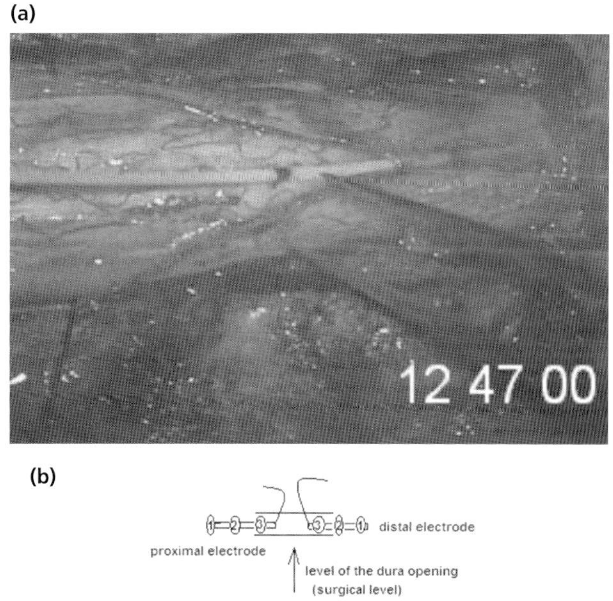


**FIGURE 1-8** Insertion of the spinal subdural electrodes for D wave recording. **a:** picture showing subdural insertion of the spinal electrode, distal to the level of surgery (right side of the screen). **b:** schematic representation of the position of the two recording spinal electrodes. The proximal electrode is placed above the level of surgery, and the recorded D waves are used as reference. The distal electrode is placed below the level of surgery for the active recordings of D waves.

The CST branches off at different spinal segments. Most of the projections are at the cervical and lumbar enlargements, containing a large population of alpha motor neurons involved in the innervations of upper and lower limb muscles. Interestingly, even if some neurons terminate directly by a synapse with the anterior horn cells (alpha motor neurons), most of them terminate by making synaptic contact with the internuncial neurons on lamina VII. The axons of the alpha motor neurons then exit the spinal cord via the ventral roots and innervate their respective myotomes via a peripheral motor nerve.

### B. Neurophysiologic-Neuroanatomic Correlation

*B1. D waves.* They represent the depolarizing electrical volley generated by direct stimulation of the corticospinal neurons that travel distally through an asynaptic pathway (i.e. CST). Because their generation takes place proximal to the level of the alpha motor neurons,

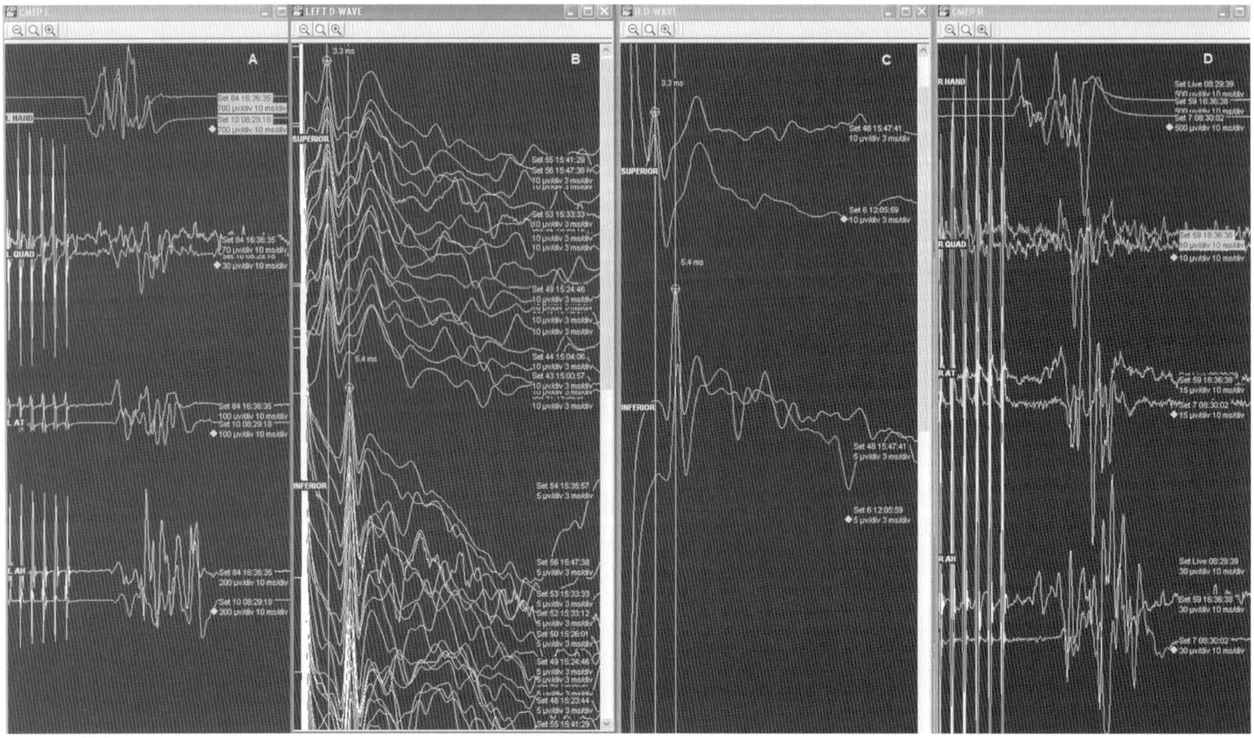


**FIGURE 1-9** tcMEPs (mMEPs and D waves) recordings under TIVA. Panel **A:** Left tcmMEPs. Panel **B:** D wave recordings in the proximal and distal subdural electrodes, when stimulation is applied over the right hemisphere. Panel **C:** D wave recordings in the proximal and distal subdural electrodes when stimulating the left hemisphere. Panel **D:** Right tcmMEPs. Notice in panels **B** and **C** the latency difference of the D waves recorded proximal and distal to the surgical level (3.3. msec versus 5.4 msec).

it relies neither on the interneuronal synaptic spinal transmission nor on the neuromuscular transmission. Thus, these waveforms are not affected by anesthetics or by neuromuscular agents. D waves are represented by an electronegative, upward deflection, seen at relatively short latencies, the shortest being at the cervical level. Because they represent a traveling depolarizing volley, they can be recorded at different spinal levels. There is thus a great variability in their latencies as they travel down the CST. This may make their identification at a certain spinal level difficult, because several peaks can be seen at different latencies, particularly in the presence of electrical artifacts. In these cases, increasing the averaging will help eliminate the artifact (see page 195). As already mentioned, unlike mMEPs, which need several pulses for their generation, each stimulation pulse is followed by a D wave. Thus, several D waves are needed in order to reach the depolarization threshold of the alpha motor neuron, which, in turn, generates the CMAP, or mMEP.

*B2. I waves.* These potentials are the result of either extrinsic or intrinsic transsynaptic excitation of the corticospinal neurons through interneuronal connections.

In the case of transcranial stimulation and even direct cortical stimulation of the motor cortex, I waves are most likely the results of cortico-cortical afferents. Additionally, other extrinsic inputs may contribute to their generation, such as thalamocortical inputs. The intrinsic inputs probably arise from local motor cortical neurons, after they are selectively stimulated. I waves are represented by negative, upward deflections of small amplitudes, following after a single-pulse stimulation triggers a D wave. Overall, they have a similar role to that of the D waves: by representing indirect excitation of the corticospinal neurons, the I depolarizing volleys bombard the spinal alpha motor neuron and contribute to the sum of EPSPs. However, in anesthetized patients, the depolarization of the alpha motor neurons relies mostly on the D waves. Because their existence depends on the activity and interconnections of the cortical interneurons, these waveforms are

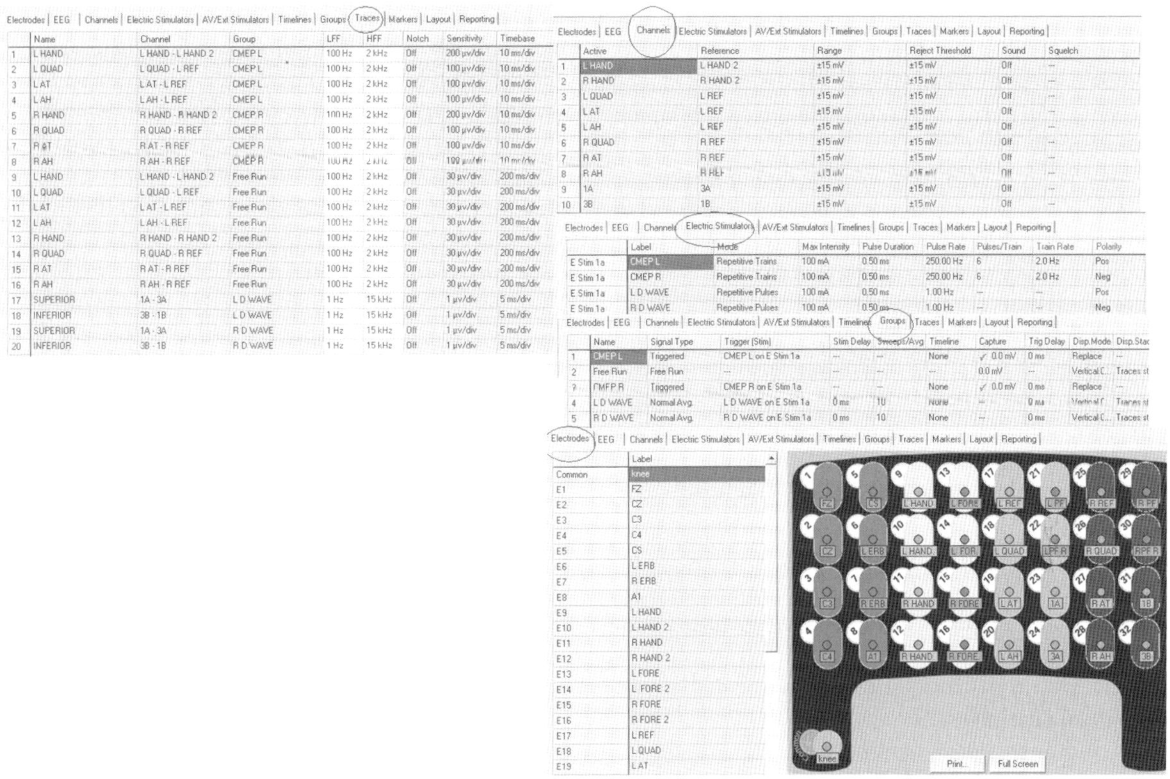

**FIGURE 1-10** Setup for tcmMEPs technique: recording electrodes, recording channels, traces, groups, and the attached stimulators.

very susceptible to anesthetics, which significantly affect the synaptic transmission. Thus, in anesthetized patients, the depolarization of the alpha motor neurons relies mostly on the D waves, and a reliable recording of the I waves is not possible. As a consequence, the I waves do not have a significant role in monitoring of MEPs and mapping procedures.

*B3. mMEPs.* As previously mentioned, the recorded muscle motor evoked response are CMAPs representing the summated activity of many muscle fibers. Robust MEPs arising in a healthy muscle, after appropriate stimulation, usually have polyphasic morphology, with high amplitudes (up to thousands of microvolts [μV]), variable durations, and variable latencies. Of course, on an average, the longer the distance is between the cortex and recorded muscle, the longer is the latency of the MEPs obtained from that muscle. Details regarding changes in morphologies and other parameters are discussed on page 185. As previously mentioned, the MEPs obtained in hand and foot muscles are, in general, more robust, have higher amplitudes, and are reliable for monitoring. Generation of mMEPs depends on appropriate depolarization of the alpha motor neuron. However, they are the result of a nonspecific depolarization of the spinal alpha motor neuron, via not only CST but also extrapyramidal tracts, such as propriospinal tract. Loss of mMEPs does not give information about injuries to specific spinal efferent and cannot predict the temporary or permanent nature of the postoperative motor deficit. Consequently, in cases of intramedullary spinal cord tumor removal, when the goal is to achieve total tumor resection, sole monitoring of the mMEPs is insufficient for determining whether the resection should stop or not (see next section and pages 202–6).

## C. Applications in Intraoperative Neurophysiology

Motor evoked responses triggered by electrical cortical stimulation are used to monitor the integrity of the descending motor pathways, from the motor cortex to the innervated muscles. More or less, their applications parallel those of SSEPs monitoring. In the last decade, MEPs monitoring has become an integral part of the spinal cord monitoring during neurosurgical, orthopaedic, and vascular procedures that could result in spinal cord damage. The necessity of adding this technique to monitoring of SSEPs derives from the fact that the two procedures assess the integrity of two different systems—the motor and the sensory pathways, respectively—not only with different functions, but also with different topographical locations at each level of neuraxias. In fact, there have been many cases of postoperative paraplegia reported after uneventful monitoring of SSEPs during spine surgery. Thus, these methods have complementary roles in ensuring a comprehensive and reliable assessment of the integrity of the neuraxis (see pages 177–78, 209–10, 225–26, and 246–47).

In addition to spinal cord monitoring, the MEPs are used to monitor more proximal parts of the CST during brainstem and supratentorial surgery (see page 303). Additionally, they can also be used in the assessment of the ventral nerve roots and peripheral motor nerves (see pages 276 and 278–79).

As already mentioned, unlike mMEPs, which are subject to changes in the presence of muscle relaxants and anesthetics, such as inhalational agents, to the point of becoming unreliable for monitoring, the neurogenic MEPs, D waves, assessing an asynaptic pathway, are not affected by anesthesia. D waves, representing strictly the electrical volleys at the level of the CST, are particularly valuable during the intramedullary spinal cord tumor removal, when they can guide the surgeon in achieving maximal tumor resection, even after disappearance of the mMEPs (see pages 202–6). Other authors (Katayama et al., 1988) investigated the use of D waves, recorded through epidural electrodes placed transcutaneously, to monitor the CST during supratentorial tumor or AVM removal. The use of D waves is preferred to that of mMEPs, because D waves specifically represent the CST and thus offer a better prediction of the potential clinical postoperative deficit.

MEPs induced by direct electrical cortical stimulation are also preferred during supratentorial surgery, in order to avoid deep stimulation of the motor pathways, distal to the level of the surgery. This procedure is also involved in mapping the cortical, subcortical, and even spinal structures involved in the motor function. Table **1-4** summarizes the uses of MEPs, triggered by electrical stimulation, in different types of monitoring and mapping procedures.

**TABLE 1-4** MEPs: Applications in intraoperative neurophysiology

| Region | Monitoring | Mapping |
|---|---|---|
| Cortical/subcortical sensory pathways | 1. Supratentorial primary or metastatic tumor resection<br>2. Supratentorial AVM resection<br>3. Brainstem surgery (e.g. tumors, Chiari malformation)<br>4. Vascular surgeries involving the carotid system (e.g. CEA, reconstructive surgery of the neck, aneurysms of the Aortic arch) | Motor cortical mapping for identification of the primary strip; subcortical mapping for identification of the corticospinal tract |
| Spinal cord | 1. Intramedullary and extramedullary intradural spinal tumor removal<br>2. Deformity correction surgeries (e.g. scoliosis)<br>3. Trauma of the spine<br>4. Other decompressive surgeries of the spine (e.g. spondylosis, extradural tumors of the spine)<br>5. Embolization of vascular tumors of the spine<br>6. Embolization of spinal AVM<br>7. Thoracoabdominal aneurysm (TAA) repair | Identification of the corticospinal tracts |
| Peripheral nervous system | 1. TAA repair (for limb ischemia)<br>2. Cervical and lumbar spine surgeries involving nerve roots (but not very sensitive in detecting damage of single roots) | Establishing the integrity of the ventral roots and certain peripheral motor nerves |

## V. FREE-RUN AND TRIGGERED EMG (fEMG AND CMAPs)

### 1. Principle

Free-run EMG (fEMG) consists of recording spontaneous muscle activity, thus allowing its live assessment. The latter depends on the integrity and function of the muscle fibers, neuromuscular junction, peripheral nerve, ventral root, alpha motor neuron, and its spinal interneuronal synapses. Although the assessment of muscles and neuromuscular junctions occurs in the outpatient lab, in intraoperative neurophysiology we use this technique as a monitoring tool for detecting surgically driven mechanical irritation of the peripheral nervous system and of the cranial nerves, hopefully before irreversible damage to these structures had occurred.

Triggered EMG consists of applying an electrical stimulus, directly on the peripheral motor nerves or roots, for eliciting CMAPs to be recorded in the corresponding muscle channels. Thus, it can be used as a mapping tool for detecting the location of peripheral or cranial nerves that may be difficult to distinguish from tumoral, fibrous, and fatty tissues. Triggered EMG can also be used in checking the functions of injured (or that are at risk for injury) nerves, roots, or trunks by assessing the electrical transmission through such structures and comparing it with a healthy (or presurgical) baseline.

### 2. Methodology

#### A. Stimulation

fEMG does not require stimulation and is based entirely on recording spontaneous and detecting irritation-driven muscle activity. Triggered EMG involves applying an electrical stimulus directly on the nervous structures.

*A1. Types of stimulating electrodes.* Both direct nerve bipolar and monopolar stimulation can be used in peripheral and cranial nerve surgery. In cases of bipolar stimulation, the surgeon uses a handheld bipolar stimulator (e.g. VIASYS) with disposable ball tips, right angle and curved tips. The latter are used as hooks to lift the nerve of interest and thus better isolate it from the surgical field, preventing the spread of current. It is very important not only that the tips of the stimulator do not touch each other but that they are also at a reasonable distance. Otherwise, appropriate stimulation will not occur. The distance between the tips varies, from 6 to 8 or 10 mm, and it could also be adjusted manually by the surgeon. In general, bipolar stimulation requires less current for stimulation and also reduces the current spread. In the case of increased amount of fluid, especially when the stimulating prongs are too close, current shunting can occur.

A monopolar handheld stimulator (e.g. MEDRONICS XOMED) can also be used, connected to the cathode (–) of a stimulator pod, while the latter is connected to the stimulator box. A sterile needle electrode is placed in the surgical field and connected to the anode (+). The monopolar stimulation is prone to current spread in the surgical field. Thus, it tends to be more sensitive and less specific, by depolarizing and detecting nerves that are not only directly stimulated but also situated adjacent to the immediately stimulated structure. Its direct application consists of ensuring that no such nervous structures are in the way of the resection path.

In the case of peripheral nerve repair surgery, when there is insufficient exposure of the nerve, or if the surgeon is reluctant to extend the incision, big nerves with a well-known trajectory (e.g. median, ulnar, radial) can be stimulated via two sterile surface electrodes applied on the skin. However, in most of our cases of peripheral nerve repair surgery, both stimulation and recording are done directly on and from the exposed nerve.

*A2. Stimulation sites.* In order to prevent damage of the nerve roots, peripheral nerves, and cranial nerves, structures to be resected during tumor removal (e.g. cauda equina, plexus, peripheral nerve, cranial nerves, or brainstem tumors) are stimulated. In cases of cervical or lumbar decompression level(s), the stimulated nerve roots correspond to the spinal level at which the surgery takes place. Similarly, in brachial plexus or peripheral nerve repair surgery, the site of stimulation is dictated by the level of injury. Prior EMG/NCS results, done in an outpatient setting, as well as the neurologic examination, can help in many circumstances to localize the level of the nerve injury. This, in turn, helps the surgeon decide on the length and extent of exposure necessary. It is important to discuss with the surgeon in advance and plan the intraoperative neurophysiology procedures accordingly. Chapter **6** includes details about the surgical preplanning (pages 284–85), as well as several figures and tables to help in choosing the appropriate nerves and sites of stimulation (Table **6-1**, page 272; Fig. **6-7**, pages 280–81).

*A3. Stimulation technique*

***a. Peripheral nervous system monitoring.*** For stimulation of peripheral nerves, roots, and nervous trunks, we generally use a bipolar stimulator. A handheld probe with hooked tips allows lifting the nerve and isolating it from the adjacent structures. This is an important step in limiting the current spread to other structures, particularly in cases of a wet surgical field, when the current may bypass the damaged portion of the nerve and stimulate the healthy part of the nerve, distal to the lesion, thus offering a false-positive response. However, we usually rely on a probe with straight tips. This successfully stimulates the nerve or root, if the tips are kept straight on it and if the nerve is well isolated from the other structures in the surgical field. The stimulus is triggered at 2.11 Hz and pulse duration of 0.1–0.3 msec. For nerve root stimulation we expect a threshold from 2–6 mA, at a pulse width of 0.2 msec. Chronic radiculopathies in spondylosis, existence of severe, diffuse neuropathies in diseases such as diabetes or chronic renal failure, or posttraumatic nerve injury can result in an abnormally high threshold and may require higher intensities and pulse durations. Also, in pediatric surgeries, with patients under the age of 2 years, before complete myelination has occurred, the threshold can be higher for the same fiber diameter.

When stimulating smaller peripheral nerves, we use repetitive pulses, at the same frequency, but with smaller widths, anywhere between 0.05 and 0.2 msec. In general, motor fibers require shorter pulses. The threshold intensity necessary to trigger a positive response also varies with the fiber size, pulse width, and integrity of the nerve, with most of the responses obtained between 0.5 to 3 mA. Again, an injured nerve or small regenerating fibers require a higher stimulus amplitude, but we avoid increasing the intensity by more than 5 mA for 0.2 msec and 20 mA for 0.05 msec pulses because of the risk of current spread and increase in stimulus artifact.

***b. Cranial nerves monitoring.*** For cranial nerve stimulation, we generally use the monopolar technique. Our stimulation technique consists of delivering constant current, with repetitive square wave pulses of 0.1–0.2 msec, intensity 0.1–1 mA, frequency 4 Hz, averaging 4–8 trials. Stimulus intensity can be increased in small increments, of 0.1 mA, until a response is observed or until a maximum allowable stimulus intensity is achieved. In all cases, one must keep in mind the potential to injure directly stimulated tissue. However, using the monopolar technique, we were able to reach the nerve depolarization threshold at stimulus intensities as low as 0.1 mA and never had to increase the intensity higher than 1 mA. Of course, the threshold value depends on a variety of factors, such as the state of the nerve (healthy versus injured), as well as on the other stimulus parameters. Likewise, it should be remembered that the risk of injury is a function not just of current intensity but of the charge delivered per unit of time and per unit of electrode surface area, thus depending on the pulse duration and pulse frequency, as well as on the surface of the stimulating electrodes.

Along these lines, Chang et al. (1999) described a technique of stimulating cranial nerves nuclei where he used 10–12 mA (stimulated in increments of 0.1 mA, starting at 0.1 mA) as a upper limit of stimulation. However, the pulse durations used were 0.02–0.05 msec. Conversely, Kimura (1989) found that stimulation at less than 0.05 msec was ineffective, whereas stimulation with a pulse of >1 msec in width was not always well tolerated.

Standard prong-type bipolar electrodes may also be used with a gap of around 1 mm, or concentric bipolar electrodes may be employed (Schlake et al., 1999).

### B. Recording

*B1. Recording electrodes, sites, and channels.* The recording of CMAP responses in triggered EMG and fEMG is done via electrodes placed at the level of the muscles innervated by the roots or peripheral nerves of interest. Thus, the recording technique is somewhat similar to that of mMEPs recordings in tcMEPs monitoring. For details regarding types of electrode, recording sites, and recording technique, see pages 18–23. Under special circumstances, the muscles selection for peripherally triggered CMAPs or fEMG needs to be more specific in identifying a certain nerve or myotome. In such cases, using two recording electrodes per muscle can increase the specificity of the recording. Thus, a recording channel represents one instead of two muscles. In order to decrease the amount of ambient noise introduced in the recording, we prefer the use of subdermal rather than surface stick-on electrodes.

According to the level and type of surgery, different muscle groups are chosen for monitoring. Table **1-3** shows the main muscles from the upper and lower limbs used in surgeries involving cervical and low thoracic high lumbar levels. Triggered EMG is also used in peripheral nerve repair, together with nerve-to-nerve recording (see pages 33–37). On pages 280–81, Fig. **6-7**, and Table **6-1** (page 272), we provide examples of the muscles innervated by several peripheral nerves. This information becomes useful when the neurophysiologist and the surgeon decide which muscles to use for CMAP recordings, depending on the location of the injury within a specific nerve. Last, as already mentioned, fEMG and triggered EMG are essential parts of cranial nerve monitoring. Table **1-5** lists the muscles used for monitoring and mapping of different cranial nerves; Fig. **1-11** shows their insertional points for the recording electrodes. Specifics on recording techniques for cranial nerves monitoring and mapping are given in Chapter 7 (pages 299–324).

*B2. Recorded potentials.* Depending on the technique, the recorded muscle activity can be either a CMAP, similar to the mMEPs described in section **IV** (triggered EMG), or spontaneous motor unit potentials (fEMG). Both types are further detailed on pages 31–32.

*B3. Recording technique.* Although the recording sites and recording channels are the same for fEMG and triggered EMG, the recording parameters differ. For fEMG we use a time base of 100 msec/div (cranial nerves) or 200 msec/div (peripheral nervous system) and a sensitivity of 50 μV/div. For triggered CMAPs, we use a time base of 5 msec/div for both upper and lower limbs muscles and 3–5 msec/div for cranial nerves muscles. We usually have the sensitivity set at 30–50 μV/div, but these can vary. In general CMAP responses vary in amplitude quite a bit, and they can be as high as a few millivolts (mV). The muscle activity has a significant amount of high-frequency oscillations—thus, HFF can be set up in the kilohertz (kHz) range. We use an LFF of 3 Hz and an HFF of 3 kHz (although higher HFF can also be used, up to 10 kHz). Because of the significant amplitudes of the CMAPs, the rejection windows have to be increased, so that high voltage activity is not rejected. Fig. **1-12** shows the recording and stimulation program setup.

## 3. Interpretation

### A. Anatomy

For an accurate setup and interpretation of the fEMG and triggered EMG results, it is necessary to have a good understanding of the anatomy of the nervous pathways that require monitoring and mapping. As an example, the decision regarding which muscles to record from in cervical and lumbar spine surgeries is based on the myotomal innervation of each muscle (see Table **1-3**). Another example is that detailed knowledge of the anatomy of brachial or lumbosacral plexus is essential in performing reliable intraoperative neurophysiology during exploration and repair in posttraumatic plexopathy. As already

**TABLE 1-5** Muscles used for monitoring and mapping of the cranial nerves

| Cranial nerve | Muscle(s) used for recording |
|---|---|
| Oculomotor (III): motor portion | Medial rectus (extraocular muscle [eom]) |
| Trochlear (IV) | Superior oblique (eom) |
| Trigeminal (V): motor portion | Masseter |
| Abducens (VI) | Lateral rectus (eom) |
| Facial (VII) | Orbicularis oris<br>Orbicularis oculi<br>Mentalis<br>Frontalis |
| Glossopharyngeal (IX): | |
| motor portion | Stylopharyngeus |
| sensory/afferent part of the gag reflex | Soft palate |
| Vagus (X): | |
| recurrent laryngeal nerve | Laryngeal (vocal cord, posterior arytenoids muscles) |
| pharyngeal branch (motor/efferent part of the gag reflex) | Pharyngeal muscles/soft palate |
| Accessory (XI) | Trapezius |
| Hypoglossal (XII) | Tongue |

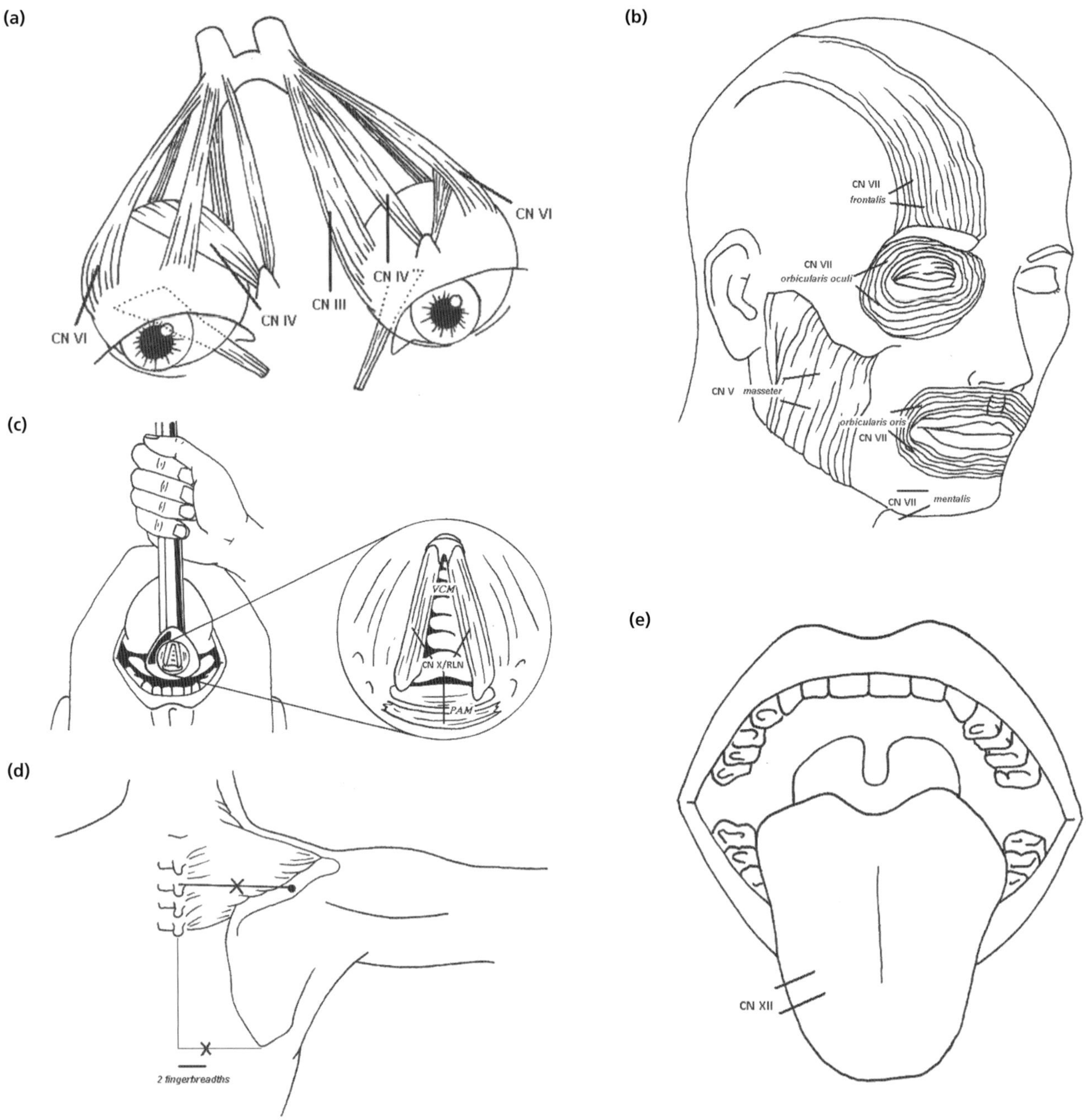


**FIGURE 1-11** Muscles used in cranial nerves monitoring and mapping procedures, with the insertional points for the needle recording electrodes. **a:** Medial rectus: CN III innervated; Superior oblique: CN IV innervated; Lateral rectus: CN VI innervated. The electrode placement for the extraocular muscles is usually done by an ophthalmologist. **b:** Masseter: CN V innervated. Insertion: two fingerbreadths posterior to its anterior margin and one fingerbreadth superior to the inferior side of the mandible. Maneuver: clenching of the teeth. CN VII innervated: Orbicularis oculi. Insertion: at the lateral portion of the eye fossa, while the needle makes a 25–30 degree angle with the skin, it is inserted in a medial and downward and upward direction. Orbicularis oris. Insertion: lateral to the angle of the mouth (one fingerbreadth), while the needle makes a 20-degree angle with the skin, it is directed toward the upper and lower lip. Frontalis. Insertion: superior to the orbital margin (one fingerbreadth) and lateral from the midline (two fingerbreadths). **c:** Vocal cord and posterior arytenoids muscles: CN X innervated (RLN). For electrode placement, see (pages 318–20). **d:** Trapezius (middle): Innervated by the spinal portion of the accessory nerve ( CN XI) and portions from C3 and C4. Insertion: midpoint on a line drawn between the midpoint of spine of scapula and spinous process of vertebra at the same level. Maneuver: adduct scapula by elevation of arm. For the cranial portion of CN XI, it is recommended to place the needle in the descending (or lower) portion of the trapezius muscle (not shown). Insertion: on a line from the inferior angle of the scapula and perpendicular to the spine, two fingerbreadths lateral from the spine. Maneuver: extension of the arm over the head. **e:** Hypoglossal muscle: CN XII innervated. Insertion: anterior two thirds of the tongue.

mentioned, Chapter **6** offers schematic representations of the brachial and lumbosacral plexus, as well as of several upper and lower limbs peripheral nerves and their innervated muscles, which are most commonly involved in peripheral nerve surgery (Fig. **6-7**, pages 280–81). I include such schemes as part of Chapter **6** rather than in this chapter because I consider them to be essential for the presurgical planning and for customizing certain monitoring techniques that, in turn, are essential for and specific to peripheral nervous system monitoring. Similarly, in Chapter 7, an increased emphasis is placed on the neuroanatomy of the brainstem, cranial nuclei, nerve root entries, and cerebello-pontine angle (CPA) (Fig. **7-1**, page 300). This knowledge is essential for the surgeon to decide on a certain surgical approach and trajectory and for the neurophysiologist to identify the nervous structures at risk and thus in need of monitoring.

## *B. Types of Neurophysiologic Responses*

*B1. fEMG activity.* This type of recording shows spontaneous firing of certain muscles, represented by spontaneous motor unit potentials (MUP). Unless a significant neuromuscular blockage is achieved via the use of muscle relaxants, muscles present spontaneous activity that occurs at random (Fig. **1-13**). The lighter the anesthesia is, the more significant is such spontaneous activity. All of our intraoperative recordings, regardless of the modality used, include a raw data window. This displays a free-run activity, recorded either from scalp or from muscle channels. As explained in Fig. **1-13**, in the case of free-run scalp recording, filter settings allow either a good sampling of the scalp muscle activity (e.g. LFF = 30 Hz and HFF = 1 kHz) or of the EEG activity (LFF = 1 Hz, HFF = 70 Hz). Of course, time base and sensitivity settings are also important. Similarly, live fEMG data are recorded from muscle channels. This raw data window allows an appreciation of the depth of anesthesia. In cases of light anesthesia, self-arousals can occur, which have the appearance of more muscle activity in the fEMG channels, sometimes as a consequence of voluntary contraction. Quiet EMG channels are seen in the cases of iatrogenic neuromuscular blockage. This should further help in recording evoked potentials such as SSEPs (in particular, SSEPs recorded in the neck channel) or brainstem auditory evoked potentials (BAEPs), which are known to have a small amplitude, by increasing the signal-to-noise ratio.

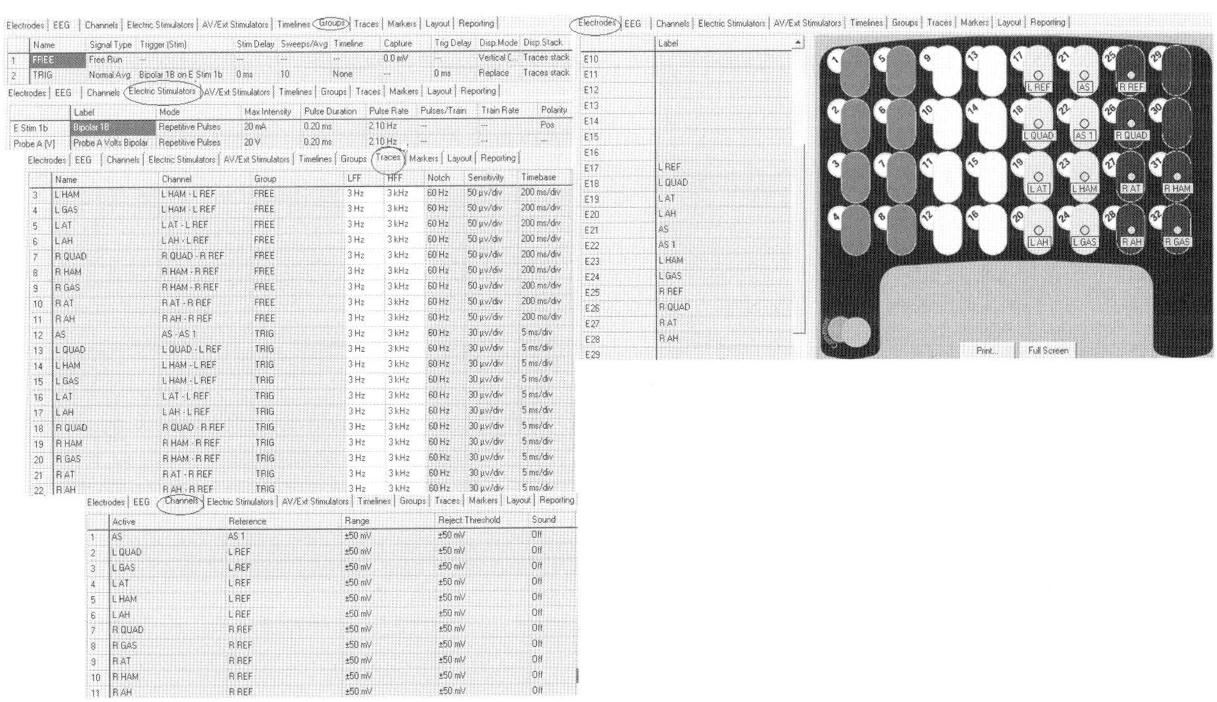


**FIGURE 1-12** Setup for free-run and triggered EMG technique: recording electrodes, channels, traces, groups, and the assigned stimulators.

However, the main role of fEMG is the appreciation of abnormal responses. These may signal irritation of the sensory and/or motor components of a peripheral or cranial nerve and are further detailed on pages 273–76, 278–79, and 308, respectively.

*B2. Triggered EMG activity.* This activity is represented by CMAPs, which are a sum of motor action potentials that arise in several muscle fibers. Thus, they tend to have more variable morphology, amplitudes, durations and latencies. In general, the morphology becomes more complex, the amplitudes higher, their latencies shorter, and their durations longer, with increases in stimulus amplitudes. However, morphology of the CMAPs may be still less complex than that of mMEPs triggered in the same muscle by direct or transcranial electrical stimulation. This seems to be due to a more restricted depolarization of a smaller number of muscle fibers. In fact, biphasic and triphasic CMAPs are frequently seen with peripheral stimulation in healthy muscles. Their characteristics can also differ from one muscle to another, based on the myotomal innervations of the muscles (thus on the distance between the stimulating and recording sites), on their function and strength. Fig. **1-14** is an example of CMAPs triggered by stimulation of lumbar roots. Similarly, Fig. **7-3** (page 306) shows examples of triggered CMAPs in the innervated muscles of the cranial nerves. For further details regarding CMAP responses and specifically how are they used in intraoperative neurophysiology, see Chapters **6** and **7** (pages 267–324).

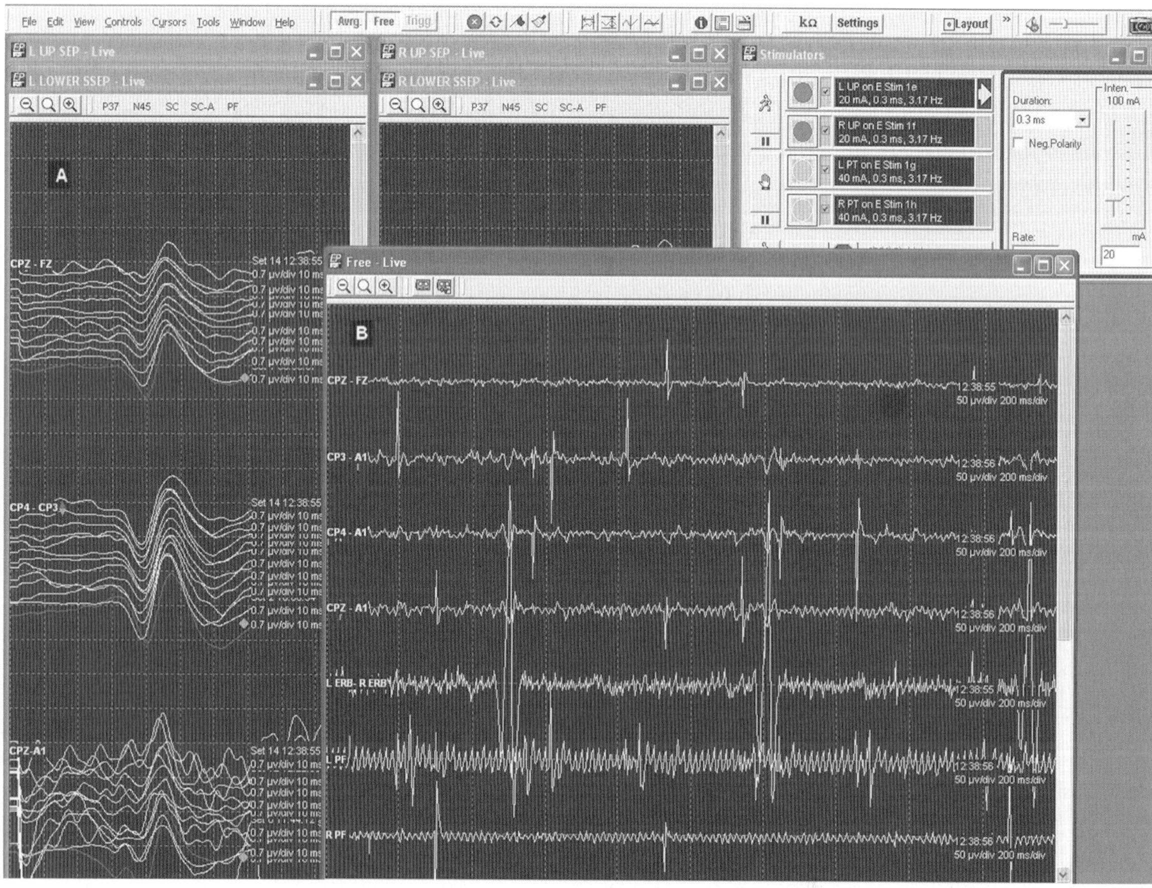


**FIGURE 1-13** Example of spontaneous muscle activity, seen in the raw data window (panel **B**), during SSEPs recordings (panel **A**). In panel **B**, notice the muscle spikes occurring randomly in all channels. The recording, done via scalp electrodes, can be used for appreciating EEG, free EMG activity, as well as the presence of electrical or electrode artifact. The assessment of these activities is important for the troubleshooting process during SSEPs recording. Because these activities differ in their frequency range, emphasis on one or another is also given by the filter settings. In the picture, the recording was done with an LFF at 30 Hz and an HFF at 500 Hz, thus favoring the appreciation of scalp EMG activity and high-frequency electrical artifact.

### *C. Applications in Intraoperative Neurophysiology*

The spontaneous and triggered activity of the muscles corresponding to different myotomes can give useful information to the neurophysiologist about the function of the peripheral part of the skeletal motor system, as well as the motor function of the cranial nerves. Thus, in intraoperative neurophysiology, we use this technique to monitor, map, and preserve the integrity of the peripheral nervous system and cranial nerves. fEMG and triggered CMAPs responses are used in neurosurgical, orthopaedic, and plastic surgery that could potentially result in damage of the ventral cervical and lumbosacral nerve roots, filum terminale, brachial and lumbosacral plexus, as well as of the peripheral nerves. The main surgeries benefiting from this technique are cervical and lumbar spine decompressive surgery for spondylosis, trauma or extradural processes (e.g. bone metastasis), tumors involving the filum terminale root and peripheral nerves, exploration and repair in posttraumatic plexopathy, peripheral nerve injuries, and brainstem or posterior fossa surgeries that could result in damage of the cranial nerves and their nuclei. Table **1-6** gives a list with the main surgical procedures in which fEMG and triggered EMG can become useful. Of note, details about specific applications can be found in the corresponding chapters (e.g. fEMG and triggered EMG in cranial nerve monitoring further detailed in Chapter 7). Last, a general utility of fEMG is that of determining variations in spontaneous muscle activity, such as detecting arousal responses, assessing the effects of anesthesia (e.g. the efficiency of the neuromuscular blockage), or in troubleshooting during procedures requiring motor evoked responses, either by detecting muscle electrodes with inadequate impedances or assessing the amount of ambient noise.

## VI. NERVE-TO-NERVE OR NERVE ACTION POTENTIALS (NAPs) RECORDINGS

Stimulation of the peripheral nerves and recording of either the CMAPs (nerve-to-muscle recordings) or NAPs (nerve-to-nerve recordings) become important in cases of peripheral nervous system surgery, such as nerve repair or brachial plexus exploration. The nerve-to-muscle

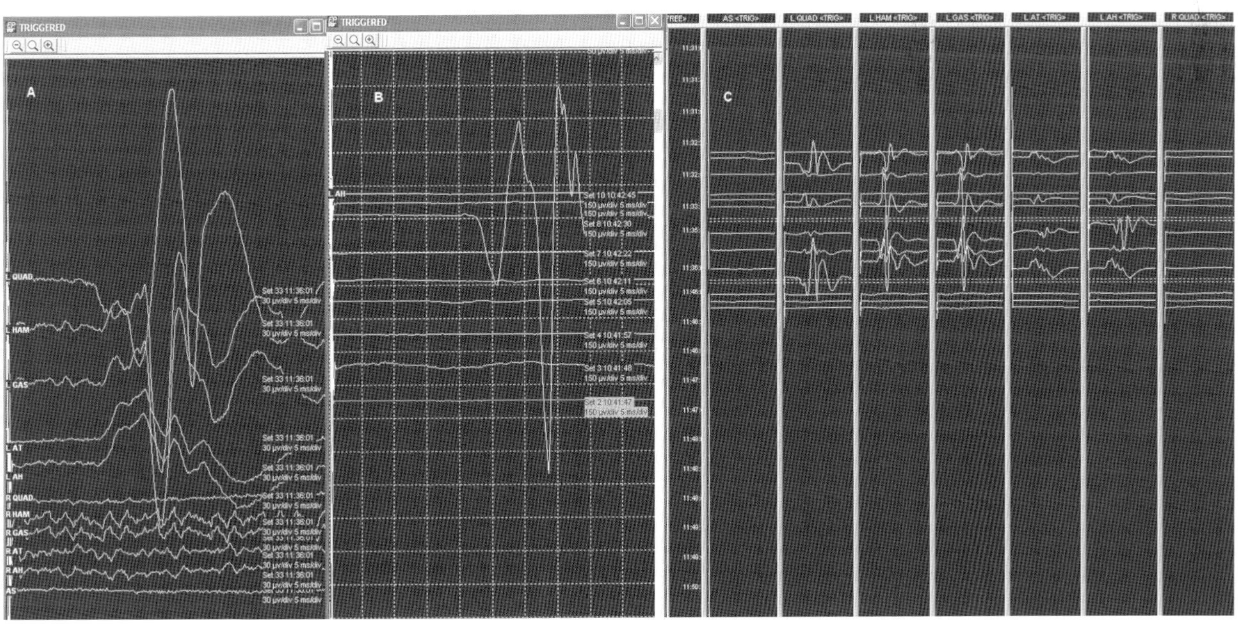


**FIGURE 1-14** Triggered EMG activity. Panel **A** shows CMAPs, occurring in all left lower limb muscles. Panel **B** shows one CMAP triggered in the left AH. Panel **C** shows the history log, displaying the activity in each muscle, presented chronologically (each muscle is represented within a column). Notice the complex morphology and high amplitude of the triggered CMAPs. The labels used for muscle channels are the same as in Fig. **1-7**. Panel A: sensitivity 30 µv/div, time base 5 msec/div. Panel B: sensitivity 150 µv/div, time base 5 msec/div.

**TABLE 1-6** FEMG and CMAPs: Applications in intraoperative neurophysiology

| Region | Monitoring (with fEMG) | Mapping (with CMAPs) |
|---|---|---|
| Peripheral nervous system | ■ Detecting spontaneous muscle activity, thus mechanical irritation<br>1. Tethered cord<br>2. Cauda equina tumors<br>3. Cervical/lumbosacral root surgery: intradural tumor removal<br>4. Cervical/lumbosacral decompressive surgery (e.g. extradural tumors, spondylosis, trauma)<br>5. Pedicle screw positioning<br>6. Posttraumatic brachial plexus/peripheral nerve repair<br>7. Brachial plexus/peripheral nerve tumor removal | ■ Triggering CMAPs, thus identifying nervous tissue<br>■ Triggering CMAPs, thus assessing the functional integrity of peripheral motor nerves<br>■ Triggering CMAPs, thus assessing the correct position of the pedicle screws<br>Same indications as for monitoring |
| Cranial nerves | ■ Detecting spontaneous muscle activity, thus mechanical irritation<br>See Table **1-8** and Chapter **7**, Table **7-1** | ■ Triggering CMAPs, thus identifying nervous tissue<br>■ Triggering CMAPs, thus identifying the cranial nerves nuclei, at the floor of the fourth ventricle, for defining a path of entry for brainstem tumor resection<br>■ Assessing the integrity of the nerves (e.g. seventh nerve) after resection of a CPA angle tumor, such as acoustic neuroma<br>■ Assessing the integrity of the recurrent laryngeal nerve (RLN) in thyroid surgery<br>Same indications as for monitoring |
| General role | 1. Assessing the depth of anesthesia (e.g. neuromuscular blockage)<br>2. Troubleshooting: detecting high impedance channels and ambient noise | |

recordings were discussed as part of triggered EMG recording (pages 27–33). This section refers to nerve-to-nerve recordings. NAPs can also be registered directly from the cochlear nerve. However, this technique is somewhat different and is further described in detail on pages 315–16.

## 1. Principle

Direct nerve electrical stimulation triggers a depolarizing volley that, if transmitted from the stimulating to the recording site, is recorded as a sensory NAP (SNAP) or a mixed NAP (MNAP). Many authors refer to it as compound NAP (CNAP), representing the sum of action potentials of individual fibers. In this chapter, for simplification, we refer to it as NAP. This confirms the functional continuity of the axon between the two sites. The stimulus amplitude necessary for generation and transmission of such potential (threshold) is higher in diseased than in healthy nervous tissue. Thus, comparison of such thresholds, between a diseased and a healthy nerve of similar size, allows appreciation of the severity as well as the location of injury. When an axon is injured and its continuity is interrupted, the distal part undergoes Wallerian degeneration, whereas the proximal part starts sprouting new axons that are smaller in diameter than the axon that was injured. Moreover, its excitability is lower than that of the axon that was injured. The conduction velocity is slower, and these fine fibers are not depolarized by pulses of short width. However, in order to become functional and to reestablish the electrical transmission, these axons need to increase in diameter. Despite the presence of a regeneration process, an effective regeneration may never occur. Thus the nerve-to-nerve recording is an important tool in assessing the appearance of an effective regeneration. The process of nerve regeneration is a slow process; it occurs over a period of several months. Thus surgical exploration needs to be postponed for several months after the injury (see also pages 284–86).

## 2. Methodology

The methodology for intraoperative nerve-to-nerve recordings differs in several aspects from that of nerve conduction studies done in the outpatient lab. The following describes the particularities of the former rather than of the latter. Nerve recordings utilize more or less the same stimulation technique as that necessary for peripherally triggered EMG.

### A. Stimulation

*A1. Types of stimulating electrodes.* Page 27 describes the bipolar handheld stimulators used in peripheral nerve recordings.

*A2. Stimulation sites.* Also, on page 27 the importance of choosing the site of stimulation appropriately, in accordance to the level of injury, is emphasized. Again, Chapter **6** (pages 267–98) offers details of topography and neuroanatomy of the brachial and lumbosacral plexus, as well as for certain peripheral nerves (see Fig. **6-7**, pages 280–81).

*A3. Stimulation technique.* Detailed description of the bipolar stimulation technique, including stimulus parameters, is given on pages 27–28. For peripheral NAP recordings, stimulation can be done in both directions (proximal to distal and distal to proximal). However, the direction of stimulation is also important. If we stimulate proximally and record distal to the lesion, in the normal direction of action potential propagation, the NAP response tends to be higher than if we stimulate in the opposite direction (stimulator kept distally and recording electrode proximally). In the latter situation, if a nerve bifurcation exists between the stimulator and recording electrodes, the stimulated nerve is smaller than the nerve where the recording is done; there are fibers that join the nerve at the bifurcation and that were not stimulated. Thus, the NAP response obtained is relatively small. However, this direction (distal to proximal) is more likely to eliminate the muscle response (CMAP), which may otherwise make the recognition of the much smaller NAP very difficult.

In order to reduce the amount of stimulus artifact, starting with short pulses is helpful. With short pulses, motor axons are more likely to be activated (Veale et al., 1973). However, higher widths may be needed for smaller sensory fibers, as well as for diseased and for fine regenerating fibers. The NAP has a fixed latency and an all–or-none response, depending on whether the stimulation is done at/above or below threshold. However, because of its small amplitude (see the following), we like to average several trials, at least 5 trials and sometimes up to 100–200. This can be difficult because the surgeon has to keep the stimulator steady, in the same place for several seconds.

More details regarding troubleshooting stimulation can be found on pages 287–88 and 290.

### B. Recording

*B1. Recording electrode, sites and channels, and recorded potentials.* We use a bipolar handheld probe (e.g. VIASYS) as recording electrodes, which has its tips bent: this helps in picking up and lifting the nerve, thus isolating it from the surrounding structures. The spacing between the tips varies according to the stimulator used and could also be gently adjusted by the surgeon. In any case, the tips of the probes need to be at a sufficient distance so that they can record a difference in potential between two regions of the nerve (with which they are in contact), from which only one is depolarized. If they are too close, they both record from an active part of the nerve, leading to a very small NAP or failure to record one. This is because, during an electrical transmission through a myelinated axon, not one but a few nodes of Ranvier are depolarized at once. Thus, the part of the axon that is active at once is relatively long (Happel and Kline, 2002). Subsequently, in the case of recording from large, myelinated fibers, the distance between the recording tips needs to be bigger than the distance between the tips for small fibers. For large fibers, we recommend at least 5 mm between the tips; for smaller fibers, the distance can be reduced to 2–5 mm.

In addition to the direct nerve recordings via the bipolar probe, when there is not enough exposure of the nerve, the recording can also be done indirectly, via surface electrodes (e.g. disposable Ag AgCl gelled disc electrodes, VIASYS or ROCHESTER) placed on the skin over the presumed location of the nerve.

A NAP is usually a biphasic potential, time-locked to the stimulus. Thus, even with an average stimulation, it appears to be still on the screen, in the same place. As the stimulus intensity increases, its amplitude increases as well, until it reaches a maximum plateau. The amplitude is, in general, small, much smaller than that of a CMAP, usually no more than 100–150 μV and sometimes as small as 2 μV. The latency is also short, usually less than 2 msec, but this varies, depending on the conduction velocity of the segment of the nerve that is studied (the more proximal this segment is, the higher the velocity is), as well as on the distance between the stimulation and recording sites (see the following). When a mixed nerve is stimulated, a CMAP, in addition to the NAP, can be obtained. This muscle potential, or simply the muscle twitches of the innervated muscle, could contaminate the recording, making the differentiation of the NAP difficult. In such circumstances, the stimulus intensity should be decreased until muscle contractions stop. Moreover, there is a possibility of falsely interpreting a CMAP as a NAP. There are several characteristics that help in differentiating between the two. A latency longer than 2 msec, amplitude higher than 1 mV, and a polyphasic morphology with a variable duration and latency should point toward a CMAP; a maximal amplitude that could not be further increased with increase stimulus intensity, as well as a frozen appearance on the screen, with fixed latency, point toward a NAP. Fig. **1-15** shows a NAP; notice the difference between a NAP and a CMAP (see Fig. **1-14**).

*B2. Recording technique.* While a stimulator is kept in place on the nerve, recording is done via a hooked bipolar handheld probe. The distance between the stimulator and the recording probes is very important in determining the expected latency of the NAP. The latter is also based on the nerve conduction velocity: NAP latency = distance(stim – rec)/conduction velocity. In general, a distance of at least 5 cm between the stimulating and recording electrodes is necessary because of the very short latencies of NAPs, which otherwise would be obscured by the stimulus artifact. The extensions of the two tips are connected in two separate electrode positions in the headbox of the amplifier. The recording channel uses the two positions. The recording parameters are 7–50 μV/div, because of the relatively small amplitudes of the NAP, with a short time base of 0.5–2 msec/div as a result of their small latencies. Of note, the recording band frequency has to include the range of frequencies contained in a NAP response. We use an LFF of 30 Hz and an HFF of 3 kHz (see Fig. **1-15**). As in the case of other neurophysiologic recordings (see pages 10–11), in order to reduce the amount of artifact, we use a larger surface electrode as common or isolated ground. Because the surface ground electrode usually has a larger surface, its application should be done in the very beginning, before the prepping and sterilizing of a large skin area. Details of troubleshooting recording can be found on pages 290–92.

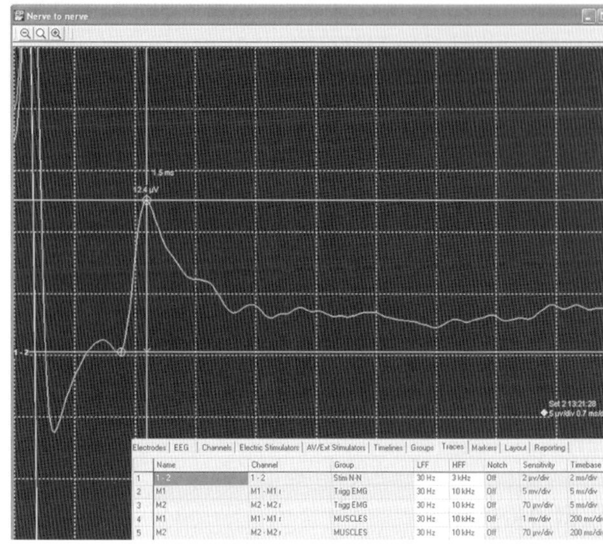


**FIGURE 1-15** Nerve action potential recording (NAP). Notice the much smaller amplitude (12.4 μV) and shorter latency (1.5 msec) compared to the CMAP responses in Fig. **1-14**. As a consequence, notice the shorter time base and higher sensitivity necessary for nerve recordings (1–2 channel) compared to triggered EMG recordings (M1–M1r channel). Time base is 0.7 msec/div; sensitivity 5 μv/div.

## 3. Interpretation

Interpretation of the nerve-to-nerve recordings needs to be done within the particular context of each case and is further detailed on pages 284–98.

### A. Applications in Intraoperative Neurophysiology

The nerve-to-nerve method is used to localize an injury site on the nerve and to assess its degree of severity. It is also used to determine whether the axonal regeneration has occurred and whether it is efficient in preserving, at least partially, the function of the nerve. This is important information sought during nerve repair surgery, when nerve grafting is done if axonal regeneration either hasn't occurred or is inadequate, without further hope for improvement. It is also used in cases of nerve tumor resection, to guide the resection of the tumor while preserving the functional nervous tissue.

As already mentioned, this method is frequently used together with nerve-to-muscle recording (triggered EMG or CMAP recording) and is further detailed on pages 290–92.

Table **1-7** summarizes the surgical procedures in which nerve-to-nerve recordings are helpful.

**TABLE 1-7** NAPs recordings: Applications in intraoperative neurophysiology

| Region | Mapping |
| --- | --- |
| Peripheral nervous system | 1. Posttraumatic brachial plexus/nerve repair<br>2. Nerve tumor resection (e.g. neuroma, schwannoma, neurofibroma)<br>3. Resection of tumoral tissue adjacent to the nerve (e.g. lipoma) |

## VII. BRAINSTEM AUDITORY EVOKED POTENTIALS (BAEPs)

### 1. Principle

BAEPs are short-latency potentials reflecting the depolarization of several structures within the auditory pathways, because they are traversed by electrical volley triggered by the peripheral stimulation of the cochlear nerve. These evoked responses are far-field potentials, being recorded by electrodes placed on the scalp, with the exception of wave I, which could be considered, in that sense, to be closer to a near-field potential. By assessing the amplitudes and latencies of these evoked responses, one can thus analyze the functional integrity of the corresponding anatomic structures. By recording the electrical transmission within the lower parts of the auditory pathways, from the cochlea to the upper pons, BAEPs are a good tool in assessing the integrity not only of the eighth nerve, but also of the brainstem structures involved in hearing, thus indirectly of the brainstem itself (as the name suggests). Of note, damage of the auditory pathways above the lower midbrain does not affect the BAEPs (see Fig. **1-16**).

### 2. Methodology

#### A. Stimulation

*A1. Types of stimulating electrodes and stimulation sites.* For recording, we use XLTEK disposable foam ear tips, connected via a front tube to a transducer, one for each ear. The foam ear tips are variable in diameter, thus ensuring a snug contact with the auditory canal.

*A2. Stimulation technique.* The stimulus is a click, produced by delivering a square wave electrical pulse to a transducer. The sound travels via a plastic tube to be delivered inside the auditory canal of the stimulated ear. The click intensity is varied and measured in peak equivalent sound pressure level (peSPL). This is the standard measure of the loudness of the sound. Stimulation is usually carried out at 100–110 dBpeSPL. It is known that the click stimulates both the ipsilateral ear as well as the contralateral ear, the latter through bone conduction. This would cause bigger BAEPs that are nonspecific as to the laterality of stimulation. Masking the contralateral ear with white noise at about 40 dB less than the stimulus intensity is therefore carried out. We can use three types of stimulus polarities. Rarefaction polarity helps in identifying wave I, as well as separate waves IV and V. Condensation polarity click helps in identifying wave V. We generally use alternating polarity clicks. This polarity is useful in the case of significant cochlear microphonics and stimulus artifact that make identification of wave I difficult. These first switch their polarity with reversal of the stimulus polarity, thus canceling each other out in the averaging process, whereas the latter (wave I) does not. Because of their very small amplitude to the fact that these recordings are, in many situations, contaminated with muscle artifact, we average about 1,000–2,000 trials. Of course, this number varies, according to the strength of the signal and the amount of noise present. The frequency applied varies as well, from one institution to the other, but it is usually between 10 and 40 Hz. As with the SSEPs, the advantage of high frequency is faster feedback to the surgeon, although increasing the rate can have a significant impact on the morphology and amplitudes of the waveforms and can thus be a problem in case of low-amplitude, abnormal waveforms. If the waveforms have good amplitudes, we use 20.17 Hz; if the waveforms are abnormal, we decrease the rate stepwise, until the highest acceptable rate is achieved (12.17 Hz is the rate that we prefer in such cases). In order to avoid amplification of the 60-Hz artifact, a rate that is not evenly divisible by 2, 3, or 6 should be used, with a pulse duration of 0.1 msec.

#### B. Recording

*B1. Recording electrodes, sites and channels, and recorded potentials.* The recording electrodes are subdermal needle electrodes, like those used for mMEPs recording. The electrodes are placed in the ear lobes (A1 and A2) and at Cz. The recording channels are Ai-Cz, Ai-Ac and Ac-Cz (where Ai is the ipsilateral ear and Ac is the contralateral ear). The action potential in the distal part of the cochlear nerve appears at and around the

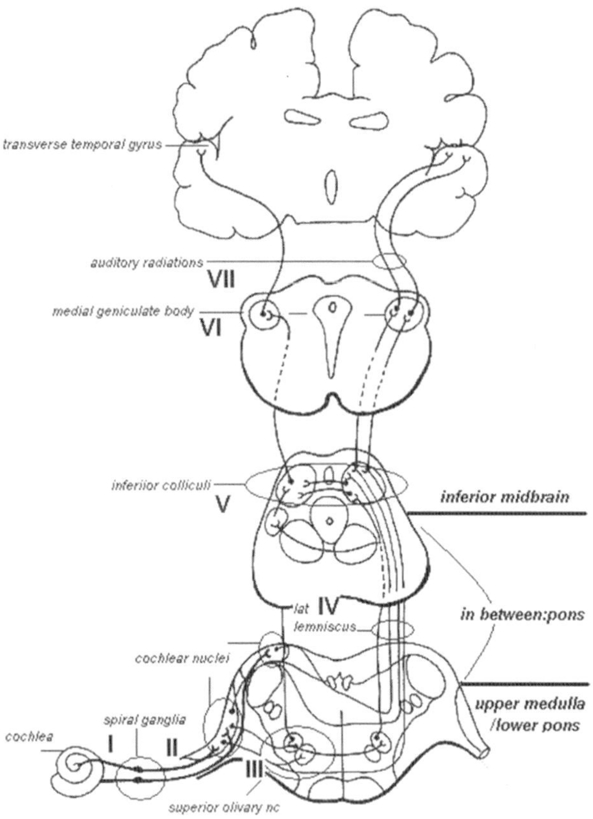


**FIGURE 1-16** Schematic representation of the auditory pathways. Notice the neurophysiologic-neuranatomic correlation between BAEPs and different levels within the auditory pathway.
Wave I: distal portion of the cochlear nerve.
Wave II: proximal portion of the cochlear nerve/cochlear nuclei.
Wave III: superior olivary complex, situated at the junction between lower and midpons (thus, this complex is not situated in the same plan with the other structures in the medulla but higher, at the level of the lower pons).
Wave IV: lateral lemniscus in the midpons.
Wave V: lateral lemniscus in the upper pons and/or inferior colliculus.
Wave VI: medial geniculate body (thalamus).
Wave VII: auditory radiations.

ipsilateral ear as a negativity. Thus, in Ai-Cz and Ai-Ac channels, wave I occurs as a negative, with the upward peak seen at about 1.5–2 msec after the stimulation. Of note, an Ai-Ac recording enhances the amplitude of wave I. Emerson et al. (1982) showed that rarefaction clicks produce a larger wave I, but Maurer (1985) found that rarefaction polarity clicks decrease its latency. As the electrical volley reaches more proximal levels within the auditory pathways, it generates waves II though V; at these latencies, all of the scalp positions show electropositivities, particularly at the vertex. Thus, in Ai-Cz, Ac-Cz, and, less often, in Ai-Ac, waves II through V are recorded, as upward negative peaks, because Ai and Ac become more electronegative than Cz at the corresponding latencies. Unlike wave I, which is more like a near-field potential, waves II through V are recorded at distance from their brainstem generators and thus are far-field potentials (see pages 39–41). Wave V is recognized as the most prominent negative peak after 5.5 msec, with a latency ranging from 5.3 to 6.3 msec (Chiappa, 1997b). It is best seen in channels Ai-Cz and Ac-Cz and is sometimes absent in Ai-Ac. Another important hint of their presence is a significant positivity in these channels following wave V, at about 7 msec (Hashimoto et al., 1981). Condensation polarity click as well as increase in stimulus intensity may help bringing up wave V. Wave IV is sometimes seen as a small negative peak, on the upward deflection preceding the peak of wave V. The recording in Ac-Cz derivation sometimes helps with separation of waves IV and V. Several investigators (Emerson et al., 1982; Maurer et al., 1980; Maurer, 1985) have found that using rarefaction polarity clicks helps with separation of waves IV and V. In normal recordings, wave III appears in all derivations as a negative upward peak between waves I and V at approximately half distance. Of course, in the OR, all waveform latencies may be longer because of the effects of anesthetics. However, the anesthetics effects on BAEPs are minimal (see page 321); thus, we do not expect significant changes in their amplitudes and morphologies. Moreover, we expect the interpeak latencies to be unchanged. Fig. **1-17a** shows normal BAEPs in an anesthetized patient.

*B2. Recording technique.* As mentioned previously, the recording is done in three channels for each side, representing a group (e.g. right BAEP); each group is assigned a display window and is connected to the ipsilateral stimulator. The recordings for each channel can be displayed in a replace, vertical stack or overlay mode. Because BAEPs are of very short latencies and small amplitudes, the following recording parameters are used: time base 1 msec (sweep of 10 msec); sensitivity 0.1–0.2 μV/division. The LFF is at 100 Hz and the HFF is 3 kHz. Fig. **1-17 b** shows the details of the recording setup.

## 3. Interpretation

### A. Anatomy

The auditory pathways start at the level of the cochlea, in the internal ear, where the auditory receptors lie within the organ of Corti. The first neuron of the pathway lies in the spiral ganglion. Its dendrites synapse with the hair cells within the organ of Corti, and its axons form the cochlear nerve. The proximal part of the cochlear nerve enters the brainstem at the CPA and ends in the ventral and dorsal cochlear nuclei, on the lateral surface of the inferior cerebellar peduncle at the level of the lower pons. From here, the trajectories of the secondary auditory pathways are quite complex. In brief, sensory input from each eighth nerve ascends both ipsilaterally and more so contralaterally, with the fibers crossing contralaterally at several levels within the pons. The next nuclear stop is in the superior olivary nucleus, at the junction between the lower and midpons. Fibers from this nucleus ascend within the lateral lemniscus through the mid- and upper pons, especially contralateral to the stimulated eighth nerve and end in the inferior colliculus, in the lower midbrain. Then the fibers project through the brachium of the inferior colliculus to the medial geniculate body in the thalamus. The auditory radiations thereafter project to the transverse temporal gyrus (see Fig. **1-16**).

### B. Neurophysiologic-Neuroanatomic Correlation

*B1. Wave I.* This waveform is the result of the volley (NAP) generated in the distal part of the eighth nerve by the click stimulus. A change in wave I can point toward inappropriate stimulation, damage of the cochlea, or of the distal part of the eighth nerve and also be followed by changes or disappearance of the all subsequent waveforms. N1 waveform of electrocochleography (ECochG) has a similar origin. This can become important when wave I cannot be recorded in a conventional manner from an electrode placed on the ipsilateral earlobe, and thus ECochG recording can be an alternative for recording the action potential of the eighth nerve. (For applications of ECochG potentials, see page 316.)

*B2. Wave II.* This potential has been traditionally considered to arise at the level of the cochlear nucleus. However, several authors (Hashimoto et al., 1981; Moller and Janetta, 1982a, 1982b; Moller et al., 1994) have recorded an action potential directly from the intracranial portion of the nerve, close to the brainstem, at a latency similar to that of wave II. Changes in this waveform signal dysfunction of the proximal part of the eighth nerve, as is the case of retraction of the eighth nerve during CPA surgeries. In fact, two components of the wave II have been described (Ananthanarayan and Durrant, 1991): IIa, arising from the proximal portion of the acoustic nerve (AN), and IIb, representing the presynaptic activity of AN terminating in the cochlear nucleus. Although wave I is still preserved, waves III and IV may show changes or disappear, with increased absolute latencies but preserved interpeak latency.

Additionally, directly recording CNAPs from the cisternal portion of the cochlear nerve can also be done, in particular when BAEPs are of small amplitude. Details regarding this technique, its advantages, and applications are given in Chapter 7, pages 315–16.

*B3. Wave III.* This evoked potential is considered to arise at the level of the superior olivary complex (Moller and Janetta, 1982b), in the lower pons. A dysfunction of the proximal part of the eighth nerve can occur because of its stretching during retraction of the cerebellum, such as during CPA tumor removal. This can result in changes to disappearance of wave III, accompanied by disappearance of wave II. Changes in wave III may also occur from a process affecting the lower part of the pons. As with the other potentials, the transmission above this level is affected, with changes or loss of waves IV and V, with increase in the absolute latency of wave V, and an increase in the I–V interpeak latency, and perhaps of III–V interpeak latency, if wave III is still recordable.

*B4. Waves IV and V.* These waveforms are thought to arise in the mid- and upper pons, as lesions of these structures affect their generation (Boller and Jacobson, 1980; Brown et al., 1981; Chiappa, 1982; Epstein et al., 1980; Lev and Sohmer, 1972; Starr and Achor, 1975, 1979; Starr and Hamilton, 1976). Moller and colleagues (1994) concluded that the tip of wave V probably generates where the lateral lemniscus terminates in the inferior colliculus. This is in accordance with the findings of other authors, who have recorded at the level of the inferior colliculi and obtained a significant positive potential (downward deflection) just after the wave V. Because the distance covered by the BAEP recordings is very small (just a few centimeters from the CPA angle to the inferior colliculus), a precise distinction of the origins of each waveform is difficult. In general, it is accepted that wave V is generated in the high pons or low midbrain, at the level of either lateral

(a)

(b)

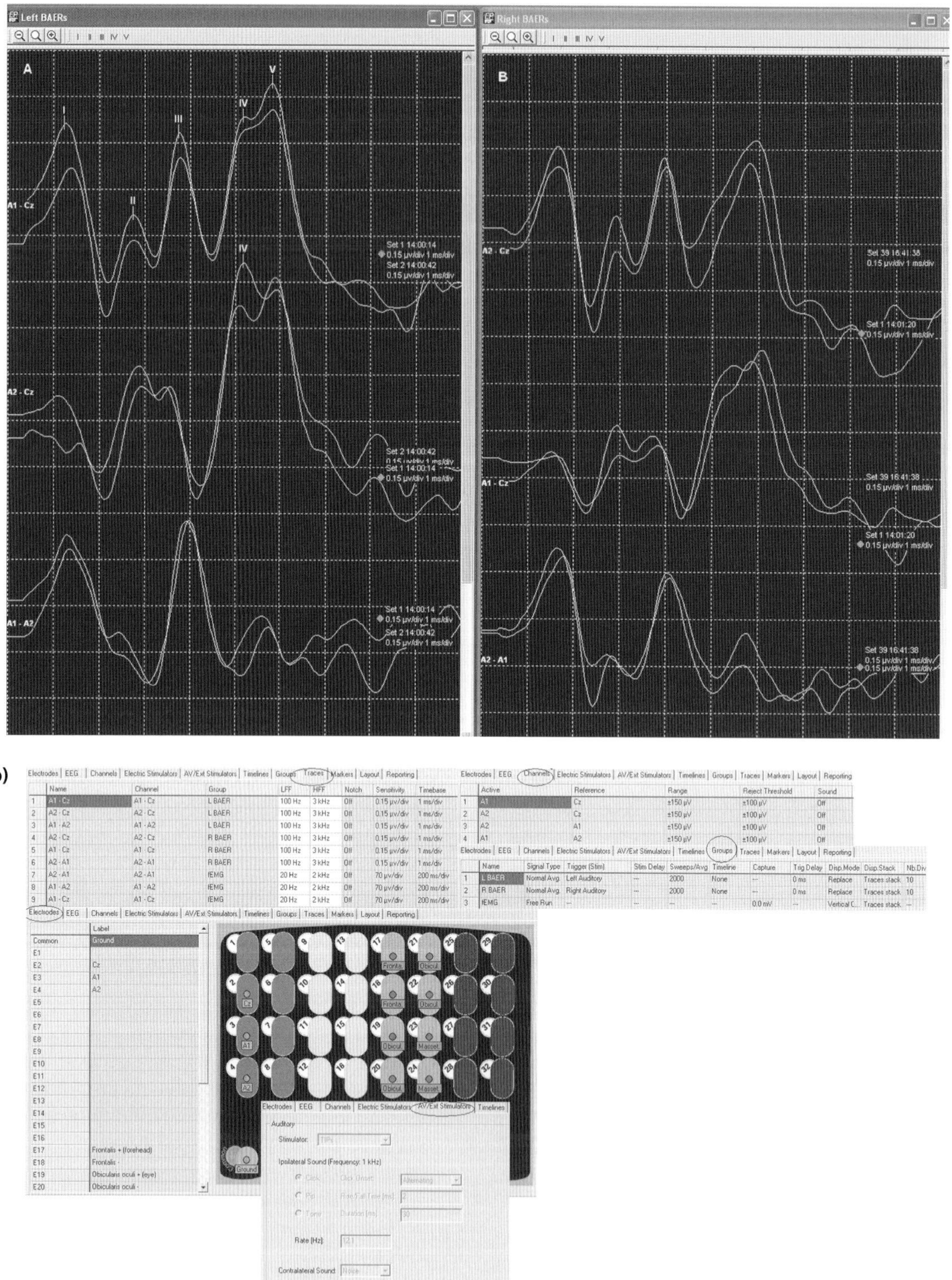


**FIGURE 1-17** Recording of BAEPs under TIVA. **a:** Normal BAEPs recordings (panel **A:** left BAEPs; panel **B:** right BAEPs). Notice the better representation of wave IV in Ac-Cz channel, whereas waves IV and V are gone in Ai-Ac channel; notice the prominence of wave I and III in Ai-Ac channel. Labels: Ac-contralateral (to the stimulation) ear; Ai-ipsilateral ear; time base 1 msec/div, sensitivity 0.15 μv/div. **b:** Setup for BAEPs technique: recording electrodes, channels, traces, groups, and stimulators.

lemniscus or inferior colliculus. This places the origin of wave IV at the midpons level. However, wave IV has a variable appearance and, because its clinical significance is in a major part related to that of wave V, not much attention is given to it in monitoring procedures. Changes in wave V can occur if damage or dysfunction arises anywhere distal to or at the level of mid- and upper pons or lower midbrain (inferior colliculus). Interestingly, an inappropriate stimulation affects wave V first, which is the first to disappear. Monitoring wave V is particularly useful during CPA surgery, when retraction of the cerebellum, and stretching of the proximal part of the eighth nerve, causes an increase in its absolute and I–V interpeak latency, as well as a change in its morphology and a decrease in its amplitude. An increase in wave V latency is usually the most common and the earliest sign of dysfunction.

*B5. Waves VI and VII.* These waveforms are believed to be generated in the medial geniculate body and auditory radiations, respectively. However, these two waveforms show an increased variability and thus are not used for clinical applications, either in the outpatient laboratory or in the OR.

In summary, the most reliable waveforms that deliver the most useful information are waves I and V; waves II and III are often waxing and waning throughout the monitoring, without clinical significance. Waves IV, VI, and VII present a significant variability and have practically no role in neuromonitoring. For further details on the use of BAEPs in monitoring of the pons and CN VIII, see pages 310–15.

### C. Applications in Intraoperative Neurophysiology

The utility of this test in intraoperative neurophysiology results from its ability to monitor the functions of the auditory pathways, from the cochlear nerve and their corresponding brainstem structures, up to the midbrain. Thus, BAEP monitoring can be used in surgical procedures that could injure the eighth nerve, including CPA surgeries such as tumor removal (e.g. acoustic schwannomas) or microvascular decompression of the facial nerve or trigeminal nerve. BAEPs can also be used in procedures that could injure the brainstem more diffusely, such as brainstem tumor removal, vascular surgery of the posterior circulation, and suboccipital decompression. Because of the close proximity of all cranial nerves as they exit the brainstem level (see Chapter **7**, Fig. **7-1**, page 300), BAEPs can be used, in combination with other cranial nerve monitoring procedures, in skull-base surgery. Table **1-8** in this chapter and Table **7-1** (page 302) in Chapter 7 summarize the uses of BAEPs in neuromonitoring.

**TABLE 1-8** BAEPs: Applications in intraoperative neurophysiology

| Region | Monitoring |
|---|---|
| CN VIII | 1. CPA tumors (e.g. acoustic neuroma)<br>2. Microvascular decompression, seventh nerve (hemifacial spasm)<br>3. Microvascular decompression, fifth nerve (trigeminal neuralgia)<br>4. Microvascular decompression of the ninth nerve (glossopharyngeal neuralgia) |
| Brainstem | 1. CPA tumors<br>2. Skull-base surgery<br>3. Suboccipital decompression (e.g. fractures/dislocations C1 vertebra, Chiari malformation)<br>4. Vascular surgeries of the posterior circulation |

## VIII. COMBINATIONS OF TESTS

Certain nervous structures or segments of the neuraxis require or are best monitored by a combination of neurophysiologic tests. Several such examples are listed.

1. Epilepsy surgery is guided by ECoG as a tool for mapping the seizure foci, and functional cortical mapping is added when the irritative zones are considered to be in close proximity to the eloquent cortex.
2. Supratentorial tumors, AVMs, or seizure foci removal frequently require functional cortical mapping employing several tests: median SSEPs for the central sulcus localization via the phase-reversal technique; mMEPs triggered by direct electrical cortical stimulation; ECoG for appreciating the baseline excitability and for monitoring of afterdischarges.
3. Brainstem surgeries, such as tumor removal, suboccipital decompression for Chiari malformation, or vascular surgeries involving the posterior circulation, may also require monitoring of the motor and sensory systems, as well as of certain cranial nerves, and thus a combination of SSEPs and MEPs monitoring with fEMG and triggered EMG for cranial nerve monitoring and mapping.
4. For other posterior fossa surgeries, especially for tumor removal (e.g. CPA or intracanalicular acoustic schwannomas), we prefer monitoring eighth, seventh, and fifth nerves together. Surgeries at the level of foramen magnum, where medulla can be easily injured, or

during skull-base surgery, involving jugular foramen, hypoglossal canal, or clivus, usually require concomitant monitoring of the lower cranial nerves, CN IX through XII, sometimes performed bilaterally, as in the case of clivus pathology . In these surgeries we rely on BAEPs for eighth nerve monitoring, as well as on fEMG and triggered EMG for the monitoring of the other cranial nerves.

5. Spine surgeries usually require both SSEPs and MEPs monitoring. In addition to SSEPs and MEPs monitoring, cervical or low thoracic, high-lumbar spine surgery may require fEMG and triggered EMG for monitoring and mapping of the nerve roots that may be at risk for damage during the surgery. A good example is an intradural, extramedullary tumor involving the conus or tethered cord, when all of the three procedures described previously are monitored. Additionally, dorsal column mapping, using SSEPs technique, is very helpful in avoiding sensory deficits caused by myelotomy, during intramedullary spinal tumors removal. D wave monitoring is important during these procedures, in order to maximize the tumor resection.
6. Thoraco-abdominal aneurysm TAA repair, involving the aortic arch, may benefit from EEG monitoring for detection of cerebral ischemia, which can be added to the SSEPs and MEPs monitoring.
7. Peripheral nervous system surgeries (e.g. brachial plexopathies, nerve repair) almost always require a combination of tests, such as SSEPs to rule out root avulsion; nerve-to-nerve and nerve-to-muscle recordings to assess the electrical transmission through the injured nerve, to locate the injury, as well as to map its extent.

Table **1-9** summarizes the surgical procedures that require several neurophysiologic tests.

## IX. SAFETY ISSUES

### 1. Characteristics of Electrical Current and Safety Concerns

The electrical current used in the OR from power lines is an alternating current (AC). Unlike direct current (DC), which flows only in one direction, AC changes direction of flow, back and forth, at a certain frequency. AC is also more harmful to the human body than DC: 3 A of the latter could pass through the human body without harmful effect, whereas only 100 mA of AC can cause

**TABLE 1-9** Combination of tests: Applications in intraoperative neurophysiology

| Surgical procedure | Combination of tests |
|---|---|
| Cervical and low thoracic/lumbar surgeries (e.g. decompressive surgery for trauma, spondylosis, extradural tumors, tethered cord, intradural extramedullary tumors such as schwannomas, etc.) | 1. SSEPs/mMEPs<br>2. Free and triggered EMG |
| Brainstem surgery (e.g. suboccipital decompression: Chiari malformation, tumor resection, vascular surgery of the posterior fossa) | 1. SSEPs/MEPs<br>2. Cranial nerve monitoring (CN IX, X , XI, and XII for medulla; V, VII, and VIII for pons; III, IV, and VI for midbrain) |
| CPA tumors, microvascular decompressions of CN V and CN VII | 1. BAEP<br>2. Monitoring of CN V and VII |
| Functional cortical mapping | 1. Median SSEPs for central sulcus localization via phase-reversal technique<br>2. MEPs triggered by direct electrical cortical stimulation for motor mapping<br>3. ECoG for appreciating the baseline cortical excitability prior to stimulation and for monitoring of afterdischarges |
| Epilepsy surgery | 1. ECoG for mapping of the irritative zones<br>2. Functional cortical mapping |
| Peripheral nervous system surgery (e.g. brachial plexus exploration, nerve repair) | 1. SSEPs to rule out root avulsion<br>2. mMEPs<br>3. nerve-to-nerve recording (NAP)<br>4. nerve-to-muscle recording (triggered CMAPs) |
| Aortic aneurysm repair | 1. SSEPs/mMEPs<br>2. EEG to monitor for cerebral ischemia |

ventricular fibrillation. The most dangerous frequency of the AC is between 10 and 200 Hz (Buczko and Mckay, 1987). Current density is the amount of current per unit area of tissue. Tissue damage (burn) occurs when the ability of capillary blood flow to dissipate the resultant heat is insufficient. Thus, this can be either the result of high current applied on relatively small surface or the result of limited blood supply of that area. The manufacturer for our equipment (XLTEK) warns against using a current density of 2 mA r.m.s/cm$^2$ or higher (e.g. if the stimulation is done at 10 Hz, with a current intensity of 100 mA and pulse duration of 1 msec, the stimulating electrode must be at least 0.5 cm$^2$). Thus, special attention has to be given when needles are used as stimulating electrodes, because of their much smaller surface. The duration of exposure to the current is also important, because the longer the pulse widths are, the higher the risk is for occurrence of burns or cardiac arrhythmias. The current path, being the route that the current follows between the entry and the exit points in the body, is most dangerous if the heart is present directly in its way.

## 2. Types of Injuries Induced by Electrical Current

Burns can occur either at the contact of stimulating electrodes with tissue, when the current density is too high (see prior discussion), or in case of leaks of high-frequency currents through alternate paths, including conductive materials in contact with human body via small surfaces (e.g. stimulating or recording electrodes, head fixation pins, etc.). In either case, the tissue damage is due to a high current density. Cardiac arrhythmias can have several mechanisms. Macroshock is caused by a high flow current, with the heart located in its path. As already mentioned, an AC at 60 Hz and 100 mA can cause ventricular fibrillation. Microshock is caused by a very small current (e.g. 200 uA, applied intramyocardially) that reaches the heart, for example, via catheters or wires of an external pacemaker. However, a single electrical stimulus can cause ventricular fibrillation if intense enough and arriving during a sensitive period in the cardiac cycle.

## 3. High-Frequency Current Leaks from the Electrosurgical Unit

Electrosurgery is the application of high-frequency electrical current that is meant to cut and coagulate biological tissue by creating a localized radio-frequency burn. An electrosurgical unit (ECU), or bovie, is used for this purpose. There are two types of ECUs that can be used, with monopolar and bipolar configurations, respectively. With the monopolar unit, the surgeon uses a pointed probe as the active electrode and a return, or dispersive electrode, with large contact surface with the patient's skin, as the reference electrode. Some refer to the latter as a grounding pad. It allows the current return to the ECU. The ECU itself has its own medical ground (an earth ground), as all electrical devices used in the OR should have, including the IOM machine (see the following). With respect to the monopolar ECU, there are several safety considerations regarding the application of the return electrode. First, it has to consist of a relatively large metallic plate or flexible metallized plastic pad, well gelled and placed over a relatively large area, ipsilateral to the site of surgery and as close as possible to it. The large surface is important for reducing the returned current density to a safe level and offering a low-resistance path for the current flow back toward the ECU. Additionally, the current density rapidly falls (quadratically) with the distance from the tip of the active electrode, whereas the heating is proportional with the square of the current density, thus occurring in a very localized region, only close to the active tip. Second, no metallic or highly conductive object should be positioned between the active and return electrode of the monopolar unit (e.g. an implanted metallic rod). It is also wise to keep the current path away from the heart, even though high frequency of the current makes it less likely to induce ventricular fibrillation than leaked current at mains frequency (Buczko and Mckay, 1987). Common areas where this electrode is placed are the patient's back and the ipsilateral thigh. Third, prior to the application of the electrode, the area should be properly cleaned and gelled for good contact. During long surgeries, the placement of the return electrode should be checked regularly.

The problem occurs when the return electrode is not appropriately placed and thus the high-frequency current finds another path of least resistance for return to an earth ground (ground loop). For example, this new path can include a recording or stimulating electrode, characterized by a small surface area, which could be situated at distance from the active probe. Thus, local burns can occur at the location of these electrodes. Worse, if the heart is in the way of the alternate path, ventricular arrhythmias can also occur.

The bipolar alternative for ECU is newer and safer. The high-frequency current flows only in between similarly sized electrodes, situated at very small distances from each other. Thus, it does not require a grounding electrode as the monopolar unit does.

## 4. Current Leaks From an Electrical Circuit to the Casing of a Medical Device

This can occur when there is faulty insulation between the wire and the casing. It can also occur as a result of inductance of a small current in the casing by an oscillating magnetic field, caused by the current flowing in the isolated wire. Third, it can be the cause of capacitive coupling (e.g. between the isolated wire and the grounded casing). These currents resulting from inductance and coupling are, in general, small, but they can still cause ventricular fibrillation if conducted directly to the heart.

## 5. Grounding in the OR and Additional Electrical Safety Warnings

In order to avoid their inclusion in alternate paths of least resistance (ground loops), patients should not be directly connected to any electrical earth grounds. The common electrode is an isolated ground, not a direct earth ground. It connects the patient to our machines in a safe way, and it has a role in artifact troubleshooting. Additional details regarding its use are found on pages 10–11.

All medical devices used in the OR are connected to a medical earth ground. However, only isolated (and not conventional) circuits should be used. This avoids the creation of a ground loop in case a person comes in contact with a faulty casing that is not well grounded (see prior section). Stimulators are isolated from ground as well as from other parts of the IOM machine. In fact, all connections between our stimulating and recording headboxes (XLTEK) and the patient are isolated. It is also recommended that the wires connecting the patient to the machine travel, if possible, suspended in the air and also that they avoid contact with any metallic surface.

This prevents a current leak through capacitive coupling. Additionally, the previously mentioned headboxes are BF devices. According to EN 606601-1, a BF device is an applied part that is isolated from the other parts of the equipment so that no current flows above a safety level (i.e. the maximum patient leakage current allowable), if an unintended voltage difference is applied to the patient.

## 6. Needle Electrodes

As mentioned, in addition to the higher risk of burns introduced by needle electrodes because of their small contact surface, they can also cause traumatic skin breaks and thus be responsible for transmission of infectious diseases.

## 7. Electrical Cortical Stimulation

Transcranial electrical cortical stimulation for MEPs monitoring has been reported to cause several types of adverse events, such as denture breaks, tongue lacerations, and even mandibular fractures, because of forceful contraction of the masseter muscles during stimulation. As a consequence, appropriate application of a bite block is important, and its position should be always checked before initiating the stimulation. Other potential risks include triggered seizures, increased intracranial pressure, displacement of vascular clips, increased current density applied to the brain in case of skull defects, and cardiac arrhythmias, in particular in the presence of pacemakers. Thus, relative contraindications for stimulation include the following: seizure disorder, presence of space occupying supratentorial lesions or significant mass effect resulting from cerebral edema, clipped AVMs, skull defects, and presence of cardiac devices such as pacemakers. However, all of these conditions are situations in which extra caution needs to be taken. Each case should be considered on an individual basis: not only the risks but also the benefits need to be weighed. In the case of direct cortical stimulation, the risks derive almost exclusively from a dangerous current density applied to the brain, as well as from the possibility of salt deposits in the case of applied monophasic pulses. In fact, caution regarding current density is necessary for any type of direct tissue stimulation, as already mentioned on pages 42–43.

## REFERENCES

Ananthanarayan AK, Durrant JD. On the origin of wave II of the auditory brain stem evoked response. Ear Hear 1991 Jun;12(3):174–179.

Boller F, Jacobson GP. Unilateral gunshot wound of the pons: Clinical, electrophysiologic, and neuroradiologic correlates. Arch Neurol 1980; 37(5):278–281.

Brown RH, Chiappa KH, Brooks EB. Brainstem auditory evoked responses in 22 patients with intrinsic brainstem lesions: Implications for clinical interpretations. Electroencephalogr Clin Neurophysiol 1981; 51:38P.

Buczko GB, McKay WPS. Electrical safety in the operating room. Can J Anesth 1987; 34(3):315–322.

Burke D, Gandevia SC, McKeon B, Skuse NF. Interactions between cutaneous and muscle afferent projections to cerebral cortex in man. Electroencephalogr Clin Neurophysiol 1982; 53:349–360.

Burke D, Skuse NF, Lethlean AK. Cutaneous and muscle afferent components of the cerebral potential evoked by electrical stimulation of human peripheral nerves. Electroencephalogr Clin Neurophysiol 1981; 51:579–588.

Carpenter MB. Tracts of the spinal cord. In: Core Text of Neuroanatomy (Carpenter MB, Ed.) 1996; pp 83–114. Baltimore: William & Wilkins.

Chang SD, Lopez JR, Steinberg GK. .Intraoperative electrical stimulation for identification of cranial nerve nuclei. Muscle Nerve 1999; 22:1538–1543.

Chiappa KH. Physiologic localization using evoked responses: Pattern shift visual, brainstem auditory and short latency somatosensory. In: New Perspectives in Cerebral Localization (Thompson RA and Green JR, Eds.) 1982; pp 63–114. New York: Raven Press.

Chiappa KH. Evoked potentials in clinical medicine. In: Clinical Neurology (Baker AB and Baker LH, Eds.) 1983; Philadelphia: JB Lippincott.

Chiappa, KH. Short-latency somatosensory evoked potentials: Methodology. In: Evoked Potentials in Clinical Medicine (Chiappa KH, Ed.) 1997a; pp. 283–340. Philadelphia: Lippincott-Raven.

Chiappa, KH. Brainstem auditory EPs: Methodology. In: Evoked Potentials in Clinical Medicine (Chiappa KH, Ed.) 1997b; pp 157–197. Philadelphia: Lippincott-Raven.

Creutzfeldt O, Houchin J. Neuronal basis of EEG waves. In: Handbook of Electroencephalography and Clinical Neurophysiology, Vol. 2, Part C (Remond A, Ed.) 1974; pp 5–55. Amsterdam: Elsevier.

Cusick JR, Myklebust J, Larson SJ, Sances JA. Spinal evoked potentials in the primate: Neural substrate. J Neurosurg 1978; 49:551–557.

Deletis V. Intraoperative neurophysiology and methodologies used to monitor the functional integrity of the motor system. In: Neurophysiology in Neurosurgery (Deletis V and Shils JL, Eds.) 2002; pp 25–51. San Diego: Academic Press.

Deletis V, Sala F. Corticospinal tract monitoring with D and I waves from the spinal cord and muscle MEPs from limb muscles. In: Intraoperative Monitoring of the Neural Function (Nuwer M, Ed.) 2008; pp 235–251. Amsterdam: Elsevier.

Emerson RG, Brooks EB, Parker SW, Chiappa KH. Effects of click polarity on brainstem auditory evoked potentials in normal subjects and patients: Unexpected sensitivity of wave V. Ann NY Acad Sci 1982; 388:710–721.

Emerson RG, Pedley TA. Generator sources of median somatosensory evoked potentials. J Clin Neurophysiol 1984; 1:159–202.

Emerson RG, Pedley TA. Effect of cervical spinal cord lesions on early components of the median nerve somatosensory evoked potential. Neurology 1986; 36:20–26.

Epstein CM, Stappenbeck R, Karp HR. Brainstem auditory evoked responses in palatal myoclonus. Ann Neurol 1980; 7:592.

Greenberg JA, Kaplan PW, Erwin CW. Somatosensory evoked potentials and the dorsal column myth. J Clin Neurophysiol 1987; 4:189–196.

Hamano T, Kaji R, Diaz AF, et al. Vibration-evoked sensory nerve action potentials derived from Pacinian corpuscles. Electroencephalogr Clin Neurophysiol 1993; 89:278–286.

Happel L, Kline D. Intraoperative neurophysiology of the peripheral nervous system. In: Neurophysiology in Neurosurgery (Delitis V and Shills J, Eds.) 2002; pp 169–195. San Diego: Academic Press.

Hashimoto I, Ishiyama Y, Yoshimoto T, Nemoto S. Brain-stem auditory-evoked potentials recorded directly from human brain-stem and thalamus. Brain 1981; 104:P841–859.

Hubbard JI, Llianas R, Quastel DM. Electrophysiological analysis of synaptic transmission. Monographs of the Physiological Society 1969; London: Edwards Arnold Ltd.

Jasper HH. The 10–20 electrode system of the International Federation. Electroencephalogr Clin Neurophysiol 1974; 10:371–375.

Jones SJ, Edgar MA, Ransford AO. Sensory nerve conduction in the human spinal cord: Epidural recordings made during scoliosis surgery. J Neurol Neurosurgery Psychiatry 1982; 45:446–451.

Katayama Y, Tsubokawa T, Maejima S, et al. Corticospinal direct response in humans: Identification of the motor cortex during intracranial surgery under general anaesthesia. J Neurol Neurosurg Psychiatry. 1988 Jan; 51(1):50–59.

Kimura J. Electrodiagnosis in diseases of nerve and muscle. In: Principles and Practice. (Kimura J, Ed.) 1989; Philadelphia: Davis.

Kobal G, Hummel C. Cerebral chemosensory evoked potentials elicited by chemical stimulation of the human olfactory and respiratory nasal mucosa. Electroencephalogr Clin Neurophysiol 1988; 71(4):241–250.

Kritchevsky M, Wiederholt WC. Short-latency evoked potentials. Arch Neurol 1978; 35(11):706–711.

Kunesch E, Knecht S, Classen J, et al. Somatosensory evoked potentials (SEPs) elicited by magnetic nerve stimulation. Electroencephalogr Clin Neurophysiol. 1993 Nov–Dec; 88(6):459–467.

Lev A, Sohmer H. Sources of averaged neural responses recorded in animal and human subjects during cochlear audiometry (electro-cochleogram). Arch Klin Exp Ohr-Nas-u Kehik Heilk 1972; 201:79–90.

Lopes da Silva F. Dynamics of EEG as signals of neuronal population: Models and theoretical considerations. In: Electroencephalography. Basic Principles, Clinical Applications and Related Fields (Niedermeyer E and Lopes da Silva F, Eds.) 1999; pp. 93–109. Baltimore: Lippincott Williams & Wilkins.

Lopes da Silva F, Van Rotterdam AB. Biophysical aspects of EEG and magnetoencephalogram generation. In: Electroencephalography. Basic Principles, Clinical Applications and Related Fields (Niedermeyer E and Lopes da Silva F, Eds.) 1999; pp. 93–109. Baltimore: Lippincott Williams &Wilkins.

Macon JB, Poletti CE, Sweet WH, et al. Conducted somatosensory evoked potentials during spinal surgery. Part 1: Clinical applications. J Neurosurg 1982; 57:354–359.

Maurer K. Uncertainties of topodiagnosis of auditory nerve and brain-stem auditory evoked potentials due to rarefaction and condensation stimuli. Electroencephalogr Clin Neurophysiol 1985; 62:135–140.

Maurer K, Schafer E, Leitner H. The effect of varying stimulus polarity (rarefaction vs. condensation) on early auditory evoked potentials. Electroencephalogr Clin Neurophysiol 1980; 50:332–334.

Moller AR, Jannetta PJ. Evoked potentials from the inferior colliculus in man. Electroencephalogr Clin Neurophysiol 1982a; 3:612–620.

Moller AR, Jannetta PJ. Auditory evoked potentials recorded intracranially from the brain stem in man. Exp Neurol 1982b; 78:144–157.

Moller AR, Jannetta PJ, Jho HD. Click-evoked responses from the cochlear nucleus: A study in human. 1994; 92(3):215–224.

Ohnishi A, O'Brien PC, Okazaki H, Dyck PJ. Morphometry of myelinated fibers of fasciculus gracilis of man. J Neurol Sci 1976; 27:163–172.

Onofrj M, Basciani M, Fulgente T, et al. Maps of somatosensory evoked potentials (SEPs) to mechanical (tapping) stimuli: Comparison with P14, N20, P22, N30 of electrically elicited SEPs. Electroencephalogr Clin Neurophysiol 1990; 77(4):314–319.

Pratt H, Politoske D, Starr A. Mechanically and electrically evoked somatosensory potentials in humans: Effects of stimulus presentation rate. Electroencephalogr Clin Neurophysiol 1980; 49:240–249.

Restuccia D, Mauguiere F. The contribution of median nerve SEPs in the functional assessment of cervical spinal cord in syringomyelia. Brain 1991; 114:361–379.

Rossini PM, Paradiso C, Zarola F, et al. Bit-mapped somatosensory evoked potentials and muscular reflex responses in man: Comparative analysis in different experimental protocols. Electroencephalogr Clin Neurophysiol 1990; 77(4):266–276.

Schlake HP, Goldbrunner R, Milewski C, et al. Technical developments in intra-operative monitoring for the preservation of cranial motor nerves and hearing in skull base surgery. Neurol Res. 1999 Jan;21(1):11–24. Review.

Shepherd GM. The Synaptic Organization of the Brain 1974; London: Oxford University Press.

Speckmann EJ, Caspers H, Elger CE. Neuronal mechanisms underlying the generation of field potentials. In: Self-Regulation of the Brain and Behavior (Elbert T, Rockstroh B, Lutzenberger W, and Birbaumer N, Eds.) 1984; pp 9–25. New York: Springer.

Starr A, Achor LJ. Auditory brain stem responses in neurological disease. Arch Neurol 1975; 32:761–768.

Starr A, Achor LJ. Anatomical and physiological origins of auditory brain stem responses (ABR). In: Human Evoked Potentials: Applications and Problems. (Lehmann D and Callaway E, Eds.) 1979; pp 415–429. New York: Plenum Press.

Starr A, Hamilton AE. Correlation between confirmed sites of neurologic lesions and abnormalities of far-field auditory brainstem responses. Electroencephalogr Clin Neurophysiol 1976; 41:595–608.

Tyner FS, Knott JR, Brem Mayer W. Electrode placement. In: Fundamentals of EEG Technology. Vol. 1. Basic Concepts and Methods. (Tyner FS, Knott JR, Brem Mayer W, Eds.) 1983; pp 136–145. Philadelphia: Lippincott Williams & Wilkins.

Veale JL, Mark RF, Rees S. Differential sensitivity of motor and sensory fibres in human ulnar nerve. J Neurol Neurosurg Psychiatry. 1973 Feb; 36(1):75–86.

Vogel P, Ruber P, Klein R. The latency difference of the tibial and sural nerve SEP: Peripheral versus central factors. Electroencephalogr Clin Neurophysiol 1986; 65:269–275.

Wall PD, Noordenbos W. Sensory functions which remain in man after complete transaction of dorsal columns. Brain 1977; 100:641–653.

Wong PKH. Technology. In: Digital EEG in Clinical Practice (Wong PKH, Ed.) 1996; pp 9–14. Philadelphia: Lippincott-Raven.

# 2

# EEG MONITORING DURING CAROTID ENDARTERECTOMY

KEITH H. CHIAPPA
MIRELA V. SIMON



## I. INTRODUCTION

A large amount of experience has been gathered over the past 30 years in using electroencephalography (EEG) to monitor brain function during carotid endarterectomies (CEAs). It has been demonstrated that the EEG is a very reliable indicator of significant cerebral ischemia during these procedures (Ballotta et al., 1997; Chiappa et al., 1979; Matsumoto et al., 1976; McGrail, 1996; Nuwer et al., 1993; Plestis et al., 1997; Pinkerton, 2002) and that it can be used confidently as a monitoring tool to guide intraoperative techniques, mainly those related to cross-clamping of the artery and the decision to use a temporary intra-arterial bypass shunt (McGrail, 1996; Pinkerton, 2002; Phillips et al., 1979; Salvian et al., 1997; Wassmann et al., 1984; Whittemore et al., 1983; Woodworth et al., 2007). There is a close correlation between the EEG and cerebral blood flow (Algotsson et al., 1990; Sundt et al., 1974), and this correlation was found to be influenced by the type of anesthetics used (Messick et al., 1987). Significant EEG changes appear within 30 seconds of cross-clamp application in those patients whose intracranial blood supply is critically dependent on the artery that was clamped. This occurs in less than 3–5% of all patients, this representing the number of patients who will need a temporary bypass shunt during the endarterectomy, which typically lasts about 20–40 minutes, depending on various technical details and surgical practices.

There has always been some confusion about the utility of EEG monitoring of CEAs (Kresowik et al., 1991). This stems from differing views of important issues that encompass the principles of EEG monitoring during CEA. We introduce these principles in the next paragraphs and expand some of them further in the remainder of this chapter. If these principles are ignored, mistakes can be made, or it can easily seem to an inexperienced team that the EEG has not been helpful.

## II. PRINCIPLES OF MONITORING OF CEAS

### 1. The Number of Adverse Events is Small

The small number of patients (3–5%) in whom the clamped carotid is a critical source of intracranial blood supply means that a surgeon might be able to perform more than 100 CEAs without shunting during the cross-clamp period before encountering a significant immediate postoperative neurologic deficit related to ischemia (and no shunt) during the clamp period. This is an infrequent enough incidence that it might be lost in the "noise" associated with outcome from a surgical procedure (i.e., morbidities not resulting from ischemia during the clamp period, such as cardiopulmonary problems).

The percentage of cases in which the EEG shows changes can be higher in a population of patients with complex intracranial and extracranial vascular disease, such as a high degree of stenosis, bilateral carotid, or multivessel involvement. Many studies have shown that occlusion of the contralateral carotid artery increases the risk of EEG changes and intraoperative ischemic events (Graham et al., 1986; McFarland et al., 1988; Phillips et al., 1979; Pinkerton, 2002; Ricotta et al., 1983; Schneider et al., 2002; Whittemore et al., 1983). Also, recent strokes (less than six weeks old) are another risk factor for intraoperative ischemia (Graham et al., 1986).

## 2. The 10–20 Rule

Experience with immediate postoperative neurologic deficits has shown that there is a "10–20 rule" in effect in EEG monitoring of CEAs. This rule refers to the clinical observations that a significant EEG change seen with cross-clamping (i.e., significant cortical ischemia) can be sustained by the cortex for about 6–10 minutes in most patients without apparent neurologic sequelae. If the EEG change is allowed to persist beyond 6–10 minutes (without release of the clamp or bypass shunting), there is an increasing incidence of immediate, postoperative significant neurologic deficits, so that, by 20 minutes duration of the significant EEG change, there is an almost 100% incidence of postoperative neurologic deficits (Chiappa et al., 1979; Jenkins et al., 1983). This is concordant with what other authors have reported: EEG changes lasting 19 minutes were associated with profound hemiparesis, which improved within a week (Kaufman et al., 1977). In assessing this incidence, note that clinical strokes may not have an immediate abnormal diffusion-weighted magnetic resonance image (MRI) (Oppenheim et al., 2000; Pedraza et al. 2002; Sylaja et al., 2008). We believe that, in these patients with negative imaging, the neurophysiologic dysfunction signaled by EEG changes and the following postoperative clinical deficits (see Fig. **2-1**) reflect transient neuronal stunning beyond the typical transient ischemic attack (TIA) time window (1 hour).

The effect of this 10–20 rule may be minimized and less apparent in some centers because the entire endarterectomy is routinely accomplished in less

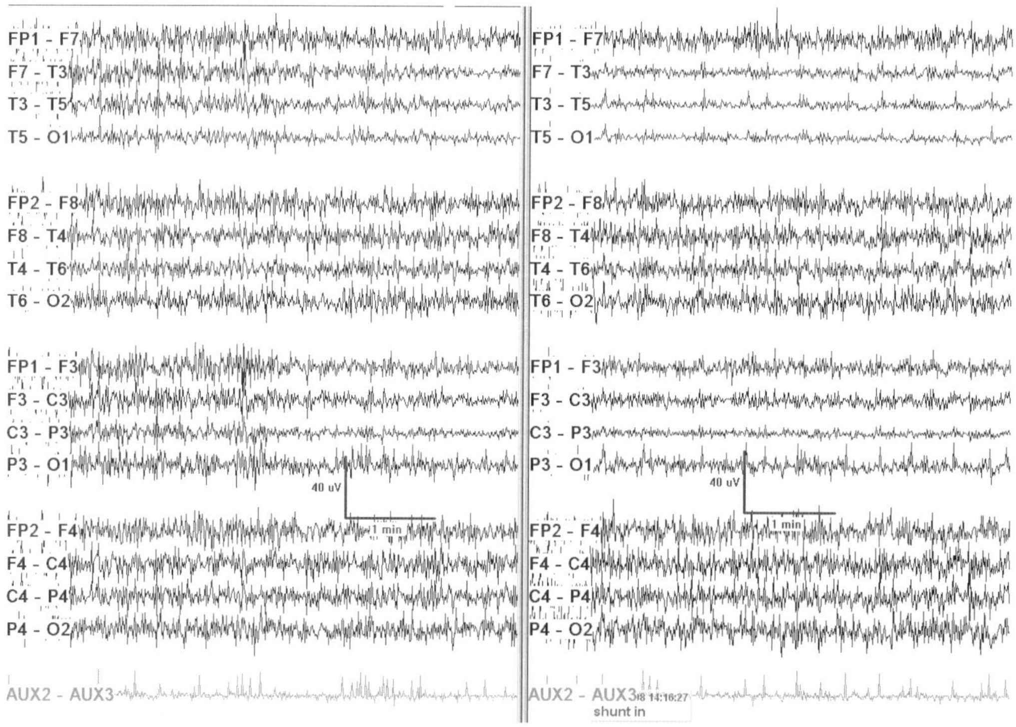


**FIGURE 2-1** The **left** panel shows the EEG associated with a clamp of the left internal carotid artery (approx. in middle of panel): there is a marked loss in amplitude over the left hemisphere. The **right** panel is taken after insertion of a temporary bypass shunt and shows no change. The patient woke up from surgery with an immediate significant neurologic deficit, but the immediate postoperative brain MRI with diffusion-weighted MRI will be normal. The calibration marks in both panels are 40 $\mu$V and 1 minute.

than 20 minutes (i.e., the "clamp time" is less than 20 minutes). Thus these patients are less likely to have a clinical problem, even if their EEG shows a significant change with the cross-clamp, because a cross-clamp maintained for 15 minutes does not have the same rate of postoperative neurologic deficits as one maintained for more than 20 minutes. Hence a center for which the clamp times are less than 20 minutes and which routinely do not use a bypass shunt will see a significant postoperative neurologic deficit even fewer times than the 1 per 100 or so CEAs mentioned in the preceding section **II, 1**, and this lesser number is even easier to lose in the "noise."

Note that some techniques used in CEAs require more than 20 minutes of clamp time (e.g., patch graft) and that, in some situations, EEG monitoring can provide data concerning cerebral blood flow, which can help the surgeon decide on the technique to be used. For example, a neurosurgeon in our group has a clamp time of about 15 minutes, and, if our assessment indicates a mild to moderate EEG change, he will not insert the shunt (Fig. **2-2**). Note that accurate gradation of the severity of the EEG changes can be problematic in patients with previous infarcts in whom lesser degree of EEG changes can have a bigger impact than expected on postoperative clinical

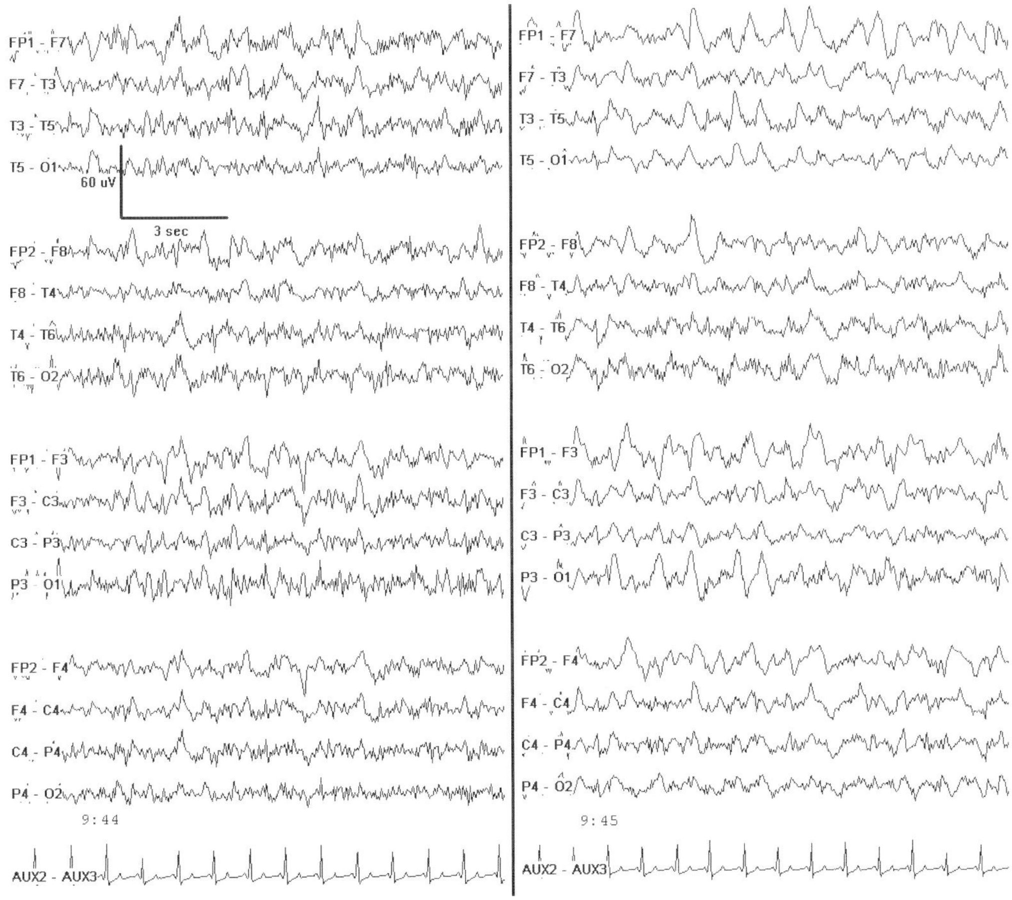


**FIGURE 2-2** Minimal EEG changes with clamp in a left carotid endarterectomy. The **left** panel shows the EEG in the few minutes prior to application of a clamp to the internal carotid artery, and the **right** panel shows the EEG a minute or so after the clamp was applied. There is some loss of amplitude of the anesthetic-induced fast activity, which is superimposed on the generalized slowing, and a mild increase in the amplitude of the generalized slowing, with a subtle asymmetry, left more pronounced than right. This would be considered a borderline change with respect to a decision to use a temporary bypass shunt. The increase in the systemic blood pressure will not reverse the EEG changes; however, due to a short expected clamp time, no shunt will be used. The patient will have no postoperative deficit. The calibration marks in both panels are 60 $\mu$V and 3 seconds.

outcome, so that a shunt may be indicated despite a modest EEG change. Thus, EEG assessment of neuronal cortical function during clamping provides significant data related to the shunt placement decision, and factors other than the clamp time also need to be considered.

## 3. Variability in the Definition of Significant EEG Change

There is considerable variation in what is termed a significant EEG change. If minor degrees of slowing associated with cross-clamping are termed "significant" in a center, the correlation between these EEG changes and postsurgical outcome becomes much poorer, and the true utility of EEG monitoring is lost in the noise. We reserve the definition of a significant EEG change to a loss in amplitude of greater than 60%, which is also always associated with a slowing in overall frequency, because the anesthetic-induced fast frequencies are the first to drop out with cortical ischemia (see Figs. **2-3** to **2-6**).

A relative increase in the amount of slowing in a clamped hemisphere is not necessarily taken as an imperative signal to shunt, but rather as a warning sign; the degree of attenuation of fast frequencies then needs to be carefully assessed.

## 4. Inadequate Recording Techniques

EEG amplitudes during anesthesia are considerably less than those encountered during routine EEGs. This requires that the EEG be viewed at an increased sensitivity, usually 2–3 microvolts per mm. If the recording is done at too low a sensitivity, attenuation of the fast-frequency activities, already suppressed by anesthesia, can be easily missed (see pages 69–70 and Fig. **2-7**).

Additionally, depending on the level of anesthesia and the frequency of the EEG activities being monitored for asymmetry, the interpreter's eye might be more sensitive at a different EEG paper speed (e.g., 10 mm/sec vs. 30 mm/sec). For recording details also see page 71 and Fig. **2-8**.

Some centers have used 6 channels of EEG or less for monitoring and have concluded that the EEG is not

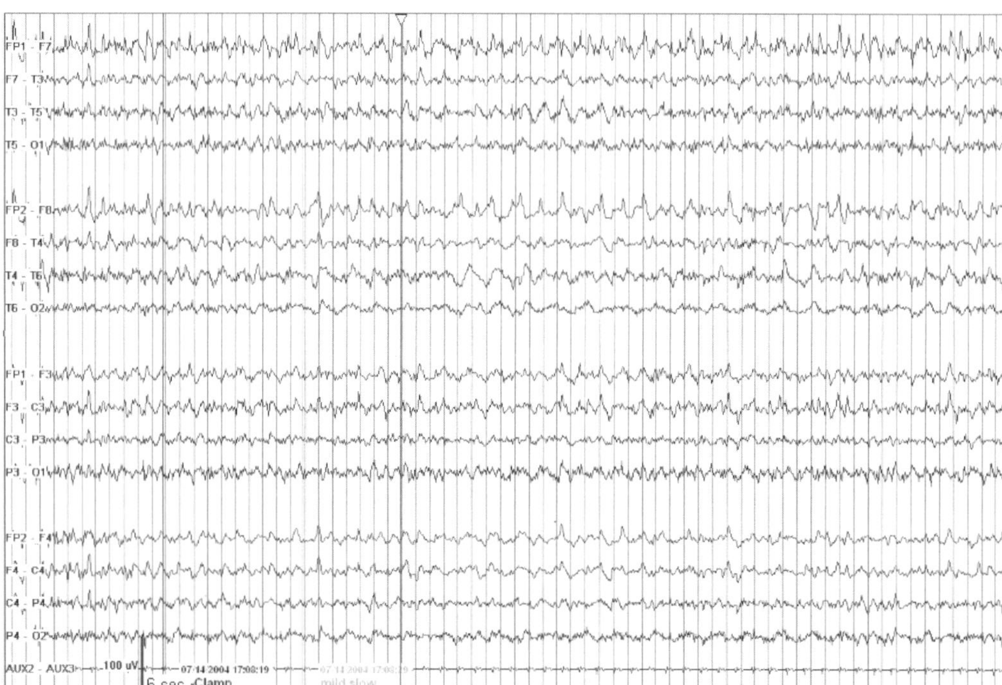


**FIGURE 2-3** Mild EEG changes following clamp of the right internal carotid artery (which occurs near the left margin at the calibration mark). There is an immediate appearance of 3–4 Hz slowing over the right hemisphere and a mild loss of the superimposed anesthetic-induced beta activity; this change is classified as mild and considered a borderline change with respect to a decision to use a temporary bypass shunt; here no shunt will be used, and the EEG changes will reverse with an increase in the systemic blood pressure. The patient will have no postoperative deficit. The calibration mark is 100 $\mu$V and 6 seconds.

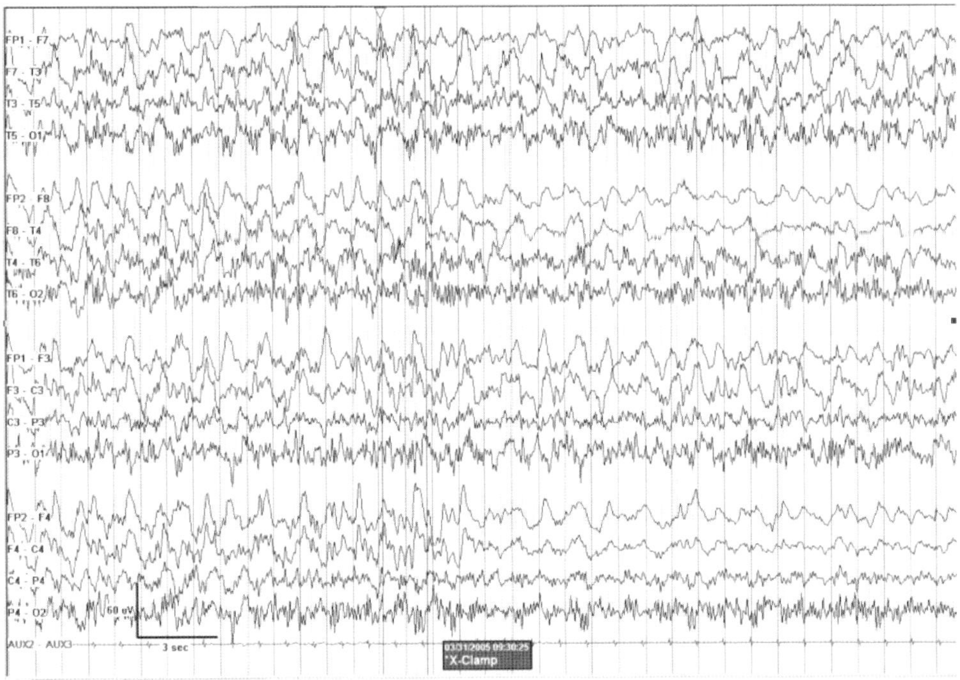


**FIGURE 2-4** Moderate to severe EEG changes associated with clamp of the right internal carotid artery (which occur at the "X-Clamp" label). There is an immediate appearance of 2–3 Hz slowing over the right hemisphere and marked loss of the superimposed anesthetic-induced beta activity; this change is classified as moderate to severe and considered a definite indication with respect to a decision to use a temporary bypass shunt. Here a shunt will be used, with return of the EEG to the preclamp baseline. The patient will have no postoperative deficit. The calibration mark is 60 $\mu$V and 3 seconds.

(a)

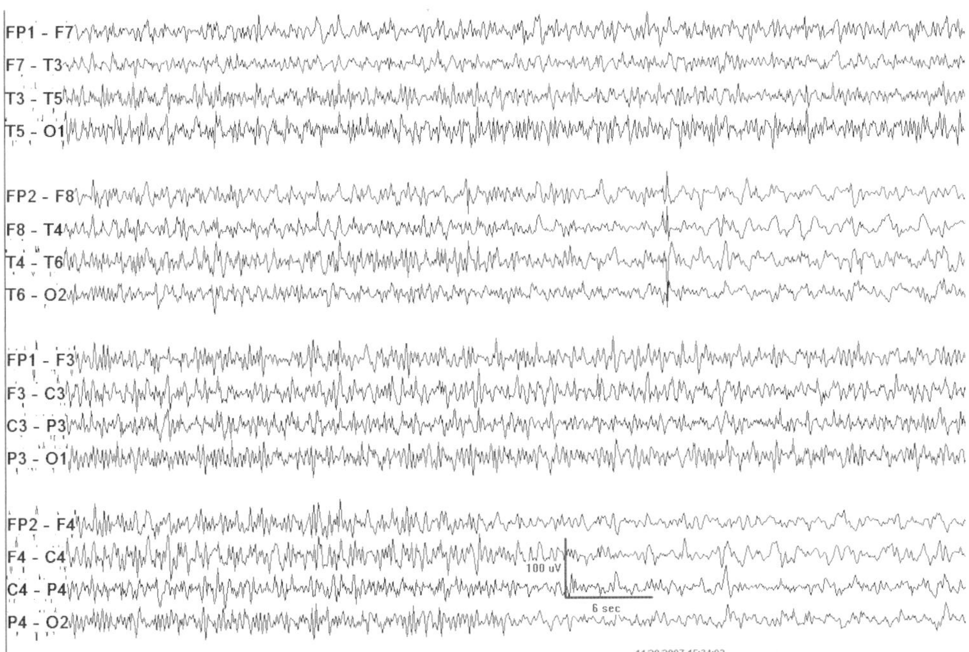


**FIGURE 2-5** Severe EEG changes after clamp of the right internal carotid artery, significantly improved by insertion of a temporary intra-arterial bypass shunt. The patient had no postoperative deficit. The calibration marks in all slides are 100 $\mu$V and 6 seconds. a: The preclamp baseline shows normal anesthetic-induced activities, without obvious asymmetries. The clamp of the carotid artery occurs at the "X-CLAMP" label. There is an immediate appearance of 1–2 Hz slowing over the right hemisphere and marked loss of the superimposed anesthetic-induced beta activity.

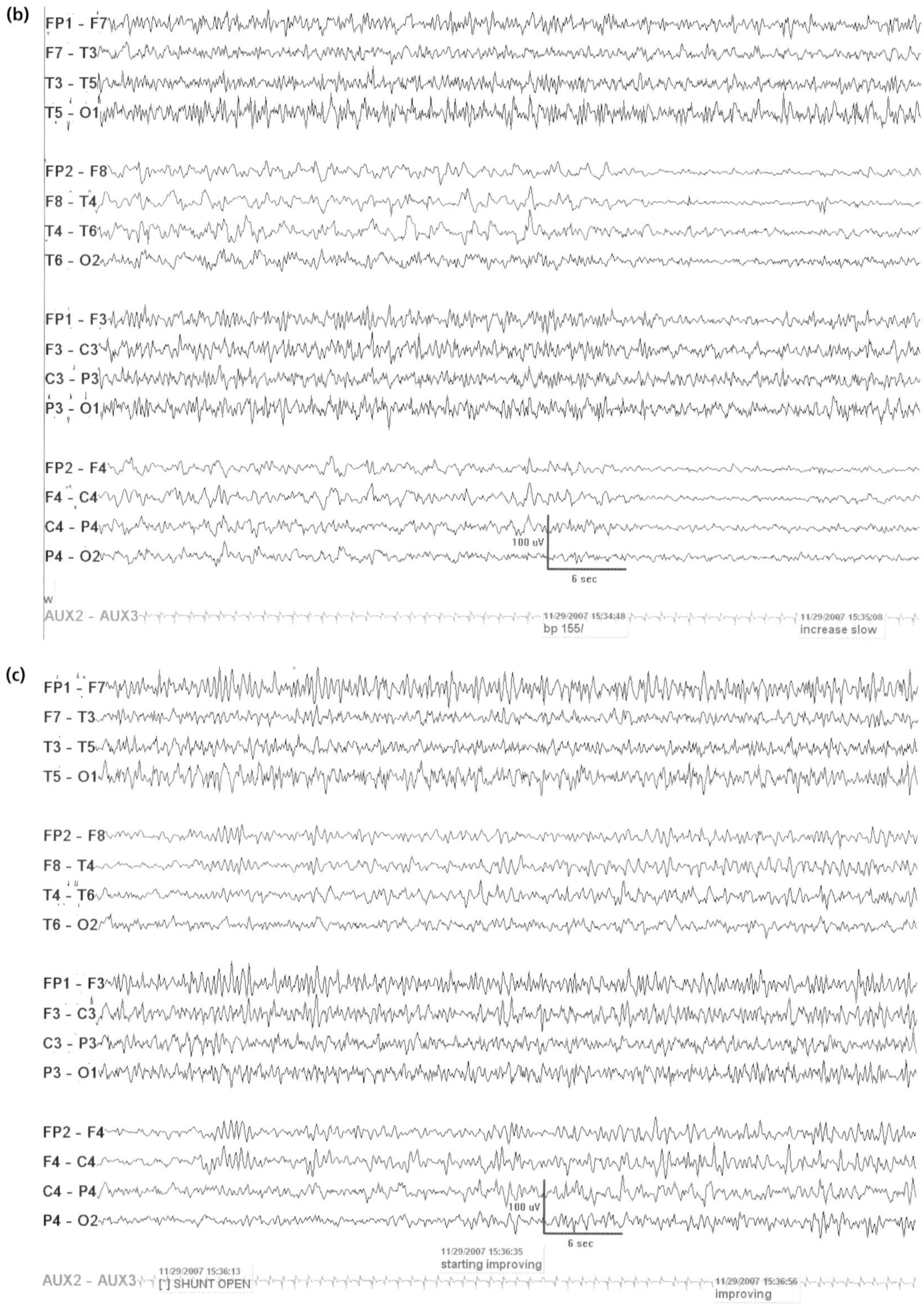


**FIGURE 2-5** *(Continued)* **b:** The postclamp right hemispheric slowing progresses to marked attenuation of all activities on the same side; this change is classified as severe and considered a definite indication with respect to a decision to use a temporary bypass shunt. **c:** Here a shunt is opened ("SHUNT OPEN" label), and there is a progressive improvement of the right hemisphere EEG activity, in particular of the fast-frequency activities.

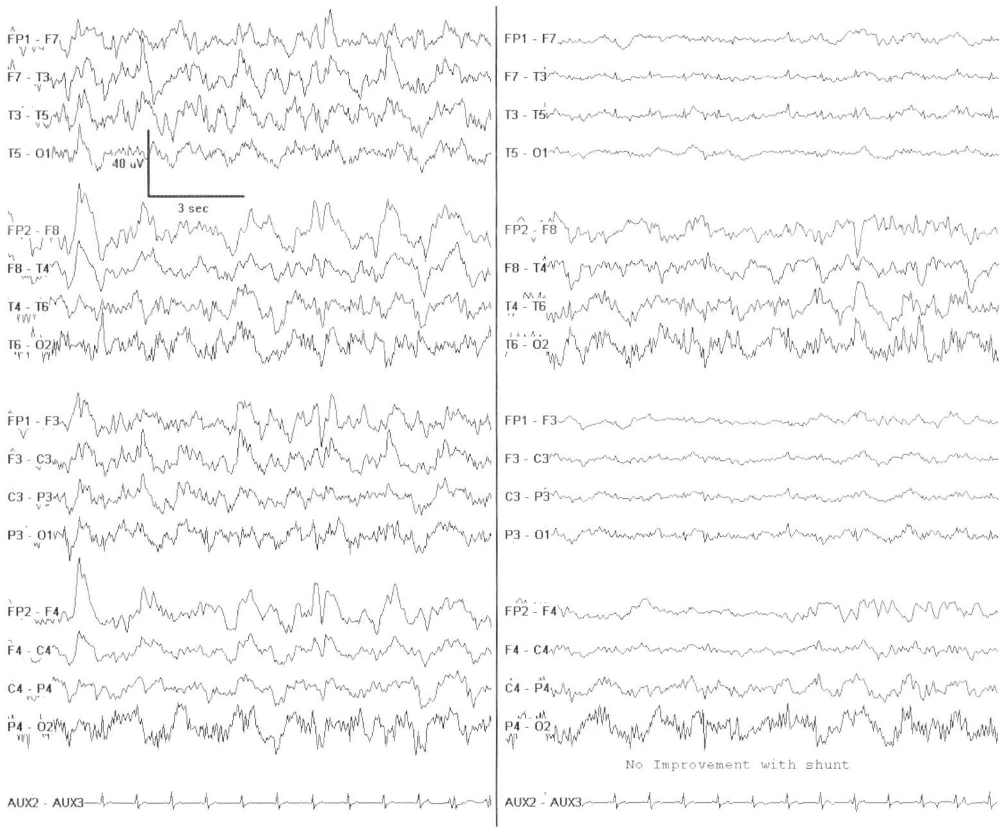


**FIGURE 2-6** Severe ipsilateral EEG changes with clamping of the left carotid artery, not improved by shunt placement. The **left** panel shows a symmetric EEG background, in the few minutes prior to application of a clamp to the left internal carotid artery. Thirty seconds after clamping of the left carotid artery, the EEG shows severe left hemispheric slowing, with attenuation of the faster frequencies on this side. The change is classified as severe and considered a definite indication with respect to a decision to use a temporary bypass shunt. In this case, however, the postclamp EEG appearance is unchanged by insertion and opening of a shunt (**right** panel taken a few minutes after shunt opening). This indicates that inadequate blood flow is being provided by the shunt, and a shunt revision or artery reopening is required. The calibration marks in both panels are 40 $\mu$V and 3 seconds.

a sensitive tool. Others have concluded that the EEG is less sensitive than somatosensory-evoked potentials (SSEPs) or other monitoring techniques (e.g., transcranial Doppler [TCD]) after comparing limited-channel EEG recordings (Arnold et al., 1997). Our prior studies (Chiappa et al., 1979; Jenkins et al., 1993) have described the importance of utilization of correct recording techniques and concluded that a 6-channel EEG recording is inadequate. Our present technique requires 16-channel EEG recording for adequate sampling of the cerebral hemispheres.

## 5. Limited EEG Recording Time

Most EEG changes are due to cerebral hypoperfusion during cross-clamp of the carotid artery. However, there are several other mechanisms acting outside the clamp time through which the patient can suffer a cerebrovascular event (Blondeau et al., 1982; de Borst et al., 2001; McKinsey et al., 1996; Mola et al., 1986; Pistolese et al., 1997). We have seen significant EEG changes related to various events: (1) positioning of the head prior to incision (Fig. **2-9**); (2) exposure of the artery, as a result of either cardiac bradyarrhythmias secondary to mechanical disturbance of the carotid sinus (Figs. **2-10**, **2-11**) or dislocation of a thrombus from a loose plaque (Fig. **2-12**); (3) clamp release, related to dislocation and release of a thrombus or other loose material (Fig. **2-13**); and (4) after reopening of the arteries and during closure of the operative site, secondary to rethrombosis there, causing either a low flow state or a thromboembolic event. In these cases, the patient can awaken from anesthesia with a significant

(a)

(b)

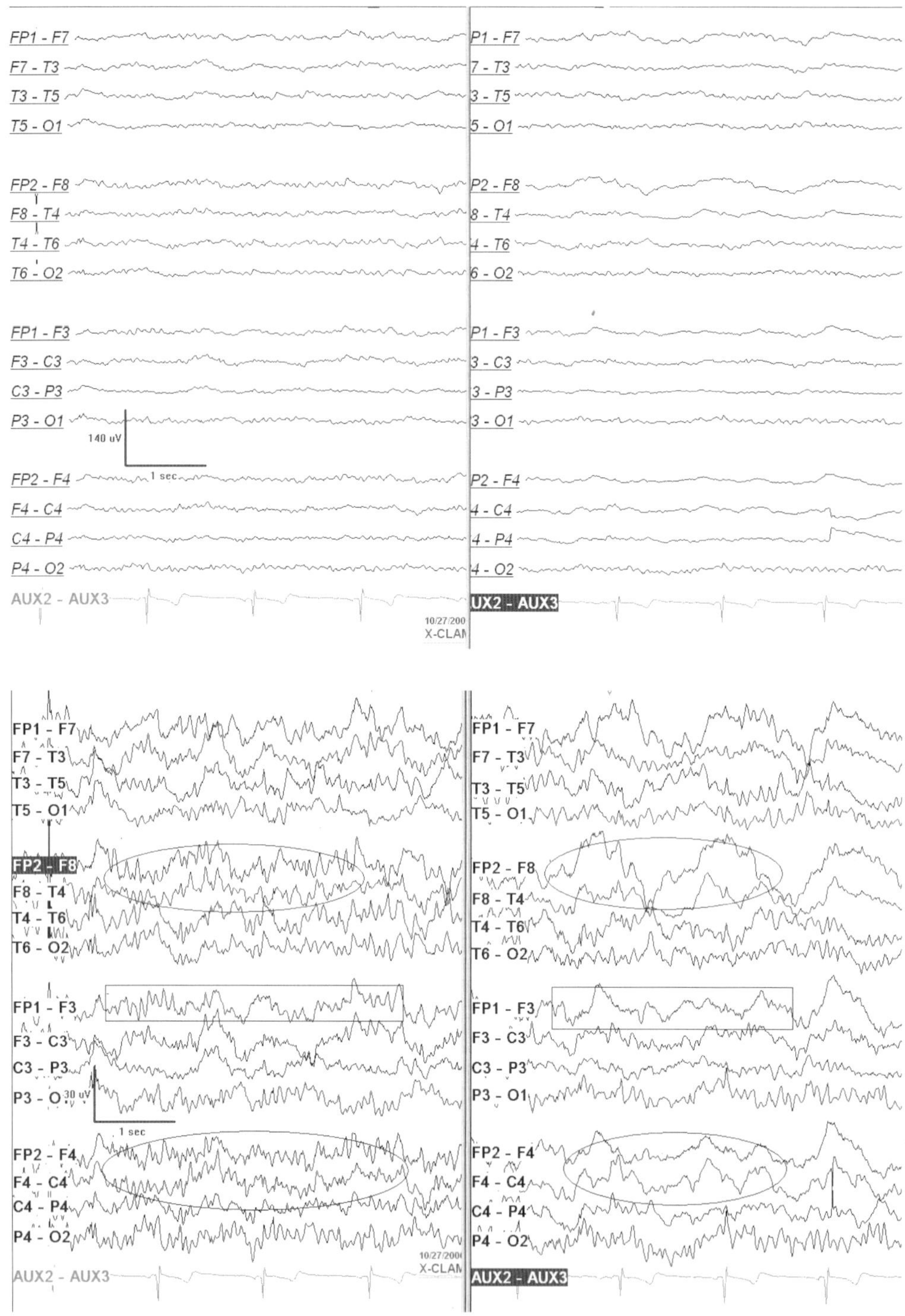

FIGURE 2-7 These EEG samples from a patient undergoing a right carotid endarterectomy illustrate the effects of different sensitivity (display magnifications or gain) on the visibility of the subtle changes that occurred in this patient with clamping of the artery. In both figures, the **left** panel shows EEG several seconds prior to clamp ("X-Clamp" label). The **right** panel shows EEG recording 19 seconds after the clamp. **a:** The display is done at 7 μV/mm sensitivity and 30 mm/sec time base (calibration mark is 140 μV and 1 second). The postinduction baseline contralateral (left) hemispheric slowing, as well as the EEG changes induced by clamping the right carotid artery, are difficult to appreciate. **b:** The display is done at 3 μV/mm sensitivity and 30 mm/sec time base (calibration mark is 30 μV and 1 second). Notice the EEG postinduction baseline asymmetry, with attenuation of the faster frequencies on the left. After the clamp, there is further attenuation of the bifrontally predominant fast-frequency activities, as well as an accentuation of the 1–2 Hz delta slowing, more evident on the right (circle) than on the left (rectangle).

(a)

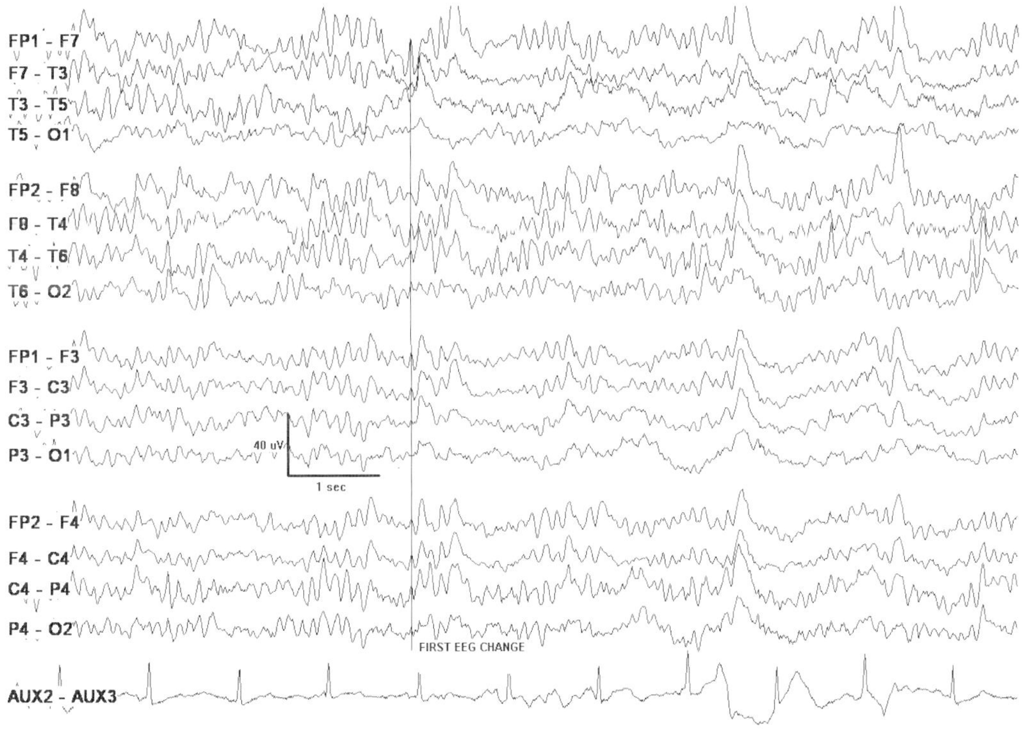


(b)

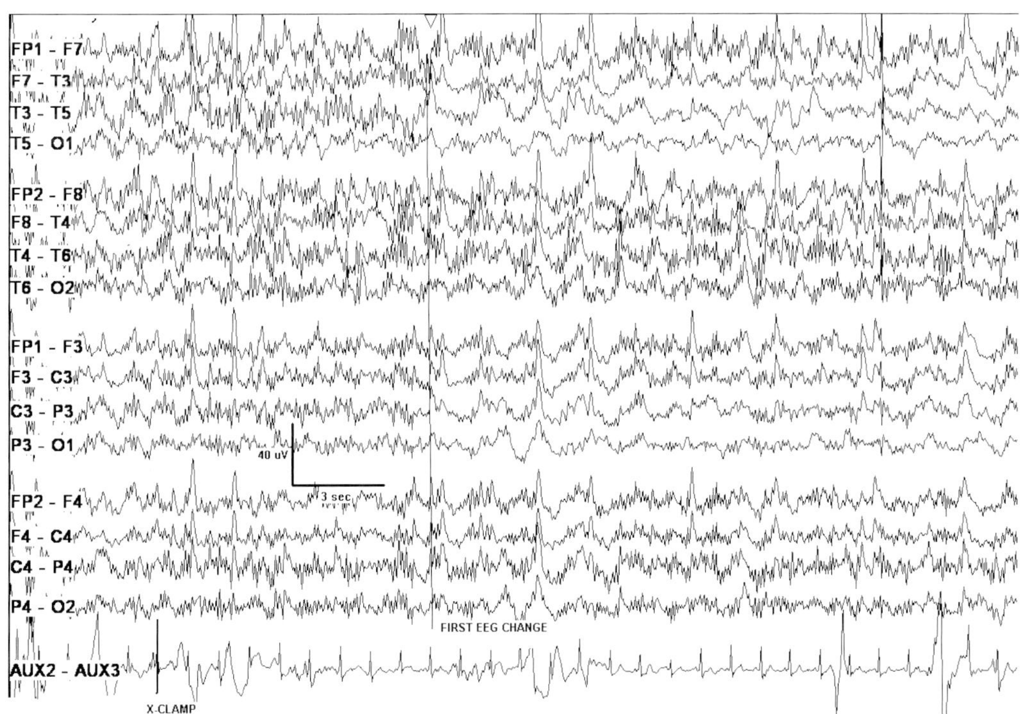


**FIGURE 2-8** EEG during a left carotid endarterectomy, showing the effects of changes in the time base in use. The calibration marks are 40 $\mu$V and 1 second in **a,** 40 $\mu$V and 3 seconds in **b**, and 40 $\mu$V and 6 seconds in **c**. Note that, at a longer time base (slower paper speed, e.g. 10 mm/sec in **b** and 5 mm/sec in **c**), the wash out of the 7–8 Hz anesthesia-induced left hemisphere activities is more evident, including the accentuation of the 0.5–1 Hz left hemisphere delta activity. On the other hand, this 7–8 Hz activity becomes similar in appearance to muscle artifact, and the frontal sharp waves also have an artifactual appearance.

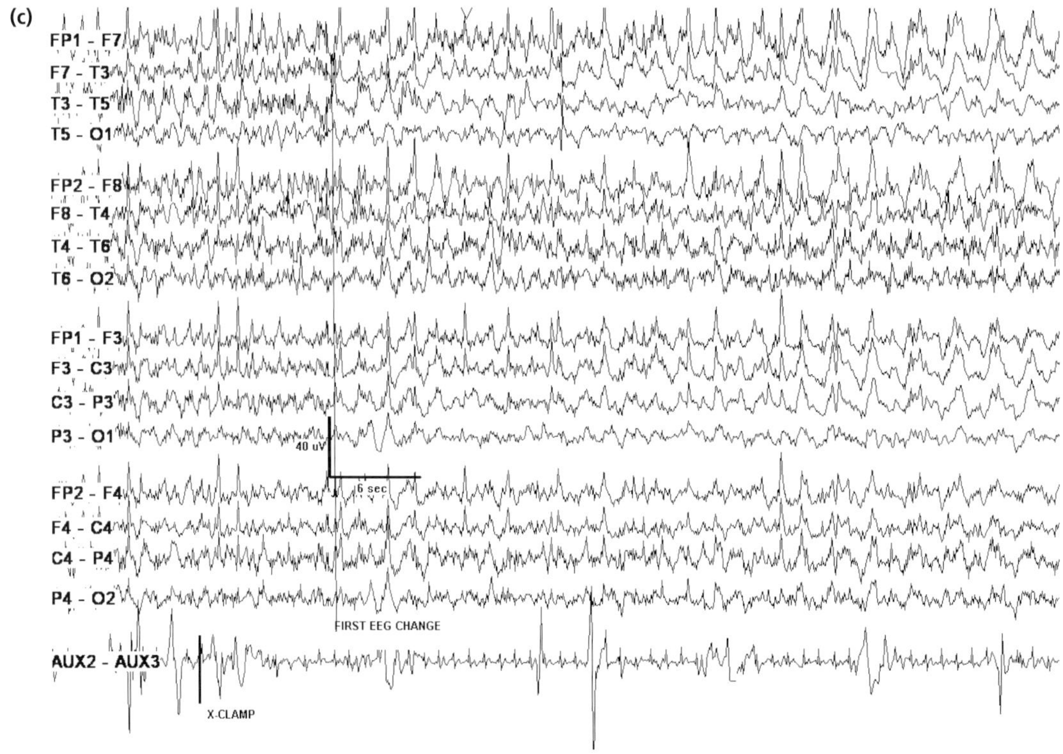


**FIGURE 2-8** *(Continued)*

deficit, yet the EEG shows no change during the cross-clamp period (as expected, because there was no problem with cortical ischemia at that time). In general, rethrombosis at the surgical site happens gradually, and thus the neurophysiologic dysfunction is reflected in EEG changes only at the end of surgery, during closure. Unfortunately, critical obstruction of the carotid may occur hours after the procedure ends and EEG recording stops (Fig. **2-14**). If the EEG was not being recorded during these events, all could be incorrectly assigned to an "incorrect EEG monitoring procedure" category (see page 90 and Fig. **2-14**).

Thus, we believe that EEG monitoring must be started prior to anesthesia, and it must not be stopped immediately after the cross-clamp is released at the completion of the endarterectomy but must be maintained until the patient is awake and responding to commands.

## 6. Embolic Strokes are Missed

Embolic strokes may occur at any time during a CEA (see pages 53–56), with the source of the embolus most commonly the artery itself or some result of the surgical procedure (loose plaque, etc.), but sometimes the aorta or the heart. These strokes may be restricted in size and thus difficult to recognize in the EEG, even with a full complement of electrodes on the scalp, and even harder to recognize if double-distance electrode placement and a limited number of channels are used (Arnold et al., 1997; Chiappa et al., 1979). If the EEG changes are not recognized during the procedure, which is possible even for an experienced EEG user, the patient awakens with a significant neurologic deficit but the EEG is reported to show no change. We do not count this as a failure in the EEG technique, which is intended to monitor for low flow states involving most of the hemisphere at risk; the EEG changes in the latter case, as opposed to those produced by a regional change, would be much more readily apparent in the EEG.

Even if the EEG is not the ideal monitoring tool for detecting small thromboembolic events, there are no better alternatives. Similar criticisms have been applied to SSEP monitoring, and the microemboli detected by TCD proved not to correlate well with the clinical outcome (Jansen et al., 1993; Rowed et al., 2004). Significantly, our experience includes cases of cerebral embolism during which the EEG not only detected the

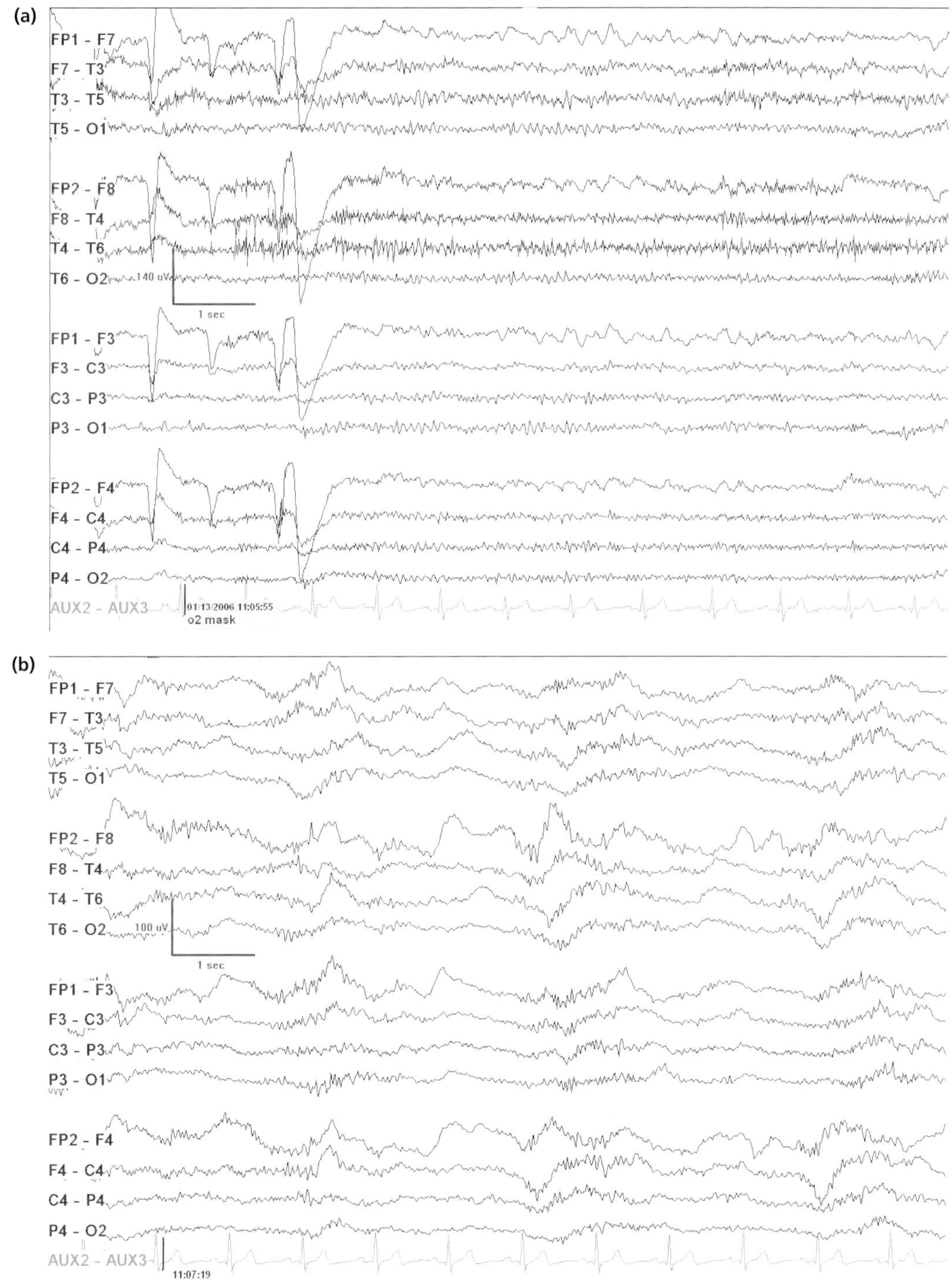


**FIGURE 2-9** EEG changes occurring after induction, in a patient undergoing left carotid endarterectomy. The ipsilateral slowing occurred after the intubation, positioning, and induction. Because these were done in rapid succession, no reliable EEG analysis was possible in between and during these events; thus no definite determination of the cause of the EEG change could be made. We have come across cases where anesthesia induction brought up an EEG asymmetry that was not appreciated in the awake recording. However, in this case, the severity of this asymmetry pointed toward a different mechanism: acute thromboembolism as a result of neck manipulation during positioning and intubation. This patient woke up with a new right hemiparesis. **a:** Normal, symmetric preinduction baseline. The calibration mark is 140 $\mu$V and 1 second. **b:** The immediate postinduction baseline shows a mild burst-suppression pattern, which precludes a thorough EEG analysis; no definite asymmetries are evident. The calibration mark is 100 $\mu$V and 1 second.

(c)

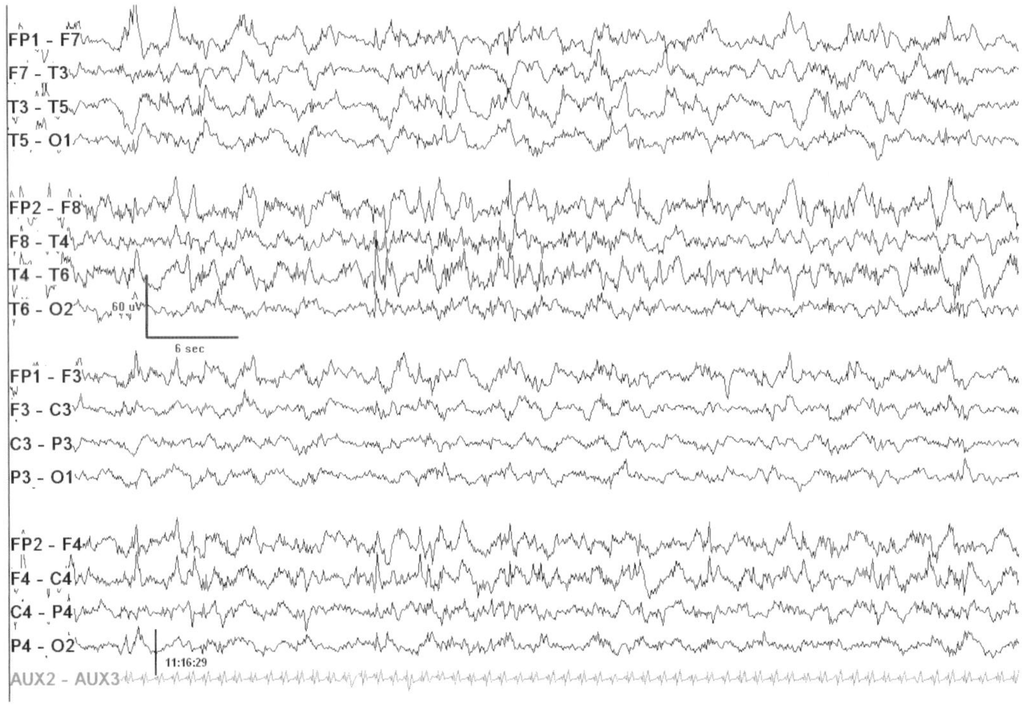


**FIGURE 2-9** *(Continued)* **c:** Once the concentrations of the anesthetics are decreased, a significant background asymmetry is evident at a slower paper speed, showing an attenuation of the faster-frequency activities over the left hemispheric regions. The calibration mark is 60 $\mu$V and 6 seconds.

(a)

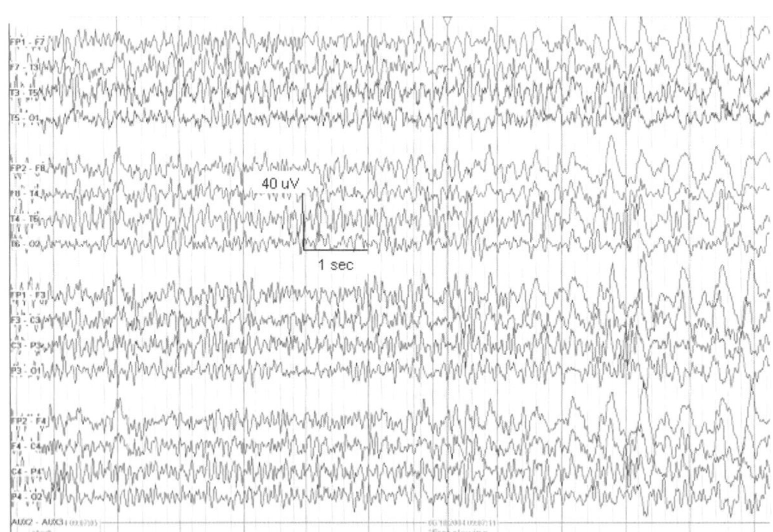


**FIGURE 2-10** EEG during a carotid endarterectomy, in which the EKG monitor channel (bottom trace) is showing an asystole produced by increased vagal tone that is secondary to mechanical manipulation of an unanesthetized carotid body (at the carotid bifurcation); the recording is shown at different paper speeds (30 mm/sec in **a** and 10 mm/sec in **b**). **a:** The last QRS complex occurs at the extreme left margin of the figure ("start" label). Approximately 6 seconds later, the EEG begins to show generalized slowing. The calibration mark is 40 $\mu$V and 1 second.

(b)

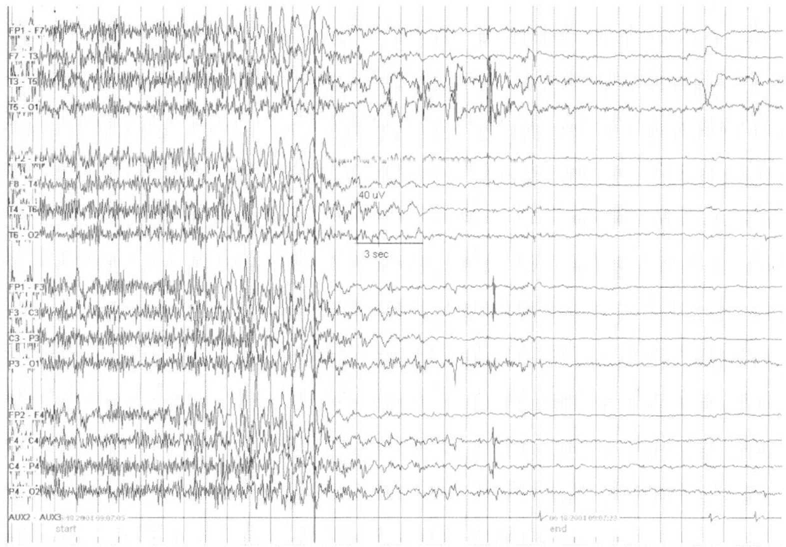


**FIGURE 2-10** *(Continued)* **b:** The last QRS complex occurs at the extreme left margin of the figure. Approximately 6 seconds later, the EEG begins to show generalized slowing. After about 6 seconds of this generalized slowing, there is a rapid loss of amplitude over 5 seconds, with an isoelectric EEG being reached about 16 seconds after the onset of asystole. A few QRS complexes are visible on the extreme right of the page as cardiac function returns ("end" label). The calibration mark is 40 $\mu$V and 3 seconds.

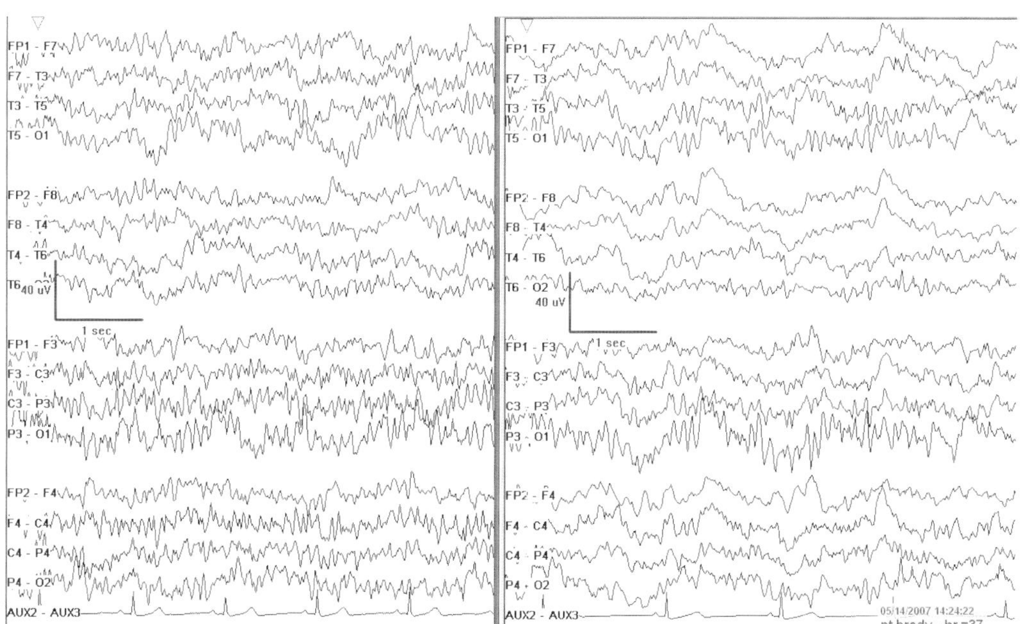


**FIGURE 2-11** EEG during a right carotid endarterectomy, with baseline ipsilateral slowing. The **left** panel shows the EEG at a heart rate of about 55 beats per min, whereas the **right** panel shows a heart rate of 37, with significant loss of the superimposed beta and alpha activities over the right hemisphere and an increase in the amplitude and prevalence of delta slowing on the same side. Calibration mark is 40 $\mu$V and 1 second.

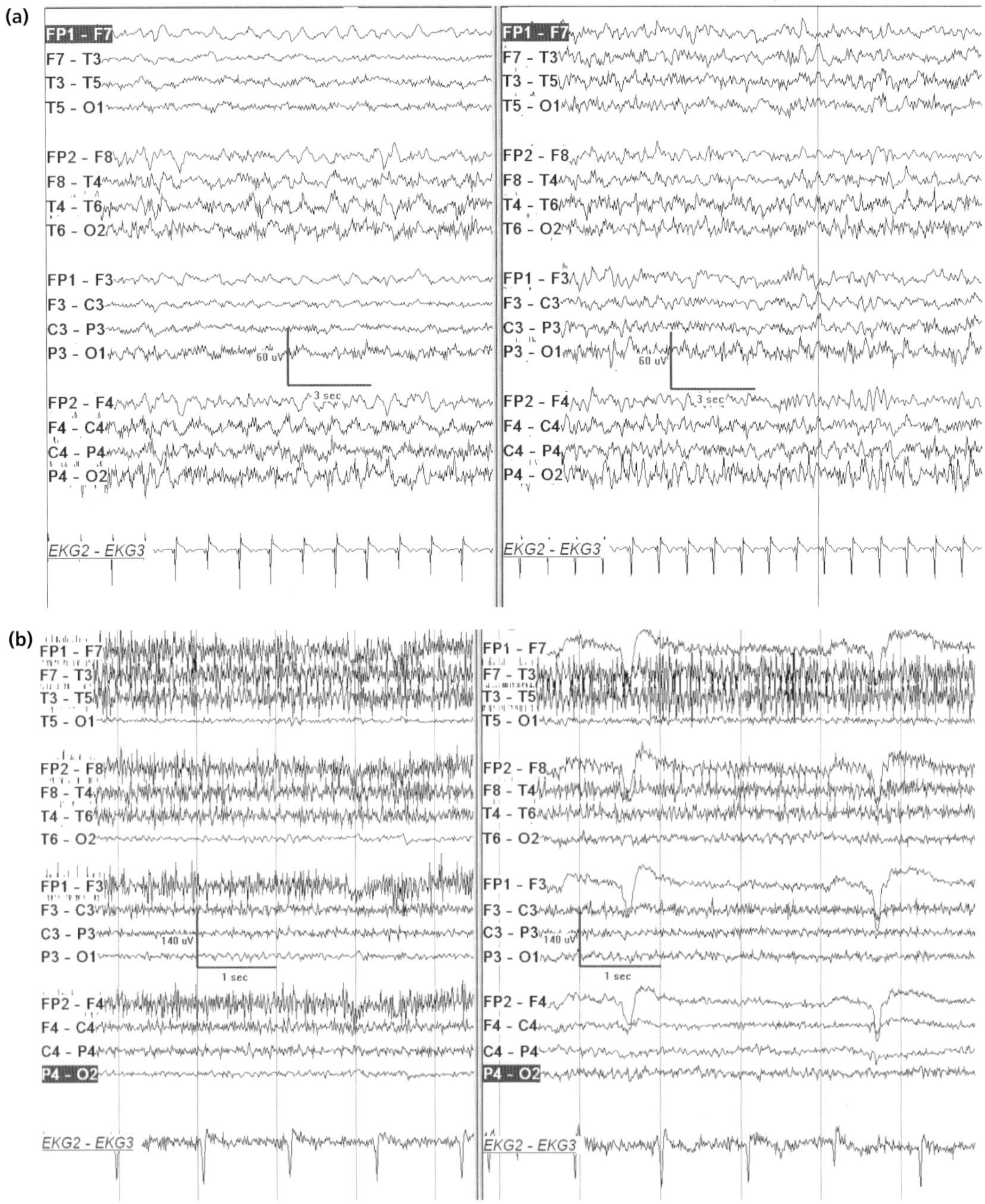


**FIGURE 2-12** EEGs recorded from a patient undergoing left carotid endarterectomy. A sudden, severe ipsilateral hemispheric slowing, with attenuation of the faster frequencies, which occurred while exposing and manipulating the left carotid artery and prior to clamp. Increase in the systolic blood pressure and insertion of a temporary intra-arterial bypass shunt did not improve this slowing. Ten minutes after administration of intra-arterial tPA (thromboplastin activator), there was a significant improvement of the EEG changes, with subsequent return to the postinduction baseline. The patient woke up neurologically intact; diffusion-weighted brain MRI, done immediately after the surgery, was normal. **a:** The **left** panel shows the EEG minutes after an intra-arterial bypass shunt was placed; notice the persistence of significant left hemispheric slowing, with marked loss of the anesthetic-induced beta activity over this side. The **right** panel shows the EEG approximately 10 minutes after administration of intra-arterial tPA; notice the significant improvement of the left-sided slowing. The calibration marks in both panels are 60 $\mu$V and 3 seconds. **b:** The **left** panel shows the symmetric awake preinduction baseline. The **right** panel shows the symmetric awake baseline at the end of the surgery after the tPA administration. The calibration marks in both panels are 60 $\mu$V and 3 seconds.

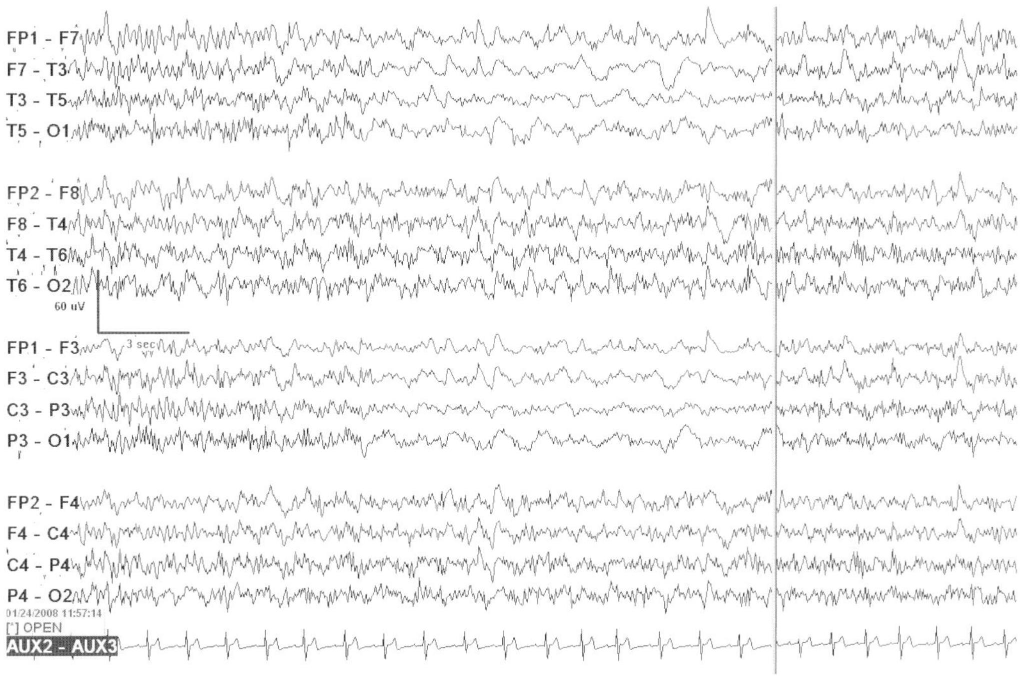


**FIGURE 2-13** EEGs recorded from a patient during a left carotid endarterectomy, showing changes associated with release of the clamp of the internal carotid artery, presumably related to dislocation of a thrombus or other loose material from the operative site. The **left** panel shows the appearance of asymmetric slowing over the left hemisphere, within 10 seconds of the release of the clamp; this asymmetry lasts about 10 minutes, until the systemic systolic blood pressure is increased to 200 mmHg. The **right** panel shows the improvement of the EEG asymmetry, with return to the baseline. We believe that the increase in the blood pressure helped to dissolve the thrombus and reinstate the appropriate flow in the left internal carotid. The calibration marks in both panels are 60 $\mu$V and 3 seconds.

neurophysiologic dysfunction but also guided the treatment (Figs. **2-12**, **2-13**, **2-15**).

## 7. Watershed Infarcts are Missed

Very rarely a patient has a low flow state in the at-risk hemisphere subsequent to clamping, which causes a watershed territory infarct. In these patients the remaining areas of cortex in that hemisphere receive sufficient blood flow to maintain function, and their EEG activity, generated from areas surrounding the involved watershed cortical regions, masks the decreased amplitude of EEG in the ischemic regions. Thus, these patients may have a postoperative neurologic deficit, despite there having been no definite change in the EEG during the clamp period. However, when the affected areas are larger, EEG changes can be detected (Fig. **2-14**), so that special attention needs to be given to persistence of asymmetries after episodes of systemic hypotension, especially in the presence of significant stenosis in the carotid systems.

## 8. Preexisting Cerebrovascular Accident (CVA)

A similar situation in which EEG changes can be difficult to detect is encountered during CEAs in patients who have had prior strokes and thus have an asymmetric EEG at baseline. In such cases, a minor change in the EEG can be difficult to appreciate but can have a major physiologic effect (Fig. **2-16**). It may be that, in these two circumstances (see this page), continuous trend analysis (graphic plots) of EEG power and frequency is a more sensitive indicator of EEG changes in this condition, but it occurs so rarely that it is difficult to study.

## 9. Variety of Monitoring Techniques

There are several other techniques for monitoring blood flow in the involved hemisphere, such as TCD, SSEPs, carotid artery stump pressures (SP), transcranial cerebral oximetry, and quantitative EEG analysis

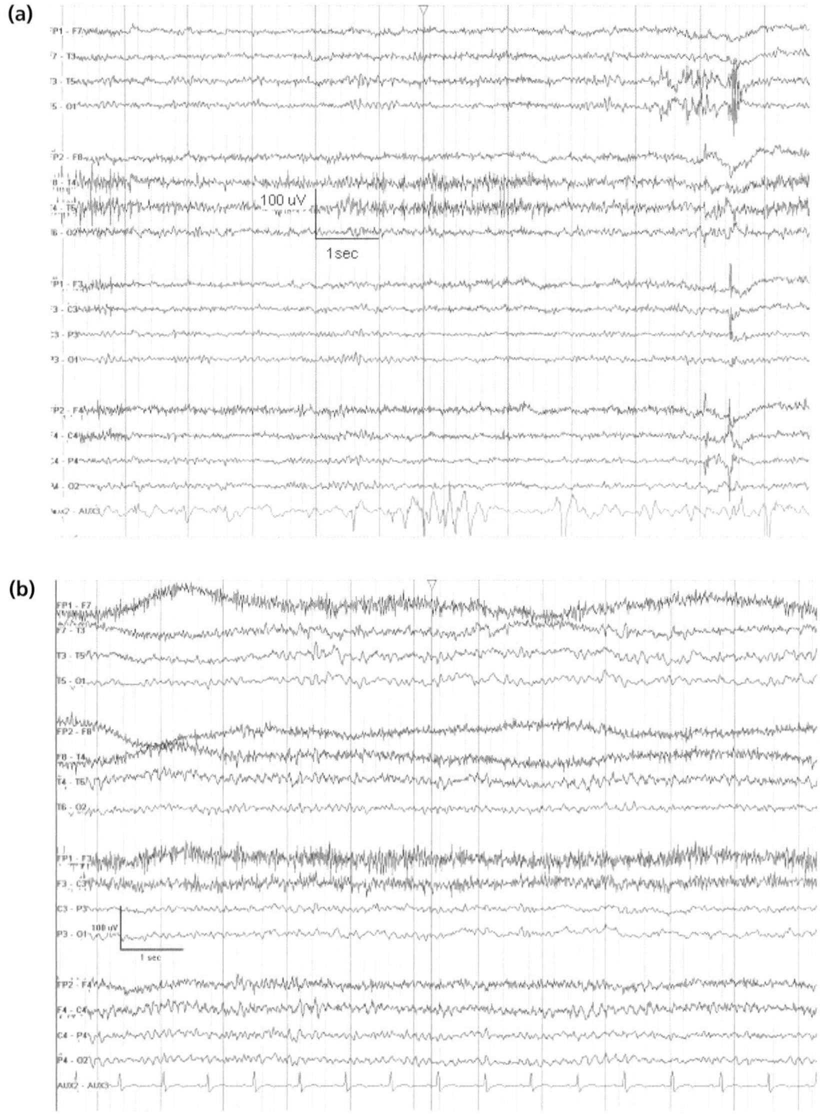


**FIGURE 2-14** EEGs from a patient who underwent left carotid endarterectomy for asymptomatic, high-degree carotid stenosis. The procedure was uneventful, and the EEG remained symmetric throughout the surgery. The patient awoke from surgery neurologically intact. However, 4 hours later, while in the recovery room, she developed new onset right hemiparesis and dysphasia. The carotid Doppler detected no flow in distal left internal carotid artery; thus, suspecting rethrombosis at the surgical site, the surgeon reintervened. The calibration marks in both slides are 100 $\mu$V and 1 second. **a:** Recording at the end of the first surgical procedure, with the patient extubated, moving all extremities and following commands. The EEG is showing a normal drowsy-awake background, without asymmetries. **b:** Five hours later, while the patient is experiencing new symptoms. The EEG shows the drowsy-awake preinduction baseline, before the surgical reintervention. Notice the presence of the new left hemispheric slowing, maximal temporally.

(Brain Symmetry Index [BSI]). The amount of experience with a particular neurophysiologic monitoring technique can play a decisive role in determining what method will be more useful in a given operating room.

### *A. TCD*

TCD is being used in many centers as a monitoring technique for cerebral ischemia because of its capacity to measure the drop in blood velocity in the middle cerebral artery (MCA) secondary to cross clamping. However,

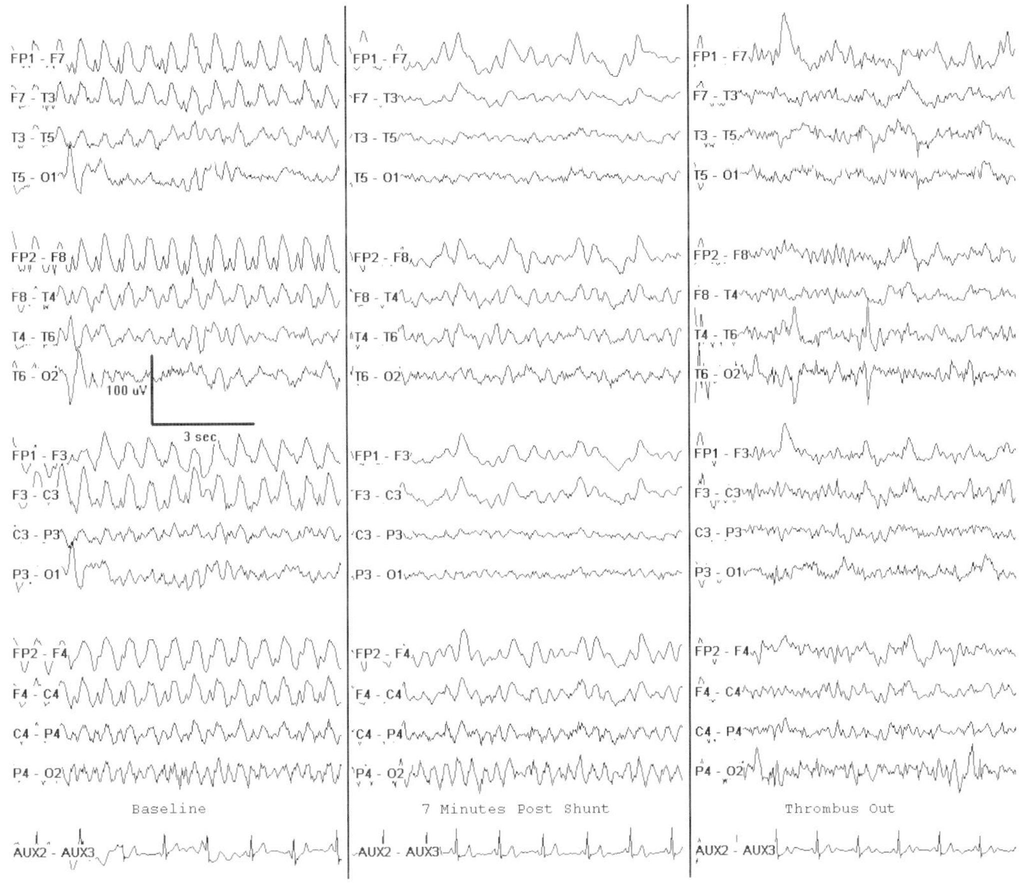


**FIGURE 2-15** EEG during left carotid endarterectomy in a patient who suffered a thromboembolic event at the clamping of the left carotid artery. In the postinduction preclamp period, the EEG showed a symmetric background (**left** panel). Clamping of the left carotid artery was followed by the sudden appearance of left hemispheric slowing, which remained unchanged after the surgeon placed a temporary intra-arterial bypass shunt. The EEG change persisted, despite shunt revision and increase in the systemic blood pressure (**middle** panel), but the surgeon decided to complete the endarterectomy; 10 minutes later the procedure was done, and the arteries were opened. However, the EEG remained asymmetric, and, despite excellent flow in the external and common carotid arteries, the internal carotid did not have a Doppler pulse. Thus the surgeon, suspecting a mechanical occlusion of the distal internal carotid artery, likely by a thrombus, reopened the artery and checked for backflow. While the proximal clamp was on, the anesthesiologist increased the systemic systolic blood pressure up to 250 mmHg. The backbleed washed out a large thrombus from the distal carotid artery. Thereafter, good retrograde flow was obtained. After the closure of the arteriotomy, all the clamps were removed, and all of the vessels had excellent flow by Doppler. Concomitantly, the EEG changes improved promptly (**right** panel), with a return to the postinduction preclamp baseline within a couple of minutes. The patient awoke from surgery neurologically intact. The calibration marks in all panels are 100 up and 3 seconds.

there is not a good correlation between regional cerebral blood flow (rCBF) and MCA blood velocity (MCAV) (Halsey et al., 1986; Halsey et al., 1989), and prior established criteria linking a decrease in the MCAV of >70% during clamping to selective shunting (Staszkiewicz et al., 1997) proved unreliable (Lacroix et al., 1999). Similarly, others did not find good correlation between changes in MCAV and development of neurologic symptoms (Bornstein et al., 1996). TCD has also been described as a more sensitive technique in detecting microemboli (Aackerstaff et al., 2004; Fiori et al., 1997; Gaunt et al., 1998; Jansen et al., 1993; Rowed et al., 2004; Spencer, 1997) and hyperperfusional states (Ogasawara et al., 2005). However, other studies revealed no significant correlation between the microemboli and postoperative neurologic deficit (Jansen et al., 1993; Rowed et al., 2004). Unlike EEG and SSEPs, TCD cannot offer a direct measure of the impact of a blood velocity drop

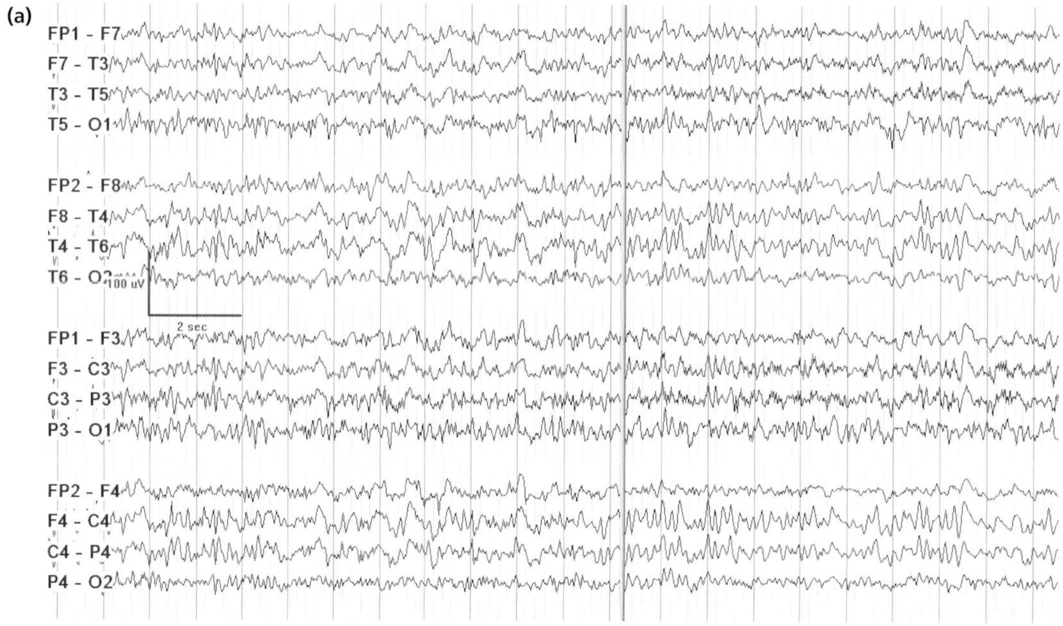


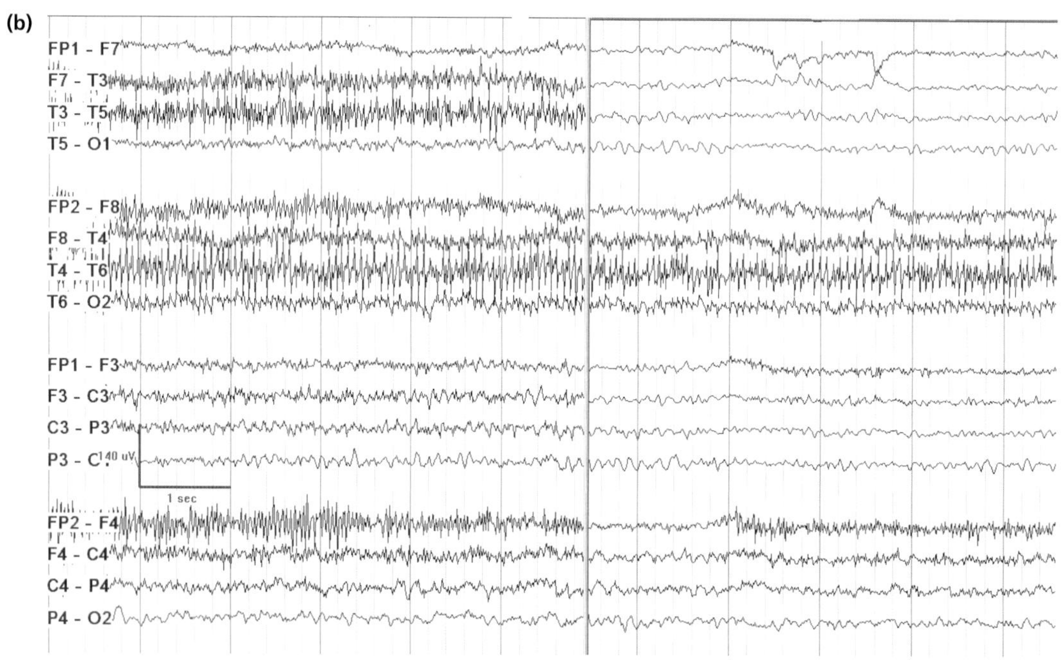


**FIGURE 2-16** EEGs from a patient with recent stroke in the territory of the right middle cerebral artery, who underwent right carotid endarterectomy (Rt CEA). The EEG recording remained unchanged during the entire procedure. However, the patient woke up with worsening of her left hemibody deficits. A postoperative brain MRI study using diffusion-weighted imaging sequences showed an increase in size of the old infarct. **a:** The **left** panel shows a mildly asymmetric postinduction EEG baseline, with more pronounced slowing over the right hemisphere. The **right** panel shows EEG recording during clamp time, with no obvious changes from baseline. The calibration marks in both panels are 100 $\mu$V and 2 seconds. **b:** The **left** panel shows preinduction EEG baseline. The **right** panel shows the drowsy-awake recording at the end of the procedure, after extubation. No obvious worsening of the baseline EEG asymmetry is seen; however, the initially symmetric, bitemporally predominant muscle artifact is now asymmetric, present on the right but not on the left, which is a worrisome sign pointing toward a possible loss of muscle tone over the left scalp. The calibration marks in both panels are 40 $\mu$V and 1 second, and the high-frequency filter (HFF) is at 70 Hz.

(c)

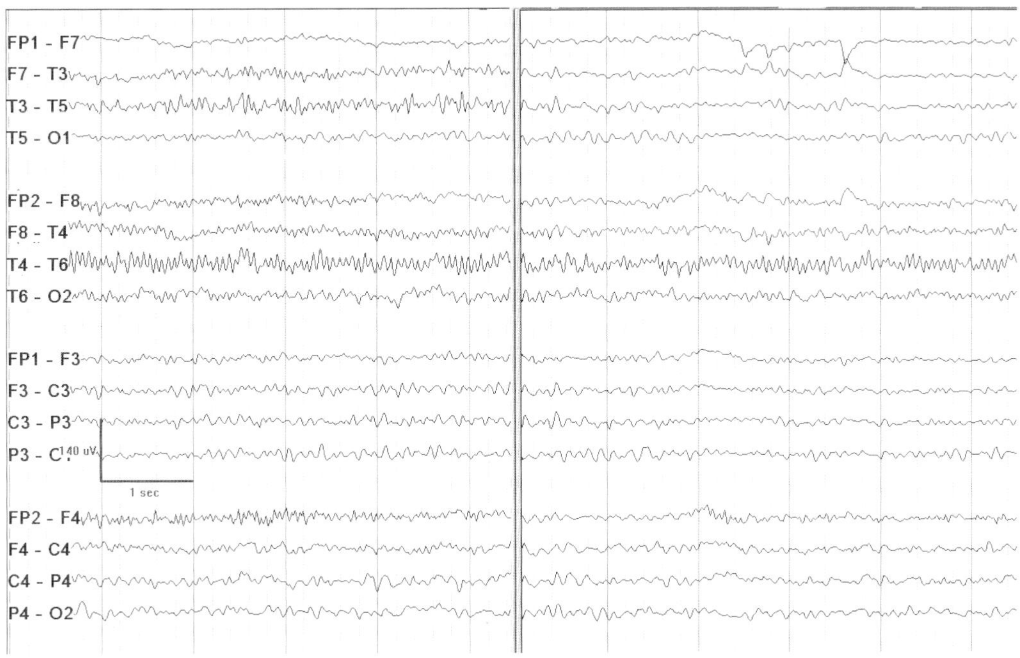


**FIGURE 2-16** *(Continued)* **c:** Same slide as in **b,** this time with HFF at 15 Hz, to attempt filtering of the muscle artifact and thus to better appreciate lower-frequencies activities (underlying slowing). No further new EEG asymmetries are detected.

in the MCA or of microemboli on neuronal function. In addition, there are technical limitations to its use reported in certain patients (up to 74% in Dinkel et al., 1994; 13% in McCarthy et al., 2001; and 21% in Moritz et al., 2007).

### *B. SP*

SP refers to the pressure in the distal internal carotid artery, which can be measured intraoperatively with an intraluminal Fogerty balloon catheter, used instead of the distal clamp. It is a very good indicator of the arterial pressure but a poor predictor of the presence of cerebral ischemia as discussed in the following. SP measurements have a suboptimal sensitivity and specificity in detecting cerebral ischemia (Brewster et al., 1980; Modica et al., 1989) and a poor correlation with EEG changes seen during CEA (Botes et al., 2007), and they are a worse outcome predictor than EEG (Fletcher et al., 1998). Similar to TCD monitoring, there is no general consensus regarding the SP cutoff threshold for selective shunting (Botes et al., 2007; Jacob et al., 2007; Modica et al., 1989); thus, SP values were found unreliable in predicting the need for shunting (Belardi et al., 2003). Finally, this type of monitoring is prone to technical errors, such as inaccurate calibration of the transducer.

The TCD and SP techniques are limited in their utility by the wide interindividual variability of the threshold for ischemia that is determined by the variability of the autoregulatory threshold for cerebral blood flow. The latter is in turn influenced by multiple factors such as the presence of (1) cerebral vasoconstrictors (e.g., hypocarbia), (2) vasodilators (volatile agents, hypercarbia), (3) chronic hypertension, (4) intracranial arterial disease, (5) different cortical metabolic requirements (with decreased requirements, with drugs such as barbiturates, or with hypothermia), and (6) variability of intracranial pressure.

### *C. Cerebral Oximetry*

As with the use of SP, with which cerebral oxygen saturation correlates well (El-Dawlatly, 2004; Lee et al., 2008), there is not a well-established threshold for selective shunting using cerebral oximetry (Carlin et al., 1998; Takeda et al., 2000; Vets et al., 2004), and studies comparing this latter method to neurophysiologic monitoring via SSEPs in detecting cerebral ischemia of clinical significance showed an inferior sensitivity and specificity

(Duffy et al., 1997), presumably as a result of interindividual and intraindividual variability of the brain tissue resistance to cerebral hypoxia.

### *D. SSEPs*

SSEPs directly measure function in the at-risk tissue (the cerebral cortex) and, like the EEG, act as a "final common denominator" of cerebral perfusion and oxygenation. Thus, SSEPs are considered a reliable method for CEA monitoring (Lam et al., 1991; Lang et al., 2000; Rowed et al., 2004). The SEP criteria used to declare significant changes are a decrease in N19 amplitude of more than 50% and an increase in the central conduction time (P/N13–N19/P22 interpeak latency) of more than 1 msec (Florence et al., 2004; Rowed et al., 2004). However, as with EEG, attempts have been made to grade the changes in SSEPs in order to increase the specificity of SSEPs (Guerit et al., 1997). Median nerve SSEPs are most often used, although the addition of posterior tibial nerve SSEPs is considered useful by some authors (Manninen et al., 2004). We do not consider SSEPs a better monitoring tool than EEG for the following reasons: (1) registration of SSEPs requires more set-up time and equipment (for stimulation and recording) and takes at least 1 minute to acquire a new data set (signal averaging is required for waveform extraction); (2) SSEPs are less sensitive to cerebral ischemia (Haupt et al., 1994), especially in the early stages of decreased cerebral blood flow (16–22 ml/gm/min), and the EEG becomes isoelectric at cerebral blood flow values about 20% higher than those giving rise to disappearance of SSEPs (Nuwer, 1993); (3) SSEPs have a recording area smaller than that of EEG and so are even less sensitive in detecting embolic events or watershed infarcts; (4) N19 and P22 are significantly affected by certain anesthetic agents (e.g., nitrous oxide and halogenated agents) that are frequently used during CEA; and (5) SSEPs cannot give additional information available in the EEG, such as depth of anesthesia and arousal responses.

### *E. Quantitative EEG Analysis*

Quantitative EEG analysis is recommended by other authors in order to extend the visual EEG analysis, based on the opinion that spectral EEG analysis is more sensitive in detecting changes in different frequency bands and thus can signal cerebral ischemia at an earlier point. As with any method, we believe that there is a learning curve in interpreting such data, because EEG changes occur in many circumstances and what makes a method reliable is the ability to discern which changes are due to cerebral ischemia and which are not. In general, the more sensitive the test is, the more likely it is that there will be an increase in false positive results. Visual EEG analysis is sensitive enough for most clinical circumstances, and the false negative cases are related not to a failure of the eye to detect a change, but rather to the spatial resolution. There is a good correlation between visual EEG changes requiring shunting and a BSI >= 0.06 (van Putten et al., 2004). However, there are some situations where the additional use of quantitative EEG analysis may prove helpful (see page 61).

## 10. Nonselective Shunting

There are surgeons who shunt in every CEA and claim that this renders neuromonitoring irrelevant. We disagree with such an approach for the following reasons: (1) The shunt is not always easily inserted, e.g., when there is a high bifurcation of the carotid system. (2) Shunting is not a risk-free procedure. TCD studies during shunt insertion have detected a high number of microemboli (Wolf et al., 2004) that are related to intraoperative complications (Aackerstaff et al., 1996; Salvian et al., 1997). This can be caused by dislocation of thrombus from an ulcerated plaque, by further intimal injury with increased potential for thrombus formation, or by air embolization (McKinsey et al., 1996; Fig. **2-17**). (3) Shunting prolongs the surgical procedure. (4) Shunts can be inadequate in size and positioning, and they can kink and otherwise malfunction. In these cases, absent continuous EEG monitoring, inadequate cerebral blood flow could not be detected (Artru et al., 1989; Figs. **2-18**, **2-19**, and **2-36**). (5) Shunting does not preclude the occurrence of other intraoperative complications that can be detected by EEG monitoring and thus treated early (e.g., emboli, thrombosis at the operative site, watershed infarction).

# III. METHODS

## 1. Electrode Application

The techniques for intraoperative EEG monitoring of carotid endarterectomies (CEAs) have routine EEG performance as their basis, and the IOM technologists should be competent in the performance of routine EEG studies. Therefore, the necessary record-

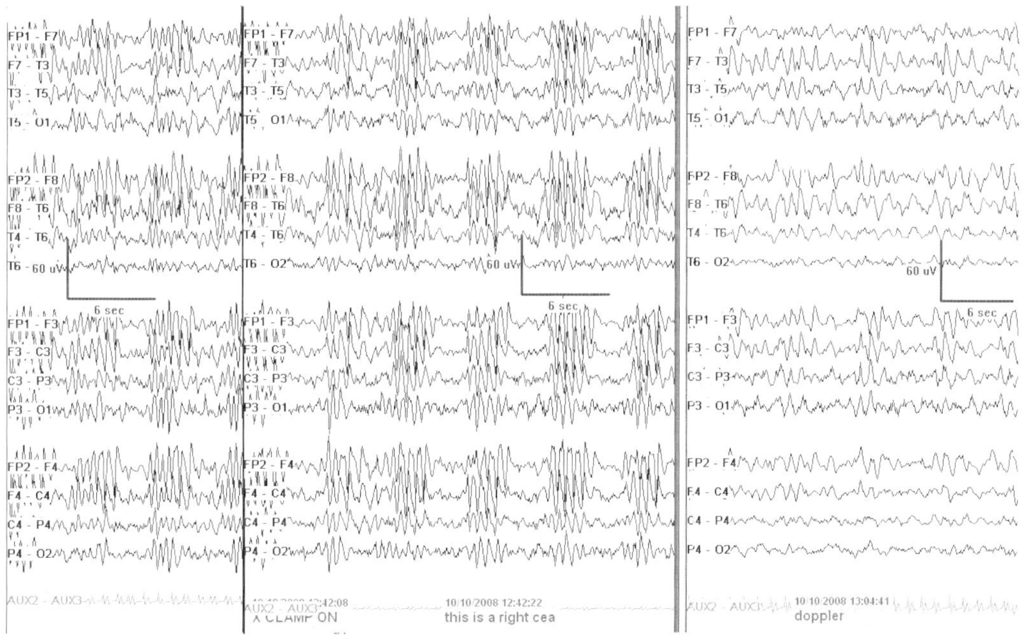

**FIGURE 2-17** EEGs samples from patient who underwent right carotid endarterectomy. The EEG prior to clamp and in the immediate postclamp period showed mild, intermittent asymmetry of the generalized slowing, with shifting predominance between the two hemispheres; however, a mild burst suppression pattern was also evident which made difficult a detailed EEG analysis. Thus, despite a definite worsening of the slowing that could have been attributed to the clamp, the surgeon elected to place an intra-arterial bypass shunt. This was promptly followed by the development of a consistent EEG asymmetry, with more slowing over the right hemisphere. The EEG change did not improve, despite an increase in the blood pressure. The patient woke up with a new left hemiparesis. The calibration marks in all panels are 60 $\mu$V and 6 seconds. The **left** panel shows the postinduction, preclamp baseline. Notice the mild asymmetry in the anesthetic-induced slowing, more prominent over the contralateral left temporal regions. The **middle** panel shows the postclamp EEG background, within the first half minute after the clamp; no changes from the preclamp baseline are present. The **right** panel shows the EEG after shunt placement: notice the new, consistent asymmetry, with more delta slowing and attenuation of the faster-frequency activities over the right hemispheric regions.

ing equipment, supplies, and electrode placement and application are not discussed here. A full set of EEG electrodes is applied in the International 10–20 locations with electrode impedances below 5,000 ohms, and an additional electrocardiogram (EKG) monitor channel is acquired, usually from an electrode applied to the clavicle opposite to the operative site (referred to a scalp electrode). The electrodes are usually applied with the patient awake.

## 2. Montages

The display is a bipolar longitudinal "double banana" montage (Fp1-F7, F7-T3, T3-T5, T5-01, Fp2-F8, F8-T4, T4-T6, T6-O2, Fp1-F3, F3-C3, C3-P3, P3-O1, Fp2-F4, F4-C4, C4-P4, P4-O2, EKG). Note that homologous areas are displayed contiguously (left temporal, right temporal, left parasaggital, right parasaggital) rather than as a hemisphere grouping (left temporal, left parasaggital, right parasaggital, right temporal). This is done because the important comparison is side to side in homologous areas, and this organization of the montage makes it easiest for the eyes and visual system to perform this task, rather than having to compare the four channels at the top of the page with the four at the bottom (for the temporal lobe comparison).

## 3. Filters and Grounds

The filter bandpass is 1 to 70 Hz, with the 60-Hz notch filter applied. The common mode rejection capability of modern EEG amplifiers is usually sufficient to exclude 60-cycle (main frequency) artifact to an extent that allows reliable EEG interpretation. If 60-cycle artifact is excessive and interferes with EEG interpretation, the usual diagnostic maneuvers need to be performed to

(a)

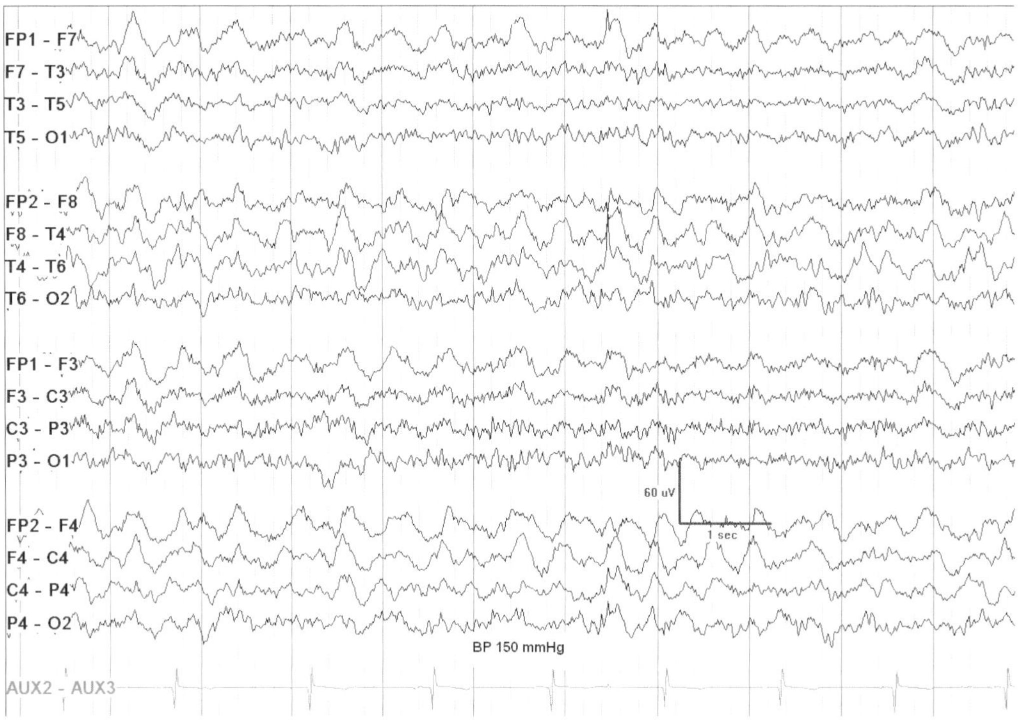


(b)

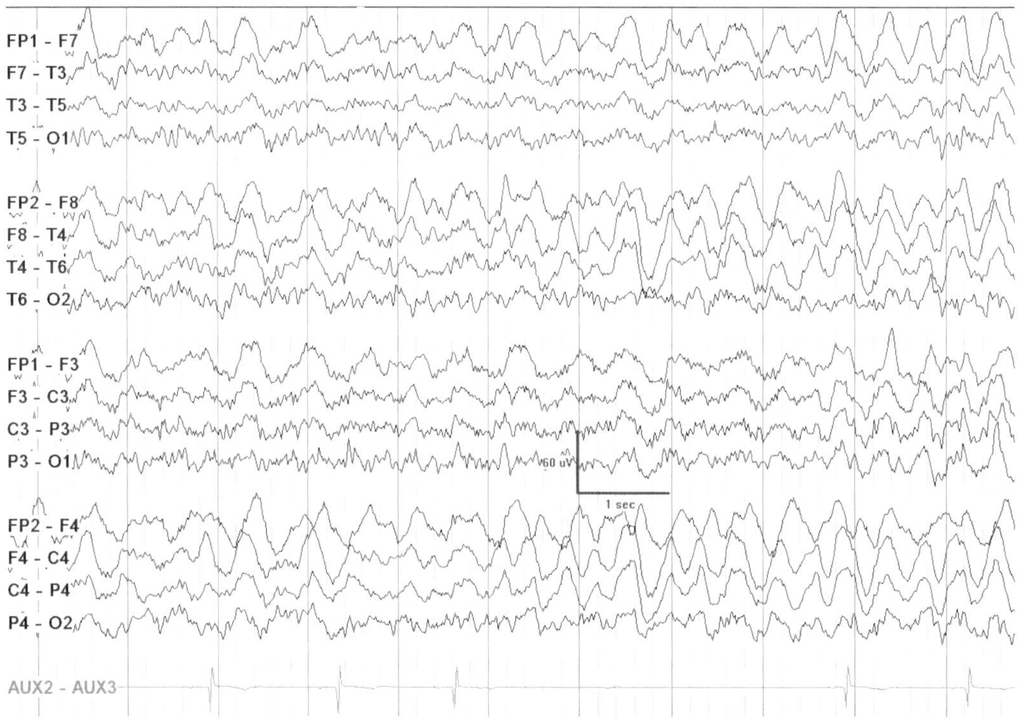


**FIGURE 2-24** EEG samples from a patient with prior cerebral infarct in the territory of the right middle cerebral artery, now undergoing right carotid endarterectomy. The calibration marks in all slides are 60 $\mu$V and 1 second. **a:** EEG background at 150 mmHg systolic blood pressure. There is a baseline EEG asymmetry, with prominent delta slowing over the right hemisphere, with decreased amplitude of the superimposed beta activity in the same regions. **b:** A 3-second pause in the heart beat signals the presence of arrhythmia, followed by a drop in the systemic blood pressure. Notice the asymmetric increase in delta slowing, mainly over the right hemisphere.

(c)

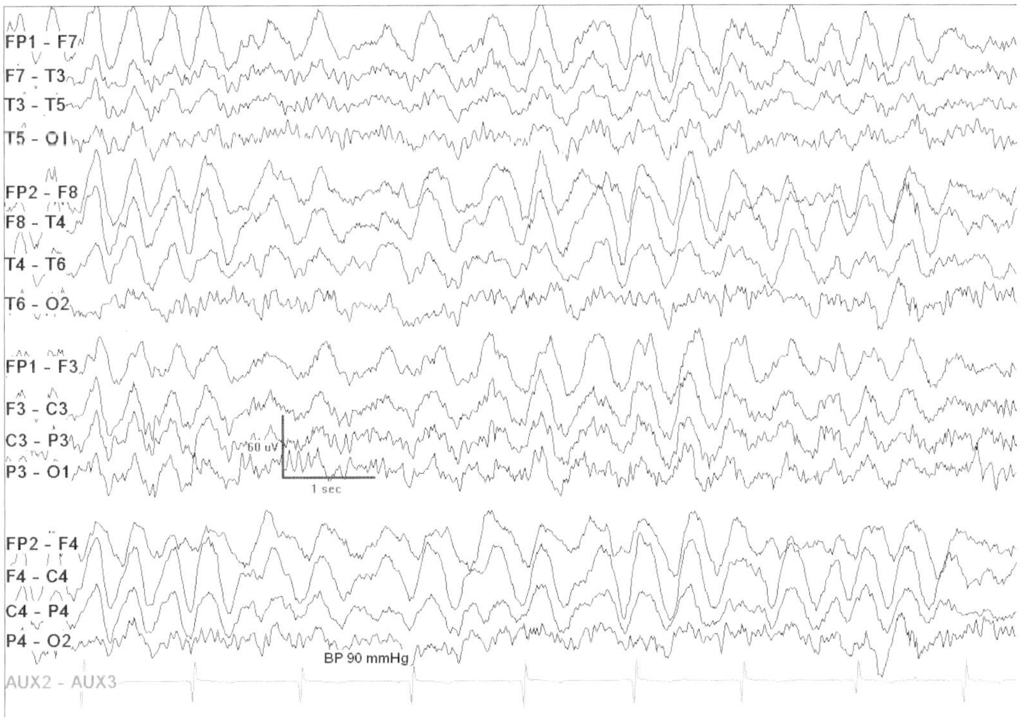


**FIGURE 2-24** *(Continued)* **c**: The systolic blood pressure is now 90 mmHg, and the EEG shows further increase in the generalized slowing, with rhythmic delta activity maximal in the anterior head regions. The asymmetry in the fast-frequency activities (which are attenuated on the right) is evident.

## 3. Anesthetic Changes

Induction of anesthesia usually involves a bolus of propofol or a short-acting barbiturate, such as thiopental, and an opiate (usually remifentanyl). The anesthesia is further maintained with an infusion of the two drugs, to which nitrous oxide and a halogenated agent (most commonly sevoflurane, isoflurane, or desflurane) may be added. These drugs are quickly administrated one after another during the induction and maintained at a constant infusion afterward. They routinely produce a rapid succession of EEG patterns: (1) mild diffuse slowing, with superimposed diffuse fast activities; (2) several seconds thereafter, a burst of high amplitude, diffuse delta slowing, followed by (3) a burst-suppression pattern, which usually resolves in a few minutes and is followed by (4) a continuous background, consisting of 2–6 Hz, slow activity at 10–20 $\mu$V. This latter pattern is fairly constant once the anesthetic agents reach a steady state, and it is usually at this the time that significant baseline postinduction background asymmetries are noticed. Depending on the type and doses of anesthetic agents in use, variable amounts of faster frequencies at 8–12 Hz and 10–15 $\mu$V are superimposed on the generalized slowing. The amplitude of these faster rhythms often varies sinusoidally over the course of a few seconds, giving the appearance of spindle patterns. It is these faster frequencies that disappear first when cortical ischemia occurs.

The amplitude of all of these activities varies over time according to anesthetic depth and the mixture of anesthetic agents in use, both of which vary over the course of an operation (Figs. **2-25**, **2-26**). Notification of these changes may or may not be supplied by the anesthesiologist, but this knowledge can be very helpful in understanding the etiology of changes in the EEG, so that good communication between the anesthesiologist and the neurophysiology staff becomes very important. For example, a sudden increase in sevoflurane concentration is followed within minutes by sudden, high-amplitude, paroxysmal diffuse slowing, with bi-anterior predominance that could mimic diffuse EEG changes related to a sudden systemic blood pressure drop, low flow related to clamping, or an arousal. We also see this pattern related to reopening of the arteries, when the anesthesiologist may be using higher concentrations of halogenated

In short, any anesthetic changes should be avoided close to key moments in the surgery, such as immediately prior to cross-clamp, during the entire clamp time, and at the time of clamp release.

An initial symmetric awake EEG background can become asymmetric with induction, or a mild asymmetry can become much more apparent when a postinduction stable background is achieved, in particular in symptomatic patients (Fig. **2-28**). Induction is a challenging time during surgery, because other mechanisms can account for EEG asymmetries occurring at this time (Fig. **2-9**) (e.g. intubation, head positioning, etc.; see page 84). Good communication with the anesthesiologist is essential, especially concerning the timing of induction and boluses of medication administered.

Another problematic EEG pattern consists of bursts of high-amplitude diffuse slowing that is seen immediately after the arteries are reopened. These can raise the question of sudden cerebral ischemia, blood pressure changes, or increase in the serum anesthetic levels (especially for halogenated agents). In many cases, the etiology of these bursts remains unknown, but it is possible that they represent a version of a paradoxical arousal pattern.

Similarly, recognizing normal arousal patterns in the EEG (e.g., attenuated background, increased muscle activity, presence of alpha activity) can be equally useful to the anesthesiologist, because inadequate anesthetic depth can be associated with intraoperative awareness (Fig. **2-29**).

## 4. Blood Pressure Changes

Changes in blood pressure affect the overall amplitude and frequency content of the EEG, and changes in this parameter often need to be obtained by querying the anesthesiologist or viewing the vital signs monitor display if it is visible. There is considerable inter- and intraindividual EEG variability with blood pressure. Because blood pressure fluctuates quite a bit during the entire procedure, it is good practice to record not one but several preclamp EEG baselines at different blood pressures values. This is especially important when the blood pressure-related background changes are asymmetric (e.g., in a previously compromised hemisphere). Also a given systemic blood pressure that provided adequate perfusion prior to clamping can prove inadequate during the clamp, especially on the side that is supplied by a stenotic carotid system. Therefore, when changes occur during clamp, the first maneuver should be to increase the blood pressure, which may reverse the change (Fig. **2-30**).

In general, EEG changes with blood pressure tend to be diffuse (Fig. **2-31**), unless there is prior unilateral hemispheric injury or unilateral/asymmetric stenosis of the anterior circulations (Fig. **2-23**).

Most of the EEG changes related to blood pressure occur with a decrease in values. However, the increase in perfusion pressure with reopening of the arteries carries the potential risk of a hyperperfusion state, although this is usually a delayed phenomenon (see page 89).

## 5. What can Go Wrong and When

As previously mentioned, it is essential that EEG monitoring is continued from the time that the patient is anesthetized until the patient is awake and cooperating with a neurologic examination at the end of the procedure, not only during clamping. This continuous monitoring is performed because adverse events can occur throughout the entire procedure time, and most of the neurologic complications of this procedure are unrelated to the cerebral ischemia secondary to cross-clamping (Blondeau et al., 1982; de Borst et al., 2001; McKinsey et al., 1996; Mola et al., 1986; Pistolese et al., 1997). When these events are detected via their EEG changes, the surgeon and the anesthesiologist should be notified at once.

### *A. Anytime*

Embolic events from the surgical site can occur at any time during the surgery, such as during head positioning, artery exposure or manipulation, shunt placement, clamp removal, and from rethrombosis at the surgical site. TCD can detect microemboli and has been used to assess the risk of microembolization during different steps of the procedure, with different types of plaques and with different degrees of arterial stenosis, as well as the correlation between such microemboli and the clinical outcome (Aackerstaff et al., 1996; Abbott et al., 2007; Babikian et al., 1997; Cantelmo et al., 1998; de Borst et al., 2001; Gaunt et al., 1996; Horn et al., 2005; Laman et al., 2002; Verhoeven et al., 2005). In general, dissection, shunting, and the immediate postoperative phases, as well as the presence of loose, fresh plaques and a high degree of stenosis, seem to be associated with clinically significant embolization. To these events one can add embolization from the aortic arch and the heart.

Additionally, it is important to acknowledge that strokes that are due to thromboembolism, as with strokes

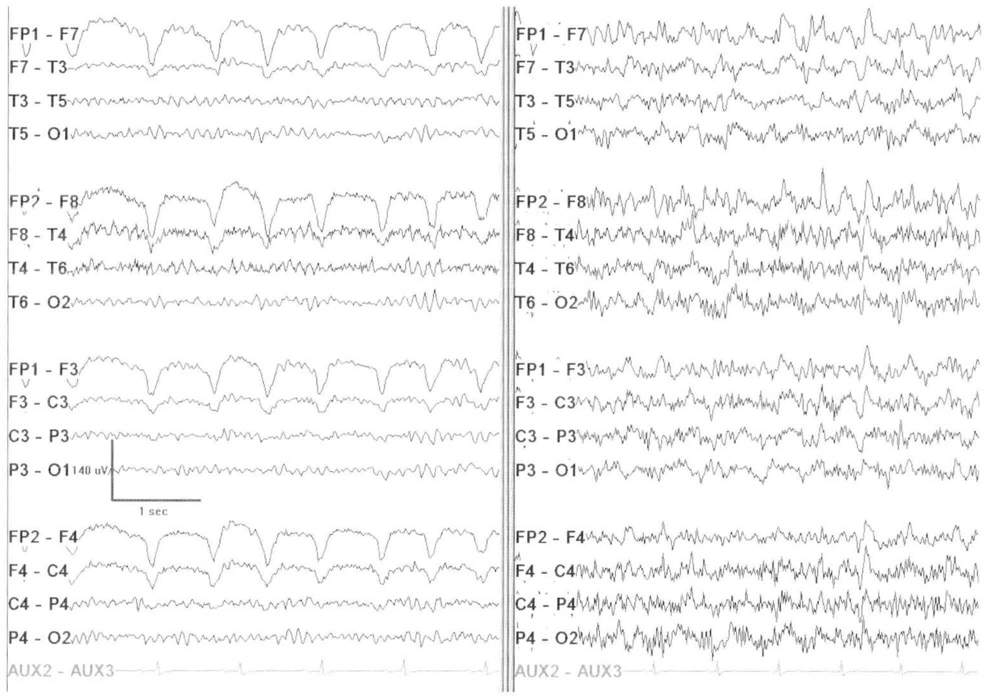


**FIGURE 2-28** EEG recorded in the awake state (**left** panel) and during induction of anesthesia (**right** panel) for a carotid endarterectomy, showing the appearance of a significant asymmetry after induction and before any surgical manipulations. The awake EEG shows normal alpha activity at 9–10 Hz and 20–25 μV, with frontal eyeblink artifact. The EEG during anesthesia shows less anesthetic-induced beta activity over the left hemisphere, with a greater than 50% amplitude asymmetry, and a small increase in the amplitude of the delta slowing on the same side. The calibration marks in both panels are 140 μV and 1 second.

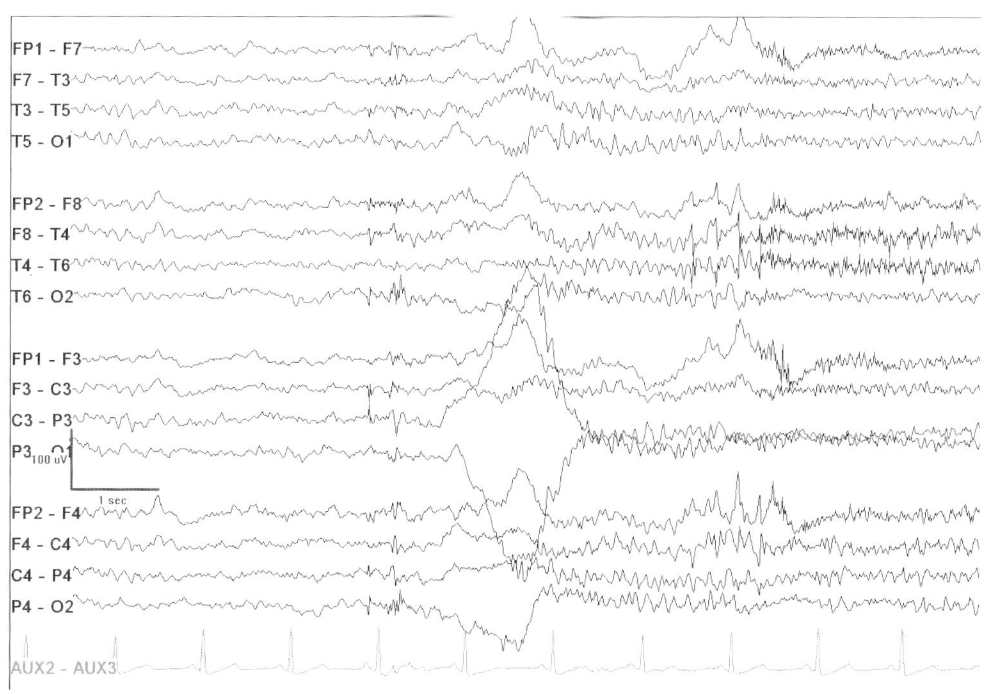


**FIGURE 2-29** EEG showing an intraoperative arousal, likely as a result of external stimulation during a period of light anesthesia. Notice how the diffuse delta slowing seen in the half left of the figure is suddenly replaced by an overall increase in the background frequencies and a decrease in delta slowing; notice the appearance of a posterior dominant alpha activity, of diffuse beta activities as well as of muscle artifact. The calibration mark is 100 μV and 1 second.

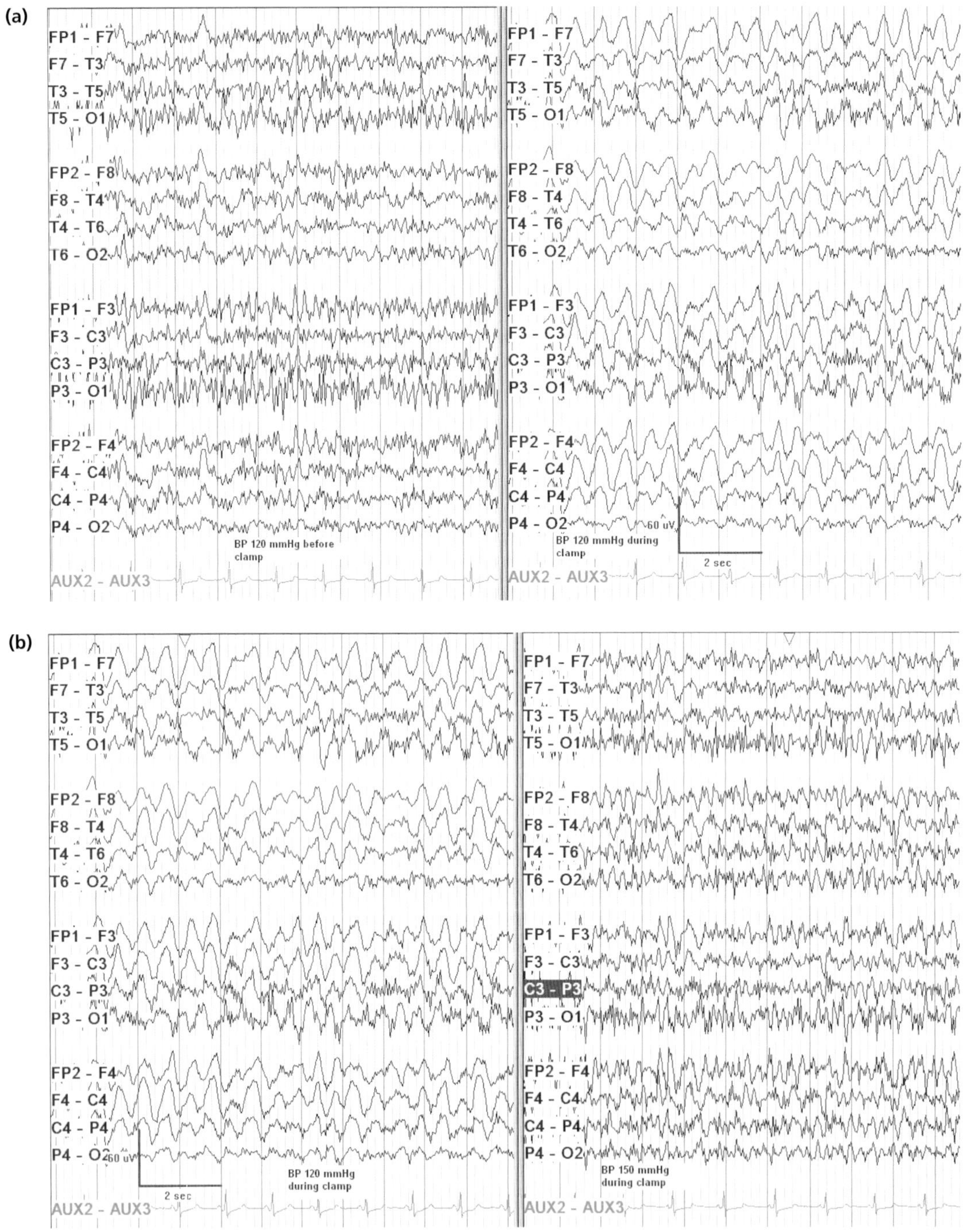


**FIGURE 2-30** EEGs during right carotid endarterectomy in a patient with right hemispheric stroke. This example shows that a certain blood pressure value that was sufficient for good cerebral perfusion before clamp may become insufficient during clamp time. The calibration marks in all panels are 60 $\mu$V and 2 seconds. **a:** The **left** panel shows the postinduction preclamp baseline: notice the mild asymmetry of the anesthetic-induced slowing, more pronounced on the right. The **right** panel shows EEG recording 1 minute after clamping the right carotid artery: notice the overall increase in the diffuse delta slowing with attenuation of the faster frequencies, with changes more pronounced on the right. The systolic blood pressure is the same in both left and right panels (120 mmHg). **b:** The **left** panel is the same as the right panel in slide **a.** The **right** panel shows EEG after the systolic blood pressure was increased to 150 mmHg; there is a reversal of the slowing and even improvement of the mild preclamp EEG asymmetry which was present at 120 mmHg.

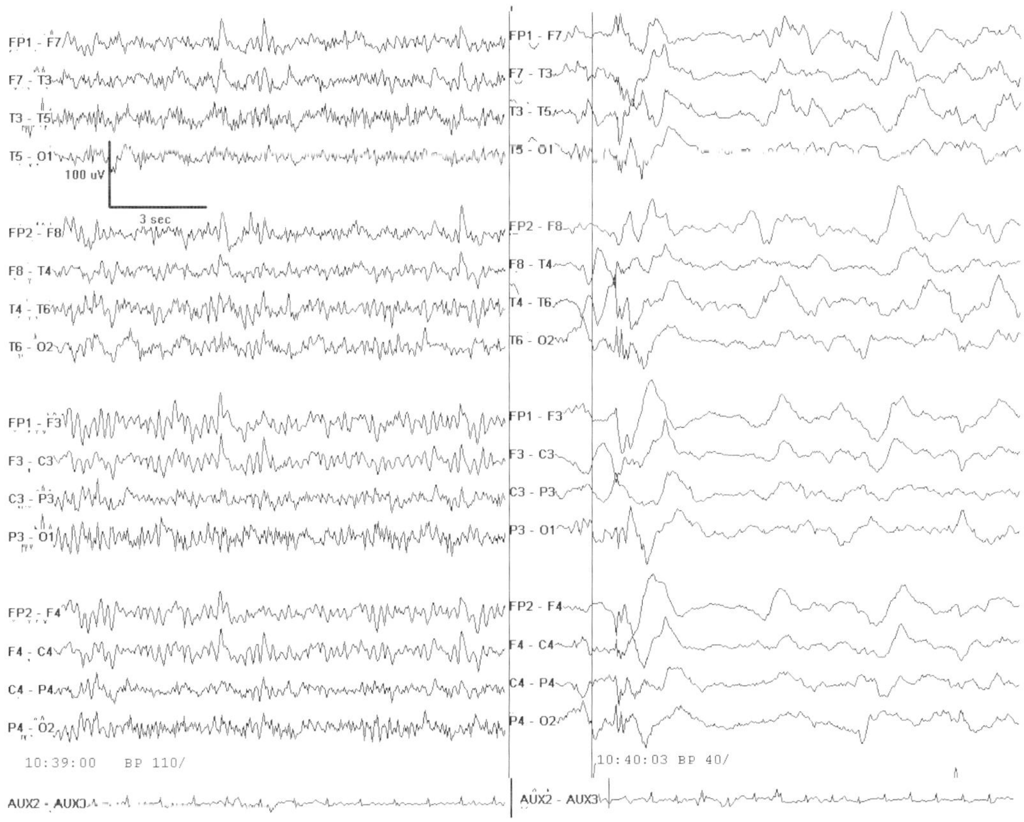


**FIGURE 2-31** EEGs recorded from a patient during a carotid endarterectomy and showing changes associated with a decrease in blood pressure. The **left** panel shows the EEG at a systolic blood pressure of 110 mmHg , consisting of anesthetic-induced mixed frequencies at 20–35 μV. In the **right** panel, after a sudden fall, the systolic blood pressure is 40 mmHg, and the EEG now shows loss of most faster frequencies and a marked increase in the generalized delta slowing, maximal in anterior head regions. The calibration marks in both panels are 100 μV and 3 seconds.

related to low flow states during clamping, can occur not only in the anterior circulation ipsilateral to the surgical site, but also in the contralateral anterior and ipsilateral posterior circulations, when there is a significant reversal of flow via collaterals (see also page 85).

Prompt recognition of the etiology of EEG changes (thromboembolism) leads to further useful investigations (SP, angiograms) and treatments, such as increasing the blood pressure (which can result in clot lysis), surgical thrombectomy, or clot-dissolving drugs. The latter treatment has been successfully used in thromboembolism complicating different stages of CEA (Barr et al., 1995; Comerota et al., 1996; Perler et al., 2000; Fig. **2-12**).

Systemic blood pressure changes can also occur at any time during the procedure. The relative hypotension can reduce cerebral blood flow to a point where cortical function is compromised, possibly to a critical level for neuronal survival. The anesthetist can increase the blood pressure, and the EEG changes can resolve without the need of further intervention (such as shunting or further exploration of the cause of EEG change). Thus, unless there is a contraindication, such as a suspected abnormally high cerebral perfusion, a blood pressure increase should be the first step taken when EEG changes occur. If a hypoperfusional state during clamping is the cause, the increase in blood pressure contributes to better brain perfusion while an intra-arterial bypass shunt is placed. If the mechanism is embolic, a blood pressure increase could help with thrombus resolution or elimination. We have seen the latter occurrence in a few cases when an increase in the SP associated with backbleeding washed out the thrombus in a retrograde direction from the distal internal carotid artery (ICA) (Fig. **2-15**).

Also, note that EEG changes related to systemic blood pressure are not necessarily diffuse. As with other causes of low flow (e.g., clamping, inadequate shunt flow), the changes tend to occur first in the more susceptible vascular territory, which could be either ipsilateral or contralateral to the clamp, especially if there is a prior brain insult.

Once the arteries are open and the end of the procedure is near, the anesthesiologist decreases the anesthetic concentrations and the patient is woken up and extubated. Unfortunately, there is much movement and muscle artifact that intermittently obscures the record during this time, and the patient is still in a drowsy state characterized by mild diffuse slowing, so that it can be quite difficult to compare the awake, preinduction EEG background (baseline) with the "awake" postsurgical background; in such cases, we are particularly looking for presence of new asymmetries. Despite these problems, new asymmetries must be sought.

## B. EEG Changes Prior to Clamping

The carotid artery sometimes occludes unexpectedly in the early part of the procedure, prior to cross-clamping. This may be caused by the following:

*B1. Hyperextension of the neck during intubation and positioning prior to the initial incision.* These maneuvers can seriously compromise carotid blood flow, especially in cases of a tight stenosis or secondary to dislocation of local thrombi. There are rare cases in which the EEG background becomes asymmetric at the time of induction. Thus, it is important to have a stepwise approach to induction, intubation, and positioning to allow time for EEG interpretation during and between these steps. Otherwise, in the case of a change in the background from symmetric awake to asymmetric postinduction, it becomes very difficult to determine its cause. The etiology is suggested by a marked postinduction asymmetry with a prior symmetric awake background. In most cases, any differences between the awake and postinduction background is rather mild, because a significant preexisting condition (signaled by a markedly asymmetric postinduction EEG) should be reflected in the awake EEG as well. If repositioning of the head does not improve the EEG changes and it is certain that the postinduction background was initially symmetric before head turning, an embolic event should be highly suspected (Fig. **2-9**).

*B2. Exposing the carotid artery.* This is an important step toward the preparation for clamping, and it occurs above and below the operative site, usually at the bifurcation of the common carotid into the internal and external branches. This location is quite variable and sometimes may be high enough in the neck to present the surgeon with a technical problem in making the clamp placement above the endarterectomy. During this phase of the procedure, EEG changes may be produced by systemic changes in circulation *that are due to mechanical stimulation of the carotid body* at the carotid bifurcation, leading to vagal nerve-mediated bradyarrhythmias and hypotension. Depending on the severity of the arrhythmias, the EEG changes can be asymmetric, ipsilateral to the more stenotic side (Fig. **2-11**), or diffuse (Fig. **2-10**).

*B3. Manipulation of the artery.* This maneuver can result in a thrombosis at that site or an embolic event *that is due to dislocation of a thrombus* from a loose plaque (Fig. **2-12**). TCD studies have shown that a thromboembolic event occurring during dissection and exposure of the artery is more likely to have adverse clinical consequences (Aackerstaff et al., 1996; Gaunt et al., 1996; Verhoeven et al., 2005; Wolf et al., 2004).

## C. EEG Changes Related to Clamping

Close attention should be paid to the features of the EEG in the minutes leading up to the clamp being applied to the carotid artery. This baseline is compared to the EEG seen in the minutes after the clamp is applied. The anesthesiologist should not make changes in the anesthetic regimen or blood pressure over this time period because that could introduce a confusing variable into the EEG interpretation. If all of the preceding steps have been carefully managed, the neurophysiology monitoring team is ready for the critical phase of the surgery, which is when the carotid artery is clamped to allow the artery to be opened for the performance of the endarterectomy. A bolus of heparin is usually given to the patient a few minutes before the cross-clamp is applied. The common carotid artery is then clamped somewhere below the bifurcation. The surgeon may or may not inform the neurophysiology team when the clamp is applied. Sometimes the lag in notification is due to the surgeon's wish to ensure a "blinded" report on EEG changes. As previously discussed, there are also many reasons for the EEG to change significantly at times that are not related to the cross-clamp itself, so that any such EEG changes need to be reported promptly to the surgeon whenever they occur.

*C1. Hypoperfusional state.* In more than 95% of patients, the Circle of Willis and the pattern of blood flow to the brain up the four major neck arteries is such that a single carotid artery can be occluded and the patient suffers no deficit, either temporary or permanent. Thus, there is no change in the EEG in the large majority of patients with application of the carotid clamp, and the surgery can proceed safely without installation of a temporary bypass shunt.

In a small percentage of patients, usually within 30 seconds of the occlusion of the carotid artery and as early as 10–12 seconds, there is a significant change in the EEG, with attenuation of the anesthetic-induced fast activities and an increase in slow frequencies over the involved hemisphere. In these situations, the surgeon and the anesthesiologist are promptly informed. The anesthesiologist attempts to raise the systemic blood pressure, while the surgeon proceeds to install a temporary bypass shunt for the period the carotid system will be clamped. Opening of intra-arterial shunt that reestablishes an adequate cerebral perfusion should be shortly followed by an improvement of the EEG changes with return to the postinduction baseline. In some situations, however, the improvement of the EEG changes is only partial and takes a longer time. In such cases, the surgeon usually attempts the reposition or changes the shunt (see the following section).

In many situations, the initial clamp is only a test clamp. Once EEG changes were present with the initial clamp, one should expect the same sequences of changes with each subsequent clamp of the artery.

Clearly, the most significant question relates to the definition of a significant change in the EEG. As discussed on page 50, there is considerable variation between neurophysiologists in what is termed a significant EEG change. The problem is that, if minor degrees of slowing associated with cross-clamping are termed significant in a center, the correlation between these EEG changes and postsurgical outcome becomes much poorer, so that the true utility of EEG monitoring is lost in the noise, becomes unclear to the surgeon, and results in unnecessary shunts being employed, with the risks and increase in time associated with their use. We reserve the definition of a significant EEG change to a loss in amplitude of greater than 60%, which is also always associated with a slowing in overall frequency because the anesthetic-induced fast frequencies are the first to drop out with cortical ischemia. There are several types of EEG changes related to cerebral ischemia: (1) ipsilateral (to the operated side) voltage attenuation of the faster-frequency activities; (2) ipsilateral decrease in the overall frequencies on that side (attenuation of the fast frequencies and increase in slowing); (3) bilateral voltage attenuation; and (4) bilateral voltage attenuation and bilateral increase in slowing (Chiappa et al., 1979). Ipsilateral voltage attenuation of the fast frequencies are the most common change seen (Chiappa et al., 1979; McFarland et al. 1988).

These changes are usually unilateral, which makes the determination of their etiology quite simple, especially when they occur in close temporal relationship to the clamping of the carotid (Fig. **2-5**). Rarely, they occur bilaterally, usually in patients who have a preoperative occlusion in the carotid artery on the side opposite to the current surgical location (Fig. **2-32**). In these cases, the blood supply to the contralateral hemisphere is dependent on the collateral flow (via Circle of Willis) from the ipsilateral ICA, which now is clamped. This may make the determination of the etiology of the change more difficult, because similar bilateral changes can be seen with a drop in blood pressure or a sudden change in an anesthetic agent or its serum concentration, but these can usually be ruled out quickly. In practice, most cases of bilateral EEG changes are asymmetric, with the hemisphere receiving the least blood supply and/or more susceptible to ischemic damage showing the greatest change. This is usually the hemisphere that is ipsilateral to the clamp, unless the contralateral hemisphere has an increased sensitivity to ischemic changes (e.g., from prior events; Fig. **2-19**).

A low flow state, with potential risk for neurologic damage in the posterior circulation territory, is also encountered when there is significant stenosis or hypoplasia of the ipsilateral vertebral artery (e.g., fetal origin of the vertebral artery) with a large ipsilateral posterior communicating artery. In these cases the blood supply for the ipsilateral posterior circulation relies on the clamped ICA.

Rarely, changes that are ipsilateral to the clamp superimposed on a prior contralateral hemispheric insult can result in a "pseudonormalization" of an asymmetric background (Fig. **2-32**).

When there is preclamp ipsilateral EEG slowing at baseline, further worsening of this slowing or its severity can be difficult to appreciate (see page 61 and Fig. **2-16**).

When a significant change occurs, the 10–20 rule comes into play (see pages 48–50). Hence a center for which the clamp times are less than 20 minutes and that routinely do not use a bypass shunt see a significant postoperative neurologic deficit even fewer times than the 1 of 100 or so CEAs (see pages 47–48), a number that is even easier to lose in the noise. Most surgeons are well

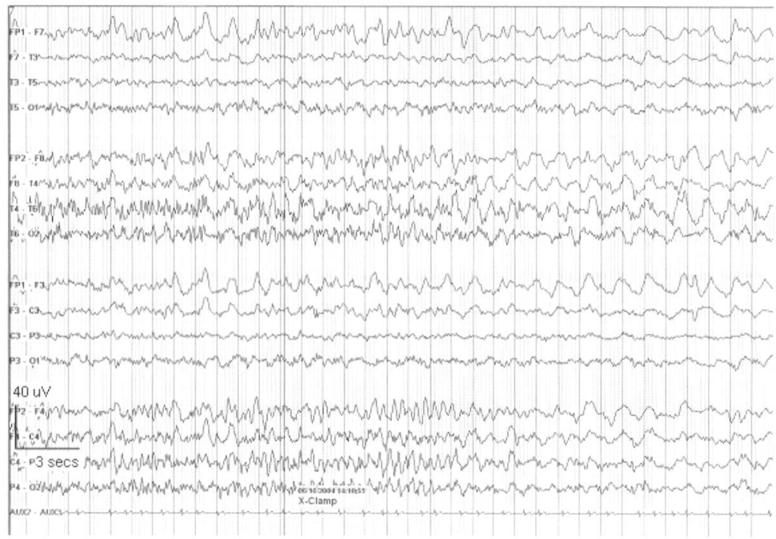


**FIGURE 2-32** EEG during a right carotid endarterectomy, showing 12 seconds of recording prior to clamping of the right internal carotid artery (at "X-Clamp" label) and 23 seconds after clamping. Prior to clamping, there is a marked asymmetry, with the anesthetic-induced faster frequencies visible over the right hemisphere and essentially absent on the left, whereas the slower frequencies are more symmetric. About 9 seconds after the clamp is applied, there is a severe loss of faster frequencies and an increase in the amount and amplitude of delta slowing over the right hemisphere; the left hemisphere also shows a mild attenuation of the faster frequencies. This recording shows diffuse EEG changes with clamp, yet asymmetric, with the most dramatic changes present over the ipsilateral hemisphere. The calibration mark is 40 $\mu$V and 3 seconds.

aware of the risks and timing involved with significant EEG changes, although some may need instruction and encouragement if they are not experienced with EEG monitoring in these cases.

The first response to a significant EEG change should be an attempt to increase the systemic blood pressure. Because most of the clamp-related changes are due to a low flow state, this action might suffice, reversing the changes or improving the slowing to an "acceptable" level (see page 80, and Fig. **2-30**).

A blood pressure increase usually helps mild EEG changes, whereas severe changes require shunt placement. In these latter cases, while the anesthesiologist is increasing the blood pressure, the surgeon proceeds with placement of a temporary intra-arterial bypass shunt. The 10–20 rule gives information about how fast a shunt insertion needs to be done. It is generally accepted that 7–10 minutes of delay is safe, because, in order to minimize the risk of embolization, it is important that the shunt insertion be done after plaque removal (Deriu et al., 1999). When the shunt is opened, the EEG usually returns to baseline within 1 to 2 minutes, and as early as 10 to 12 seconds (Figs. **2-5**, **2-33**, **2-34**), although some patients may take up to 5 minutes. There should be a progressive change and improvement in the EEG over this time period; otherwise, the lack of improvement is a signal of something wrong with the shunt: it is not providing sufficient blood flow and needs to be revised or repositioned. If the EEG shows a return of the anesthetic-induced activities to at least 85% of the preclamp baseline values, the cerebral cortex has sufficient blood flow to maintain function and the surgery can proceed beyond the 20-minute limit.

There are situations, however, when the significant changes occur gradually, resulting from worsening of minimal changes seen immediately after the clamp and that, it was initially thought, could be managed with just an increase in the systemic blood pressure. In such situations, when significant changes are "called" 10–15 minutes in the clamp (or later), the challenge resides in making the decision whether to insert a shunt or not. If the endarterectomy is advanced and the opening of the arteries will occur shortly, the surgeon rushes to finish the procedure without shunt placement and with an additional increase in the blood pressure (Fig. **2-35**). If more than 15–20 minutes of artery clamping is still to be expected, the surgeon may opt for a shunt placement (see the following). In order to assist in making this decision, the surgeon may also check for

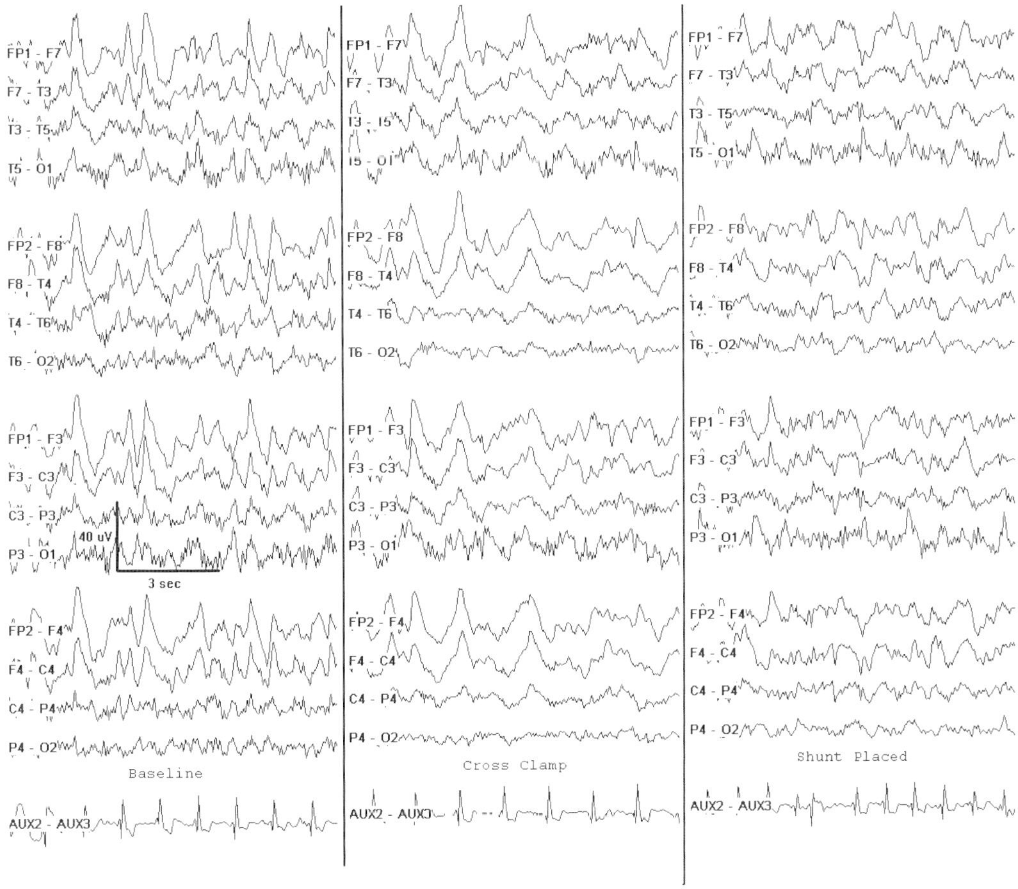


**FIGURE 2-33** EEG during a right carotid endarterectomy (CEA). The **left** panel shows the postinduction preclamp EEG baseline; there is symmetric, continuous, irregular delta slowing maximal anteriorly up to 30 $\mu$V, with superimposed fast frequencies of lower amplitude, which are mildly asymmetric, somewhat attenuated on the right. This slowing does not change much after clamping of the right ICA, but there is a significant amplitude decrement of the fast frequencies over both hemispheres, which is much more marked on the right (**middle** panel). Opening of a temporary bypass shunt (**right** panel) results in a return of some of the fast frequencies over the right hemisphere to the preclamp baseline, indicating that the procedure can go forward. The calibration marks in all panels are 40 $\mu$V and 3 seconds.

backbleeding, as a decreased retrograde flow is usually a reliable additional indication of the paucity of the collateral flow and increased risk for significant cerebral hypoperfusion.

*C2. Shunt-related problems.* Insertion of a temporary intra-arterial bypass shunt is not always followed by improvement of the EEG. This can be due to inadequate shunt placement or inadequate size (Fig. **2-18**). In such cases, repositioning of the shunt or changing the shunt may be followed by appropriate blood supply.

Similarly, if a temporary bypass shunt is employed during the time of cross-clamp, the shunt may occlude without the surgeon's knowledge. This occurs if the shunt orifices are blocked by the arterial walls or if the shunt tubing is compressed or kinked during the procedure (Fig. **2-36**). In all cases of shunt malfunction, the problem could be missed without EEG monitoring.

*C3. Embolization.* Failure of improvement of the EEG changes with adequate shunting can indicate a different mechanism of injury, such as embolization, which can occur either before clamp time (Fig. **2-12**) or at clamp time, because of dislocation of the thrombotic material by the mechanical act of clamping, especially in case of friable or ulcerated plaques (Fig. **2-15**).

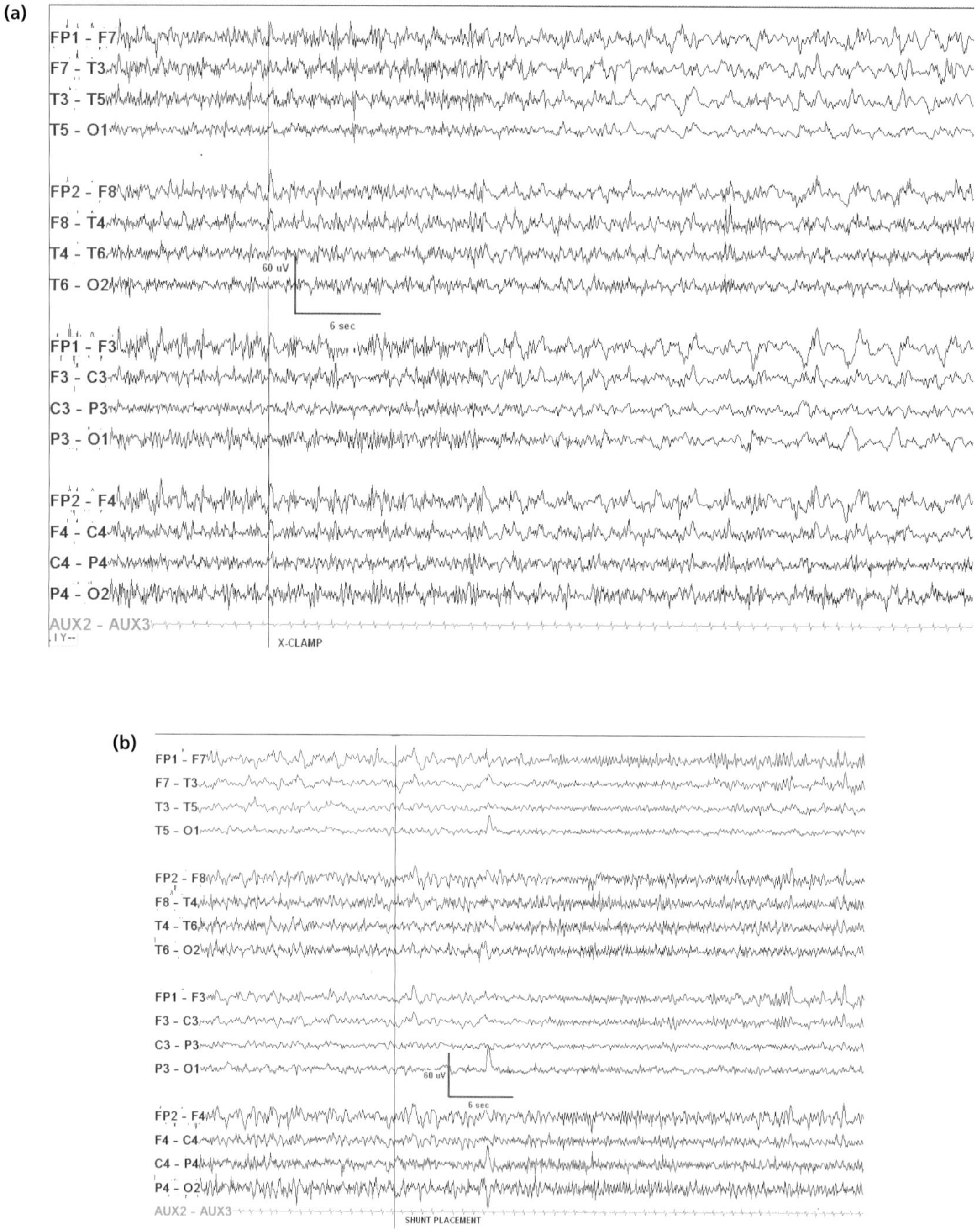


**FIGURE 2-34** The EEGs show ipsilateral hemispheric changes caused by clamping of the left carotid artery, and recovery of these changes with the insertion of a temporary intra-arterial bypass shunt. The calibration marks in both slides are 60 $\mu$V and 6 seconds. **a:** About 14 seconds after clamping of the left carotid artery, the left hemisphere shows irregular, delta slowing with attenuation of the faster frequencies on this side. **b:** About 12 seconds after insertion of a temporary intra-arterial bypass shunt, there is an increase in the anesthetic-induced fast-frequency activities over the left hemispheric regions, concomitant with the attenuation of the ipsilateral slowing.

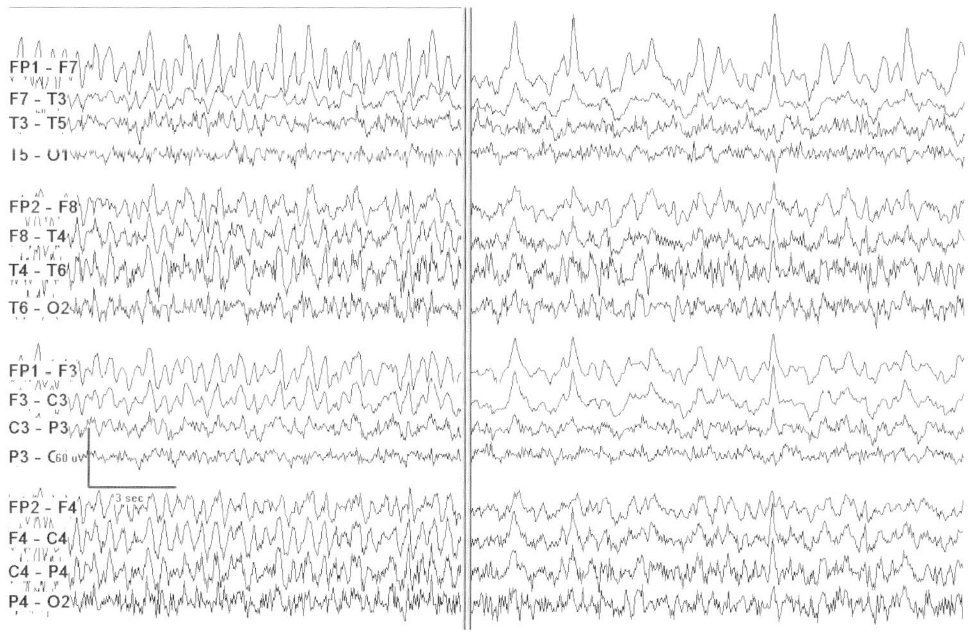


**FIGURE 2-35** EEG during left carotid endarterectomy, showing an increase in the postinduction baseline left hemispheric slowing, which occurred about half an hour into the clamp. During this time the blood pressure was noticed to be lower than before. The systemic blood pressure was increased, and the surgeon opened the left carotid artery in about 5 minutes after completion of the endarterectomy. No intra-arterial shunt was placed. The patient woke up with no new neurologic deficit. The **left** panel shows the baseline asymmetric left hemispheric slowing, unchanged for about 30 minutes after the clamp. The **right** panel shows an increase in the left hemispheric slowing, occurring toward the end of the clamping period. The calibration marks in both panels are 60 $\mu$V and 3 seconds.

## D. EEG Changes Just before Clamp Removal

*D1. Occlusion of the shunt for removal.* These are expected EEG changes in patients who had slowing with clamping prior to shunt placement. They are of little significance, because the arteries will be reopened within a minute afterward.

*D2. Backbleeding.* This is performed immediately prior to the closure of the arteriotomy, to clean the distal artery of debris, and before reestablishing the normal anterograde flow with the unclamping (Fig. **2-37**).

## E. EEG Changes Just after Clamp Removal

*E1. Thromboembolism.* Clamp removal is associated with an increased risk of microembolization (Wolf et al., 2004). Thus, sudden EEG changes occurring with the opening of the arteries should raise this question and a consideration of further investigation (e.g., arteriogram) and therapeutic options, such as increase in systemic blood pressure (Fig. **2-13**), embolectomy, and clot lysis.

*E2. Changes of uncertain etiology.* When the surgeon is ready to reopen the arteries, the anesthesiologist decreases the systemic blood pressure in an attempt to moderate the hemispheric reperfusion pressure, which could otherwise result in a hemorrhagic event. As mentioned on pages 75–80, sometimes the anesthesiologist achieves this goal by increasing the concentration of halogenated agents (such as sevoflurane). This technique introduces problems in the EEG interpretation (Fig. **2-27**) because the sudden burst of high-amplitude slowing induced by the anesthetic agent could also signal a paradoxical arousal response or, more important, a thromboembolic event.

It is very unlikely that these EEG changes are the consequence of an immediate postclamp release hyperperfusional state, unless sudden brain damage has already occurred (e.g., an intracerebral hemorrhage). The cerebral hyperperfusion syndrome is a delayed complication of CEAs and is usually seen within several days after the surgery, with development of vasogenic edema and/or intracerebral hemorrhage (Aubert et al., 2007; Ogasawara et al., 2007; Streifler, 2004; van Mook et al., 2005). The

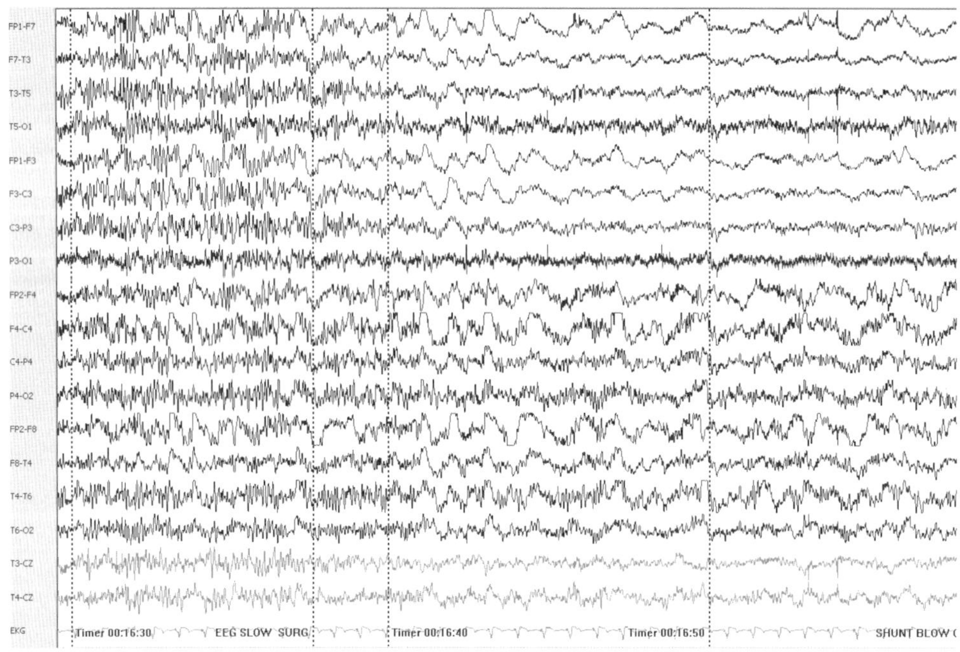


**FIGURE 2-36** EEGs from a patient undergoing left carotid endarterectomy, with a shunt placement for ipsilateral slowing, which occurred with the clamp; the shunt initially reversed the EEG changes, but later on, during clamp period, the EEG again became asymmetric, likely as a result of shunt malfunctioning. The left third of the slide shows a symmetric EEG during the clamp period and with the insertion of an intra-arterial bypass shunt; notice that starting at 00:16:40 (16 minutes into the clamp) there is a significant increase in the left hemispheric delta slowing with attenuation of the faster frequencies over this side, in particular over the anterior regions. The calibration mark is 40 $\mu$V and 3 seconds.

syndrome is usually seen in cases where there was an increase in the postsurgical cerebral blood flow that was two to four times greater than the presurgical baseline (Reigel et al., 1987; van Mook et al., 2005). Thus, it is more likely to occur in patients with significant preoperative hemodynamic impairment, acute cerebral ischemia during clamping, and impaired cerebrovascular reserve (Komoribayashi et al., 2006; Reigel et al., 1987; Russell et al., 2004). EEG recordings are more likely to show unilateral periodic epileptiform discharges (PLEDS) (Reigel et al., 1987) that are consistent with the typical clinical presentation: unilateral headache, focal neurologic deficits, and seizures, rather than diffuse, bilateral slowing. Thus, we believe that the sudden EEG changes sometimes seen with clamp release are not related to a hyperperfusional state as it is usually defined.

### *F. Late EEG Changes after Clamp Removal*

*F1. Rethrombosis at the surgical site.* After the endarterectomy has been performed, the shunt removed (if present), and arterial blood flow restored, the carotid artery sometimes occludes during the time when the operative site is being closed and before the patient is awake and able to participate in a neurologic examination. Most commonly this is caused by a thrombosis at the operative site. This can cause either a low flow state because of significant obstruction in the blood flow or dislocation of an embolus from a freshly formed clot. These mechanisms are considered by some authors as the main causes of neurologic complications associated with CEA (Abbott et al., 2007; Cantelmo et al., 1998; de Borst et al., 2001; Horn et al., 2005; Laman et al., 2002). If this thrombosis occurs while the EEG is still running, the surgeon can be alerted immediately and no time wasted in appropriate management. However, strokes through the same mechanisms can occur within the first 24 hours after the surgery (usually while the patient is still in the recovery unit). In such cases, the outcome may be worse because of either delayed diagnosis of the neurologic dysfunction (in the absence of EEG monitoring during this period and a relatively limited neurologic exam) or as a result of an increased time gap between the onset of neurologic dysfunction and appropriate intervention (Fig. **2-14**).

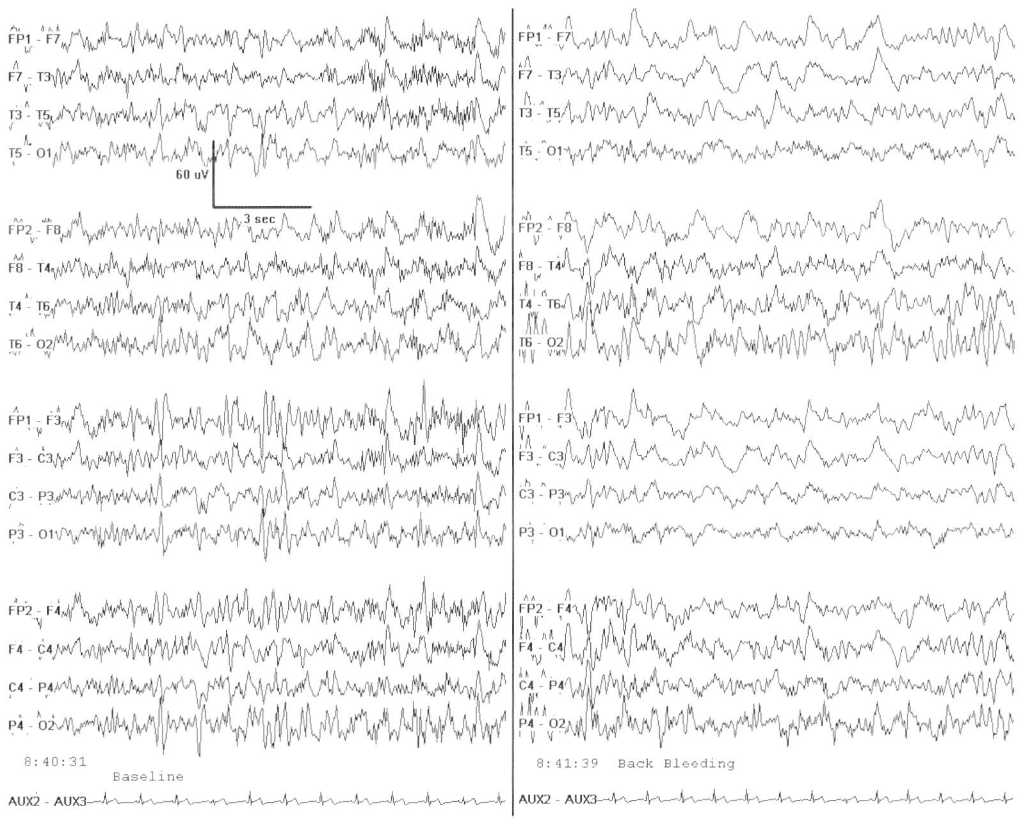

**FIGURE 2-37** EEGs recorded from a patient during a left carotid endarterectomy, showing changes associated with backbleeding immediately after completion of the endarterectomy and prior to reopening the artery to normal flow. The **left** panel shows a symmetric record with mixed-frequency, anesthetic-induced activities. One minute later, during backbleeding (**right** panel), there is a marked loss of faster frequencies and the appearance of asymmetric slowing over the left hemisphere. The calibration marks in both panels are 60 $\mu$V and 3 seconds.

## REFERENCES


Aackerstaff RG, Jansen C, Moll FL. Carotid endarterectomy and intraoperative emboli detection: Correlation of clinical, transcranial Doppler, and magnetic resonance findings. Echocardiography 1996; 13(5):543–550.

Aackerstaff RG, Moll FL, van de Pavoordt ED, et al. Carotid endarterectomy, postoperative hyperperfusion, and cerebral haemorrhage. Eur J Vasc Endovasc Surg 2004; 28(6): 679–680.

Abbott AL, Levi CR, Stork JL, et al. Timing of clinically significant microembolism after carotid endarterectomy. Cerebrovasc Dis 2007; 23(5–6):362–367.

Algotsson L, Messeter K, Rehncrona S, et al. Cerebral hemodynamic changes and electroencephalography during carotid endarterectomy. J Clin Anesth 1990; 2(3):143–151.

Arnold M, Sturzenegger M, Schaffler L, Seiler RW. Continuous intraoperative monitoring of middle cerebral artery blood flow velocities and electroencephalography during carotid endarterectomy. A comparison of the two methods to detect cerebral ischemia. Stroke 1997; 28(7):1345–1350.

Artru AA, Strandness DE. Delayed carotid shunt occlusion detected by electroencephalographic monitoring. J Clin Monit 1989; 5(2):119–122.

Aubert S, Sellal F, Rouyer O, et al. Reperfusion syndrome with cerebral vasogenic edema after carotid artery endarterectomy. Rev Neurol 2007; 163(8–9):840–844.

Babikian VL, Wijman CA, Hyde C, et al. Cerebral microembolism and early recurrent cerebral or retinal ischemic events. Stroke 1997; 28(7):1314–1318.

Ballotta E, Dagiau G, Saladini M, et al. Results of electroencephalographic monitoring during 369 consecutive carotid artery revascularizations. Eur Neurol 1997; 37(1):43–47.

Barr JD, Horowitz MB, Mathis JM, et al. Intraoperative urokinase infusion for embolic stroke during carotid endarterectomy. Neurosurgery 1995; 36(3):606–611.

Belardi P, Lucertini G, Ermirio D. Stump pressure and transcranial Doppler for predicting shunting in carotid endarterectomy. Eur J Vasc Endovasc Surg 2003; 25(2):164–167.

Blondeau C, Hebert F, Ducasble G, et al. Electroencephalographic surveillance during carotid endarterectomy. Rev Electroencephalogr Neurophysiol Clin 1982; 12(3):238–242.

Bornstein NM, Rossi GB, Treves TA, Shifrin EG. Is transcranial Doppler effective in avoiding the hazards of carotid surgery? Cardiovasc Surg 1996; 4(3):335–337.

Botes K, Le Roux DA, Van Marle J. Cerebral monitoring during carotid endarterectomy: A comparison between electroencephalography, transcranial cerebral oximetry and carotid stump pressure. S Afr J Surg 2007; 45(2):43–46.

Brewster DC, O'Hara PJ, Darling RC, Hallett JW Jr. Relationship of intraoperative EEG monitoring and stump pressure measurements during carotid endarterectomy. Circulation 1980; 62(2 pt 2):14–17.

Cantelmo NL, Babikian VL, Samaraweera RN, et al. Cerebral microembolism and ischemic changes associated with carotid endarterectomy. J Vasc Surg 1998; 27(6):1024–30; discussion 1030–1031.

Carlin RE, McGraw DJ, Calimlim JR, Mascia MF. The use of near-infrared cerebral oximetry in awake carotid endarterectomy. J Clin Anesth 1998; 10(2):109–113.

Chiappa KH, Burke SR, Young RR. Results of electroencephalographic monitoring during 367 carotid endarterectomies. Use of a dedicated minicomputer. Stroke 1979; 10(4):381–388.

Comerota AJ, Eze AR. Intraoperative high-dose regional urokinase infusion for cerebrovascular occlusion after carotid endarterectomy. J Vasc Surg 1996; 24(6):1008–1016.

de Borst GJ, Moll FL, van de Pavoordt HD, et al. Stroke from carotid endarterectomy: When and how to reduce perioperative stroke rate? Eur J Vasc Endovasc Surg 2001; 21(6):484–489.

Deriu GP, Milite D, Mellone G, et al. Clamping ischemia, threshold ischemia and delayed insertion of the shunt during carotid endarterectomy with patch. J Cardiovasc Surg 1999; 40(2):249–255.

Dinkel M, Langer H, Loerler H, et al. Neuromonitoring in carotid surgery: Possibilities and limits of transcranial Doppler ultrasound. Vasa 1994; 23 (4):337–344.

Duffy C, Manninen PH, Chan A, Kearns CF. Comparison of cerebral oximeter and evoked potential monitoring in carotid endarterectomy. Can J Anaesth 1997; 44(10):1077–1081.

El-Dawlatly AA. Carotid endarterectomy: A study of cerebral oxygen saturation vs. stump pressure. Middle East J Anesthesiol 2004; 17(6):1037–1044.

Fiori L, Parenti G, Marconi F. Combined transcranial Doppler and electrophysiologic monitoring for carotid endarterectomy. J Neurosurg Anesthesiol 1997; 9(1):11–16.

Fletcher JP, Morris JG, Little JM, Kershaw LZ. EEG monitoring during carotid endarterectomy. Aust N Z J Surg 1998; 58(4):285–288.

Florence G, Guerit JM, Gueguen B. Electroencephalography (EEG) and somatosensory evoked potentials (SEP) to prevent cerebral ischemia in the operating room. Neurophysiol Clin 2004; 34(1):17–32.

Gaunt ME, Brown L, Hartshorne T, et al. Unstable carotid plaques: Preoperative identification and association with intraoperative embolisation detected by transcranial Doppler. Eur J Vasc Endovasc Surg 1996; 11(1):78–82.

Gaunt ME, Rimmer T, Smith JL, et al. The effect of perioperative embolisation on visual function in patients undergoing carotid endarterectomy. Eur J Vasc Endovasc Surg 1998; 16(3):231–237.

Graham AM, Gewertz BL, Zarins CK. Predicting cerebral ischemia during carotid endarterectomy. Arch Surg 1986; 121(5):595–598.

Grego F, Antonello M, Lepidi S, et al. Is contralateral carotid artery occlusion a risk factor for carotid endarterectomy? Ann Vasc Surg 2005; 1996:882–889.

Guerit JM, Witdoeckt C, de Tourtchaninoff M, et al. Somatosensory evoked potential monitoring in carotid surgery. Relationships between qualitative SEP alterations and intraoperative events. Electroencephalogr Clin Neurophysiol 1997; 104(6):459–469.

Halsey JH, McDowell HA, Gelman S. Transcranial Doppler and rCBF compared in carotid endarterectomy. Stroke 1986; 12(6):1206–1208.

Halsey JH, McDowell HA, Gelman S, Morawetz RB. Blood velocity in the middle cerebral artery and regional cerebral blood flow during carotid endarterectomy. Stroke 1989; 20(1):53–58.

Haupt WF, Erasmi-Korber H, Lanfermann H. Intraoperative recording of parietal SEP can miss hemodynamic infarction during carotid endarterectomy: A case study. Electroencephalogr Clin Neurophysiol 1994; 92(1):86–88.

Horn J, Naylor AR, Laman DM, et al. Identification of patients at risk for ischaemic cerebral complications after carotid endarterectomy with TCD monitoring. Eur J Vasc Endovasc Surg 2005; 30(3):270–274.

Jacob T, Hingorani A, Ascher E. Carotid artery stump pressure (CASP) in 1135 consecutive endarterectomies under general anesthesia: An old method that survived the test of times. J Cardiovasc Surg 2007; 48(6):677–681.

Jansen C, Moll FL, Vermeulen FE, et al. Continuous transcranial Doppler ultrasonography and electroencephalography during carotid endarterectomy: A multimodal monitoring system to detect intraoperative ischemia. Ann Vasc Surg 1993; 7(1):95–101.

Jansen C, Vriens EM, Eikelboom BC, et al. Carotid endarterectomy with transcranial Doppler and electroencephalographic monitoring. A prospective study in 130 operations. Stroke 1993; 24(5):665–669.

Jenkins G, Chiappa KH, Young RR. Practical aspects of EEG monitoring during carotid endarterectomies. Am J EEG Technol 1983; 23:191–203.

Kaufman HH, Reilly EL, Porecha HP, et al. Cerebral ischemia during carotid endarterectomy with severe but reversible changes. Surg Neurol 1977; 7(4):195–198.

Komoribayashi N, Ogasawara K, Kobayashi M, et al. Cerebral hyperperfusion after carotid endarterectomy is associated with preoperative hemodynamic impairment and intraoperative cerebral ischemia. J Cerebr Blood Flow Metab 2006; 26(7):878–884.

Kresowik TF, Worsey MJ, Khoury MD, et al. Limitations of electroencephalographic monitoring in the detection of cerebral ischemia accompanying carotid endarterectomy. J Vasc Surg 1991; 13(3):439–443.

Lacroix H, Beyens G, Van Hemelrijck J, et al. Is transcranial Doppler useful in the detection of internal carotid artery cross-clamp intolerance? Cardiovasc Surg 1999; 7(2):203–207.

Lam AM, Manninen PH, Ferguson GG, Nantau W. Monitoring electrophysiologic function during carotid endarterectomy: A comparison of somatosensory evoked potentials and conventional electroencephalogram. Anesthesiology 1991; 75(1):15–21.

Laman DM, Wieneke GH, van Dujin H, van Huffelen AC. High embolic rate early after carotid endarterectomy is associated with early cerebrovascular complications, especially in women. J Vasc Surg 2002; 36(2):278–284.

Lang W, Dinkel M. Cerebral ischemia during carotid clamping: Diagnosis and prevention. Zentralbl Chir 2000; 125(3):243–250.

Lee TS, Hines GL, Feuerman M. Significant correlation between cerebral oximetry and carotid stump pressure during carotid endarterectomy. Ann Vasc Surg 2008; 22(1):58–62.

Manninen P, Sarjeant R, Joshi M. Posterior tibial nerve and median nerve somatosensory evoked potential monitoring during carotid endarterectomy. Can J Anaesth 2004; 51(9):937–941.

Matsumoto GH, Baker JD, Watson CW, et al. EEG surveillance as a means of extending operability in high risk carotid endarterectomy. Stroke 1976; 7(6):554–559.

McCarthy RJ, McCabe AE, Walker R, Horrocks M. The value of transcranial Doppler in predicting cerebral ischaemia during carotid endarterectomy. Eur J Vasc Endovasc Surg 2001; 21(5):408–412.

McFarland HR, Pinkerton JA Jr, Frye D. Continuous electroencephalographic monitoring during carotid endarterectomy J Cardiovasc Surg 1988; 29(1):12–18.

McGrail KM. Intraoperative use of electroencephalography as an assessment of cerebral blood flow. Neurosurg Clin N Am 1996; 7(4):685–692.

McKinsey JF, Desai TR, Bassiouny HS, et al. Mechanisms of neurologic deficits and mortality with carotid endarterectomy. Arch Surg 1996; 131:526–532.

Messick JM Jr, Casement B, Sharbrough FW, et al. Correlation of regional cerebral blood flow( rCBF) with EEG changes during isoflurane anesthesia for carotid endarterectomy; critical rCBF. Anesthesiology 1987; 66(3):344–349.

Modica PA, Tempelhoff R. A comparison of computerized EEG with internal carotid artery stump pressure for detection of ischemia during carotid endarterectomy. J Neurosurg Anesthesiol 1989; 1(3):211–218.

Mola M, Collice M, Levati A. Continuous intraoperative electroencephalographic monitoring in carotid endarterectomy. Eur Neurol 1986; 25(1):53–60.

Moritz S, Kasprzak P, Arlt M, et al. Accuracy of cerebral monitoring in detecting cerebral ischemia during carotid endartrectomy: A comparison of transcranial Doppler sonography, near-infrared spectroscopy, stump pressure, and somatosensory evoked potentials. Anesthesiology 2007; 107(4):563–569.

Nuwer MR. Intraoperative electroencephalography. J. Clin Neurophysiol 1993; 10(4):437–444.

Ogasawara K, Inoue T, Kobayashi M, et al. Cerebral hyperperfusion following carotid endarterectomy: Diagnostic utility of intraoperative transcranial Doppler ultrasonography compared with single-photon emission computed tomography study. AJNR AM J Neuroradiol 2005; 26(2):252–257.

Ogasawara K, Sakai N, Kuroiwa T, et al. Intracranial hemorrhage associated with cerebral hyperperfusion syndrome following carotid endarterectomy and carotid artery stenting: Retrospective review of 4494 patients. J Neurosurg 2007; 107(6):130–1136.

Oppenheim C, Stanescu R, Dormont D, et al. False-negative diffusion-weighted MR findings in acute ischemic stroke. AJNR Am J Neuroradiol 2000; 21(8):1434–1440.

Pedraza S, Osuna MT, Davalos A, et al. False-negative diffusion in acute ischemic stroke. Rev Neurol 2002; 34(12): 1127–1129.

Perler BA, Murphy K, Sternbach Y, et al. Immediate postoperative thrombolytic therapy: An aggressive strategy for neurologic salvage when cerebral thromboembolism complicates carotid endarterectomy. J Vasc Surg 2000; 31(5):1033–1037.

Phillips MR, Johnson WC, Scott RM, et al. Carotid endarterectomy in the presence of contralateral carotid occlusion: The role of EEG and intraluminal shunting. Arch Surg 1979; 114(11):1232–123.

Pinkerton JA Jr. EEG as a criterion for shunt need in carotid endarterectomy. Ann Vasc Surg 2002; 16(6):756–761.

Pistolese GR, Appolloni A, Ronchey S, et al. Update on cerebral monitoring and protective methods. Ann Ital Chir 1997; 68(4):441–451.

Plestis KA, Loubser P, Mizrahi EM, et al. Continuous electroencephalographic monitoring and selective shunting reduces neurologic morbidity rates in carotid endarterectomy. J Vasc Surg 1997; 25(4):620–628.

Reigel MM, Hollier LH, Sundt TM Jr, et al. Cerebral hyperfusion syndrome: A cause of neurologic dysfunction after carotid endarterectomy. J Vasc Surg 1987; 5(4):628–634.

Ricotta JJ, Charlton MH, DeWeese JA. Determining criteria for shunt placement during carotid endarterectomy. EEG versus backpressure. Ann Surg 1983; 198(5):642–645.

Rowed DW, Houlden DA, Burkholder LM, Taylor AB. Comparison of monitoring techniques for intraoperative cerebral ischemia. Can J Neurol Sci 2004; 31(3):347–356.

Russell DA, Gough MJ. Intracerebral haemorrhage following carotid endarterectomy. Eur J Vasc Endovasc Surg 2004; 28(2):115–123.

Salvian AJ, Taylor DC, Hsiang YN, et al. Selective shunting with EEG monitoring is safer than routine shunting for carotid endarterectomy. Cardiovasc Surg 1997; 5(5):481–485.

Schneider JR, Droste JS, Schindler N, et al. Carotid endarterectomy with routine electroencephalography and selective shunting: Influence of contralateral internal carotid artery occlusion and utility in prevention of perioperative strokes. J Vasc Surg 2002; 35(6):1114–1122.

Spencer MP. Transcranial Doppler monitoring and causes of stroke from carotid endarterectomy. Stroke 1997; 28 (4):685–691.

Staszkiewicz W, Antepowicz W, Madycki G, et al. Does intraoperative transcranial Doppler monitoring improve results of carotid endarterectomy? Wiad Lek 1997; 50(suppl 1, pt 2):45–47.

Streifler JY. Hyperperfusion Syndrome. Harefuah 2004; 143(7):466–469.

Sundt TM, Jr, Sharbrough FW, Anderson RE, Michenfelder JD. Cerebral blood flow measurements and electroencephalograms during carotid endarterectomy. J Neurosurg 1974; 41(3):310–320.

Sylaja PN, Coutts SB, Krol A, et al. When to expect negative diffusion-weighted images in stroke and transient ischemic attack. Stroke 2008; epub ahead of print.

Takeda N, Fujita K, Katayama S, Tamaki N. Cerebral oximetry for the detection of cerebral ischemia during temporary carotid artery occlusion. Neurol Med Chir 2000; 40(11):557–562.

van Mook WN, Rennenberg RJ, Schurink GW, et al. Cerebral hyperperfusion syndrome. Lancet Neurol 2005; 4(12):877–888.

Van Putten MJ, Peters JM, Mulder SM, et al. A brain symmetry index (BSI) for online EEG monitoring in carotid endarterectomy. Clin Neurophysiol 2004; 115(5):1189–1194.

Verhoeven BA, de Vries JP, Pasterkamp G, et al. Carotid atherosclerotic plaque characteristics are associated with microembolization during carotid endarterectomy and procedural outcome. Stroke 2005; 36(8):1735–1740.

Vets P, ten Broecke P, Adriaensen H, et al. Cerebral oximetry in patients undergoing carotid endarterectomy: preliminary results. Anaesthesiol Belg. 2004;55(3):215–220.

Wassmann H, Fischdick G, Jain KK. Cerebral protection during carotid endarterectomy: EEG monitoring as a guide to the use of intraluminal shunts. Acta Neurochir 1984; 71(1–2):99–108.

Whittemore AD, Kauffman JL, Kohler TR, Mannick JA. Routine electroencephalographic (EEG) monitoring during carotid endarterectomy. Ann Surg 1983; 197(6):707–713.

Wolf O, Heider P, Heinz M, et al. Microembolic signals detected by transcranial Doppler sonography during carotid endarterectomy and correlation with serial diffusion-weighted imaging. Stroke 2004; 35 (11):373–375.

Woodworth GF, McGirt MJ, Than KD, et al. Selective versus routine intraoperative shunting during carotid endarterectomy: A multivariate outcome analysis. Neurosurgery 2007; 61(6):1176–1177.

# 3

# ELECTROCORTICOGRAPHY

MIRELA V. SIMON, JASON L. GERRARD,
AND EMAD N. ESKANDAR



## I. INTRODUCTION

Intraoperative electroencephalogram recording, or electrocorticography (ECoG), has been used for more than five decades to provide a wide array of vital information to neurosurgical teams. ECoG can be used for mapping as well as for monitoring electrical cortical activity. It is commonly used during the surgical treatment of epilepsy syndromes and is particularly useful for the localization or mapping of epileptogenic brain tissue, both the irritative zones and the epileptic foci. ECoG is also helpful in determining the prognosis of the surgical outcome by analyzing any residual epileptiform activity. To this end, this technique is currently being used by the majority of epilepsy centers within the United States and Canada. Another important role of ECoG consists in monitoring stimulation-triggered epileptiform discharges or afterdischarges (AD) during electrophysiologic functional mapping, which is further detailed on pages 157–73. In many instances, ECoG is used in conjunction with functional mapping in epilepsy surgery, with an attempt to map the epileptic foci and assess their topographical relationship with the eloquent language or sensorimotor cortex. In such cases, ECoG is used first as a mapping tool for epileptic foci and subsequently as a monitoring tool for the stimulation-triggered ADs during mapping of the eloquent cortex. This chapter, however, addresses only its first utility.

## II. HISTORICAL PERSPECTIVE

Intraoperative electroencephalographic recordings were initially pioneered in human patients by Hans Berger (1929), who placed electrodes on the dural surface of patients with skull defects. In the 1950s and 1960s, several investigators began using increasingly refined measures to study the local field potentials of various cortical locations, as well as the pathological responses of epileptogenic tissues (Ajmone Marsan et al., 1958; Bancaud et al., 1965; Walker et al., 1961). As the prominent role of the medial temporal lobe in seizures came to light, stereotactically placed depth electrodes came into use in order to study these deeper structures (Ajmone Marsan, 1980; Crandall et al., 1963). These intracranial recordings greatly expanded the study of seizures and cortical mapping and became integral to the surgical treatment of intractable epilepsy.

Early ECoG recordings showed that the interictal discharges seen on electroencephalogram (EEG) recordings were better localized with intraoperative electrodes. In addition, electrophysiologic responses to low-intensity stimulation were found to be variable between cortical networks and were pathologically enhanced in epileptic tissue such that rhythmic spikes and waves were produced (Ajmone Marson, 1980; Ajmone Marsan et al., 1973; Bates, 1963; Bengzon et al., 1968; Gloor, 1975).

In recent years, the ever-increasing sophistication of imaging techniques used for functional mapping (functional magnetic resonance imaging [fMRI], positron emission tomography [PET]), as well as improved intracranial and extracranial electrophysiologic recording techniques, such as magnetoencephalography (MEG), subdural electrodes, and chronic depth electrodes, used during the preresection study of epileptic patients, has reduced the overall role of ECoG to an extent. The vast majority of epilepsy centers, however, continue to use intraoperative ECoG to help identify the location and boundaries of the epileptogenic zone and to guide the resection borders during seizure surgery.

## III. THE CONTROVERSIAL ROLE OF ECOG IN EPILEPSY SURGERY

There is much controversy in the literature regarding the utility of ECoG recording in epilepsy surgery for the following reasons.

### 1. ECoG relies on a short intraoperative recording

This further limits the amount of information that can be obtained. Even though invasive recordings in the epilepsy monitoring unit (EMU) offer more detailed information regarding the seizure foci, there are several situations, as detailed on pages 98–99, when ECoG is preferred either as a substitute to invasive EMU recordings or as an additive.

Even though we appreciate the utility of prolonged intraoperative recordings, as practiced by the Montreal school, in most of the situations such extended recordings are not feasible. In order to optimize the information obtained during short ECoG recordings, preoperative planning is essential. First, the neurophysiologist has to get acquainted with the case, with prior neuroimaging and EEG results, as well as with the patient's clinical presentation. The neurophysiologist and neurosurgeon need to set a specific goal or goals to be achieved by the intraoperative recordings (e.g., What is the extent of the irritative zone? Are there several independent irritative zones? Which are the lead-in foci? Is functional mapping required as well?). Second, the recording electrodes and montages need to be determined as soon as the surgeon has a good appreciation of the size of the exposed cortex. We suggest checking the technical setup at least once before the actual recording. For a systematic troubleshooting, we found it quite useful to have a check-up list with the succession of the set-up steps. Third, the neurophysiologist and anesthesiologist should agree in advance on the anesthetic regimen required for ECoG recording and, possibly, functional mapping, in case the latter procedure is also needed. Because these procedures are affected by different anesthetics, good coordination between the neurophysiology and anesthesia team is essential for a smooth transition from one procedure to the other, as further detailed on page 126. Also, during ECoG, an epileptogenic drug should be available if iatrogenic activation of the irritative zone is considered (see pages 121–24). In case a significant amount of epileptiform activity is present during ECoG, after the recording is completed and before proceeding to functional mapping, the team may decide to administer intravenously a loading dose of an antiepileptic drug (AED), in order to prevent stimulation-triggered seizures. For the same purpose, ice-cold saline should also be available. Fourth, two screen windows should be used: one containing the live study and the other offering a review capability. Fifth, we find using the graphic representation of the location of the recording electrodes to be useful for further appreciating the topography of the irritative zones, their electrical fields, propagation, and evolution, as well as the distances between the epileptogenic foci and eloquent cortex. Thus, we believe that good cooperation—between neurology, neurosurgery, and the anesthesia team—and planning in advance greatly increase the chances of obtaining useful information, even during such short studies. Finally, a good experience of the interpreting neurophysiologist is essential for a reliable interpretation and analysis of the ECoG recordings.

## 2. Interictal and Ictal Activities can Occur at Different Locations

Because the chances of capturing a spontaneous electrographic seizure are relatively small during a 30-minute intraoperative recording, one has to rely on the information obtained by analyzing interictal epileptiform activity. However, the utility of the latter has been the subject of much debate (Bernier et al., 1990; Gloor, 1975; Kuruvilla et al., 2003; Wennberg et al., 1998). Many authors believe that using interictal abnormal activity to guide the resection of an epileptic focus can be misleading, especially with multifocal epilepsy. Moreover, the reliability of interictal discharges in guiding surgical treatment has been questioned not only in multifocal but also in unifocal epilepsy. It has been known since the time of Penfield that the most actively spiking area on ECoG may not coincide with the location of a known lesion (Penfield et al., 1954). Even in seizure disorders with a recognized lesion on imaging that has been associated with seizure induction via extracranial neurophysiologic monitoring, regions with interictal spiking may be rather widespread during ECoG recordings (Awad et al., 1991; Cascino et al., 1992; Cascino, Hulihan, et al., 1993; Wennberg et al., 1999).

However, we agree with Alarcon and colleagues (1997), who specify that some of the characteristics of interictal activities can still give reliable information, with the most informative being the differences between the times of onset of paroxysmal interictal events.

## 3. The Induced Seizure Activity may Differ from the Spontaneous Ictal Event

Because the probability of witnessing a spontaneous seizure during ECoG is very small, many centers utilize methods of provoking an epileptic event. These include intravenous administration of an epileptogenic drug (e.g. alfentanyl, methohexital, remifentanyl, or even low-dose propofol), as well as electrical stimulation, and are extensively detailed on pages 121–24. Bernier et al. (1990) showed that the location of electrically induced seizures (with or without prior existing epileptiform activity) was a good indicator of the abnormally increased baseline excitability and thus of the cortical resection areas. Furthermore, some epileptologists have tried to use cortical stimulation as a tool in depicting the areas of increased cortical excitability. The onsets of such induced electrographic seizures, however, are not universally accepted as a safe replacement for the location of the onset of the spontaneous seizure activity. For example, Blume and his colleagues (2004) have not found a consistent correlation between the site of stimulation-triggered ADs and that of spontaneously appearing seizures. However, we believe that such induced ictal activity can deliver important information to the neurophysiologist and to the neurosurgeon, particularly if it is associated with the typical seizure semiology or a certain electrographic pattern of onset and propagation, characteristic of the spontaneous seizures (Fig. **3-1**).

## 4. Postoperative ECoG Recordings may not Reliably Predict the Seizure Freedom Outcome

Much controversy surrounds the use of postoperative ECoG recordings for determining prognosis. This is, in particular, due to the fact that immediately following resection, the tissue at the margin of resection may be excessively irritated, because of edema, hemorrhage, or local trauma, and thus can exhibit abnormal epileptiform and even epileptic activity. We agree with the fact that this activity, present immediately in the postresection period, should not be of concern. However, there are situations when postoperative ECoG can give important information, such as persistent activity at a distance from the resection margins (see page 102).

We believe that the approach to epilepsy surgery should be multidisciplinary and that ECoG recordings can add valuable information to planning the surgical approach. On pages 121–24 we have discussed several circumstances when ECoG can make a difference in surgical planning and implementation. Careful analysis of the recordings by an experienced neurophysiologist is essential for these purposes, because valuable information is obtained by interpreting and weighing different morphologies and patterns of evolution of the epileptiform and epileptic activity.

Last, the investigation and deliberation over the role of ECoG in localizing the ictal focus has led to a better understanding of the characteristics of various epileptiform and epileptic activities in regard to specific epilepsy types.

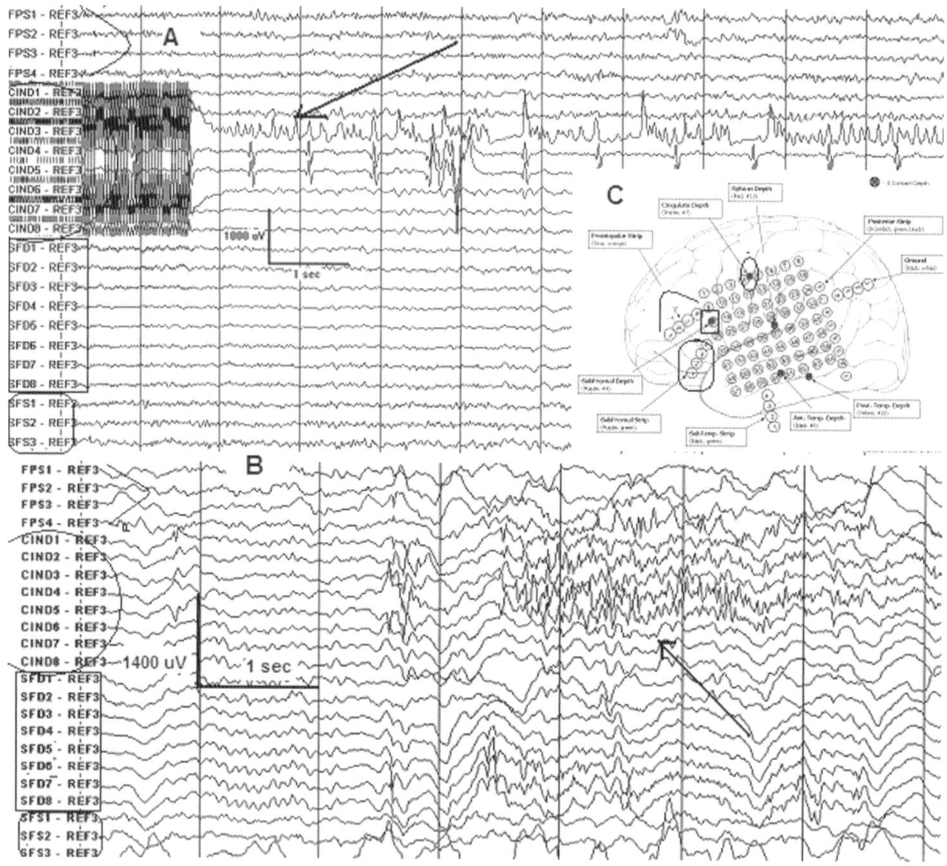


**FIGURE 3-1** Stimulation-triggered seizure, reproducing the electrographic onset and the semiology of the spontaneous seizure. Panel **A** shows the stimulation-triggered seizure, consisting of a rhythmic run of sharp 9 Hz activity (arrow), localized at contact 3 on the cingulate depth (circled), when stimulation was done at contacts 4 and 5. (Notice EKG artifact present at these contacts) Its onset and end are synchronous with the onset and offset of the typical clinical aura. The stimulation consists of repetitive pulses at 50 Hz, 1 msec pulse duration, and an intensity of 1.5 mA applied in 5-second epochs. The calibration mark is 100 $\mu$V and 1 second. Panel **B** shows a spontaneous 3-second burst of fast sharp activity (arrow) present at the cingulate contacts, after which the patient wakes up and has one of his typical clinical events. The calibration mark is 1400 $\mu$V and 1 second. Panel **C** shows the position of the implanted recording hardware. The reference electrode is one of the contacts of a subdural strip, placed upside down. The cingulate depth electrode position and its corresponding ECoG channels are circled. The position of the subfrontal depth electrode and its corresponding ECoG channels are surrounded by a rectangle. The subfrontal strip and its channels are surrounded by a rounded rectangle. The fronto-polar strip and its channels are also marked.

## IV. INDICATIONS

The main indications for ECoG mapping, as well as indications for a combination of ECoG and functional mapping, are listed in Tables **3-1** and **3-2**, respectively, and are further detailed in the following.

### 1. Indications for ECoG Mapping

#### *A. Type of Surgery*

*A1. Elective epilepsy surgery.* Almost all of our ECoG recordings are used as part of elective epilepsy surgery evaluation and usually follow a comprehensive workup. Prior to patients undergoing surgical treatment for a medically intractable seizure disorder, an extensive preoperative evaluation, including high-resolution imaging and inpatient neurophysiologic monitoring, is performed in an attempt to localize or, at the very least, lateralize the seizure source. Seizure disorders that are clearly bilateral are typically not candidates for surgical resection. Patients whose seizure source remains questionable, after initial prolonged surface video EEG monitoring, often undergo chronic implantation of subdural and/or depth electrodes to further characterize

**TABLE 3-1** Indications for ECoG mapping

| | | |
|---|---|---|
| **I Type of surgery** | Epilepsy surgery<br>1. Elective<br>2. Emergent | Nonepilepsy surgery |
| **II Goals** | Guiding the surgical approach and prognosis<br>1. Extent of Resection<br>2. Subpial transections<br>3. Location of stimulator placement<br>4. Postoperative prognosis | 1. Diagnostic-seizure activity versus direct tumor effect<br>2. Management: assessing local cortical excitability and need for AED treatment |
| **III Specific situations** | 1. Temporal lobe epilepsy<br>2. Extratemporal lesional epilepsy<br>3. Focal nonlesional epilepsy<br>4. Widespread pathology: unifocal or multifocal epilepsy? | |

their seizures. Thus, at the time of craniotomy for resection, the treatment team has a good idea regarding the laterality and region (lobe or lobes) involved in seizure generation.

ECoG recording becomes a valuable tool in four circumstances. First, there are patients in whom implantation of the hardware and long-term recordings pose an increased procedural risk, higher than the surgical resection itself. The main risks refer to complications that have been described in both pediatric (Onal et al., 2003) and adult (Wong et al., 2009) cases of invasive recording for epilepsy surgery. These include cerebral edema, subdural hematomas, blood loss, osteomyelitis, superficial wound infections, and cerebral spinal fluid leaks. Consistent correlation has been found between the size of the grid and the risks for complications. Other risks are due to the patient's inability to tolerate such procedure or to the type of seizures, where the increased confusional states (either ictal or postictal) or violent semiology would pose higher risk for injury as a result of, for example, self-explantation of the electrodes. Second, unlike the implanted hardware, which can shift over the span of several days, ECoG recordings allow a better appreciation of the local topography and location of the abnormal cortical activities. Third, ECoG gives the surgeon the opportunity to readjust its position repetitively, as needed. This is particularly important in situations when the invasive recordings are not able to localize the seizure onset precisely because of their unexpected location. We have had several cases when the interictal or ictal activities of interest were seen at the margin of the implanted grid (for the EMU recording) or, worse, arose from a region not included in the implant planning. In such cases, intraoperative ECoG, performed after the EMU session, allowed full evaluation of such activities. Fourth, ECoG is helpful in the case of lesional epilepsy, when the location of the lesion and the patient's stereotyped seizure semiology show concordant data, whereas the surface EEG recording is not helpful.

**TABLE 3-2** Indications for combination of ECoG and functional mapping

| | |
|---|---|
| **I General indication** | Whenever the irritative zones/ epileptic foci are or could be in close proximity to eloquent cortex |
| **II Goals** | Guiding the surgical approach<br>1. Extent of resection<br>2. Subpial transections |
| **III Specific situations** | 1. Left temporal foci<br>2. Frontal or parietal foci<br>3. Multifocal epilepsy<br>4. Neurodevelopmental lesions |

*A2. Emergent epilepsy surgery.* On a few occasions, however, ECoG recordings are undertaken in emergent cases, guiding the surgical treatment of ongoing seizure activity. In such circumstances, the ECoG recording guides the excision of an active ictal focus, and, from this perspective, its guidance is more precise than in the case of elective surgery, when most of the information is based on the location and behavior of interictal abnormalities. Because a good prognosis is more likely associated with a well-demarcated location of the ictal abnormalities, it is also more likely to occur in lesional (versus nonlesional) epilepsy. However, we have been involved in cases of nonlesional and even newly diagnosed epilepsy when the surgical treatment under ECoG guidance aborted the status epilepticus (Fig. **3-2**; Costello et al., 2006). Thus, we believe that the key for successful resection in such cases is the consistent and preserved focality of the ictal pattern over time (rather than the identification of a lesion on neuroimaging studies).

*A3. Nonepilepsy surgery.* ECoG can also be used in nonepilepsy surgery, such as resection of supratentorial brain tumors, as further described in the following.

## B. Goals: Epilepsy Surgery

In epilepsy surgery, the main purpose is to guide in choosing a certain surgical approach.

*B1. Establishing the extent of resection.* This is, in a major part, determined by the extent of interictal and ictal activities. However, in cases when these activities are or could be located in close proximity to eloquent cortex, mapping of the seizure foci and functional mapping should be done in combination (see pages 114–15).

*B2. Establishing the need for multiple subpial transections.* In cases where the irritative zone is in close proximity to the eloquent cortex, as well as in cases of extensive irritative zones and/or multifocal epilepsy, the resection is combined with or entirely replaced by multiple subpial transections (MSTs). This technique (Morrell et al., 1989) is used in attempts to disconnect epileptogenic brain tissues and thus to prevent the spread and organization of seizure activity, and the extent of these transections is often determined by ECoG recordings (Fig. **3-3**).

*B3. Establishing the location for stimulator implantation.* Several authors have shown that cortical electrical stimulation, when applied at certain parameters, can abort interictal and ictal abnormalities (Lesser et al., 1999; Motamedi et al., 2002; Yamamoto et al., 2002). This principle is the basis for ongoing trials, where the patients who are considered poor candidates for epilepsy surgery have cortical electrical stimulators implanted over the irritative zones. These stimulators are expected to trigger an electrical current at the seizure's onset, which can abort the abnormal ictal activity. Thus, it becomes important to locate such irritative zones accurately, to ensure a precise placement of the electrical devices. We have had cases of refractory epilepsy (e.g. multifocal epilepsy and/or close proximity of the seizure foci to eloquent cortex) in patients who are considered to be good candidates for implantation of such a device, as part of the Neuropace trial. In these cases, we used ECoG to determine the precise location of the future implant (Fig. **3-4**).

*B4. The prognostic value of post resection ECoG recordings*
There are many reports suggesting that the persistence of spiking activity after the surgical excision does not have prognostic value (Ajmone Marsan et al., 1958; Chatrian et al., 1997; Schwartz et al., 1997). Moreover, the significance of ECoG recordings after temporal lobe resection remains controversial. In fact, activation of new interictal spiking in temporal cortex following a selective amygdalohippocampectomy occurs frequently and does not have any prognostic significance (Cendes et al., 1993; Wennberg et al., 1997). Some authors, however, have suggested that persistent spike activity after completion of the surgical resection predicts unsatisfactory outcome (Bengzon et al., 1968; Gloor, 1975; McBride et al., 1991).

In the cases of extratemporal epilepsies or dual pathology, requiring resection or MST outside the standard temporal lobectomy, the prognostic value of postexcision ECoG is much higher. It is also the case of other epilepsies where the seizure foci were found overlying eloquent cortex, thus requiring MST under ECoG guidance. Orbach et al. (2001) evaluated long-term outcome after MST guided by intraoperative ECoG. Approximately 82% of the patients achieved seizure freedom within 2 years after the surgery. In frontal lobe epilepsy, the presence of residual spiking activity after resection is associated with poorer outcome with regard to seizure control (Wennberg et al., 1998), especially if epileptiform activity is distant from the resection border. Tran et al (1997) found a trend toward a relationship between postresection spikes and seizure occurrence in patients with brain tumors and seizures.

## B. Goals: Nonepilepsy Surgery

In nonepilepsy surgery, such as supratentorial tumor resection, the main role of ECoG is directed toward diagnosis and management.

*B5. Diagnosis.* In many cases of brain tumors, the patients' first clinical presentation, which, in many instances, leads to the diagnosis, is seizure activity. However, there are cases when the patients are not diagnosed with seizures, but rather present with symptomatology that is thought to be due to a space-occupying process (e.g. muscle weakness, language disturbance). However, these symptoms may represent the manifestation of recurrent or ongoing seizure activity, especially if they fluctuate. ECoG recordings at the time of resection can thus be used to characterize the local cortical excitability and its relationship, if any, with the clinical presentation (Fig. **3-5**).

*B6. Intraoperative and postoperative management.* In the case of undiagnosed seizures, the brain tumor patients are most likely not treated with AEDs at the time of the surgery. Because, in many of these cases, electrophysiologic functional mapping is performed intraoperatively, the preexistence of epileptiform or epileptic activity further increases the risks of stimulation-triggered seizure activity. Thus, ECoG recordings help assess the presence, frequency, and pattern of increased cortical excitability, thus grading the risk for stimulation-triggered seizure activity (organized, high-frequency discharges are more likely to further evolve into seizure activity) and the need for intravenous loading with an AED. Determining the location of the irritative zones in relationship to the site of stimulation helps in a better estimation of the risks of triggering ADs. As mentioned on page 102, some authors have described a trend toward a relationship between the

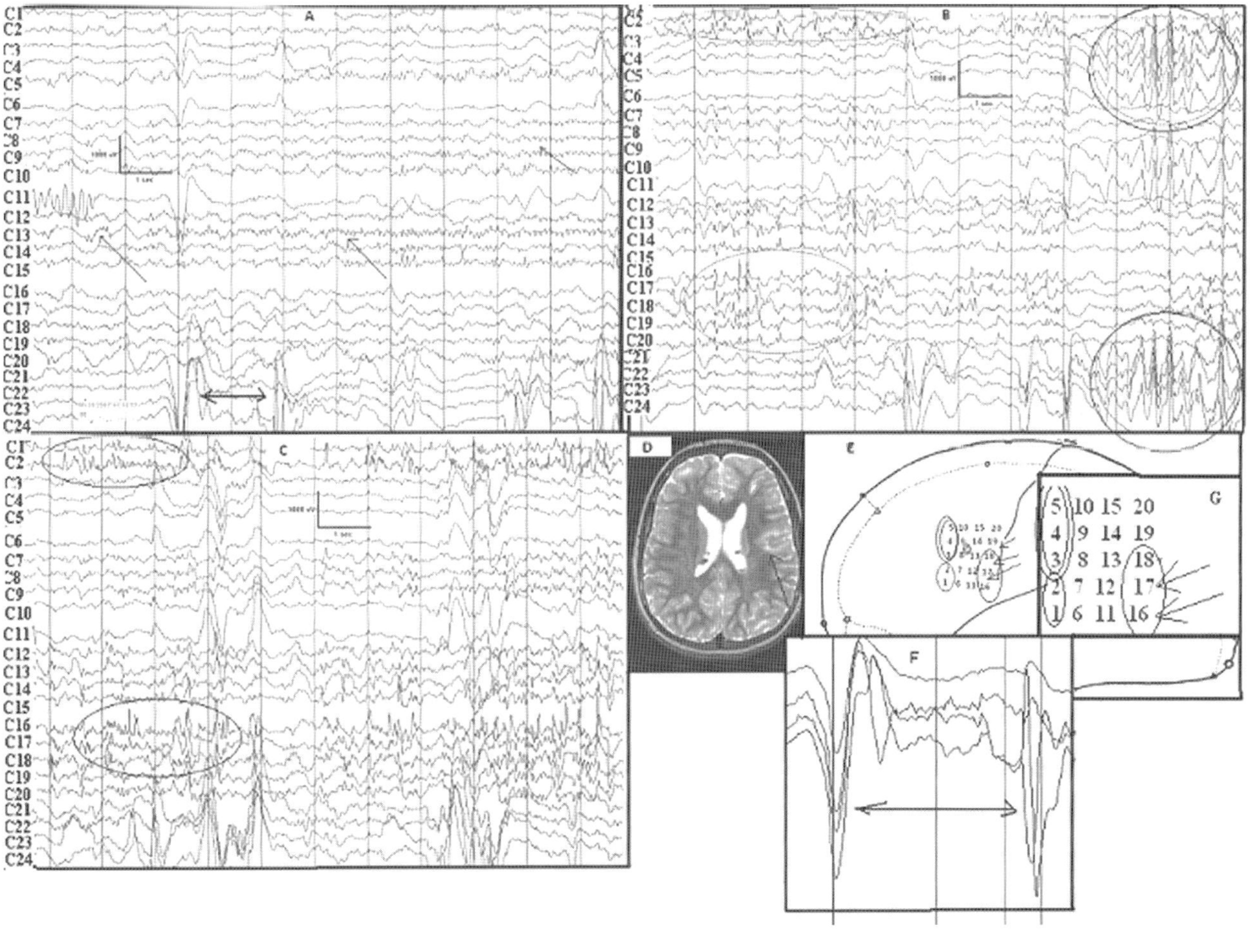


**FIGURE 3-3** Combined ECoG and functional mapping guiding the surgical treatment of multifocal epilepsy. This is the case of a 3-year-old-girl with tuberous sclerosis, motor seizures and drop attacks. The ECoG recording is done via a 20-contact (labeled C1 through C20) grid and a 4-contact (labeled C21 through C24) depth electrode, with reference on the contralateral mastoid. Panel **A** shows the onset and spread of an electrographic seizure (thin arrow), induced by the administration of 30 mcg/kg alfentany1. The depth electrode contacts show high amplitude epileptiform discharges, throughout the recording (thick arrow), with some corresponding field on grid contacts (double circled in panel **B**). Panel **B** and **C** show bursts of high amplitude, multifocal spikes, and polyspikes, seen at several grid contacts (single circles). Panel **D** shows an axial T2 MRI image, with a horse show hyperdensity in the left frontal lobe, representing the targeted tuber (arrow). Panels **E** and **G** show placement and close up of the 20 contact grid. The depth electrode is placed in between grid contacts 8, 9, 14, and 13. The most active zones are circled. The arrows point towards areas of eloquent motor cortex, as delineated by functional cortical mapping. Consequently, multiple subpial transections were applied to these regions. The calibration marks in all panels are 1000 uV and 1 second.

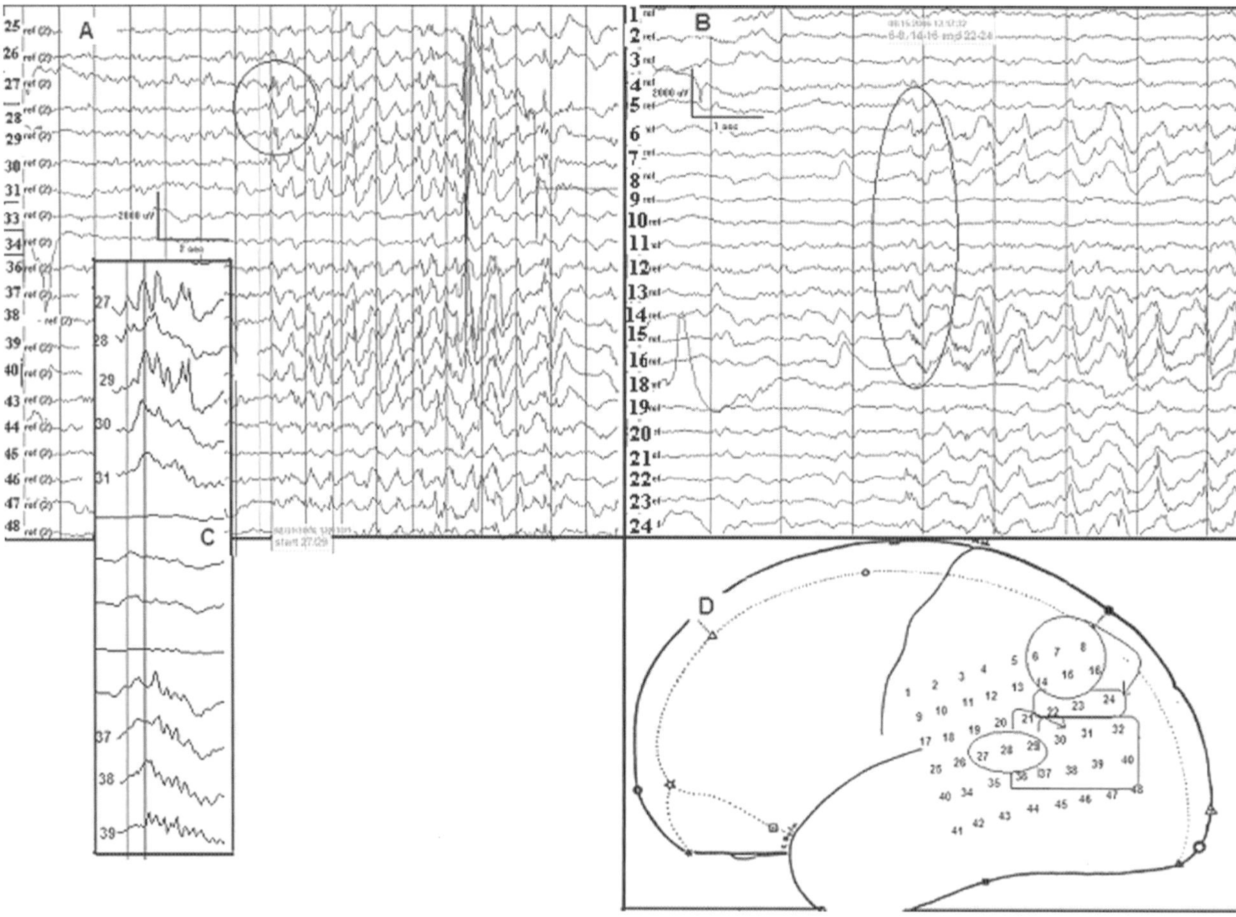

**FIGURE 3-4** Cortical stimulator placement, guided by ECOG recordings in multifocal epilepsy. This is the case of a 21-year-old man with intractable epilepsy and two types of seizures: one starting with a visual aura and the other one with language disturbance. The surface EEG showed frequent, broad field left posterior quadrant interictal discharges, as well as more focal independent left parietal and left posterior temporal sharps. The ictal onset was also spread over the left posterior quadrant regions. Because of the multifocality of interictal discharges and their location in a region likely presenting visual and language functions, as well as presence of different semiology types, a cortical stimulator implantation was preferred over surgical resection. The placement of the stimulators was done under ECoG guidance. The recordings showed two distinct irritative zones: one situated toward the left parieto-occipital region and the other in the posterior part of the left superior temporal gyrus. Thus, two stimulators were placed separately in these regions. Postoperatively, the patient described a clear reduction of his ictal events. Panel **A** shows a burst of epileptiform activity originating at contacts 27, 28, 29 (circled), located over the posterior part of left superior temporal gyrus. However, there is almost instantaneous spread toward contacts 30, 31, and 36–40. Panel **B** shows a burst occurring independently from the one described in panel **A**, more superior and posterior on the left hemispheric convexity, toward the left parieto-occipital junction. Notice the near synchronous activity at contacts 6–8, 14–16(circled), and, to a lesser extent, at 22–24. Panel **C** is a close up of the onset of the burst captured in panel **A** and shows a small lead-in of the activity seen at contacts 27 and 29. Panel **D** shows the placement of the 48-contact grid and the approximate location of the lead-in irritative zones (circled), with immediate spread of the interictal/ictal activity to the adjacent contacts, situated at the left temporo-parieto-occipital junction (rectangle).

presence of postresection spikes and seizure occurrence in brain tumor patients. Thus, assessing the location of epileptiform and epileptic activity in relationship to the tumor location also helps in deciding if the patient should receive AEDs for a longer period of time after the surgery. This is due to the fact that ECoG recordings rarely, if ever, guide the extent of the resection in order to incorporate such abnormal epileptiform activity beyond the tumor margin, particularly when such activities are situated at a distance from tumoral tissue (Fig. **3-6**).

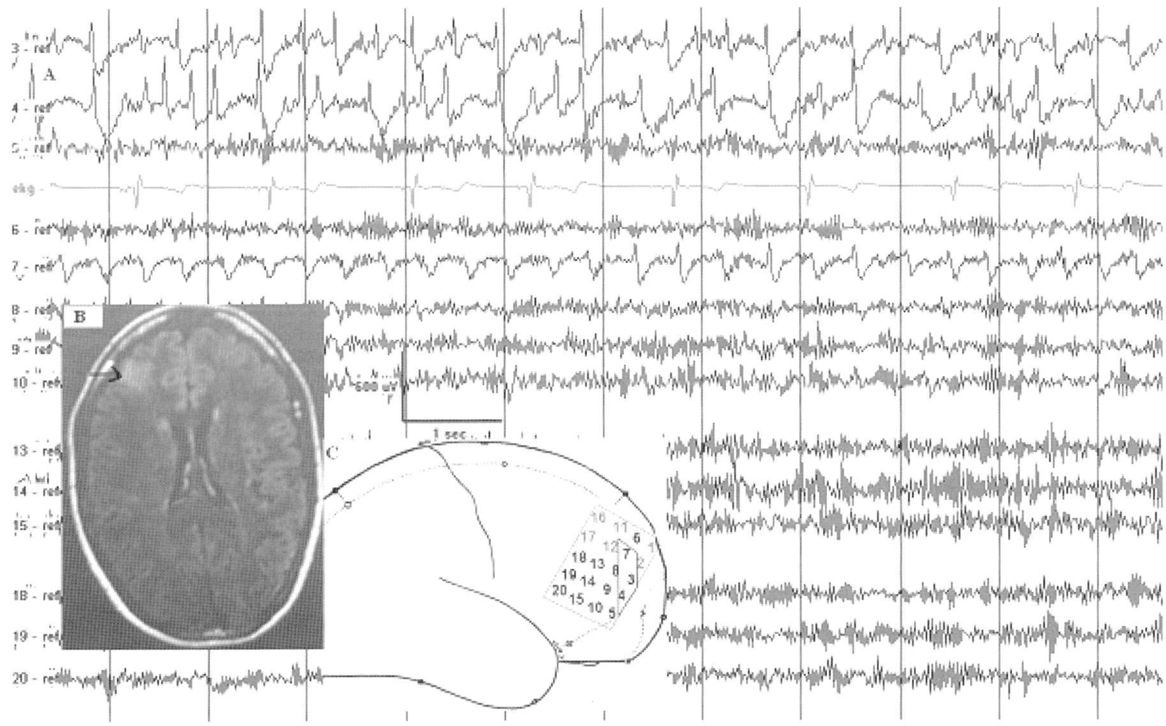


**FIGURE 3-5** The diagnostic value of ECoG recording in nonepilepsy surgery. This is a case of a 30-year-old woman who presented with headache and discrete episodes characterized by an inability to concentrate and a poorly described abnormal sensation. Brain MRI revealed a right frontal lesion, suggestive of a low-grade glioma. The etiology of her spells remained uncertain, and she was not treated with AEDs at the time of lesion resection. In an attempt to determine whether these spells represent seizures, evaluation of local cortical excitability was done with intraoperative ECoG. The ECoG revealed rhythmic runs of epileptiform discharges, prompting intravenous load with an AED (Keppra), which was continued after surgery. Postoperatively, her spells stopped. The ECoG recording is done via a 14-contact grid (see panel **C**), in a referential montage. The reference electrode is placed on the contralateral mastoid. Panel **A** reveals rhythmic brief spikes at contacts 3, 4, and 7, up to 500 $\mu$V, seen in almost continuous runs, up to 2–4 Hz. Also, notice the significant amount of diffuse beta activities, likely an anesthetic effect. The calibration mark is 600 $\mu$V and 1 second. Panel **B** shows an axial FLAIR MRI image, revealing a small right frontal hyperdense area. Panel **C** shows a drawing of the customized grid placement. The gray contacts were discarded by the surgeon. The recording was done via the black contacts.

## C. Specific Situations

*C1. Temporal lobe epilepsy.* As with the other circumstances, the contribution of ECoG recordings in the surgical treatment of temporal lobe epilepsy is most likely related to the overall surgical plan and suspected seizure foci that are derived from the extensive preoperative evaluation. Many centers argue that the utility of ECoG in directing resection in a standard temporal lobectomy (Falconer at al., 1955) is minimal (Engel, 1987; Spencer et al., 1992), and there remains considerable debate regarding the use of ECoG recordings to shape the extent of temporal lobe resection in order to achieve a better postoperative outcome (Ajmone Marsan, 1980; Engel, 1987; Graf et al., 1984; Jennum et al., 1993; Ojemann, 1980; Primrose et al., 1992; Spencer et al., 1992; Stefan et al., 1991). We agree that, in general, intraoperative ECoG data are not needed for standard anterior temporal lobectomy.

Some centers, however, do use intraoperative ECoG to tailor the temporal lobe resection in specific situations (Chatrian et al.,1997; Engel, 1997), for more selective resections of the temporal lobe structures.

1. The extent of hippocampal resection can be tailored by the intraoperative ECoG data (Jooma et al., 1995; Kanner et al., 1995; McKhann et al., 2000; Nayel et al., 1991; Stefan et al., 1991), which have been shown to improve postoperative outcomes. In fact, the most common cause of surgical failure in the

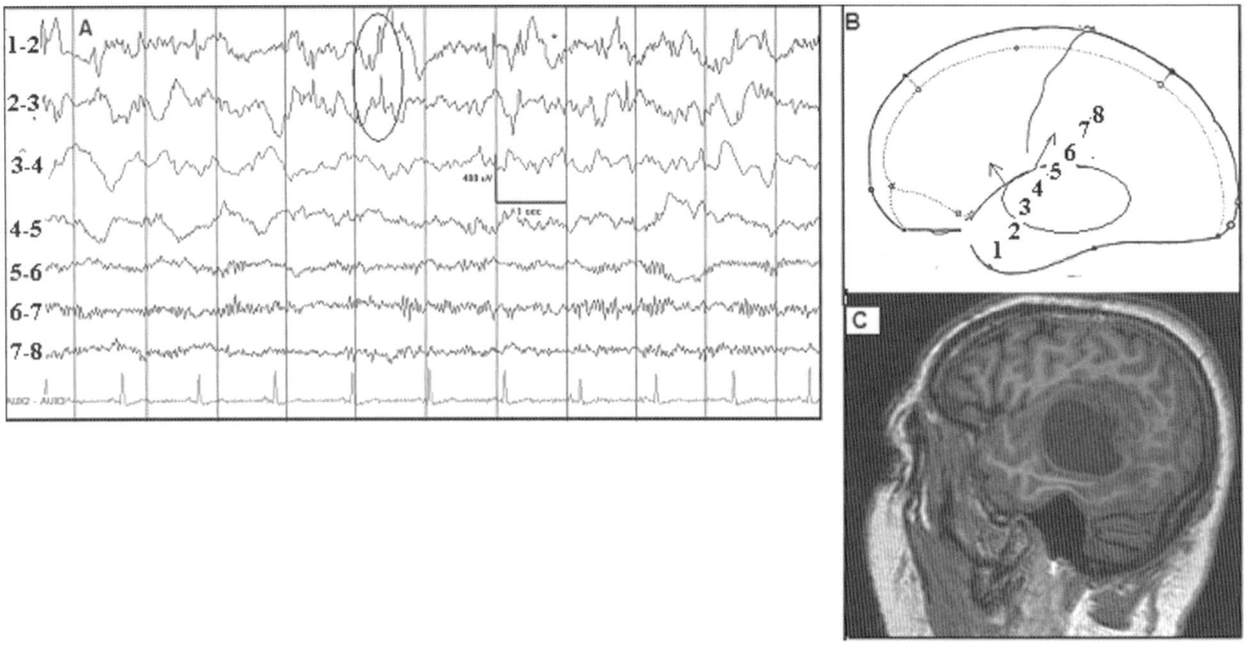


**FIGURE 3-6** ECoG recording guiding intraoperative and postoperative management in nonepilepsy surgery. This is the case of a 68-year-old woman who developed discrete episodes of confusion and fluctuating language difficulties and was found to have a left temporal tumor, with subsequent left-sided craniotomy and tumor resection. ECoG recording showed epileptiform activity in the anterior temporal tip, relatively away from the more posteriorly located tumoral tissue. It was decided to load the patient with AEDs, before the intraoperative cortical stimulation for language mapping, because of its epileptogenic effects. The patient underwent lesionectomy, instead of temporal lobectomy, which failed to remove the irritative zone. It was therefore decided to continue AEDs postoperatively. Panel **A** shows ECoG recording done via an 8-contact strip, in a bipolar montage, with reference electrode placed on the contralateral mastoid. Notice brief spikes (circled) present at contacts 1 and 2, up to 300 $\mu$V and 1 Hz. Also, notice the relative slower, 1–4 Hz, 200–600 $\mu$V activities present in the first 4–5 channels, as well as the beta frequencies present in the last 3 channels. Panel **B** shows the approximate position of the strip, relative to the tumor location (circled). Notice that contacts 6, 7, and 8 are overlying relatively normal parietal cortex, although some mass effect from the tumor is pushing the Sylvian fissure superiorly (arrows). Contacts 3, 4, and 5 are overlying temporal region infiltrated by the tumor. Contacts 1 and 2 are placed close to the temporal tip, adjacent to the anterior margin of the tumor. Notice the good correlation between the location of different contacts relative to the location of the tumor and the cortical activities recorded in those regions. Panel **C** represents sagittal SPGR MRI sequences showing the location of the left temporal tumor and of the surrounding vasogenic edema.

treatment of temporal lobe epilepsy is believed to be inadequate hippocampal resection. We had several similar cases when ECoG recording was helpful in determining the posterior extent of the resection of the mesiotemporal structures, especially when there was a discrepancy between the locations of their interictal and ictal activities in the anterior-posterior axis. In such situations, we used two or three depth electrodes to sample the activities at the level of amygdala, middle, or posterior part of the hippocampus (Fig. **3-7**).

2. In patients with temporal lobe epilepsy, the interictal abnormalities on ECoG often involve both the mesial temporal lobe structures and cortical structures, either simultaneously or independently, when data are available from all of these structures (Devinsky et al., 1992; Engel et al., 1975; Graf et al., 1984; Kajtor et al., 1958; Stefan et al., 1991; Walker, 1967). The relationship or independence of these epileptiform activities can provide information regarding whether the seizure generator lies in the medial temporal structures, temporal cortex, or both. ECoG recordings can help further define the discrepancy between abnormal electrical activity in the mesiotemporal structures and temporal neocortex, respectively. In such instances, using a small grid for the neocortical activity and a depth electrode could tailor the extent of the resection, which may encompass more neocortex and fewer mesiotemporal structures or vice versa.

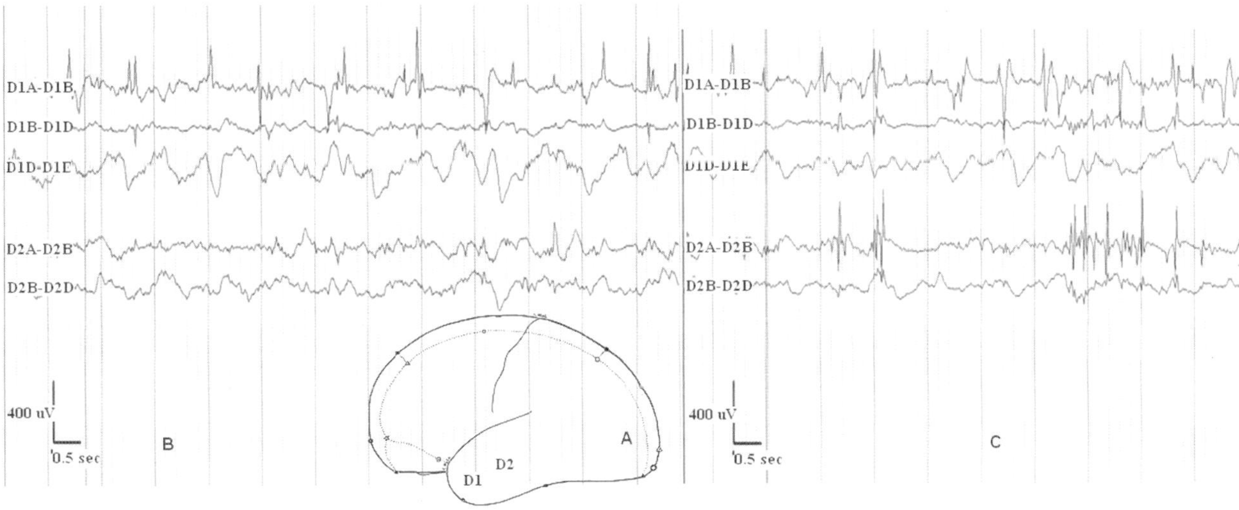


**FIGURE 3-7** Left temporal lobectomy guided by ECoG recording. The long-term surface EEG recording in this situation was significant for frequent interictal left anterior temporal spikes (T1, F7) and stereotyped electrographic seizures onset in the mid-posterior temporal regions (T3, T5). The brain MRI showed signal changes and atrophy of the left amygdala, but normal-looking left hippocampus. ECoG recording done intraoperatively confirmed the posterior extent of the irritative zone within the left mesiotemporal structures. Panel **A** shows the placement of two 4-contact (labeled A, B, D, and E) depth electrodes, labeled D1 and D2. Contact D1A was at the level of the left amygdala, and contact D2A was at the level of the left hippocampus; contacts D1E and D2E were the most superficial. Panel **B** shows ECoG recording via the two depth electrodes, in a bipolar montage. The reference electrode was on the contralateral mastoid. Notice frequent interictal spikes, up to 400 $\mu$V or so, present at the level of amygdala and seen in channel D1A–D1B. At the same time, very infrequent spikes are present at the level of hippocampus (D2 recording). Panel **C** shows the ECoG recording a couple of minutes after administration of an intravenous bolus of 30 mcg/kg alfentanyl. Notice a significant increase in the interictal spikes at the level of hippocampus (channel D2A–D2B) and only a modest increase, if any, in the spikes at the level of amygdala. The calibration marks in panels **B** and **C** are 400 $\mu$V and 0.5 second.

A particular situation arises in the case of a left (dominant) temporal lobectomy, when the resection of abnormal temporal neocortex in the posterior part of the superior temporal gyrus should be guided by both ECoG and language mapping. In the case of epileptiform or epileptic focus or foci identified within eloquent language cortex, subpial transections should be considered (see page 102).

3. At times, however, interictal epileptiform activities can be observed in adjacent lobes (frontal and/or parietal), as well as within the temporal lobe (Cendes et al., 1993; Niedermeyer, 1982; Quesney et al., 1993; Rasmussen, 1983; Tuunainen et al. 1994; Walker, 1967). In such cases, additional surgical intervention, outside of the boundaries of the standard anterior temporal lobectomy and in areas that retain significant sensorimotor function, again requires concomitant ECoG and functional mapping and, possibly, MSTs instead of a complete resection. This is especially relevant in patients with additional interictal or ictal activity in fronto-parietal regions (Cendes et al., 1993; Levesque et al., 1991). In such cases of planned subpial transections, ECoG data become more important.

*C2. Extratemporal lesional epilepsy.* The extratemporal epilepsies constitute a much more heterogeneous group, which differs considerably from the better known temporal lesional epilepsy (TLE). Extratemporal epilepsies are defined by the presumed site of seizure origin (i.e., frontal or parietal), although some territorial overlaps can occur (Quesney et al., 1992). Intraoperative ECoG recordings are valuable tools for the mapping of interictal epileptogenic zone(s), particularly in tailored surgical resections (Quesney et al., 1998) and during the use of MSTs. In patients with cortical dysplastic lesions, the most common cause of non-TLE, ECoG provides vital information regarding the location of the lesion, as well as the extent of resection around the lesion required for cure. These lesions often have an electrographic correlate consisting of rhythmic electrographic discharges, coinciding with the anatomical location of the dysplastic lesion on magnetic resonance imaging (MRI). Thus, ECoG is a reliable tool to indicate the presence of underlying dysplastic cortex (Palmini et al., 1995) and may also be used to determine when the extent of resection is adequate for seizure treatment. It has been shown that excision of all lesional epileptiform activity in the form of the so-called continuous rhythmic spiking correlates with better postoperative outcomes (Gambardella et al, 1996; Palmini et al., 1995). In fact, the very active interictal patterns are so pathognomonic for cortical dysplasia (see page 121) that, if ECoG recordings thought to be from a dysplastic lesion are "quiet," either the lesion is not a cortical dysplasia or the recording is not done from the lesion. We came across a case where ECoG recording from a presumed dysplastic cortical lesion failed to demonstrate any significant interictal or ictal activity. The patient continued to have seizures after excision of the "quiet" zone. On further assessment of the neuroimaging and neurophysiologic data, it was felt that the lesion was not resected. At reintervention, repeat ECoG showed significant interictal abnormalities originating from the lesion. Its excision rendered the patient seizure free. (Fig. **3-8**). Because cortical dysplasias are such

(a)

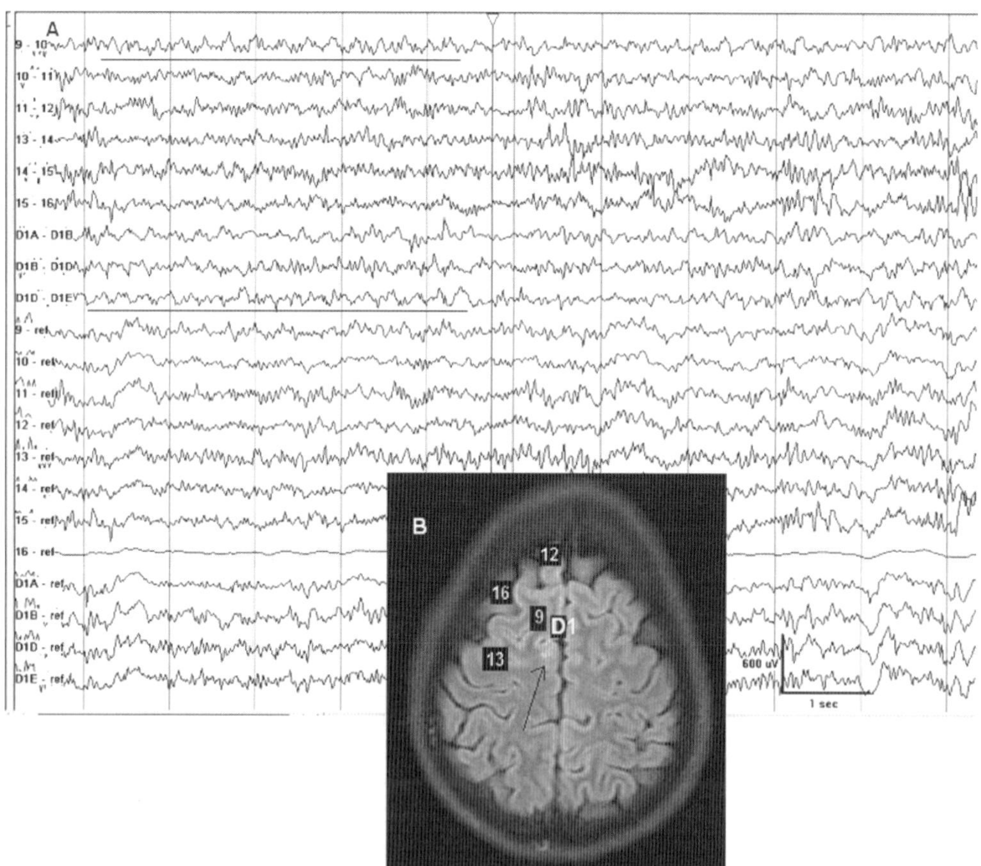


**FIGURE 3-8** ECoG recording form cortical dysplasia: the absence of epileptiform/epileptic activity should question either the location of the lesion or its pathology. This is the case of a 21-year-old man with long-standing history of stereotyped, frequent (~15 per night) nocturnal events. They consisted of an aura of left thigh tingling, followed by stiffness of the lower extremities. These events were diagnosed as frontal lobe seizures. However, they had been refractory to AED treatment. The interictal surface EEG was normal and the ictal EEG was obscured by artifact. A review of the brain MRI study revealed a questionable hyperintense lesion in the right superior frontal gyrus, which was considered to be the cause for the patient's seizures. Lesionectomy under ECoG guidance was performed. Unfortunately, the first positions of the ECoG electrodes were mislocated from the lesion, and the recording was unremarkable. The patient underwent a limited resection, which did not eliminate the dysplastic lesion.

active epileptogenic lesions, ECoG recordings showing the typical and well-demarcated interictal patterns offer concordant data to weak MRI findings where the lesion is not well defined. Thus, in the presence of stereotyped seizure semiology in concordance with the location of the presumed lesion, the surgical intervention can be guided solely by neuroimaging studies, their co-registration and ECoG recordings, even in the absence of relevant data from extracranial EEG monitoring and avoiding implantation of hardware for phase II invasive recording. (Fig. **3-9**). Similarly, resection of tubers in tuberous sclerosis is another typical case of lesional epilepsy where ECoG recordings help determine the extent of the resection. It is known that, in many circumstances, the irritative zone is the neocortex surrounding the tuber, rather than the tuber itself (Major et al., 2009; Fig. **3-10**), and thus the surgical approach is guided with the help of ECoG recording. In general, the irritative zone is characterized by multifocal epileptiform discharges and can be rather extensive. Depending on its location and proximity to the eloquent cortex, resection, subpial transections, or a combination of both can be entertained (Fig. **3-3**). There are also other lesions that do not consist of epileptogenic tissue themselves, but which cause irritation of the surrounding neocortex; with the help of ECoG, resection of such lesions can be tailored to include the irritative zones.

**(b)**

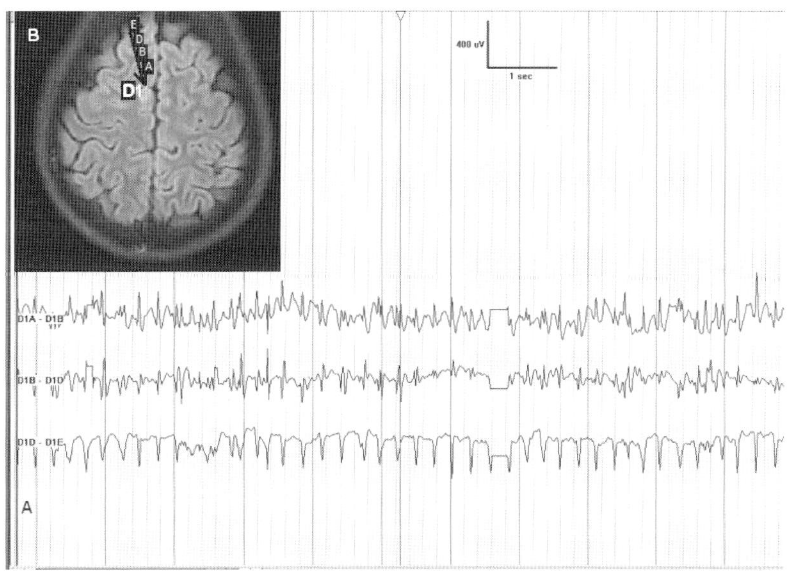


**FIGURE 3-8** *(Continued)* The patient continued to have seizures postoperatively. It was felt that the lesion was still present at the posterior margin of the initial resection. The patient underwent surgical reintervention under ECoG guidance. This time the depth electrode recording showed a significant amount of epileptiform and epileptic activity characteristic of a cortical dysplasia. Resection of the lesion led to seizure freedom. The patient was tapered off his AEDs and has remained seizure free since then. **a:** Panel **A** shows ECoG recording at the first surgical intervention, showing bipolar as well as referential recording from the two 4-contact strips, as well as from the 4-contact depth electrode. The reference electrode is a sterile needle placed in the surgical field. Notice the 4–5 Hz rhythmic activity seen in channels 9–10 and D1D–D1E, with depth contacts D and E being the most superficial (underlined). Also notice the close proximity of D1 to contact 9 (see panel **B**). However, even with such reasonable field, no definite epileptiform or epileptic activity was seen. The calibration mark is 600 $\mu$V and 1 second. Panel **B** shows an axial FLAIR MRI sequence, revealing a hyperintense lesion (arrow) at the level of the superior frontal gyrus. The labels show placement of the contacts of two subdural strip electrodes, 9 through 12 and 13 through 16, respectively. Label D1 shows placement of a 4-contact (labeled A, B, D, and E) depth electrode, mistakenly positioned anterior to the lesion. **b:** Panel **A** showing ECoG recording at the second surgical intervention, via a 4-contact (labeled D1A, D1B, D1D, and D1E) depth electrode (D1), in a referential montage, with the reference electrode placed in the surgical field. The recording reveals almost continuous spiking in channels D1A–D1B and D1B–D1D. The spikes and polyspikes are 100–200 $\mu$V and up to 200 msec apart. The most superficial channel, D1D–D1E, is showing 4-Hz, 200-$\mu$V rhythmic, notched theta activity. Both these interictal and ictal patterns are characteristic for cortical dysplastic regions. Panel **B** shows axial FLAIR MRI sequences, demonstrating the second position of the depth electrode, during the surgical reintervention, this time traversing the lesion.

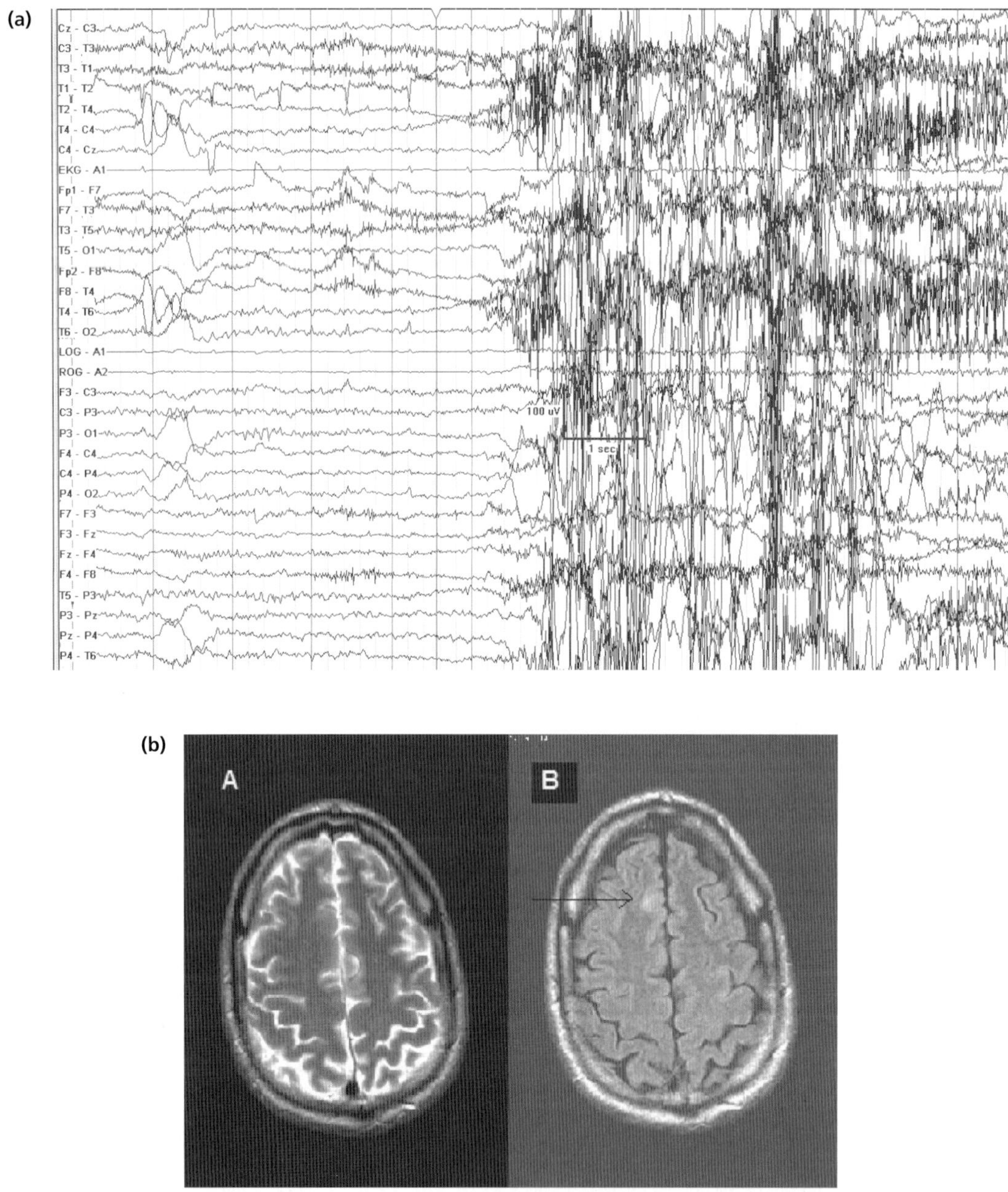


**FIGURE 3-9** Characteristic features of ECoG recording in cortical dysplasia. This is the case of a 40-year-old man with a 20-year history of highly stereotyped, nocturnal clinical events. These were brief in duration and involved violent leg movements, accompanied with a guttural vocalization. They were not followed by postictal confusion or dysphasia. The interictal surface EEG was normal, and the ictal surface EEG was markedly obscured by artifact. The brain MRI study was initially interpreted as normal. The patient was diagnosed with a sleep disorder, although frontal lobe epilepsy could not entirely be ruled out. The episodes became more frequent and started occurring even during the day. On careful review of the MRI, a possible cortical dysplastic lesion in the right superior frontal gyrus was found, and the diagnosis of frontal lobe epilepsy was reconsidered. The patient underwent right-sided craniotomy, with lesionectomy under ECoG guidance. The recording from the lesion via depth electrodes revealed significant epileptiform and epileptic discharges, characteristic for cortical dysplasia. The pathology confirmed the diagnosis. The patient became seizure free following the procedure. **a:** Ictal surface EEG in a bipolar recording. The left half is showing symmetric awake background. The right half is showing sudden ictal onset, characterized by significant movement, with electrode and muscle artifact obscuring the record. The calibration mark is 100 $\mu$V and 1 second. **b:** Two axial MRI images: panel **A**-axial T2 sequence showing no obvious lesions; panel **B**-axial FLAIR sequence showing a small area of increased signal in the right superior frontal gyrus (arrow).

(c)

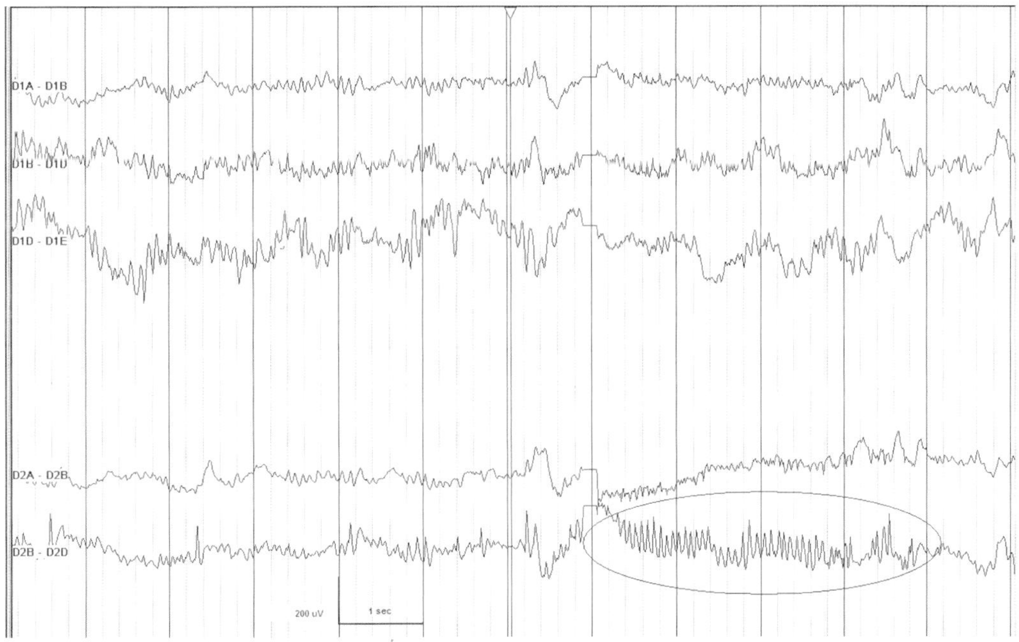


**FIGURE 3-9** *(Continued)* **c:** ECoG recording via two 4-contact (labeled A, B, D, and E) depth electrodes, D1 and D2. The electrodes are placed perpendicular to the cortical surface, through the region thought to harbor the cortical dysplasia, with contact A being the deepest. The recording is done in a bipolar montage. The reference electrode is placed in the surgical field. Notice the rhythmic, 4-second run of sharp activity, seen in channel D2B–D2D (circled) and fragmented similar activities at D1 electrode contacts. This run of sharp alpha/beta activity is one of the electrographic signatures of cortical dysplasias. The calibration mark is 200 $\mu$V and 1 second.

ECoG recordings can be particularly helpful in cases of lesional frontal lobe epilepsy, when significant amount of movement, often associated with these type of seizures, causes movement and muscle artifact, obscuring the surface EEG recordings.

So far, we have addressed cases pertaining to well-demarcated, focal lesions. Of course, lesional epilepsy can present with extensive or multiple lesions; this is further discussed on pages 112–14. In all of the extratemporal epilepsies, but particularly in frontal lobe epilepsy (FLE), large epileptogenic zones can be seen. Therefore, accurately estimating the boundaries of the surgically relevant epileptogenic is an important intraoperative consideration. In one study, a group of 60 patients with nontumoral FLE underwent operative treatment, guided by ECoG recordings. In this group, the distribution and abundance of interictal spiking recorded on ECoG was the most powerful prognostic indicator of postsurgical outcome with respect to seizure control (Wennberg et al., 1998). Patients with epileptiform activity localized to the one or a few gyri were significantly more likely to have a good outcome from surgical resection than those with more widespread epileptiform activity.

*C3. Focal nonlesional epilepsy.* Epilepsy surgery is an elective procedure for treatment of refractory epilepsy. However, we used resection of the seizure focus in a few patients under urgent circumstances as the ultimate management resource to treat status epilepticus (Fig. **3-2**). In both occasions, ECoG was the recording tool that helped delineate the seizure focus. Of note, in one of these cases, no lesion was identified on neuroimaging studies, and in neither of these cases was there a history of epilepsy. One of the challenges of such ECoG recording is the fact that it usually follows days after aggressive pharmacological treatment. Thus, the cortical activity is usually suppressed and so are the guiding epileptiform discharges. As noticed in the case presented in Fig. **3-2**, after 3 weeks of anesthetics infusion and multiple AED treatment, the interictal activity had a blunt morphology, even as the ECoG was recorded after the propofol was stopped. In such refractory cases, it may be worthwhile intervening sooner rather than later,

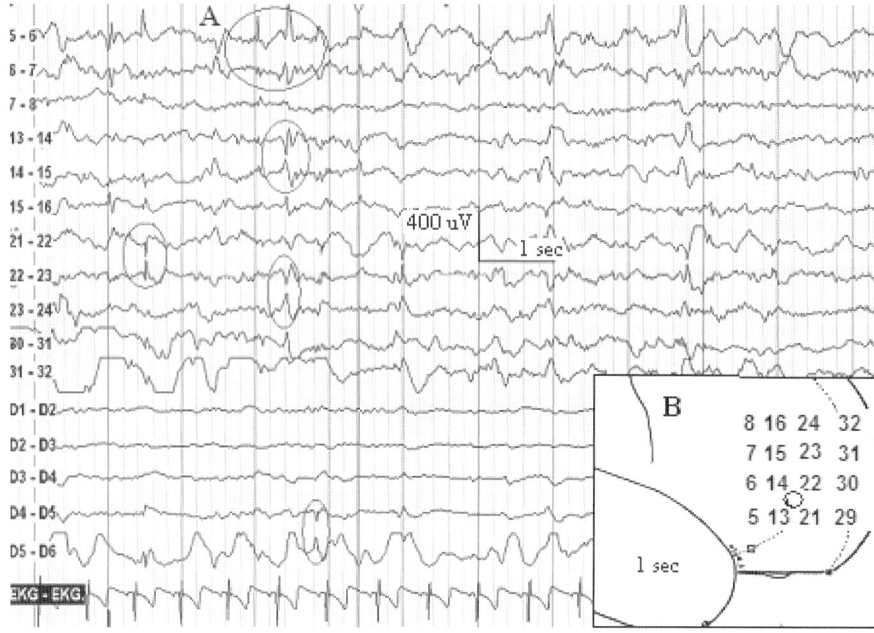


**FIGURE 3-10** ECoG recording in lesional epilepsy, determining the extent of the irritative zone. This is the case of a 3-year-old girl with medically intractable seizures and tuberous sclerosis. Her seizures were attributed to a right frontal tuber. Thus, she underwent right frontal craniotomy for lesionectomy under ECoG guidance. The recording depicted an irritative zone extending to the surrounding cortex and less to the tuber itself. As a consequence, the surgeon resected not only the tuber but also the adjacent cortical irritative zones depicted by the ECoG recording. Postsurgically, the patient gained very good seizure control. Panel **A** shows the ECoG recording in a bipolar montage. The recording was done via a 4 by 4 contact subdural grid, cut from a 65-contact grid and placed over the right inferior frontal gyrus, overlying the location of the tuber and its surrounding cortex. A 6-contact (labeled D1 through D6) depth electrode was placed through the right frontal tuber, with contact 6 being the most superficial. The depth electrode was adjacent to contacts 21, 22, 13, and 14, as shown in panel **B.** Notice the presence of multifocal independent spikes located in the cortex surrounding the tuber (circled). Notice also the absence of epileptiform discharges at the deeper contacts of the depth electrode and the relative attenuation of the background in these regions. The calibration mark is 400 $\mu$V and 1 second.

when the focality and morphology of such discharges are still preserved.

We have also used ECoG recordings to guide resection in other nonlesional but elective cases of epilepsy surgery, when the EEG data pointed toward a very focal electrographic abnormality. We believe that, in some of these cases, the inability to identify a lesion on neuroimaging studies is the result of limited resolution rather a result of the absence of focal pathology.

*C4. Widespread pathology: Unifocal or multifocal epilepsy?* There are cases of epilepsy when the causal pathology can cause widespread cortical damage, sometimes unpredictable, as, for example, in postinfectious and posttraumatic epilepsies. Similarly, in cases of large neurodevelopmental abnormalities, such as polymicrogyria, more than a limited region can be involved; sometimes the entire hemisphere is involved. In other situations, the initial primary focus or foci can induce a dual pathology, for example, inducing secondary mesiotemporal sclerosis from repetitive, longstanding seizure activity. The widespread pathology is a warning sign for multifocal epilepsy, and these patients are not good candidates, in general, for epilepsy surgery. However, there are other cases when, despite the existence of multiple interictal foci, only one of them is the ictal focus. In such cases, multifocal interictal abnormalities could be responsible for "activating" the seizure focus; thus, they can have a secondary role in seizure onset, without being the primary ictal focus. In other cases, the ictal discharges, as well as the seizure onset, can have a broad-surface EEG representation, involving an extensive area (e.g., anterior quadrant). This makes it difficult to localize the interictal and ictal abnormalities precisely, particularly when the pathological abnormalities involve the entire region of the electrical field. In these situations, the consistent, stereotyped electroclinical features of the seizure activity, which have remained unchanged over a prolonged period, are a reassuring sign of possible unifocal epilepsy. It is also possible that, despite widespread cortical pathology, some of the regions involved are sufficiently damaged to have a lower, rather than higher, cortical excitability. In many situations, these areas show low-amplitude activities and an increase in the delta slowing (Fig. **3-11**).

Epilepsy surgery should be considered in select patients when a single focus is suspected, particularly when medical treatment proves insufficient or when seizures are debilitating or life-threatening. In such cases, phase II invasive video-EEG recording is usually the rule. In certain instances, the patient may not be implanted because of higher risks that are associated with implantation. In these cases, neuroimaging data, such as brain MRI, ictal single photon emission computer tomography (SPECT), and seizure semiology can deliver concordant data that point toward a region most likely involved into the seizure onset (e.g. mesiotemporal sclerosis). Subsequently, the decision of at least palliative, if not curative, resection of this region can be made. In these circumstances, intraoperative ECoG can add valuable information regarding the surgical approach, offering more precise and specific

(a)

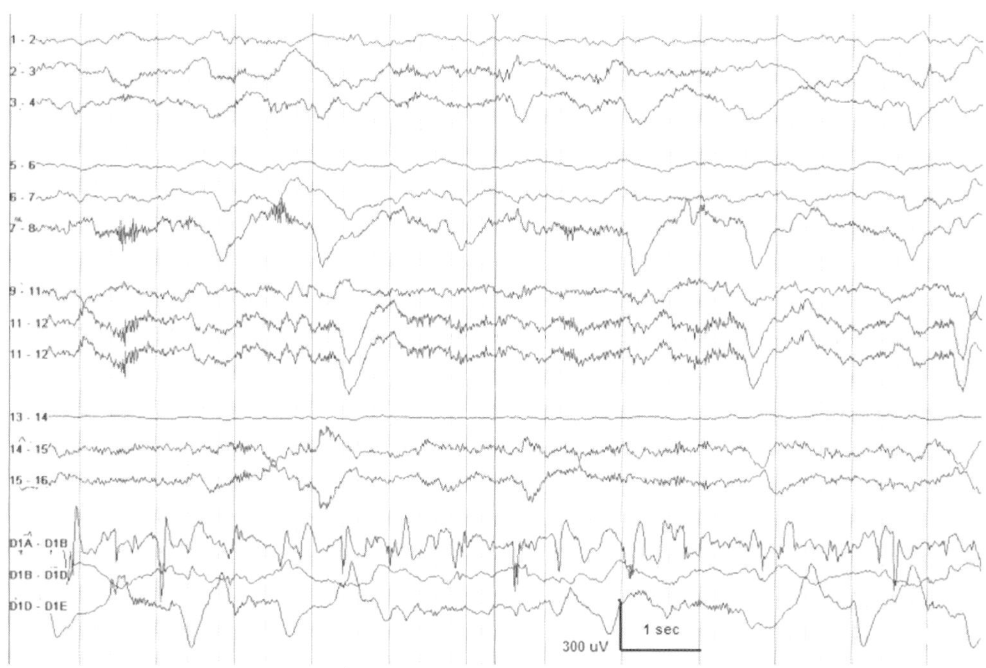


**FIGURE 3-11** **a:** ECoG recording for determining the number of irritative zones. This is the case of a 46-year-old man with history of posttraumatic brain injury and intractable epilepsy. His seizures had similar semiology: they were complex partial, with stereotyped behavior, lip smacking, secondary generalization, and postictal dysphasia. The video surface-EEG monitoring captured six seizures, with broad left fronto-temporal onset and maximal amplitude at T1. They were also frequent interictal sharps present within the same regions, also maximal at T1. The brain MRI showed a markedly damaged left temporal lobe, particularly involving the left temporal tip and mesiotemporal structures, as well as damaged orbito-frontal and frontal regions. Despite the lack of precise localization of the ictal electrographic onset, based on the stereotyped temporal semiology of his seizures and the markedly damaged mesiotemporal structures, the patient was offered a palliative left temporal lobectomy. At the time of the surgery, ECoG recording was performed in an attempt to better localize the irritative zones by assessing the excitability of all injured regions and deciding if additional subpial transections were needed. The ECoG recording showed diffuse slowing but no epileptiform discharges over the injured orbito-frontal, frontal, and temporal tip regions, whereas there were frequent spikes present at the amygdala. The patient underwent left temporal lobectomy and achieved seizure freedom postoperatively. **a:** ECoG recording via Montreal frame, with electrodes placed as shown in slide **b.** A 4-contact depth electrode was placed perpendicular, with contact A reaching the level of the amygdala. The recording is done in a bipolar montage, with the reference electrode placed on the contralateral mastoid. Notice the relative attenuation of the fast frequencies and the increased slowing present over the injured regions, with sharply contoured delta waves at contacts 3, 4, 7, 8, 11, and 12. Also notice the frequent spikes seen at the level of amygdala (contact 1A). The calibration mark is 300 $\mu$V and 1 second.

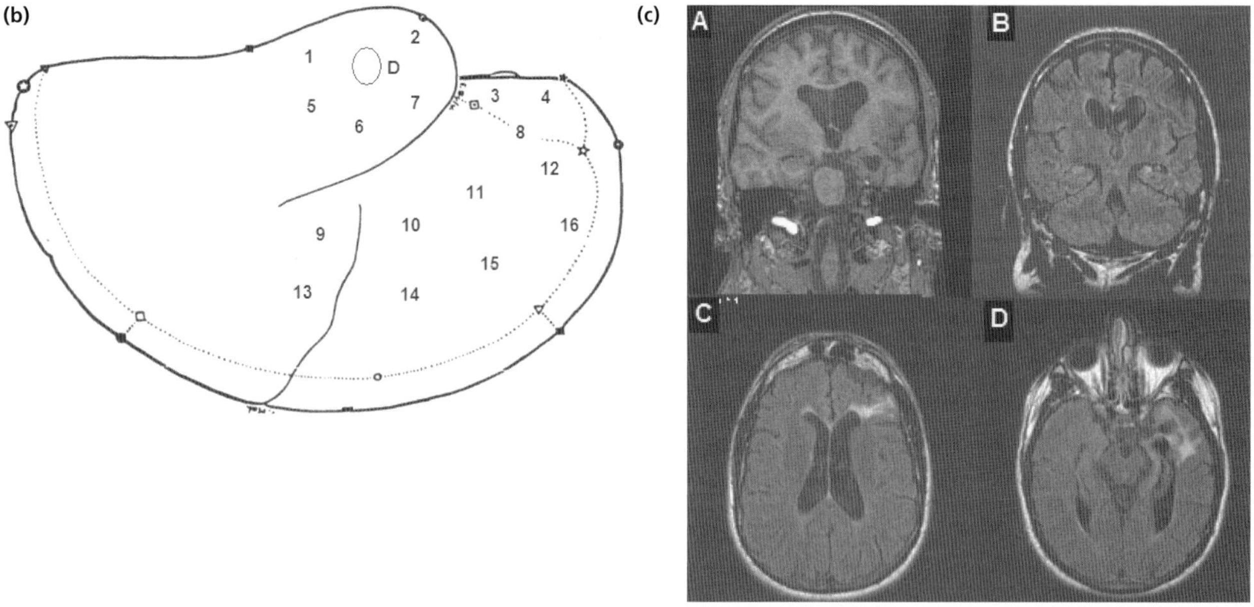


**FIGURE 3-11** *(Continued)* **b:** Representation of the electrode position. The goal was to sample a wide area, especially the injured regions. The location of the depth electrode is circled. **c:** Several brain MRI sequences showing the injured regions. Panel **A** represents a coronal T1 sequence, showing marked atrophy of the left mesiotemporal structures. Panel **B** represents a coronal FLAIR sequence, showing increase in signal of the left mesiotemporal structures. Both findings are consistent with left mesiotemporal sclerosis. Panel **C** represents an axial FLAIR sequence, which reveals a hyperintense left frontal lesion. Panel **D** represents an axial FLAIR sequence, demonstrating a hyperintense lesion at the left temporal tip.

information about the interictal and ictal activities. These include information on the number of interictal foci; presence of their independent activities, if any; whether the activity of one focus has a consistent lead-in; as well as the behavior of different interictal foci to seizure-inducing methods (see pages 121–24). Thus, they can guide the extent of resection, and/or its combination with or replacement by MSTs, for limiting the spread and organization of interictal activity.

## 2. Indications for Concomitant Use of ECoG and Functional Mapping

The use of both ECoG and functional mapping may be necessary in patients with parietal, frontal, or temporal epilepsy, when functional cortical mapping is used to determine whether the location of epileptogenic tissue overlaps with important functional cortex, including sensorimotor or language regions (Keles et al., 2004). A particular situation occurs in the case of neurodevelopmental lesions. Even with well-defined neurodevelopmental lesions, plasticity-induced cortical reorganization could result in the functional cortical regions having unexpected locations, which may allow more or less extensive resection than was initially anticipated. Even if the targeted epileptogenic lesion seems to be located at reasonable distance from the eloquent cortex, in many cases of neurodevelopmental disorders, the development of an entire region, lobe, or hemisphere may be abnormal. Thus, the location of eloquent cortex in these situations may be unpredictable (Fig. **3-3**). Consequently, ECoG should be used in combination with functional mapping, while mapping seizure foci in patients with neurodevelopmental lesions.

However, the use of both methods during the surgery is not done simultaneously, but rather in succession, because of the requirements of different anesthetic regimens. We prefer to perform first the ECoG recording and then to continue with functional mapping. For further details, see page 126.

Recently, preoperative functional imaging with fMRI and/or PET or MEG (Toga et al., 2002) has become more common for mapping cortical function prior to any planned resection. Cortical regions, adjacent to a lesion planned for resection, that are believed to play a role in eloquent function(s) are often identified preoperatively using these functional imaging techniques. There remains, however, some legitimate concern regarding the exact neurophysiologic correlate of increased signal on function imaging maps. In addition, electrophysiologic recordings have been shown to improve intraoperative targeting, even in patients undergoing preoperative functional imaging (Pirotte et al., 2005).

## V. METHODOLOGY

The technique of ECoG typically involves the use of several recording electrodes, placed either in light contact with the cortex, exposed during a craniotomy, or located in the vicinity of deeper structures, situated underneath the visualized cortex.

Historically, rigid electrodes with ball-shaped tips were used to achieve contact between the electrode and brain tissue. These systems, known as Medussa or Montreal frame (Fig. **3-12**), require individual placement of each of the 16–20 electrodes over the cortical region, while their other end is connected to a wooden frame at the craniotomy site. The Montreal frame has the advantage of allowing adjustment of the distance between the contacts, when the presumed area of interest proves to be smaller or larger than initially appreciated (Fig. **3-2**). It also allows visualization of and access to the cortex in between the contacts. This becomes important in cases of concomitant functional mapping.

However, because the subdural recording electrodes used for ECoG recordings must be able to make light contact with the cortex that is exposed by the craniotomy, flexible electrode packages are preferred over rigid electrodes. More recently, flexible plastic grids and strips of electrodes have become widely available for intracranial electrophysiology. These grids of disc electrodes, implanted within thin clear plastic sheets, such as those used as subdural electrodes, have become very popular. A grid of electrodes can be quickly and easily placed over a large region of exposed cortex and easily moved or adjusted if needed. Care must be taken, however, that the grid or strip does not unknowingly shift or move during ECoG recordings. With the availability of these plastic grids full of contacts, any number of electrodes can be used: from a strip of electrodes up to a complete grid, including several rows of contacts.

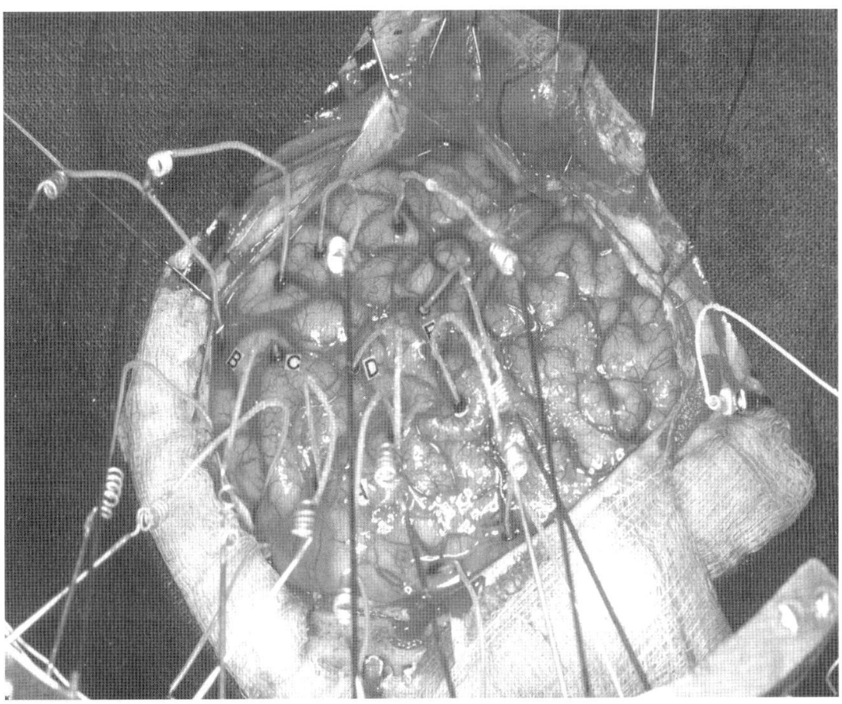


**FIGURE 3-12** Medussa or Montreal frame consists of 16 electrodes with ball-shaped tips, placed by the surgeon in direct contact with the cortical surface. The opposite ends are connected to a wooden frame fixed by the surgeon at the craniotomy site.

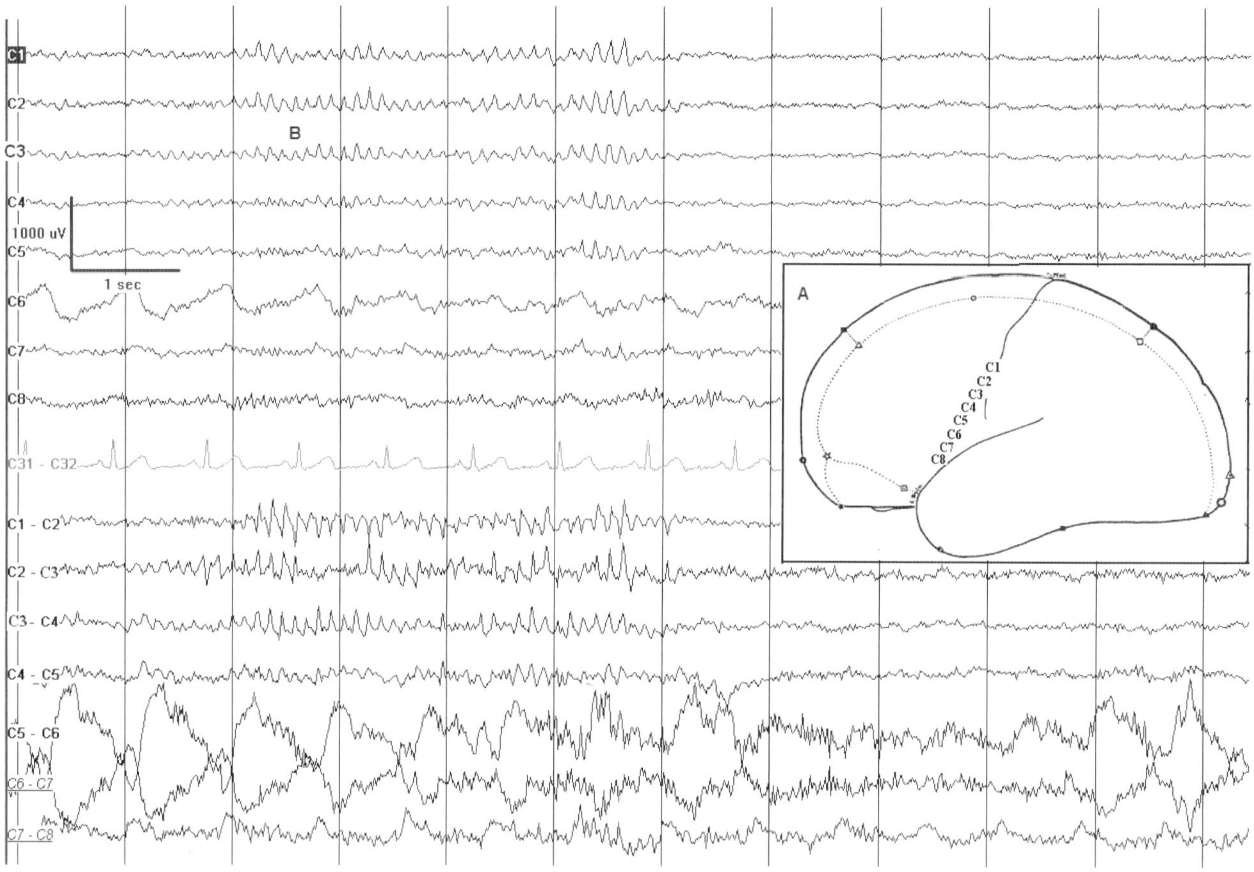


**FIGURE 3-14** ECoG recording showing a mu rhythm that needs to be differentiated from epileptiform activity. The ECoG was recorded via an 8-contact strip during awake left fronto-temporal craniotomy for epilepsy surgery. Panel **A** shows placement of the 8-contact strip, labeled C1, C2, C3, C4, C5, C6, C7, and C8. Notice the location of contacts C1 through C5, overlying the motor strip. Panel **B** shows the ECoG recording via the strip described previously, in a bipolar and referential montage, with the reference electrode placed on the contralateral mastoid. Notice the 8–9 Hz, 200–300 μV rhythmic sharp activity present at contacts C1 through C5, representing a mu rhythm. The patient was asked to open and close her right fist, which led to disappearance of this activity. Notice its sharp morphology, with the phase reversal seen at contact C2. The calibration mark is 1000 μV and 1 second.

frequencies, but others lower the frequencies. On the other hand, areas with preexisting lesions (e.g., from trauma, old infarct, tumor) show overall low amplitudes (attenuation) and low-frequency activities (slowing). (Fig. **3-15**). The irritative or epileptic zones, in such cases, may be not within the damaged tissue but located elsewhere. We have encountered this situation in cases of dual pathology, in particular when the mesiotemporal structures become involved, sometimes years after the initial injury (Fig. **3-11**). Similarly, in cases of tumor resections, the epileptiform activities have been seen adjacent or at distance from the actual lesions (Figs. **3-6, 3-15 a**). With depth electrode recordings, the presence of delta slowing seen at the deepest contacts usually suggests recording from white matter.

The interpretation of ECoG recordings in the anesthetized patient is challenged by the effects of anesthetics on cortical activities. Depending on their type and their concentration, the ECoG background may show different patterns. The main goal is to avoid a burst-suppression pattern, which is not adequate for mapping epileptiform and epileptic foci. The bursts usually contain high-amplitude, very sharp activity that can be easily mistaken as epileptiform activity, such as spikes and polyspikes (see Fig. **3-16 A, B**; and Fig. **4-14**, pages 163–64). On the other hand, the flat epochs offer no activity to analyze. Thus, in order to analyze the ECoG accurately, a continuous background should first be obtained. We usually achieve this by decreasing the propofol infusion rate, or turning it off entirely, while adding another anesthetic.

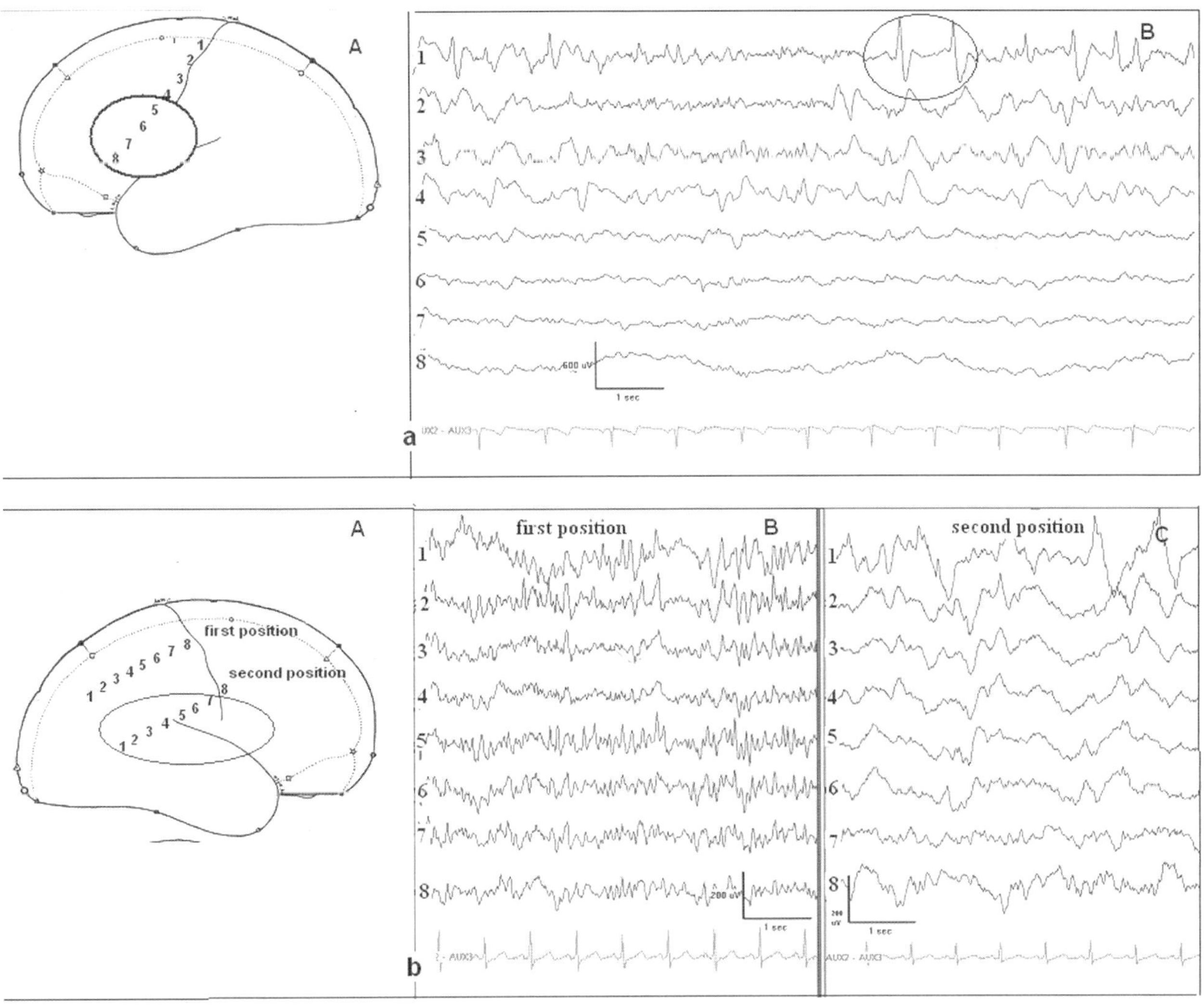


**FIGURE 3-15** ECoG recordings showing marked differences in the cortical activities between the regions invaded by tumors and distant, unaffected areas. **a:** ECoG recording during a left inferior frontal glioma resection. The patient is a 58-year-old woman who presented with intermittent confusion and language disturbance. Panel **A** shows the approximate placement of the 8-contact strip in relationship to the tumor location (circled). Notice that contacts 1 through 4 are situated more superior on the left hemispheric convexity, whereas contacts 5 though 8 are overlying the diseased cortex. Panel **B** shows a referential ECoG recording (reference electrode on the contralateral mastoid). There is a significant attenuation of the overall amplitudes at contacts 5 through 8, most likely directly related to the presence of the tumor. The activities recorded at contacts away from the tumor have overall higher amplitudes. At the most distant contact (1), several high-amplitude, spikes are present (circle). The calibration mark is 600 $\mu$V and 1 second. **b:** ECoG recordings from a 73-year-old woman presenting with a high-grade glioma involving the right perisylvian region and temporal lobe. Panel **A** shows the placement of the 8-contact strip. The first position overlies an area that is not affected by the tumor. The second position is overlying the perisylvian region, significantly affected by the presence of the tumor. Panel **B** shows the ECoG background during the first strip position. Notice the presence of diffuse theta, alpha, and beta frequencies. Panel **C** shows the ECoG background during the second strip position. Notice an overall attenuation to disappearance of fast activities and the exacerbation of the delta slowing. The recording is done in a referential montage, with the reference electrode on the contralateral mastoid. The calibration marks in both panels are 200 $\mu$V and 1 second, respectively.

An alternative is to switch from propofol and remifentanyl to a combination of nitrous oxide and a halogenated agent, such as isoflurane.

Even with a continuous background, one has to be aware that many anesthetics, particularly at low doses, cause an abundance of fast-frequencies activities, which may have a sharp contour (Fig. **3-16 C**). These fast activities need to be differentiated from "buzzes" of epileptiform or epileptic activity, which are typical of certain structures such as hippocampus or cortical dysplasia. One useful hint is that anesthetic-induced fast activities tend to be diffuse and continuous and they have no evolution, whereas the buzzes of abnormal activity are present within the epileptogenic zones and are short lived, or they further evolve and organize. The cortical depressive effect of anesthetics causes an overall attenuation of the activities. Although this is a problem during scalp recordings (see pages 50–53 and 69–70), it should not affect the interpretation of an ECoG recording that is done directly from the cortical surface.

We use propofol and dexmedetomidine, at sedative doses, with ECoG recordings during awake craniotomies. Both these medications can be stopped 15–20 minutes before the ECoG recordings.

For further details regarding anesthetic effects on ECoG recordings, see pages 124–27.

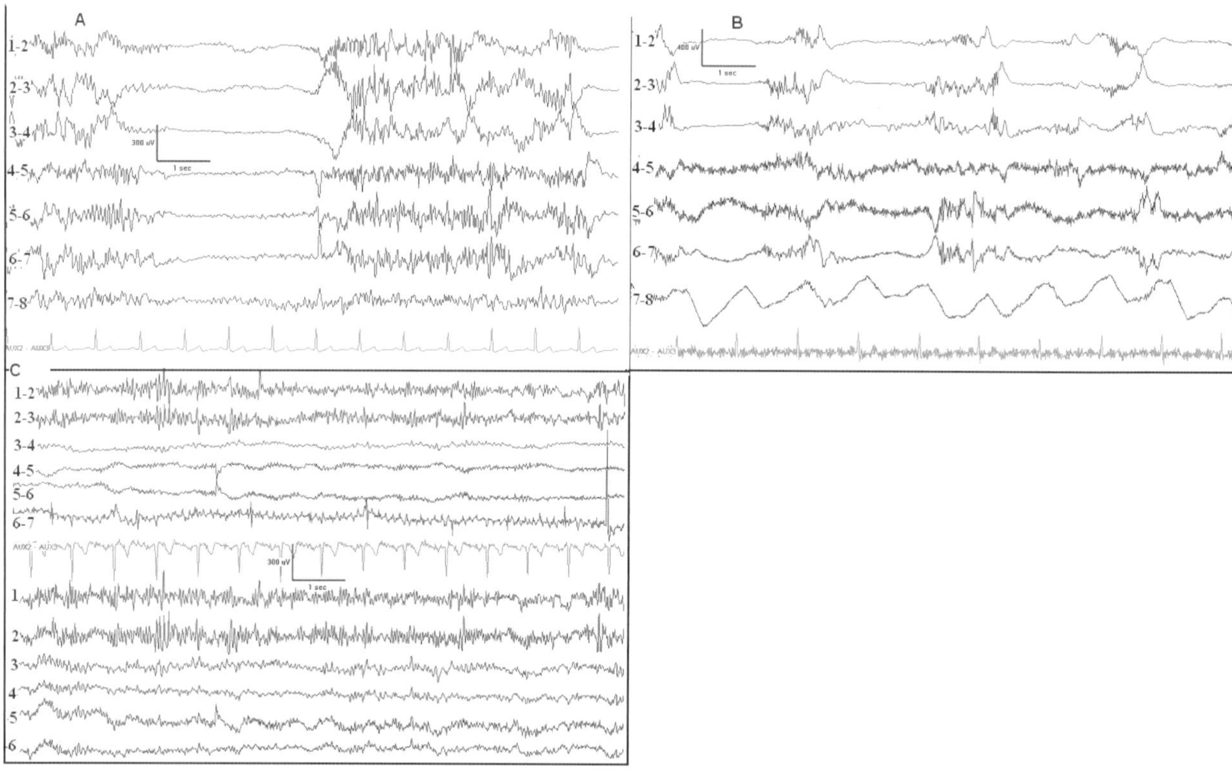


**FIGURE 3-16** ECoG recordings showing normal, anesthetic-induced background activities that could be misinterpreted as epileptiform/epileptic activities. The recording in all panels is done via an 8-contact strip, with the reference electrode on the contralateral mastoid. Panel **A** demonstrates a burst-suppression pattern in a 55-year-old man, during infusion with propofol 70 mcg/kg/min and remifentanyl 0.5 mcg/kg/min. The bursts contain sharp, 50 to 300 $\mu$V beta activity intermixed with slower frequencies. The calibration mark is 300 $\mu$V and 1 second. Panel **B** is also showing a burst-suppression pattern that is different from the one described in panel **A**. The recording is done in a 55-year-old woman, during an infusion of propofol 65 mcg/kg/min and remifentanyl at 0.12 mcg/kg/min. This time the flats and bursts are almost equal in duration (approx. 1 second). Again, notice the very sharp morphology of the bursts, resembling polyspikes. The calibration mark is 400 $\mu$V and 1 second. Panel **C** shows ECoG recording in an awake patient, after propofol and remifentanyl were stopped. Notice a significant amount of 50 to 300 $\mu$V sharp beta activities. In all of the panels, it is very difficult to distinguish reliably real epileptiform activity from normal sharp fast activities, particularly within the bursts. The recording in all panels is done via an 8-contact strip, with the reference electrode placed on the contralateral mastoid. The calibration mark is 300 $\mu$V and 1 second.

## 2. Epileptiform and Epileptic Activity

### *A. Morphology*

Observing the morphology of the interictal activity can provide precious information about the distance of the recording from a focus. Direct cortical recording from an irritative zone usually shows brief epileptiform discharges (spikes, polyspikes) rather than sharps, with higher rather than lower amplitudes. A blunt, small-amplitude AD should raise the following questions: (1) Is the AD arising from a damaged cortical area, exhibiting an overall attenuation and slowing (see pages 117–20)? (2) Are its shape and amplitude attributable to an anesthetic effect? (3) Is it arising from a region at a distance from the recording electrode?

The type of discharge seen also depends on the tissue of origin. For example, cortical dysplasias are notorious for very focal buzzes of low-amplitude, fast activity, or very frequent spikes that organize quickly in rhythmic alpha or theta runs (Fig. **3-9 c**). Continuous spiking has been found to be highly associated with cortical dysplasias (Ferrier et al., 2006). As a matter of fact, the absence of abundant interictal activity, during recording from a presumably cortical dysplastic lesion, should raise questions about the etiology of the lesion or its correct localization (Fig. **3-8**). Similarly, mesiotemporal structures usually generate focal, low amplitude, fast-frequency buzzes, mostly at seizure onset. Alternatively, high amplitude spikes and polyspikes are also frequent.

As already mentioned on pages 112–14, the presence of multifocal spikes, with or without similar morphology, usually signals widespread pathology, such as in the case of posttraumatic or postinfectious epilepsies or in the case of extensive neurodevelopmental lesions (Fig. **3-3**). Although, in general, multifocality is a warning sign for the success of surgery, there are situations where presence of typical, highly stereotyped seizures is encouraging. In such cases, capturing either a spontaneous or an induced typical ictal event is invaluable in determining which of the interictal foci is also the ictal one. However, in cases when several interictal foci are present, it may be impossible to achieve complete resection of the abnormal tissue. These are in general good indications of concomitant functional mapping, as well as consideration of MSTs under ECoG guidance. This approach prevents the propagation and organization of the multifocal epileptiform discharges, while preserving the eloquent cortex.

In addition to presence of the typical epileptiform activity of sharp-waves or spikes, the presence of other slow waves, such as delta waves, may have some value in seizure localization in lesional epilepsy (Gloor et al., 1977; Schaul et al., 1981). There have been reports of increased delta wave activity in TLE as well (Panet-Raymond et al., 1990).

### *B. Frequency*

There are few aspects regarding the frequency of interictal epileptiform discharges that we consider worth mentioning. First, their frequency is not necessarily correlated with their tendency to organize in a habitual ictal event. We have encountered many cases when the "quieter" focus became the ictal focus (Fig. **3-7**). Second, important information is gained in the EMU by observing the evolution in frequency and patterns of different interictal activities as the AEDs are tapered down over the course of several days. This information is further available in patients who are undergoing additional intraoperative ECoG. Thus, during the intraoperative recordings, special attention is paid to those interictal discharges that show the most dynamism and evolution during the EMU session, as the medication adjustment takes place. Third, continuous, high-frequency spiking is a characteristic sign for cortical dysplasias.

### *C. Synchrony*

In the case of multifocal interictal discharges, one of the most valuable information is given by the existent lag, if any, in their onset. Synchronous discharges do not help, but those which consistently show a lead-in, when compared to the others, may point toward the area of interest (Alarcon et al., 1997).

### *D. Ictal Activity*

In general it is rare to capture a spontaneous ictal event during an ECoG recording. However, if this is the case, analysis of the seizure onset, as done in EMU, should be the goal. The ECoG analysis poses its own challenges, because the time is limited and the surgeon requires a prompt answer. We usually use two windows: one for the life recording and the other for reviewing. If the patient is awake, it is important to correlate the timing of the electrographic with the clinical onsets. If no spontaneous seizure is seen, we recommend trying to induce a habitual ictal event.

### *E. Methods for Inducing an Ictal Event*

Whereas such methods can be useful in any ECoG recording done for the purpose of localization of seizure focus, there are several instances when they are particularly helpful, because no reliable localizing information is delivered by existing interictal activity. First, these methods can help identify regions of abnormally increased cortical excitability, which show normal electrical activity at the time of the recording. Second, they can be used to better define the already existing interictal activity, particularly in situations when the morphology, amplitude, and frequency of the latter have been greatly affected by the cortically depressive effects of different anesthetics (Fig. **3-2**). Their third utility is in the cases with multifocal epileptiform discharges. In these situations, an activating method could help identify those discharges that are more prone to further organize and propagate. Interestingly enough, the seizure focus may not be the one displaying the most frequent interictal discharges before the method is administered (Fig. **3-7**).

For the purpose of triggering a run of epileptiform discharges, an electrographic seizure or, at best, an electroclinical event, an epileptogenic drug, or direct electrical cortical stimulation can be used. As mentioned previously, during intraoperative interictal recordings, the irritative zones may exhibit abnormal epileptiform activity. Alternatively, some of these structures may show normal electrical activity and thus may be difficult to recognize. However, these epileptogenic structures may have disproportionately large and rhythmic responses to low-intensity electrical stimulation, and thus the latter may help in delineating them, by bringing up abnormal activity.

This increased excitability results in the production of large field potentials after the cessation of electrical stimulation (ADs). More so, these ADs can further organize into an electrographic or electroclinical seizure. As previously mentioned, the specific role of epileptiform activity, including ADs, in the exact localization of the ictal generator or epileptic tissue has been the subject of much debate (Bernier et al., 1990; Gloor, 1975; Kuruvilla et al., 2003; Wennberg et al., 1998).

Therefore, ECoG recordings are most frequently limited to the spontaneous interictal electrophysiology, such as spikes and sharp-waves, along with the stimulation-induced ADs. As in the case of spontaneous interictal discharges, the specific role of such stimulation-triggered epileptiform discharges, in the exact localization of the ictal generator or epileptic tissue, has been controversial (see page 97).

However, when stimulation is done for long enough epochs, at sufficiently high stimulus amplitude, these ADs can further organize into an electrographic or electroclinical seizure (see pages 164–66). In an ideal situation, the electrical stimulation in the region of epileptogenic tissue should simultaneously generate a run of ADs or stimulation-induced electrographic seizure producing the patient's typical aura, early ictal manifestations, or even the patient's characteristic seizure phenotype. In the presence of this ideal ECoG finding, it could be rather convincing that the epileptic zone has been identified. Even though it is difficult to reproduce the patient's typical seizure semiology with electrical stimulation, this situation was achieved in several centers (e.g. Chassagnon et al., 2008), including ours (Fig. **3-1**). As previously mentioned, several authors have found a good correlation between the localization of such triggered seizures and that of spontaneous ictal events (Bernier et al., 1990), although others have not (Blume et al., 2004). For further details regarding stimulation-triggered epileptiform and epileptic activity, see pages 157–73.

Another method for inducing an ictal event is via intravenous administration of an epileptogenic drug such as alfentanyl or methohexital. For these purposes our preference has been to use alfentanyl. We have used this agent frequently and thus have more substantial experience with its effects than with other drugs. On the other hand, we think that alfentanyl selectively increases the cortical excitability, particularly of the irritative zones and not so much of the healthy tissue. Thus, the focality of such induced epileptiform or epileptic activity is still preserved. We usually administer 30 mcg/kg intravenous bolus and expect to see an effect within the next few minutes. If necessary, two additional similar boluses can be given. The use of alfentanyl as a provoking method has been extensively described in the literature, as has its epileptogenic effect (Manninen et al., 1999; Ross et al., 2001). We have benefited from its use during ECoG recordings in many situations (Fig. **3-2 c, d;** and Fig. **3-7**). In those cases where the administration of alfentanyl does not make a difference, subsequent administration of an intravenous bolus of methohexital, which has a strong epileptogenic effect, can be helpful. In concordance with the findings of other authors (Fiol et al., 1990), we think that methohexital is less selective than alfentanyl and causes an overall increase in excitability and multifocal epileptiform discharges (Fig. **3-17**).

Ketamine, etomidate, and propofol, at lower doses, and remifentanyl, at higher doses, have been described to increase the cortical excitability. However, we do not have adequate experience with their use for these purposes. For further details regarding the cortical effects of these drugs, see pages 124–27.

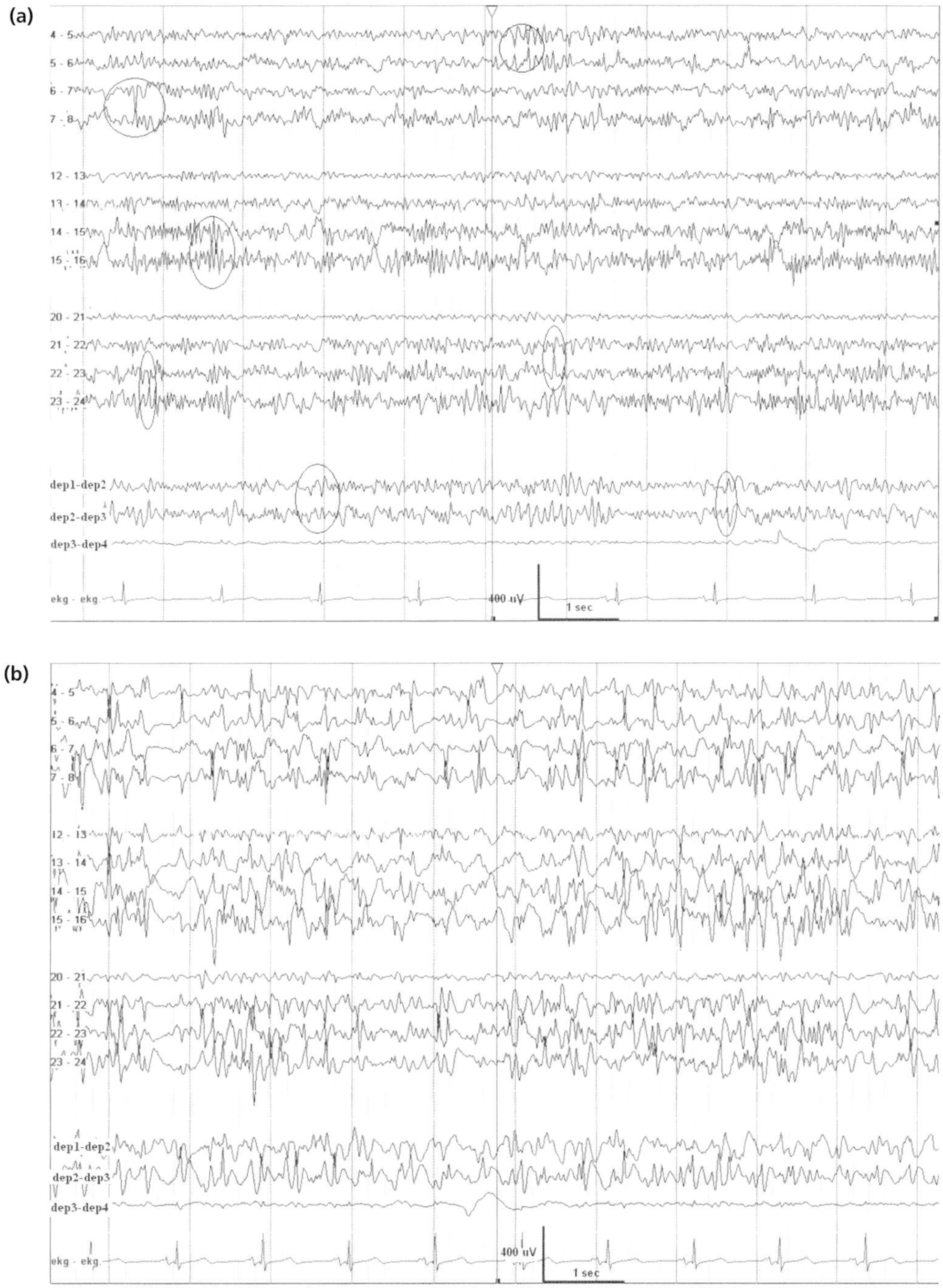


**FIGURE 3-17** The epileptogenic effect of methohexital. Patient undergoing right frontal lesionectomy under ECoG guidance. The recording is done via three 5-contact strips (labeled 4 through 8; 12 through 16; 20 through 24), placed parallel to each other within anterior-posterior axis, overlying the frontal lobe. A 4-contact (labeled dep1 through dep4) depth electrode is placed perpendicular to the cortical surface, through the lesion. Calibration marks in both slides are 400 $\mu$V and 1 second. **a:** ECoG recording in a bipolar montage (reference on the contralateral mastoid) before administration of methohexital. Notice that the background is relatively quiet. A few independent spikes are possibly present, with phase reversal at grid contacts 5, 7, 15, 22, and 23, as well as at contact 3 on the depth electrode (circled). However, there is a significant amount of anesthetic-induced, sharp fast activities, which makes the identification of epileptiform discharges difficult. **b:** ECoG recording 2 minutes following intravenous administration of methohexital. Notice a marked increase in the multifocal epileptiform discharges. They are seen up to 200 $\mu$V, at irregular intervals, independently, with phase reversal at multiple contacts, including 5, 7, 13, 14, 15, 22, and 23, as well as contact 3 on the depth electrode. The recording is showing a relatively diffuse rather than focal increase in the cortical excitability.

Using such inducing methods can occasionally trigger a significant electroclinical seizure. In the case of awake craniotomies, this can be in particular challenging, especially when dealing with a generalized tonic clonic seizure (see pages 132–33). Ice-cold saline irrigation of the cortical surface can be in particular helpful in aborting the seizure activity (see pages 170–73).

## 3. Differentiating Brain Activity from Artifact

There are several circumstances in which differentiating artifact from brain activity becomes challenging.

### *A. Pulsation Artifact*

This type of artifact is frequently seen in intraoperative ECoG recordings and may result from close proximity of the recording contacts to a pulsatile vessel, from pulsatile movements of the brain itself, or rhythmic movements of the wires connected to the electrodes, translating in ballistogram artifact. It is represented by rhythmic 1–2 Hz delta activity, which can be falsely interpreted as rhythmic slowing, or even as ictal activity. However, this activity is time locked to the QRS complexes and sometimes T waves (see also pages 166–67 and Fig. **4-17**, page 170). In order to determine this relationship, dragging the electrocardiogram (EKG) channel next to the channel displaying this artifact can be helpful. If the contacts showing the artifact are situated on a pulsatile cortical vessel, repositioning the recording strip or grid helps. In such situation, the low signal-to-noise ratio is determined not only by an increased amount of pulsation artifact but also by decreased signal.

### *B. EKG Artifact*

EKG artifact can easily be taken as brief, rhythmic spikes though the synchrony between the QRS complexes, and the suspected discharges can be a key in making this distinction (Fig. **3-18**). There can be situations in which the epileptiform discharges occur at similar frequency with the QRS complexes but are not time-locked to them (see Fig. **4-12**, page 163). In such situations, irrigation with cold saline water may help in the differentiation, because this is expected to abort the epileptiform activity.

### *C. Electrical Artifact*

A 60-Hz artifact, present intermittently, can easily be taken as "buzz" of abnormal, epileptiform activity, especially when it is located at the same contacts with known epileptiform activity. The absence of a reasonable field or the presence of similar activity in nonscalp channels, such as EKG channel, is helpful in determining its artifactual nature (see Fig. **4-10 b**, page 160). In other situations, particularly during cortical stimulation using Penfield method, applying the 60-Hz Notch can be helpful. At other times, virtual channels can show runs similar in frequency with epileptic runs. Again, the lack of a reasonable electrical field, as well as the monotonous appearance, without further evolution helps in establishing its artifactual nature (see Fig. **4-15 b**, page 167). In the case of significant ambient electrical artifact that interferes with the ECoG interpretation, it becomes important to identify the source of artifact, usually an electrical machine situated in its close proximity. The source of artifact should be unplugged for the time of the ECoG recording if possible (e.g. the surgeon's headlights, the bed, the venodyne boots). In order to reduce the ambient noise, it is also helpful to isolate the recording cables and wires by grouping them together and covering them with an isolating pad.

### *D. Electrode Artifact*

An unstable, loose contact can introduce a large amount of artifact, mimicking epileptiform discharges in the recording. It can also display disproportionate amplitudes of all activities, compared to the similar activities recorded at adjacent contacts (see Fig. **4-17 b**, page 170). Thus, we believe that it is essential to check and, if necessary, improve the impedances of the recording electrodes (see pages 115–17).

## VII. ANESTHESIA CONSIDERATIONS

In order to perform ECoG properly, drugs that alter or inhibit cortical neuronal function for any substantial amount of time must not be used during the surgery. Anesthetics, especially general anesthesia, have a profound effect on cortical activity and thereby on ECoG recordings, which can show a disappointing lack of activity in the area of exploration because of the effect of general anesthesia (Quesney et al., 1993; Sloan, 1998). It is important, therefore, that the pharmacological agents used during the ECoG recording phase of the procedure have minimal depressant effects on the electrical activity of the brain. We aim for a continuous background rather than a burst-suppression pattern,

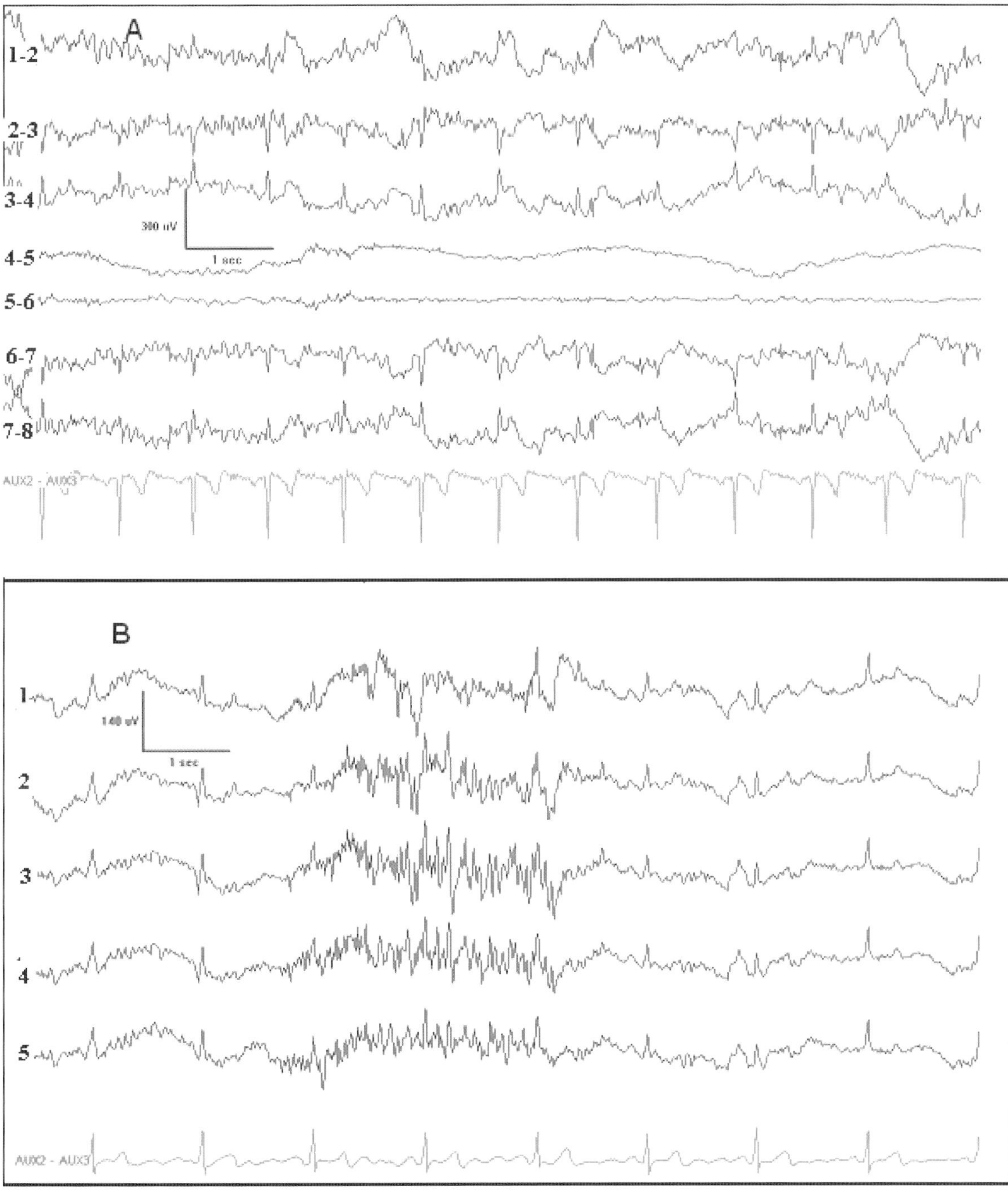


**FIGURE 3-18** Two ECoG recordings showing rhythmic EKG artifact. Panel **A** shows a recording via an 8-contact strip, in a bipolar montage, with reference electrode on the contralateral mastoid. Notice that the rhythmic EKG artifact, time-locked to the QRS complexes could be taken as rhythmic sharps at about 1 Hz, with phase reversal at contacts 3 and 7. The calibration mark is 300 $\mu$V and 1 second. Panel **B** shows a recording via a 5-contact strip, in a referential montage, with reference electrode placed on the contralateral mastoid. Notice the morphology and amplitudes of the artifact, equipotential in all channels and time-locked to the QRS complexes. The artifact is more evident during the flats in the burst-suppression pattern. The calibration mark is 140 $\mu$V and 1 second.

which would not be appropriate for ECoG analysis (see pages 117–20). The halogenated inhalational anesthetics, such as nitrous oxide, halothane, enflurane, isoflurane, desflurane, and sevoflurane, which are typically used for general anesthetic, produce a significant dose-related reduction in amplitude and frequency of cerebral activity (McDonald et al., 2000). However, studies have shown that the concomitant use of nitrous oxide with isoflurane attenuates the depth of an EEG burst-suppression pattern. It is also our experience that using a combination of nitrous oxide and halogenated agent facilitates the recording of a continuous background, with an acceptable suppression of the cortical activities, which does not have a major impact on the reliability of the information delivered by our recordings.

On the other hand, the use of large doses of propofol during these procedures invariably causes a certain degree of burst suppression and hinders the ECoG results. Thus, we do not recommend the use of propofol as the cornerstone for TIVA during ECoG for mapping the epileptic foci. However, its use in combination with other anesthetics allows titration of the propofol concentration to a level where the ECoG recordings display a continuous background. Moreover, the use of propofol, at low doses, may help in identifying the irritative zones by increasing their cortical excitability (Hewitt et al., 1999; Smith et al., 1996).

The awake craniotomy technique is important for intraoperative ECoG recordings. This technique requires a skilled neuroanesthesiologist and a good communication within the operative team. In most cases, a sleep-awake-sleep method is used by administering short-acting anesthetics and discontinuing these agents 10–15 minutes before ECoG recordings. In the past decade, propofol has become a popular agent for sedation for these procedures. Although propofol does have a dose-dependent suppressive effect on EEG, the rapid metabolism of this agent makes it a good sedative for awake craniotomies, without further effect on the ECoG recordings (Soriano et al., 2000). In addition, preliminary reports on a new sedative called dexmedetomidine suggest that this sedative may be also very useful for awake craniotomies and appears to have little effect on ECoG recordings at low doses (Souter et al., 2007). Thus, these agents can be used as for sedation during the initial and terminal stages of the operation without significantly affecting the intraoperative ECoG recordings. However, due to limited data, we do recommend caution when dexmedetomidine is cited in combination with other anesthetics (see page 332).

A special situation remains the concomitant use of ECoG and functional mapping. As already mentioned, the two procedures are done successively, not simultaneously, during the surgical procedure. As mentioned previously, during the ECoG phase, we prefer the use of inhalational agents or, in case of awake craniotomy, of sedative medication, such as dexmedetomidine. However, once the irritative zones are identified and we proceed with functional mapping, we request a change in the anesthetic regimen from inhalational agents, known to affect significantly the generation of sensori and motor-evoked potentials, to TIVA (propofol and remifentanyl). As mentioned on page 96, this transition require a good coordination between the neurology and anesthesia teams.

Analgesia in cases requiring ECoG recording is largely accomplished by local anesthesia, placed into the scalp prior to head fixation and incision. In addition, most narcotics have little influence on ECoG spiking activity (Rampil, 1997). The most notable exception to this is alfentanyl (Cascino et al., 1993). This is considered by several neurophysiologists (including us), as the drug of choice for inducing an electrographic seizure during ECoG recordings (see pages 121–24). At high doses, intravenous narcotics are also known to increase the epileptiform activity seen during epilepsy surgery and may thereby facilitate the localization of the epileptogenic zone.

A newer narcotic is remifentanyl, an ultrashort-acting opiod with a half-life of less than 10 minutes, which may also facilitate neurologic testing immediately after the resection or operation. In higher doses (2 mcg/kg/hr), this has been shown to increase significantly the frequency of spikes in the epileptogenic zone while suppressing the activity in the surrounding normal brain tissue (Wass et al., 2001). In lower doses (0.6 mcg/kg/hr), remifentanyl was shown to have a minimal effect on the ECoG spike activity in awake patients (Herrick et al., 2002).

Other agents have been used as activating agents during ECoG recordings. Low-dose methohexital has been used in the past to enhance epileptiform activity during ECoG recordings as part of a surgery to remove seizure foci (Rampil, 1997). Methohexital is an ultrashort-acting barbiturate anesthetic, which is known to activate epileptiform activity only in epileptogenic brain tissue (Hufnagel et al., 1992; Modica et al., 1990; Morrell, 1985; Musella et al., 1971; Wennberg et al., 1997; Wyler et al., 1987). As already specified on pages 121–24, our experience with the use of methohexital as an inducing agent during ECoG recordings shows that it obscures the specificity of the seizure focus at interest because of its high potential to increase the cortical excitability in both abnormal and normal tissue. Etomidate and ketamine have a cortical excitatory effect, by enhancing the synaptic transmission. Thus, at low doses, they may induce seizures in the epileptic patient.

For more details regarding the effects of different anesthetics on cortical activities and thus on the ECoG recordings, see Chapter **8** (pages 327–36).

## VIII. CONCLUDING REMARKS

Intraoperative ECoG is a widely used and well-established tool for the surgical treatment of intractable epilepsies. This technique provides real-time information regarding the specific locations of epileptiform activity that may be associated with the seizure foci or epileptogenic zone. There are differing opinions in the literature regarding the usefulness of cortically recorded interictal spiking in TLE, especially when a standard anterior temporal lobectomy is the planned surgical treatment. Most experts, however, agree that ECoG is very useful in planning surgical resections in patients with epilepsies other than typical medial TLE, such as frontal epilepsy or seizure syndromes caused by cortical dysplasia or lesions without clearly visualized anatomic landmarks, multifocal epilepsies, or in determining the extent of epileptogenic brain tissue to be removed in the surgical treatment of epilepsy. In addition, cortical subpial transections done to disconnect seizure prone networks are typically performed under ECoG guidance.

We believe the surgical treatment of epilepsy requires a multidisciplinary approach, and thus the additional information gathered by ECoG recordings may have a significant role in its success.

## REFERENCES


Ajmone Marsan C. Depth electrography and electrocorticography. In: Electrodiagnosis in Clinical Neurology (Aminoff MJ, Ed.) 1980; pp 167–196. New York: Churchill Livingstone.

Ajmone-Marsan C. Focal electrical stimulation. In: Experimental Models of Epilepsy (Purpura DP, Penry JK, Tower DD, et al., Eds.); pp 147–172. New York: Raven Press.

Ajmone Marsan C, Baldwin M. Electrocorticography. In: Temporal Lobe Epilepsy (Baldwin M, Bailey P, Eds.) 1958; pp 368–395. Springfield, IL: Charles C Thomas.

Ajmone Marsan C, O'Connor M. Electrocorticography. In: Handbook of Electroencephalography and Clinical Neurophysiology (Remond A, Ed.) Vol 10C, 1973; pp 3–49. Amsterdam: Elsevier.

Alarcon G, Garcia Seoane JJ, Binnie CD, et al. Origin and propagation of interictal discharges in the acute electrocorticogram: Implications for pathophysiology and surgical treatment of temporal lobe epilepsy. Brain 1997; 120(Pt 12):2259–2282.

Awad IA, Rosenfeld J, Ahl J, et al. Intractable epilepsy and structural lesions of the brain: Mapping, resection strategies, and seizure outcome. 1991; Epilepsia 32:179–186.

Bancaud J, Talairach J, Morel P, et al. Generalized epileptic seizures elicited by electrical stimulation of the frontal lobe in man. Electroencephalogr Clin Neurophysiol 1965; 37:275–282.

Bates JAV. Special investigation techniques: Indwelling electrodes and electrocorticography. In: Electroencephalography: A Symposium on Its Various Aspects (Hill H, Parr G, Eds.) 1963; pp 429–470. New York: Macmillan.

Bengzon ARA, Rasmussen T, Gloor P, et al. Prognostic factors in surgical treatment of temporal lobe epilepsy. Neurology 1968; 18:717–731.

Berger H. Über das Elektrenkephalogramm des Menschen. Arch Psychiatr Nervenkr 1929; 87:527–570. (Translated by Gloor P. Electroenceph Clin Neurophysiol Suppl 1969; 28:37–73.)

Bernier GP, Richer F, Giard, N. Electrical stimulation of the human brain in epilepsy. Epilepsia 1990; 31:513–520.

Blume WT, Jones DC, Pathak P. Properties of after-discharges from cortical electrical stimulation in focal epilepsies. Clin Neurophys 2004; 115:982–989.

Cascino GD, Hulihan JF, Sharbrough FW, Kelly PJ. Parietal lobe lesional epilepsy: Electroclinical correlation and operative outcome. Epilepsia 1993; 34:522–527.

Cascino GD, Kelly PJ, Sharbrough FW, et al. Long-term follow-up of stereotactic lesionectomy in partial epilepsy: Predictive factors and electroencephalographic results. Epilepsia 1992; 33:639–644.

Cascino GD, So EL, Sharbrough FW, et al. Alfentanil-induced epileptiform activity in patients with partial epilepsy. J Clinical Neurophysiology 1993; 10(4):520–525.

Cendes F, Cook MJ, Watson C, et al. Identification of dual pathology in patients with lesional epilepsy. Epilepsia 1993; 34(6):126.

Chassagnon S, Minotti L, Kremer S, et al. Somatosensory, motor, and reaching/grasping responses to direct electrical stimulation of the human cingulate motor areas. J Neurosurg 2008; 109(4):593–604.

Chatrian GE, Quesney LF. Intraoperative electrocorticography. In: Epilepsy: A Comprehensive Textbook (Engel J Jr, Pedley TA, Eds.) 1997; pp 1749–1765. Philadelphia: Lippincott-Raven Publishers.

Costello DJ, Simon MV, Eskandar EN, et al. Efficacy of surgery for cryptogenic refractory focal status epilepticus. Arch Neurol 2006; 63(6):895 901.

Crandall PH, Walter RW, Rand RW. Clinical application of studies on stereotactically implanted electrodes in temporal lobe epilepsies. J Neurosurg 1963; 20:827–840.

Devinsky O, Canevini MP, Sato S, et al. Quantitative electrocorticography in patients undergoing temporal lobectomy. J Epilepsy 1992; 5:178–185.

Engel J Jr. Approaches to localization of the epileptogenic lesion. In: Surgical Treatment of the Epilepsies (Engel J Jr, Ed.) 1987; pp. 75–100. New York: Raven Press.

Engel J Jr. Comment on the predictive value of intraoperative electrocorticography in resections for limbic epilepsy associated with mesial temporal sclerosis. Neurosurgery 1997; 40(2):310–311.

Engel J Jr, Driver MV, Falconer MA. Electrophysiological correlates of pathology and surgical results in temporal lobe epilepsy. Brain 1975; 98:129–156.

Falconer MA, Meyer A, Hill D, et al. Treatment of temporal lobe epilepsy by temporal lobectomy: A survey of findings and results. Lancet 1955; 1:827–835.

Ferrier CH, Aronica E, Leijten FS, et al. Electrocorticographic discharge patterns in glioneuronal tumors and focal cortical dysplasia. Epilepsia 2006; 47(9):1477–1486.

Fiol ME, Torres F, Gates JR, Maxwell R. Methohexital (Brevital) effect on electrocorticogram may be misleading. Epilepsia 1990; 31(5):524–528.

Gambardella A, Palmini A, Andermann F, et al. Usefulness of focal rhythmic discharges on scalp EEG of patients with focal cortical dysplasia and intractable epilepsy. Electroencephalogr Clin Neurophysiol. 1996;98(4):243–249.

Gloor P. Contributions of electroencephalography and electrocorticography to the neurosurgical treatment of the epilepsies. In: Neurosurgical Management of Epilepsies (Purpura, DP, Penry, JK, Walker, RD, Eds.) 1975; pp 59–105. New York: Raven Press.

Gloor P, Ball G, Schaul N. Brain lesions that produce delta waves in the EEG. Neurology 1977; 7:326–333.

Graf M, Niedermeyer E, Schiemann J, et al. Electrocorticography: Information derived from intraoperative recordings during seizure surgery. Clin Electroencephalography 1984; 15:83–91.

Herrick IA, Craen RA, Blume WT, et al. Sedative doses of remifentanil have minimal effect on ECoG spike activity during awake epilepsy surgery. J Neurosurg Anesthesiol 2002; 14(1):55–58.

Hewitt PB, Chu DL, Polkey CE, Binnie CD. Effect of propofol on the electrocorticogram in epileptic patients undergoing cortical resection. Br J Anaesth 1999; 82(2):199–202.

Hufnagel A, Burr W, Elger C, et al. Localization of the epileptic focus during methohexital-induced anesthesia. Epilepsia 1992; 33:271–274.

Jasper HH. Electrocorticogram in man. Electroencephalogr Clin Neurophysiol Suppl 1949; 2:16–29.

Jennum P, Dhuna A, Davies K, et al. Outcome of respective surgery for intractable partial epilepsy guided by subdural electrode arrays. Acta Neurol Scand 1993; 87:434–437.

Jooma R, Yeh HS, Privitera MD, et al. Seizure control and extent of mesial temporal resection. Acta Neurochirurgica 1995; 133:44–49.

Kajtor F, Hullay J, Farago L, Haberland K. Electrical activity of the hippocampus of patients with temporal lobe epilepsy. AMA Arch Neurol Psychiatry 1958; 80:25–38.

Kanner AM, Kaydanova Y, de Toledo-Morrell L, et al. Tailored anterior temporal lobectomy: Relation between extent of resection of mesial structures and postsurgical seizure outcome. Arch Neurol 1995; 52:173–178.

Keles GE, Lundin DA, Lamborn KR, et al. Intraoperative subcortical stimulation mapping for hemispherical perirolandic gliomas located within or adjacent to the descending motor pathways: Evaluation of morbidity and assessment of functional outcome in 294 patients. J Neurosurgery 2004; 100(3):369–375.

Kuruvilla A, Flink R. Intraopertive electrocorticography in epilepsy surgery: Useful or not? Seizure 2003; 12:577–584.

Lesser RP, Kim SH, Beyderman L, et al. Brief bursts of pulse stimulation terminate afterdischarges caused by cortical stimulation. Neurology 1999;53(9):2073–2081.

Levesque MF, Nakasato N, Vinters HV, Babb TL. Surgical treatment of limbic epilepsy associated with extrahippocampal lesions: The problem of dual pathology. J Neurosurg 1991; 75:364–370.

Major P, Rakowski S, Simon MV, et al. Are cortical tubers epileptogenic? Evidence from electrocorticography. Epilepsia 2009; 50(1):147–154.

Manninen PH, Burke SJ, Wennberg R, et al. Intraoperative localization of an epileptogenic focus with alfentanil and fentanyl. Anesth Analg 1999; 88(5):1101–1106.

McBride MC, Binnie CD, Janota I, Polkey CE. Predictive value of intraoperative electrocorticograms in resective epilepsy surgery. Ann Neurol 1991; 30:526–532.

McDonald DB, Pillay N. Intraoperative electrocorticography in temporal lobe epilepsy surgery. The Canad J Neurologic Sci 2000; 27:S85–S91.

McKhann G M II, Schoenfeld-McNeill J, Born DE, et al. Intraoperative hippocampal electrocorticography to predict the extent of hippocampal resection in temporal lobe epilepsy surgery. J Neurosurgery 2000; 93:44–52.

Modica PA, Tempelhoff R, White PF. Pro- and anticonvulsant effects of anesthetics (Part II). Anesth Analg 1990; 70(4):433–444.

Morrell F. Secondary epileptogenesis in man. Arch. Neural 1985; 42:318–335.

Morrell F, Whisler WW, Bleck TP. Multiple subpial transections: A new approach to the surgical management of focal epilepsy. J Neurosurg 1989; 70:231–239.

Motamedi GK, Lesser RP, Miglioretti DL, et al. Optimizing parameters for terminating cortical afterdischarges with pulse stimulation. Epilepsia 2002;43(8):836–846.

Musella L, Wilder B, Schmidt R. Electroencephalographic activation with intravenous methohexital in psychomotor epilepsy. Neurology 1971; 21:594–602.

Nayel MH, Awad IA, Luders H. Extent of mesiobasal resection determines outcome after temporal lobectomy for intractable complex partial seizures. Neurosurgery 1991; 29:55–61.

Niedermeyer E. Electrocorticography. In: Electroencephalography (Niedermeyer E, Lopes da Silva F, Eds.) 1982; pp 537–541. Munich: Urban & Schwarzenberg.

Ojemann GA. Basic mechanisms implicated in surgical treatments of epilepsy. In: Window to Brain Mechanisms (Lockard JS, Ward AA Jr, Eds.) 1980; pp 261–277. New York: Raven Press.

Onal C, Otsubo H, Araki T, et al. Complications of invasive subdural grid monitoring in children with epilepsy. J Neurosurg 2003; 98(5):1017–1026.

Orbach D, Romanelli P, Devinsky O, Doyle W. Late seizure recurrence after multiple subpial transections. Epilepsia 2001; 42(10):1316–1319.

Palmini A, Gambardella A, Andermann F, et al. Intrinsic epileptogenicity of human dysplastic cortex as suggested by corticography and surgical results. Ann Neurol 1995; 37(4):476–487.

Panet-Raymond D, Gotman J. Can slow waves in the electrocorticogram (ECoG) help localize epileptic foci? Electroenceph Clin Neurophysiol 1990; 75:464–473.

Penfield W, Jasper H. Epilepsy and the Functional Anatomy of the Human Brain. 1954; Boston: Little Brown.

Penfield W, Kristiansen K. Epileptic Seizure Patterns. 1951; Springfield, IL: Charles C Thomas.

Pirotte B, Neugroschl C, Metens T, et al. Comparison of functional MR imaging guidance to electrical cortical mapping for targeting selective motor cortex areas in neuropathic pain: A study based on intraoperative stereotactic navigation. American J Neuroradiol 2005; 26(9):2256–2266.

Primrose DC, Ojemann GA. Outcome of respective surgery for temporal lobe epilepsy. In: Epilepsy Surgery (Lüders H, Ed.) 1992; pp 601–611. New York: Raven Press.

Quesney LF, Constain M, Rasmussen T, et al. How large are frontal lobe epileptogenic zones? EEG, ECoG and SEEG evidence. In: Advances in Neurology: Frontal Lobe Seizures and Epilepsies, Vol. 57 (Chauvel P, Delgado-Escueta VA, Halgren F, Bancaud J, Eds.) 1992; pp 311–323. New York: Raven Press.

Quesney LF, Niedermeyer, E. Electrocorticography. In: Electroencephalography: Basic Principles, Clinical Applications and Related Fields (Niedermeyer E, Lopes da Silva F, Eds.) 1993; pp 695–699. Baltimore, MD: Williams and Wilkins.

Quesney LF, Niedermeyer E. Electrocorticography. In: Electroencephalography: Basic Principles, Clinical Applications, and Related Fields (Niedermeyer E, Lopes da Silva F, Eds.) 2004; pp 769–775. Baltimore: Lippincott Williams & Wilkins.

Quesney LF, Wennberg R, Olivier A, Rasmussen T. ECoG findings in extratemporal epilepsy: The MNI experience. Electroenceph Clin Neurophysiol 1998(Suppl); 48:44–57.

Rampil IJ. Electroencephalogram. In: Textbook of Neuroanesthesia with Neurosurgical and Neuroscience Perspectives (Albin, MA, Ed.) 1997; pp 193–220. New York: McGraw-Hill.

Rasmussen T. Surgical treatment of complex partial seizures: Results, lessons and problems. Epilepsia 1983; 24:S65–S75.

Ross J, Kearse LA Jr, Barlow MK, et al. Alfentanil-induced epileptiform activity: A simultaneous surface and depth electroencephalographic study in complex partial epilepsy. Epilepsia 2001;42(2):220–225.

Schaul N, Gloor P, Gotman J. The EEG in deep midline lesions. Neurology 1981; 31:157–167.

Schwartz TH, Bazil CW, Walczak TS, et al. The predictive value of intraoperative electrocorticography in resections for limbic epilepsy associated with mesial temporal sclerosis. Neurosurg 1997; 40:302–311.

Sloan TB. Anesthetic effects on electrophysiologic recordings. J Clin Neurophysiol 1998; 15(3):217–226.

Smith M, Smith SJ, Scott CA, Harkness WF. Activation of the electrocorticogram by propofol during surgery for epilepsy. Br J Anaesth 1996; 76(4):499–502.

Soriano SG, Eldredge EA, Wang FK, et al. The effect of propofol on intraoperative electrocorticography and cortical stimulation during awake craniotomies in children. Paediatr Anaesth 2000; 10(1):29–34.

Souter MJ, Rozet I, Ojemann JG, et al. Dexmedetomidine sedation during awake craniotomy for seizure resection: Effects on electrocorticography. J Neurosurg Anesthesiol 2007; 19(1):38–44.

Spencer DD, Inserni J. Temporal lobectomy. In: Epilepsy Surgery (Luders HO, Ed.) 1992, pp 533–545. New York: Raven Press.

Stefan H, Quesney LF, Abou-Khalil B, Olivier A. Electrocorticography in temporal lobe epilepsy surgery. Acta Neurol Scand 1991; 83:65–72.

Toga AW, Mazziota JC. Brain Mapping: The Methods, 2nd ed. 2002; New York: Academic Press.

Tran TA, Spencer SS, Javidan M, et al. Significance of spikes recorded on intraoperative electrocorticography in patients with brain tumor and epilepsy. Epilepsia 1997; 38(10):1132–1139.

Tuunainen A, Nousiainen U, Mervaala E, et al. Postoperative EEG and electrocorticography: Relation to clinical outcome in patients with temporal lobe surgery. Epilepsia 1994; 35:1165–1173.

Walker AE. Temporal lobectomy. J Neurosurg 1967; 26:642–649.

Walker AE, Marshall C. Stimulation and depth recording in man. In: Electrical Stimulation of the Brain. (Sheer DE, Ed.) 1961; pp 498–518. Austin: University of Texas Press.

Wass CT, Grady RE, Fessler AJ, et al. The effects of remifentanil on epileptiform discharges during intraoperative electrocorticography in patients undergoing epilepsy surgery. Epilepsia 2001; 42(10):1340–1344.

Wennberg R, Quesney LF, Lozano A, et al. Role of electrocorticography at surgery for lesion-related frontal lobe epilepsy. The Canad J Neurologic Sc 1999; 26:33–39.

Wennberg RA, Quesney LF, Olivier A, et al. Post-excision residual spiking after frontal lobe removal: Outcome. Electroenceph Clin Neurophysiol 1998; 48:97–104.

Wennberg RA, Quesney LF, Olivier A, Dubeau F. Induction of burst-suppression and activation of epileptiform activity after methohexital and selective amygdalo-hippocampectomy. Electroenceph Clin Neurophysiol 1997; 102:443–451.

Wennberg RA ,Quesney LF, Olivier A, Rasmussen T. Electrocorticography and outcome in frontal lobe epilepsy. Electroenceph Clin Neurophysiol 1998; 106:357–368.

Wong CH, Birkett J, Byth K, et al. Risk factors for complications during intracranial electrode recording in presurgical evaluation of drug resistant partial epilepsy. Acta Neurochir (Wien) 2009; 151(1):37–50.

Wyler A, Richey E, Atkinson R, Hermann B. Methohexital activation of epileptogenic foci during acute electrocorticography. Epilepsia 1987; 28:490–494.

Yamamoto J, Ikeda A, Satow T, et al. Low-frequency electric cortical stimulation has an inhibitory effect on epileptic focus in mesial temporal lobe epilepsy. Epilepsia. 2002;43(5):491–495.

# 4

# FUNCTIONAL CORTICAL MAPPING

**MIRELA V. SIMON, DONALD C. SHIELDS, AND EMAD N. ESKANDAR**



## I. INTRODUCTION

Within the field of neurologic surgery, it is currently accepted that maximal resection of the supratentorial gliomas in both adult and pediatric population is associated with a higher survival rate and an increased quality of life (Albright et al., 1995; Ammirati et al., 1987; Berger et al, 1989; Berger, 1996; Guilburd et al., 1989; Nitta et al., 1995; Sala et al., 2002; Scerrati et al., 1996; Yang et al., 1999). Likewise, in cases of surgical treatment of epilepsy, incomplete resection of the seizure focus is followed by unsatisfactory reduction of the number of epileptic events (Bonilha et al., 2004; Kim et al., 2008; Paolicchi et al., 2000).

In order to achieve complete or near complete lesional resection without damage of healthy nervous tissue and thus to avoid postoperatory deficits, accurate delineation of pathological tissue from functionally normal cortex is necessary. In many occasions, this is difficult to achieve, mainly for two reasons: first, in certain cases, visual inspection of the cortex during surgery fails to delineate accurately the lesional margins, particularly in cases of infiltrative pathology and in cases of microscopic abnormalities without definite macroscopic anatomic distortion (e.g., cortical dysplasia). Second, several authors have emphasized that diseased brain tissue can still retain clinically important function (Ojeman et al., 1996; Schiffbauer et al., 2001; Skirboll et al., 1996). As a consequence, surgical resection strictly based on visual inspection of distorted anatomy poses a real risk, even in cases when distinction between diseased and healthy cortex is relatively easy to make. Thus, perioperative functional cortical mapping is often required for resection of neurosurgical lesions in the vicinity of eloquent cortex in order to preserve cognitive function (Skirboll et al., 1996) while resecting the maximal amount of pathological cortex/lesion. Multiple modalities, including functional magnetic resonance imaging (fMRI), positron emissions tomography, magnetic source imaging, and electrophysiologic mapping, have been utilized for this purpose. More so, several studies have compared the efficacy of each of these methods in accurately mapping eloquent cortex (e.g. comparison between fMRI and intraoperative neurophysiologic mapping-Fandino et al., 1999).

Intraoperative electrophysiologic mapping offers several advantages over other methods with similar goals, such as functional MR imaging (fMRI). First, it allows live assessment of function, with direct feedback to the neurosurgeon. Second, it allows subcortical mapping, with identification of the white matter tracts that are essential in the interconnectivity between eloquent cortical regions, thus essential in function generation; this advantage over fMRI is best seen in case of language mapping. Third, it offers an increased specificity in identification of the essential functional regions of the brain (Duffau et al., 2002; Lehéricy et al., 2000). Fourth, its results are not affected by hemodynamic changes, which can distort the spatial localization, as in the case of fMRI.

This method consists of applying an electrical current directly on the cortical surface or epidurally, designed to produce a transient activation/disruption of normal function pathways. The assessment of these pathways is done by recording the triggered muscle motor-evoked potentials (mMEPs) in the contralateral hemibody muscles for motor mapping in anesthetized or awake patients and/or evoked clinical responses for both primary sensorimotor and language mapping in awake patients.

From a historical perspective, electrical stimulation of the brain has been used for over a century in the investigation of anatomically specific brain function. Bartholow (1874) is most often credited as the first to attempt electrical stimulation of brain tissue, which was performed on a patient with a brain abscess eroding through the skull. Later, Foerster (1936) and Penfield et al. (1937) began mapping the human cerebral cortex and eventually presented detailed descriptions of human sensory and motor cortex. Penfield and his co-workers went on to use intraoperative electrical stimulation during electrocorticography (ECoG) recordings to map speech/language areas, as well as the anatomical identification of somatosensory and motor cortices (Penfield et al., 1954; Penfield et al., 1951).

Intraoperative electrophysiologic mapping of language, motor, and sensory cortex is extensively employed in resection of a wide variety of brain pathology, from tumors to neurodevelopmental lesions. When done appropriately, it is safe and very reliable. However, there are two main challenges associated with this technique. First, there is a considerable variability in thresholds between individuals (see pages 155–57). This variability becomes a dangerous source for false-negative results, when the mapping is stopped prematurely, based on the wrong assumption that the stimulated area is "nonfunctional" and thus safe for resection. Similarly, false-positive results can lead to erroneous identification of primary sensory, motor, and language areas. This can occur from inadvertent stimulation at higher intensity than the actual threshold, producing either spread of the current from the stimulated region to the primary sensorimotor and language regions or direct stimulation of supplementary motor cortex. Thus, in order to ensure reliable results, it is important that the mapping be performed at the threshold of the primary cortex when only primary functional cortical regions are stimulated. Second, direct cortical electrical stimulation has epileptogenic potential. Stimulus-triggered electrographic seizures can jeopardize the patient's safety and also interfere with the mapping results (see page 157).

In this chapter, we address the methodology, interpretation, and technical and theoretical challenges associated with the electrophysiologic mapping techniques.

## II. SURGICAL PREPARATION

### 1. The Pros and Cons of Awake Craniotomy

Intraoperative cortical mapping for motor cortex has historically been performed under general anesthesia, although conscious sedation techniques allow the clinician to achieve consistent mapping results that are not obscured by the variable depths of anesthesia (Vitaz et al., 2003). Moreover, certain mapping procedures, such as language and sensory mapping, can be done only with the patient being awake. Awake craniotomies could result in loss of adequate airway control and produce hypercapnia-induced brain swelling. Previous studies indicate that the incidence of such complications is less than 1% in controlled settings (Danks et al., 2000; Haglund et al., 1994; Meyer et al., 2001). Furthermore, the judicious use of perioperative dexamethasone, mannitol, or furosemide often provides satisfactory amelioration of unwanted cerebral swelling. On the other hand, seizure activity during awake craniotomies could be more deleterious than in cases of general anesthesia, both because of compromise of airway control and production of postictal refractory effects, which interfere with subsequent mapping results where patient's input is essential (e.g., language mapping). With respect to cortical mapping, anesthetic agents have antiepileptic properties that provide considerable protection in these individuals, but the interpretation of the stimulation results can also be affected by the depth of anesthesia (Prasad et al., 2001; Stecker et al., 1998).On the other hand, administration of general anesthesia requires higher stimulation amplitudes to arrive at the mapping threshold, sometimes more than twice that required during conscious sedation (Tan et al., 2001; Vitaz et al., 2003; see pages 155–56). This can not only affect the interpretation of the stimulation results but also increase the risk of triggering a seizure, even in the presence of a burst-suppression pattern. This electroencephalogram (EEG) pattern, even though considered a measure of the overall suppression of the cortical activities, remains a succession of inhibitory and excitatory epochs. Thus, levels of anesthesia should be decreased sufficiently during the surgery to allow effective but also safe stimulation of the stimulate cortex.

Awake craniotomy is not appropriate for all patients despite these advantages. Individuals with significant preoperative neurologic functional deficits, significant pulmonary or cardiac conditions, and those with psychiatric disorders that can compromise their potential participation in such a procedure should be considered carefully prior to surgery. The emotional maturity of children and teenagers should also be established prior to any such procedure.

Awake craniotomy procedures, including asleep-awake-asleep protocols with repeated intubation and those with only conscious sedation (without endotracheal intubation) used during portions of the surgery, vary among institutions. Centers that employ intubation often preoperatively induce with sodium thiopental or propofol. Patients can then be hyperventilated prior to durotomy. Experienced anesthesiologists are therefore required to extubate the patient while the patient's head is immobilized with cranial fixation devices. Following functional mapping and resection of the portion of cerebral cortex in question, the patient is reintubated, using a tube changer or fiber-optic laryngoscope, for the remainder of the surgical procedure. Proponents of this technique argue that intubation prior to durotomy provides for the ability to hyperventilate the patient for brain relaxation (Pouratian et al., 2004). Alternatives to this procedure, which avoid potential complications associated with intubation or extubation of patients in cranial fixation, include the use of conscious sedation during the craniotomy portion of the procedure (Danks et al., 2000). Patients can then be easily awakened for the cortical mapping portion of the case and sedated again for closure. In cases where propofol is used for sedation, infusion of this agent can be stopped approximately 10 minutes prior to the start of mapping. Versed, labetalol, and fentanyl can be administered during the awake portions of the procedure to treat anxiety, hypertension, and pain. With this technique, patients often can more quickly recover from surgery than those given general anesthesia.

### 2. Reducing the Risk of Stimulation-Triggered Seizures

Given the epileptogenic effect of direct electrical cortical stimulation, it is important to ascertain whether the patient harbors a preexisting seizure disorder. It is also important to know the type of seizure, because generalized tonic-clonic seizures (GTCZ) are more worrisome than partial seizures. Tumors of the frontal lobe or seizures from a frontal focus (such as cortical dysplasia or tumors of the frontal lobe) have a higher risk of secondary generalization because of rapid spread.

Patients with a seizure disorder should be adequately medicated with antiepileptic drugs (AEDs) and a recent serum level of the AED should be available. If the levels are subtherapeutic, additional bolus or dose of medication should be given preoperatively, preferably intravenously. In those instances where there is a high suspicion of seizure activity but the patient is not on AEDs, loading with an intravenous agent such as Keppra can be helpful. Fosphenytoin or depakote are also good options.

Ice-cold saline should be available intraoperatively for prompt irrigation of the cortex, in case of stimulation-triggered afterdischarges (AD) and seizures. This has been shown to be very effective in aborting triggered epileptiform or epileptic activity (Sartorius et al., 1998).

## III. SENSORIMOTOR MAPPING

### 1. Central Sulcus Localization

#### *A. Principle*

Sensorimotor mapping starts with accurate identification of the central sulcus. The central sulcus can be identified on the 1.5 T MRI images by using anatomic landmarks generally accepted in the neurosurgical literature (Bittar et al., 2000), such as the intersection between the superior frontal sulcus with the precentral sulcus (sagittal images); the motor hand area as the omega-shaped knob (axial images) or hooklike configuration (sagittal images). Using co-registration techniques and the stealth neuronavigational system, the surgeon correlates the anatomic location, determined by direct visual inspection within the surgical field, with the anatomic location, determined by inspection of the MRI images.

We agree with other authors (Korvenoja et al., 2006) that one of the most reliable tool for identifying the central sulcus is the use of median somatosensory evoked potentials (SSEPs phase-reversal technique). Using this method, in many instances, we have found that the presumed location of the central sulcus, as determined by visual inspection, was erroneous. This can be the result of several factors, such as interindividual anatomic variability; presence of pathology, causing distorted anatomy or dissociation between the anatomy and physiology of the region;

imprecise calibration of the neuronavigational system; inadequate brain MRI study. Thus, the "correlated" presumed location of the central sulcus, as determined with the help of neuronavigational system, should be further confirmed with the neurophysiologic method.

This technique is based on the reversal in polarity of the scalp-recorded median thalamocortical SSEPs, at the level of the central sulcus. At about 19 msec after the stimulation of the contralateral (to the hemisphere operated on) median nerve at the wrist, the postcentral primary somatosensory parietal cortex becomes more electronegative than the precentral frontal region, creating a dipole-oriented perpendicular to the central sulcus, from posterior toward anterior regions. Placing a multicontact strip in this presumed direction enables recording of the thalamocortical SSEPs, localization of their polarity shift, and thus of the central sulcus.

## B. Methodology

### *B1. Technical setup*

***a. Stimulation technique.*** The contralateral median nerve at the wrist is the most commonly used stimulation site for this method, in particular if the lesion to be resected is located in the vicinity of the hand or arm region of the contralateral motor strip. However, because the SSEPs phase reversal is commonly seen along the extent of the central sulcus, other peripheral nerves can also be stimulated. For example, in case of lesions present along the motor cortex near the interhemispheric fissure, the femoral nerve can also be stimulated for motor strip mapping. In this case needle electrodes are inserted over the contralateral inguinal ligament, approximately 15 mm lateral to the femoral artery. Thus, the femoral nerve is stimulated with increasing intensities until small twitches are noted in the quadriceps femoris. However, the most common nerve in the lower extremity used for stimulation and SSEPs recording is the posterior tibial nerve. Other peripheral sites that can be stimulated include the peroneal and ulnar nerves. Alternatively, in cases when the lesion was localized lateral to the motor hand cortex, or the latter was not exposed by the craniotomy, Kumabe and colleagues (2005) successfully used the SSEP polarity inversion observed by stimulation of the contralateral lower lip mucosa, for central sulcus localization.

For stimulation, two disposable conductive solid-gel electrodes, with attached 1.5-mm safety lead wires (Rochester, Electro-Medical Inc), are attached to the skin overlying the contralateral (to the hemisphere of interest) median nerve at the wrist, along its length. The electrical stimulus delivered is a rectangular pulse of 0.3 msec width, delivered at a frequency of 3.17 Hz, with a progressive increase in the intensity until thumb muscle twitches are observed. Usually, the stimulus amplitude needed for initiating the thumb twitch varies between 10 and 20 mA. Other stimulus parameters can also be used. For further details regarding stimulation technique see pages 1–46. Awake patients should be informed and prepared about their expected experience during stimulation. Median nerve stimulation can be unpleasant, with the patient attempting to move the stimulated hand. Minimizing such arm movements is important for noise reduction. In such cases, a gradual increase in the stimulus intensity, within the level of the patient's tolerance, should be performed until a reliable thumb twitch is obtained.

We usually average at least 200–300 stimulation trials. Averaging is essential for increasing the signal-to-noise ratio: it helps in canceling out the random noise and identifying the neurophysiologic signal. In many situations, the number of averaged trials can be less as a result of time constraints. When the recorded waveforms have good amplitudes and well-defined morphology, less averaging may suffice. However, during a noisy recording and/or when SSEPs are expected to have poor morphology, such as in the case of baseline sensory deficits (small signal-to-noise ratio), increasing the number of averaged trials gives more reliable results.

***b. Recording technique.*** For SSEPs recordings we use a platinum strip electrode (Ad-Tech subdural strip electrodes), placed by the surgeon directly on the cortex (can also be placed epidurally), in a direction perpendicular to the presumed direction of the central sulcus (Fig. **4-1**). The contact diameter is 2.3 mm, the intercontact distance is 10 mm, and the total length of the strip is about 90 mm. The stimulated peripheral nerve should dictate the location of this strip on the convexity (higher, closer to the interhemispheric fissure, or lower, closer to the hand regions). For example, when stimulating the contralateral median nerve, ideally the strip should be located at the level of the hand area on the convexity (see pages 136–40). When the femoral or posterior tibial nerves are stimulated, the recording strip should be along the superior lip of the interhemispheric fissure. The surgeon places the strip and chooses which end of the strip is anterior or posterior (contact 1 versus contact 8). It is preferable that the neurophysiologist remain unaware of the orientation and position of the strip and, in particular, of the location of the contacts relative to the presumed location of the central sulcus. Our reference electrode is either a subder-

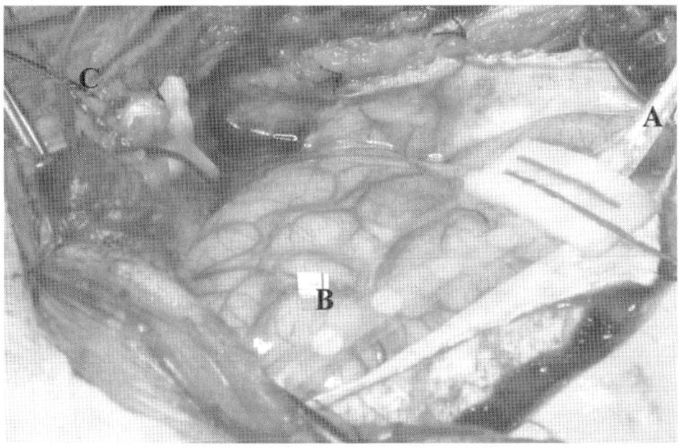


**FIGURE 4-1** Placement of a subdural 8-contact strip on the cortex. In electrophysiological functional mapping, the strip is used for recording thalamocortical SSEPs in central sulcus localization (phase-reversal technique) and ECoG recording to monitor stimulation-triggered afterdischarges. In the first case, the strip has to be placed perpendicular and crossing the presumed direction of the central sulcus. In the second case, the strip should be placed close to the stimulated regions. The contact closest to the extension cable (**A**) is contact number 8. Also, notice placement of a "flag" (**B**), for marking an area of eloquent cortex, as well as placement of a sterile subdermal needle in the field as the cathode for the monopolar stimulation (**C**).

mal needle (anesthetized patients) or a surface stick-on (awake patients) electrode, placed on the contralateral mastoid. Some institutions use a reference electrode, placed on the patient's forehead, or a needle, placed in the exposed temporalis muscle. We like to use the contralateral mastoid because we consider this location to be more or less at an equal distance from all of the recording contacts, thus ensuring a relatively fair representation of the amplitudes and polarities of the SSEPs at the eight recording contacts.

However, if functional mapping is attempted during phase II invasive recordings for epilepsy surgery evaluation, median SSEPs recording is done successively through several rows of grid contacts, which presumably intersect the central sulcus. In particular, the best option for obtaining the SSEPs phase reversal is a row oriented perpendicular on the central sulcus and also located at the hand level on the hemispheric convexity.

The usual recording parameters employed are the following: time base = 5 msec/div; sensitivity 10–20 $\mu$V/div. The recording channels include both reference and bipolar channels. The reference recordings provide a representation of the "absolute" polarities of the regions situated underneath each contact, relative only to the charge of the reference electrode, as well as an "imperfect" phase reversal at the level of the central sulcus, with the first (closest to the central sulcus) positive peak lagging behind the first (closest to the central sulcus) negative peak by a few milliseconds (see pages 136–40). The bipolar recordings give a better-defined phase reversal, with synchronous positive and negative peaks (see pages 136–40), pointing not to the central sulcus but rather to the postcentral region; the contact at which the phase reversal occurs is situated on the most electronegative region, i.e., somatosensory parietal region. The bipolar montage needs to have one channel crossing the central sulcus and each channel containing one contact from the previous channel (i.e., 1–2, 2–3, 3–4, 4–5, 5–6, 6–7, 7–8).

*B2. Troubleshooting*

***a. 60-Hz noise and other types.*** One of the most common problems with this technique is represented by the significant amount of 60-Hz artifact that is introduced in the recording. Even when such direct cortical recordings are expected to deliver high-amplitude evoked potentials, the signal-to-noise ratio can be markedly decreased by the presence of 60-Hz noise. We find that turning "on" the 60-Hz filter notch can help. One should ensure that the signal obtained is of the maximum amplitude, in other words that the stimulation is appropriate: a good thumb twitch is a sure indication of appropriate stimulation. The stimulus frequency should be a number that does not exactly divide 60 (e.g., we use 3.17 Hz, not 3 Hz). Unstable contacts can translate in increased rejection or can decrease the signal-to-noise ratio by introducing significant amounts of noise. Impedance check should thus be performed. Irrigation of the contacts with saline usually improves the impedances. We usually ask our surgeons to secure the strip with a couple of stitches and to apply a soaked sterile gauze on top of it. As already mentioned, increasing the number of averaged trials helps increase the signal-to-noise ratio.

***b. Increased rejection.*** Inability to average is usually a result of rejection of the trials. One has to ensure that the rejection window of the recording system is wide open. These are direct cortical recordings, and the expected amplitudes of the SSEPs are much higher than the amplitudes obtained with scalp recordings; if the rejection threshold is set too low, the high-amplitude signals are rejected.

***c. No recorded SSEPs or no phase reversal.*** This is generally a result of either distorted somatosensory cortex with poor SSEPs or inadequate placement of the strip, away from the somatosensory cortex. Reposition of the strip is essential; this should be done in the anterior-posterior axis as well as in the superior-inferior axis (see pages 136–40). Additionally, rotating the strip can help in orienting it perpendicular to the central sulcus.

## C. Interpretation

As expected, the highest amplitudes of the median thalamocortical potentials are recorded on the lateral surface of the hemispheric convexity, posterior to hand motor area, in the postcentral primary somatosensory region. Thus, the strip should be placed at the level of the hand area and crossing the central sulcus, so that a phase reversal can be obtained. The somatosensory cortex becomes most electronegative at about 19 msec after stimulation of the contralateral median nerve at the wrist; an electrode at this level (G1) reflects this relative negativity to the reference electrode (G2). Our machines are wired so that the recorded waveform in a G1 negative–G2 positive channel starts with an upward deflection. Thus, the recorded N19 thalamocortical potential is represented by an upward, negative peak recorded in the postcentral region, at the level of somatosensory cortex and seen at 19 msec after stimulation of the contralateral median nerve at the wrist. Of note, the absolute latency of N19 potential does vary because of the impact of factors such as existing pathology, patient's height, or, in particular, the anesthetic effects. In order to appreciate the "absolute" polarity of the thalamocortical projections to the somatosensory cortex, we prefer to use the contralateral mastoid as the reference electrode (rather than Fz, which introduces an additional charge and makes the polarity interpretation even more difficult).

At the same time, the frontal cortical regions, situated anterior to the central sulcus, become more positive than the reference electrode. Their greatest positivity occurs at a latency a bit longer than the latency at which the postcentral parietal regions attain their highest negativity. Thus, in the reference montage, we expect to see an imperfect phase reversal, at the level of central sulcus, with the first positive peak lagging behind the first negative peak by a few milliseconds. A clearer phase reversal, possibly with more synchronous positive and negative peaks, is seen in a bipolar montage, which offers a good appreciation of the relative electronegativity of this region compared to the other regions. Thus, the phase reversal occurs at the contact situated in the postcentral region, overlying the somatosensory cortex, which is the most electronegative area at 19 msec. In order to obtain the phase reversal, the 8-contact recording strip needs to be placed perpendicular to the central sulcus and straddling across it. According to the position of the strip, with contact 1 placed most anteriorly or most posteriorly, respectively, the phase reversal obtained in the reference montage is represented by waveforms pointing to each other or pointing away from each other. At the same time, the phase reversal in bipolar montage points toward the contact situated on the primary somatosensory cortex, irrespective of the position of the strip (Fig. **4-2 a–c**).

If the recordings show low amplitudes of what is believed to represent the N19 waveform, repositioning of the strip to a different location on the hemispheric convexity may be helpful. If no N19 is identified, it is possible that the strip is located entirely anterior to the central sulcus; thus pulling the strip posteriorly should be tried. If good amplitude cortical SSEPs are recorded in all of the channels but no phase reversal is seen, either the strip is entirely posterior to the central sulcus or it is not perpendicular to the central sulcus. In this case, pushing the strip anteriorly or rotating it may help (Fig. **4-2 d**). We like to record the SSEPs blinded to the position of the strip. Once the phase reversal is obtained, an additional repositioning of the strip is done for confirmation of the first results. For example, if the first phase reversal was seen between contacts 2 and 3, the surgeon repositions the strip in the anterior-posterior axis, which leads to another pair of contacts to cross the central sulcus. Being able to identify this new pair of contacts, without knowing the new strip position, is reassuring for a correct identification of the location of the change in polarity and, thus, the location of the central sulcus. However, situations can arise when, despite multiple repositionings of the strip, a phase reversal cannot be identified. Instead, a sharp and consistent change in morphology and amplitudes of the evoked potentials is observed closest to the central sulcus. This situation can be particularly seen in cases of pericentral lesions, associated with significant changes in anatomy and function of these regions, as well as in the case of a tortuous central sulcus, when positioning the strip perpendicular to the central sulcus becomes challenging.

Despite the potential use of other peripheral nerves (see page 134), we prefer to use the median nerve for stimulation, regardless of the location of the pathology. The median SSEPs have a robust morphology and good

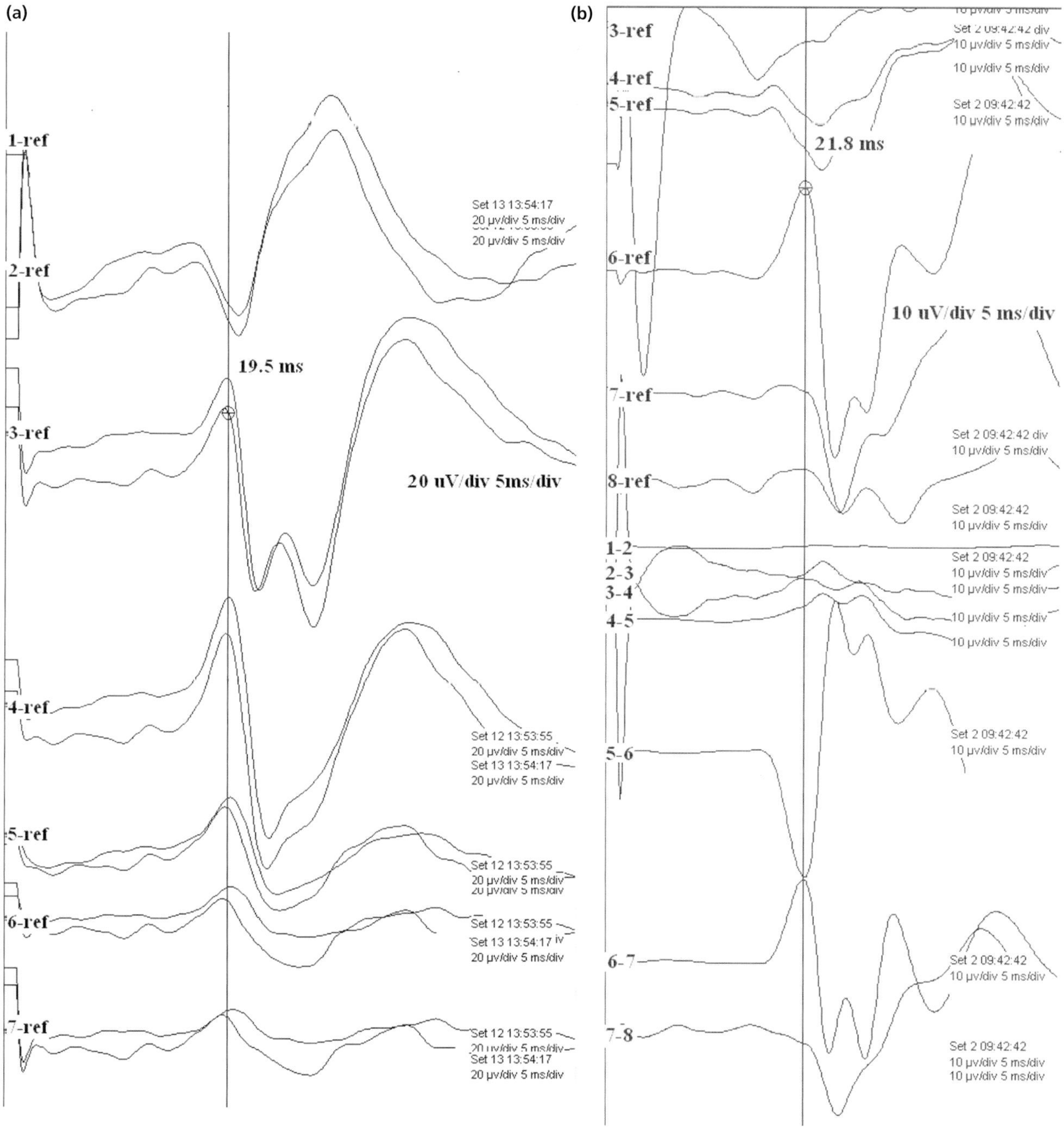


**FIGURE 4-2** Median SSEPs phase-reversal technique. **a:** Referential recording, with reference electrode on the contralateral mastoid and the strip placed with contact 1 most anterior in an anterior-posterior axis. Notice a phase reversal of the field polarity between contact 3 and contact 2. The electronegative potential seen at contact 3 represents the N19 thalamocortical potential obtained in the postcentral somatosensory cortex, at 19.5 msec after stimulation of the contralateral median nerve at the wrist. **b:** Referential and bipolar recordings from a different patient, but with similar placement of the reference electrode and orientation of the strip (contact 1 anterior). N19 is recorded at contact 6, 21.8 msec latency.

(c)

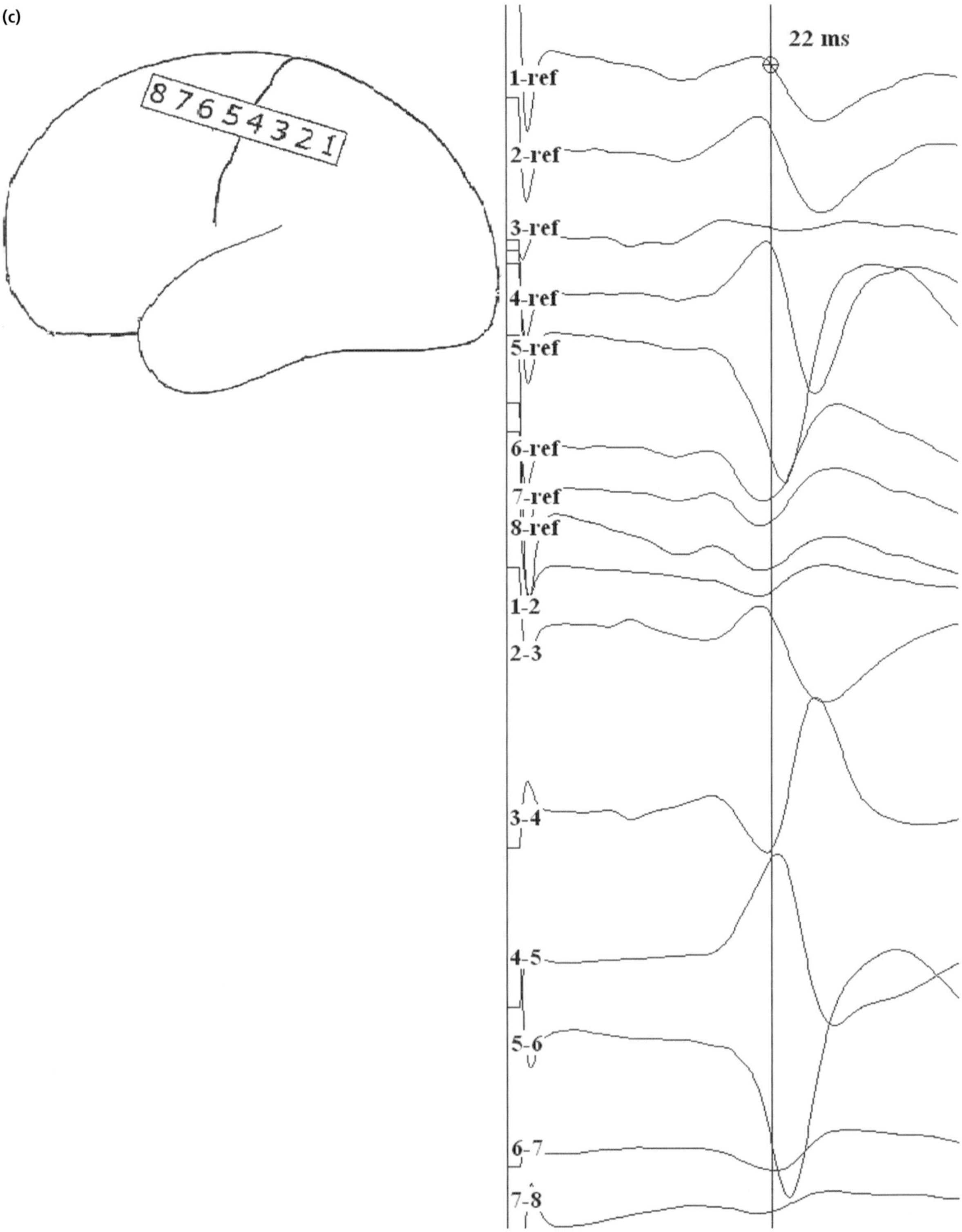


**FIGURE 4-2** *(Continued)* **c:** Referential and bipolar recordings, this time with the strip placed with contact 1 most posterior in the anterior-posterior axis. The referential montage shows a phase reversal at the level of the central sulcus, however, with the waveforms pointing away from each other.

(d)

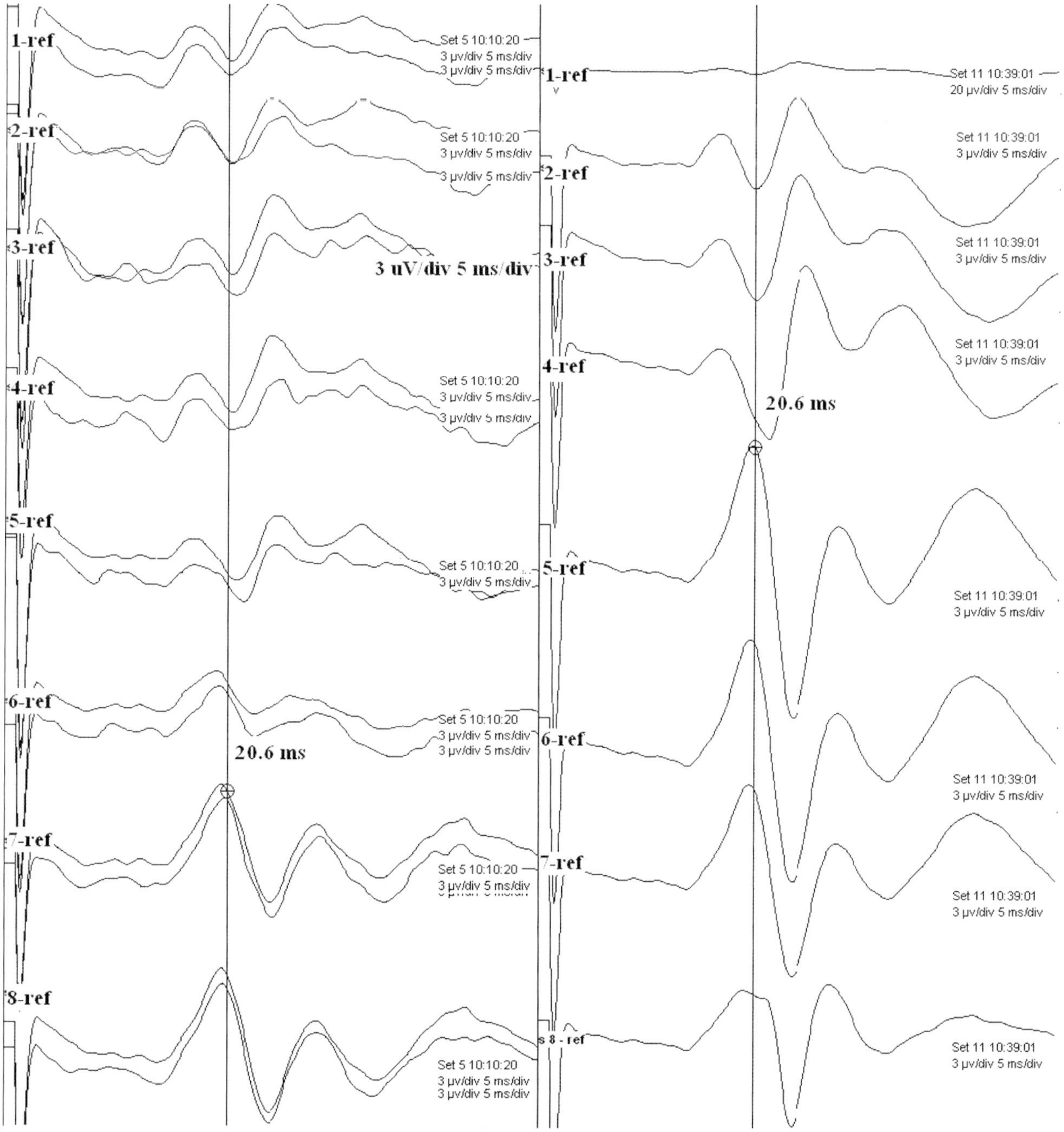


**FIGURE 4-2** *(Continued)* **d:** Reorientation of the strip can improve the morphology and amplitudes of the recorded SSEPs. Notice the poorly defined negative peak, with an absolute latency of 20.6 msec at contacts 7 and 8 (**left** panel). In the **right** panel, the strip was pushed more posterior in the anterior-posterior axis and more inferior on the lateral hemispheric surface, reaching the level of the hand area. Notice the better-formed negative peaks at contacts 5 through 8 and the phase reversal between 4 and 5 contacts. The recordings show that contact 5 is situated postcentrally and contact 4 is precentral.

the threshold for the primary cortex. Alternatively, trying to reduce the upslope or downslope of the stimulus artifact, by decreasing LFF can be helpful. A third option is to change the stimulus parameters so that the "length" of the stimulus artifact is shortened and the latter does not superimpose on the responses of the mMEPs. We have previously used repetitive trains of 4 pulses per train, train frequency of 1 Hz, pulse frequency of 500 Hz, pulse width of 0.5 msec, and obtained facial mMEPs (Fig. **4-6 b**). The intensity threshold, however, was higher by 3–4 mA than the motor mapping threshold found when the stimulation was applied at the usual stimulus parameters (Fig. **4-3**).

One should be able to distinguish between triggered neurophysiologic responses and movement artifact. We have had instances when the contralateral facial mMEPs were thought to be evoked at a latency shorter than expected (Fig. **4-6 c**). However, these proved to be movement artifact. In these cases, the cathode was placed in the ipsilateral temporalis muscle and most likely led to direct spread of the current to the peripheral part of the facial nerve and facial muscles (see pages 140–41). This spread was responsible for significant twitching of the ipsilateral facial muscles and produced movement artifact. This bilateral facial type of artifact tends to be picked more often when there is a longer distance between the recording electrodes. On the other hand, unilateral cortical stimulation can produce bilateral upper facial muscle responses. A good way to distinguish between peripherally and centrally triggered responses is to modify the stimulus parameters: only a multipulse train can trigger muscle responses when applied to the cortex, because cortical stimulation has to overcome the motorneuron threshold for depolarization (in the facial nc); this can be achieved only when several pulses per train are applied. If one changes the stimulus parameters to repetitive pulses at 2 Hz (one pulse per train) and still obtains the facial responses, the stimula-

**(a)**

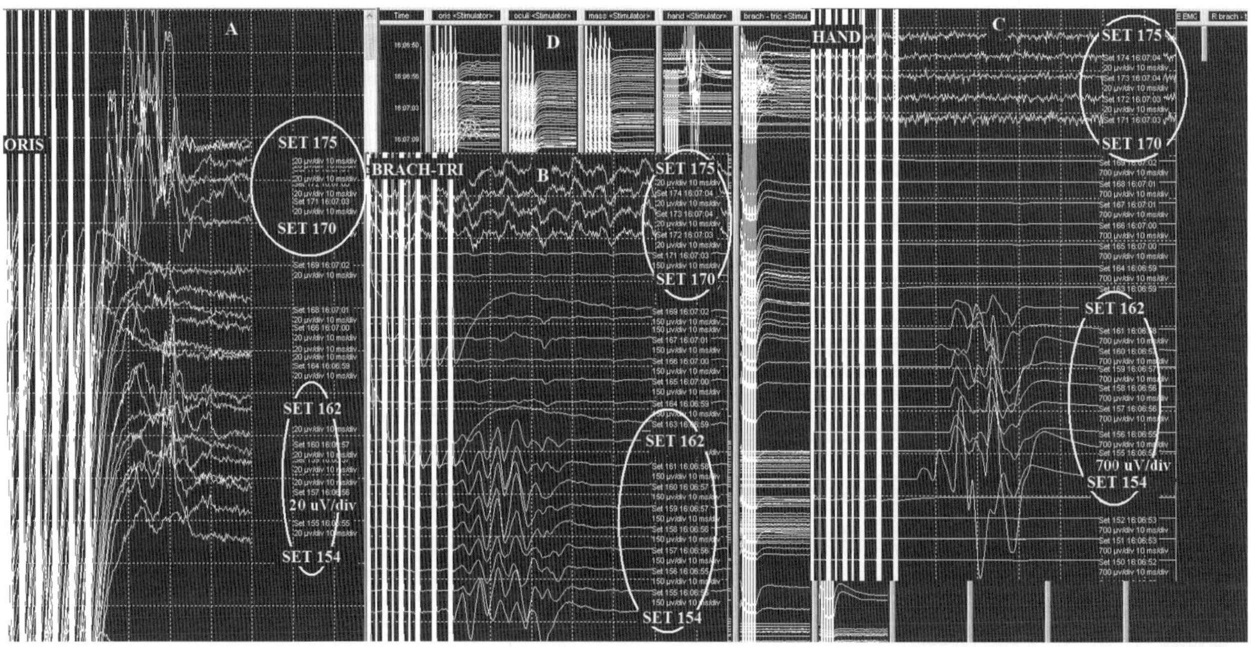


**FIGURE 4-6** Recordings of face mMEPs in functional motor mapping. **a:** In panel **A**, mMEPs seen in the orbicularis oris channel were triggered on two occasions: first time in sets 154 through 162 (bottom circle) and second time in sets 170 through 175 (top circle). Notice that, during the sets 154–162, high-amplitude mMEPs are also triggered in the brachioradialis-triceps channel (panel **B**) and even higher amplitude mMEPs in the hand channel (panel **C**). The stimulation was done at current intensity (i.e. 7 mA), which is higher than the mapping threshold. Thus, the activity seen in orbicularis oris channel is most likely due to a spread of the current from the primary motor hand region. Subsequently, the stimulus intensity was decreased, and the surgeon stimulated the primary motor face region at the primary motor mapping threshold (4 mA). This stimulation triggered mMEPs in the orbicularis oris channel (sets 170 through 175 in panel **A**), but not in the brachioradialis-triceps or hand channels (same sets in panels **B** and **C**, respectively). Notice that these facial mMEPs are of higher amplitude, more complex morphology, and shorter latencies than the mMEPs triggered in sets 154 through 162. Panel **D** in the background shows a history log.

tion must have occurred distal to the motorneuron, likely by direct spread of the current to the peripheral part of the facial nerve. Using a bipolar stimulator or moving the cathode away from the temporalis muscle may help (see pages 140–45).

***c. Noisy muscle channels.*** This could impair identification of the lower-amplitude mMEPs. As a consequence, in order to be able to distinguish the evoked motor potentials, one has to stimulate at high stimulus amplitude, probably higher than the actual primary motor mapping threshold. Thus, the latter may not be reliably identified. Checking the impedances of the muscle electrodes is a good start. All should be below 5 KOhms and relatively equal. Uneven impedances are more troublesome, sometimes, than higher but even impedances.

A special situation is the presence of high-amplitude (usually 60-Hz) artifact in the muscle channels. Unless adjacent muscle channel allows accurate additional data, one should make all efforts to eliminate the 60 Hz, such as changing the existing ground electrode or adding an additional one on that limb. The hardware on our machines has ground electrodes connected in parallel, and thus, the higher the number of grounds is, the lesser the resistance of the final ground to the current's path will be. Of course, using subdermal needle electrodes rather than surface electrodes for muscle responses may help reduce such artifact.

For troubleshooting recording of epileptiform discharges, see pages 162–63.

## C. Interpretation

As mentioned previously, motor responses can be recorded with subdermal needle or surface stick-on electrodes in the related muscle groups.

To be considered as a positive response, the motor potential should have at least 50 $\mu$V in amplitude and be reproducible in subsequent stimulations. We believe that a polyphasic morphology is a sign for reliable activation of the motor pathways. However, consistent, high amplitude, normal latency biphasic/triphasic responses are also

**(b)** **(c)**

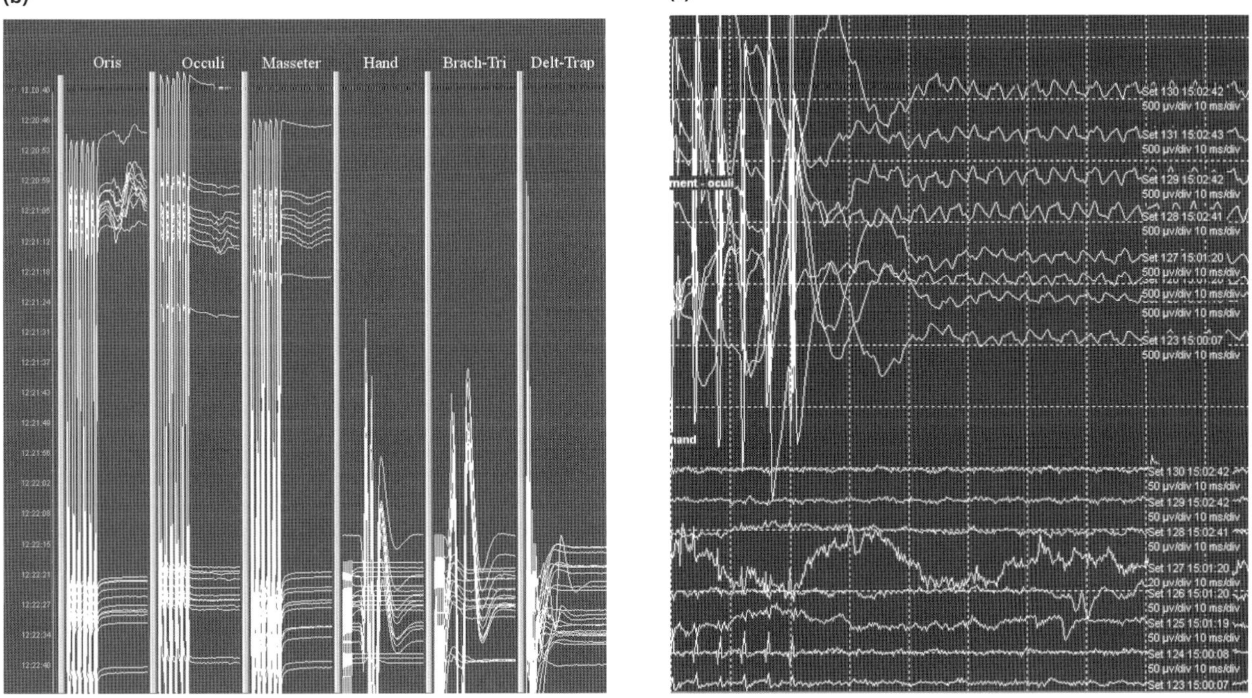


**FIGURE 4-6** *(Continued)* **b:** History log of triggered mMEPs in facial and arm muscles: orbicularis oris and oculi mMEPs, as well as arm mMEPs obtained at modified stimulus parameters: train of 4–5 pulses, train frequency of 1 Hz, pulses frequency of 500 Hz, pulse width of 0.5 msec. The recording parameters are different for the face muscle channels and arm muscle channels, with shorter time base for the face channels (10 msec/div for arm channels; 3 msec/div for face channels). **c:** Short latency muscle activity with superimposed stimulus artifact, representing facial movement artifact as a result of direct spread of the current to the peripheral part of the facial nerve and the facial muscles.

seen (Fig. **4-4 d**). In certain cases, the triggered motor responses have a long duration and inconsistent latency, but they are very reproducible. In these situations, viewing these potentials in the history window is helpful (Fig. **4-4 d**). Also, using overlay display (see Fig. **4-4 c**) is indicated. The reliability of the mapping results is further confirmed by remaining "blinded" to the cortical location of each stimulation: reproducible motor responses should be evoked every time that the surgeon stimulates the presumed eloquent cortex. Of note, normal variations of the latencies, amplitudes, and widths of the mMEPs are seen within different regions of the motor cortex. In general, there is a progressive increase in the absolute latencies of mMEPs, with shortest latencies seen in the muscles of the face, followed by the shoulder, forearm, hand, and then leg. For example, when hand MEPs are obtained at a latency similar to that of the AH MEPs, one should question the accuracy of hand area identification. However, in the case of preexisting hand weakness (e.g., brain tumor infiltrating the hand motor area), one can expect abnormally long latencies of the motor responses or abnormally high mapping thresholds (compared to the threshold of other primary motor regions).

If the stimulation is done with high stimulus amplitudes, the motor response triggered may be due to spreading of the current to cortex adjacent to the directly stimulated region. For example, when stimulating the motor strip to identify the hand region, we may obtain simultaneous responses in several muscle groups (e.g., hand and brachioradialis-triceps channels). In this instance, one has to decrease the stimulus intensity until the muscle responses are obtained in only one channel. The motor strip area that corresponds only to hand muscle activation at this new, lower stimulus amplitude should be considered as the primary motor hand region.

In other instances, stimulation at intensities higher than primary motor threshold can be also followed by localized motor-evoked responses. These may arise from direct stimulation of a supplementary motor region and thus may lead to false localization of the primary motor cortex.

As mentioned in the following (see pages 155–57), there is significant interindividual and intraindividual variability of the mapping threshold. A deep burst-suppression pattern, with long epochs of flats, a pericentral highly infiltrative tumor, baseline muscle weakness, or young age, could determine a possible high threshold. Knowledge of a correct stimulus intensity, which corresponds to the threshold, can expedite the mapping, as well as eliminate the risk of false-negative results.

There are several instances in which muscle activity can be seen in the recording muscle channels. Some of this activity may have the right latency for triggered mMEPs but is not reproducible, or it does not appear to be time-locked to the stimulus and occurs more or less randomly. One has to differentiate such spontaneous activity from the triggered muscle responses. A similar circumstance is that of seizure-triggered movements, which translate in a "jittery" muscle channel. We have seen several cases where mapping and the seizure thresholds were the same. In such situations, the mMEPs of the corresponding muscle, directly triggered by stimulation, as well as seizure-related movement of that muscle were present. The two types of muscle activity are, however, different in morphology, as well as in their relationship to the stimulation, and are usually present in different trials. For example, triggered mMEPs are first obtained, followed by the "jittery" trials (Fig. **4-5**). The ECoG at the time of the seizure-related muscle activity is analyzed to determine the spread of the electrographic seizure activity at that moment. Also, analysis of the ECoG at the time of the suspected triggered mMEP reveals whether any ADs or seizures were present. In several of our cases, the latter (ECoG at the time of mMEPs) showed no ADs or seizures and the former (ECoG at the time of seizure-related muscle artifact) showed a very localized run that was building up in the stimulated region. In such cases we did not disregard the results of the mapping.

The challenges of a noisy muscle channel have already been discussed (see page 147).

The anesthetic regimen that we use for motor mapping includes total intravenous anesthesia TIVA and no muscle relaxant. If inhalational agents or muscle relaxants are administered, interpretation of a negative response becomes challenging. In order to assess the degree of neuromuscular blockade, anesthesiologists apply TOF (train of four) to the hand muscles. However, the presence of muscle twitches at this level does not guarantee the presence of twitches in other muscles. We believe that, in order to assess reliably the effect of muscle relaxants on the muscles monitored, it is important to apply TOF to those muscles tested during the procedure. Also, one should bear in mind that TOF is not good in assessing the effects of the inhalational agents on mMEPs because most of these effects are due to cortical depression (for details regarding the effects of anesthesia, see page 173).

As a general rule, stimulated areas resulting in muscle responses should not be resected, and removal of motor-inhibitory regions should also be avoided, although resulting deficits from their removal are often temporary.

In many instances, the surgeon is constrained to perform resection of supplementary motor regions. In these cases, the patients often wake up with a motor deficit, which resolves within a few weeks following the surgery (SMA sdr). Thus, an accurate differentiation between the supplementary and primary motor regions becomes important. Unlike the cases of supplementary motor regions, damage of the primary motor cortex results in irreversible motor deficit.

### 3. Sensory Function Mapping

#### A. Principle and Interpretation

Less commonly, clinicians also employ cortical mapping during resection of lesions that are localized in the occipital-parietal lobes. Mapping of the primary somatosensory areas in the postcentral regions is usually done with the patient awake. Direct cortical stimulation of such regions is followed by positive sensory symptoms, such as tingling or pins and needles.

One could also use concomitant mapping of SSEPs, to identify the area of maximum electronegativity, represented by the highest amplitude of the N19 thalamocortical median SSEPs and the central sulcus by the median SSEPs phase-reversal technique (see pages 136–40). Subsequently, continuous stimulation of the contralateral median nerve at the wrist and recording the SSEPs via the strip allow monitoring of the integrity of such regions, as well as of the sensory pathways during supratentorial lesional resection. In the case of baseline sensory deficits or symptoms, the thalamocortical SSEPs may have poor morphology and low amplitudes; thus, recording such potentials can be challenging

Stimulation of occipital cortex may result in production of visual illusions, phosphenes, which can manifest as lights or colors that may move across the field of vision. Phosphenes from stimulation of cortex adjacent to the occipital poles are observed in the center of the visual field, whereas stimulation of parieto-occipital regions can result in complex visual illusions. Stimulation caudal to the calcarine fissure often results in phosphenes in the contralateral superior visual field or quadrant (Lesser et al., 1998).

#### B. Methodology

*B1. Technical setup*

***a. Stimulation technique.*** In order to trigger sensory or visual symptoms, stimulation can be done with either multipulse train or Penfield methods. As already mentioned, monitoring of the integrity of the large fiber sensory system is done by continuous stimulation of the contralateral median nerve at the wrist and recording of the thalamocortical SSEPs. For details regarding these techniques, see pages 134–36 and 150–51.

***b. Recording technique.*** With the patient awake, sensory and visual symptoms are the end point of the stimulation. However, as in the case of the central sulcus localization, mapping of the somatosensory cortex is done by recording the thalamocortical, usually median SSEPs over the ipsilateral hemisphere while stimulating the contralateral (e.g., median) nerve at the wrist (see pages 134–36).

***c. Recording stimulation-triggered epileptiform discharges (ADs).*** This is done as mentioned on page 162.

## IV. LANGUAGE MAPPING

#### A. Principle

The purpose of language mapping is to avoid deficits in language-processing functions after surgery. Lesion resections adjacent to the dominant Sylvian fissure, at the level of the superior temporal gyrus, in particular its posterior part and at the level of inferior frontal gyrus, are especially important indications for language mapping. The electrical stimulation of eloquent language cortex causes an interruption of language functions. Thus, during short epochs of stimulation, different aspects of the language functions should be tested.

Obviously, neurophysiologic language mapping can be done only in awake patients, which poses its own challenges (see pages 132–33). Careful *preoperative patient selection and preoperative assessment* are required. Many patients are not suited for such awake language-mapping procedure because of anxiety, immaturity, inattention, or confusion. Previous studies have not demonstrated significant differences among individuals based on age, seizure type, lesion presence, handedness, or psychiatric history (Lee et al., 2000). However, patients with elevated depression scores (Minnesota Multiphasic Personality Inventory) and significantly lower IQs are more often unable to cooperate effectively during language mapping in awake craniotomies. We believe that individuals with lesions near the primary speech centers should be carefully screened prior to surgery with neuropsychiatric testing and fMRI.

As already mentioned, electrical cortical stimulation may lower the threshold for epileptiform activity, with the possibility of triggering a seizure. In addition to interfering with accurate identification of the eloquent cortex, triggered GTCZ poses additional risks in awake patients that are related to possible compromise of airway control. Moreover, in the case of language mapping, a postictal state, characterized by dysphasia or excessive sedation and confusion, related to either medication or the seizure itself, can jeopardize further testing and compromise subsequent mapping results. Minimizing the risks of stimulation-triggered seizures is thus important and starts with the preoperative assessment (see page 133).

## B. Methodology

### *B1. Technical setup*

***a. Stimulation technique.*** For language mapping we use the bipolar Penfield stimulation technique. As in sensorimotor mapping, there is a significant variability of the stimulus parameters and techniques used for language mapping (Jayakar et al., 1992; Schäffler et al., 1996). As a general rule, in order to produce disruption of language function, the electrical impulses need to repeat, with a relatively high frequency and in a continuous fashion, during the epoch of stimulation (e.g., the multipulse train technique used in motor mapping cannot be used in language mapping). The stimulation is done with a handheld bipolar stimulator (Viasys Healthcare, Straight bipolar probe 900X143), connected to our machine. At our institution, we use an electrical current with biphasic rectangular pulse of 1 msec width as stimulus. Repetitive pulses at 60 Hz are applied directly on a cortical region (Fig. **4-3**) for short epochs (3 to 8 seconds in duration). The lengths of these epochs is, in general, determined by two factors. First is the task to be used for the patient's testing; tasks such as comprehension testing and visual recognition require more time than repetition testing or verbal fluency. The second factor is represented by the degree of baseline cortical excitability. Many of these patients also have seizures, which in fact may have prompted initial medical attention, leading to the diagnosis of brain lesion. Some of the patients with first-time seizures are suboptimally treated or, in the absence of a seizure history, may not be on any AEDs. Thus, if they harbor significant epileptiform activity, they do not have the protection of a reasonable level of AED. Application of an electrical current on such an excitable cortex poses an enhanced risk for stimulus-triggered ADs or electroclinical seizures. From our experience, a longer stimulation epoch increases the chances for the ADs to organize into an electrographic seizure. Some of our surgeons prefer that we leave the stimulator on indefinitely. This enables their control not only of location but also of length of stimulation. The neurophysiologist can recognize the initiation of the brain stimulation by watching the appearance of stimulus artifact on the ECoG.

We usually start at 1 mA and slowly increase the intensity in 0.5-mA increments, up to 15 mA or until ADs are triggered. Similar to the motor mapping technique, at each intensity, all areas of interest are stimulated in a successive fashion before the stimulus amplitude is increased to the next level. If ADs are triggered with a field beyond the stimulated area and there is language disruption, the stimulus amplitude is decreased by 0.5 mA. If ADs are triggered but no symptoms are elicited and, especially, if ADs occur at very low intensity, we continue to increase the stimulus intensity slowly. In many instances we have been lucky in "clearing" certain areas of interest for resection, in the presence of ADs, but without triggering a seizure or provoking a symptom.

We believe that a cutoff of 15 mA for calling an area nonfunctional is a somewhat conservative (as far as not missing a high mapping threshold) but safe approach. It is known that there is significant mapping threshold variability and that the Broca's threshold may be lower than the threshold of a language association cortex, such as in the posterior part of the superior temporal gyrus. It is also likely that a region infiltrated or "pushed" by the lesion may have a higher threshold than expected (see pages 155–57). In order to preserve the language function, it also becomes important to map the subcortical regions involved in the connectivity between different cortical regions involved in language function. White matter, in particular, has higher stimulation thresholds than that of the cortical regions.

On the other hand, the "cap" used for language mapping (15 mA) is lower than the one used for motor mapping (25 mA). In addition to the fact that the two stimulation techniques apply different current densities at the same stimulus intensity, the language mapping via Penfield method is done with the patients awake, whereas the motor mapping via multipulse train technique is done mostly in anesthetized patients. It is our experience that the stimulus amplitudes that are necessary to reach the stimulation thresholds are, in general, lower in the cases of language mapping than in the cases of motor mapping.

Some surgeons prefer to identify first the eloquent cortex and then to stimulate around the lesion in order to clear the area for resection. Other surgeons limit the mapping to detecting only whether the surgical path chosen for entry is clear. We usually discuss and agree with the surgeon on the plan before the surgery.

As in the cases of motor mapping, language mapping is sometimes performed prior to surgery, during the phase II monitoring of patients who are candidates for epilepsy surgery. In such cases, the stimulation is done by connecting one pair of grid (or strip) contacts at a time to the stimulation headbox (see pages 140–42).

***b. Recording for ADs.*** In stimulation with Penfield method, early detection of ADs may become problematic because of the presence of a dense 60-Hz stimulus artifact. Applying a 60-Hz notch is often helpful. For detailed technical and theoretical considerations, see pages 157–73.

*B2. Troubleshooting*

***a. Lack of stimulus artifact.*** As in the case of sensorimotor mapping, this may signal a problem with the stimulation. For example, when the two ends of the bipolar stimulator are too close to each other, the stimulation is not appropriate and the stimulus artifact is absent. Of note, we found that stimulus artifact seen with the Penfield method has a greater variability than that seen with the multipulse train. Thus, there are instances when the stimulus artifact with the Penfield method can hardly be seen with lower stimulus amplitudes.

***b. The synchronization between the cortical stimulation and patient's testing.*** This remains one of the biggest challenges. As mentioned previously, leaving the stimulator on indefinitely and allowing the surgeon to have full control of the timing of the stimulation, according to the patient's performance, helps solve this problem.

## C. Testing the Patient and Interpretation of the Results

*C1. Using function-specific tasks.* Prior to surgery, baseline verbal function must be established for verbal fluency, repetition, visual object naming, sentence comprehension, and word generation. Patients are trained to perform the same types of tasks that are used during surgery, and those with significant preoperative deficits require consideration on a case-by-case basis. Depending on the region of interest, the emphasis is put on one task versus another.

***a. Verbal fluency and repetition.*** These tasks should be tested in particular if the lesion to be resected is in close proximity to the inferior frontal region, corresponding to the *Broca's area* or the *arcuate fasciculus*. Verbal fluency can be tested by tasks such as enumeration of the months of the year, days of the week, forward and backward, counting by 2, repetition of sentences, or engaging patients in conversations about their hobbies, family, etc.

***b. Sentence comprehension.*** This can be established by showing patients two sentences that appear to be similar but contain a single word that has been substituted for another with a different or similar meaning. The patient's task is to determine whether or not both sentences carry the same meaning. This task activates *perisylvian cortices*, which participate in semantic processing and should be carefully tested in particular for resections involving *the posterior part of the superior temporal gyrus.*

***c. Mathematical deficits.*** This can also be elicited with selective stimulation of the *dominant angular gyrus.*

***d. Visual object naming.*** This is tested by showing the patients pictures or line drawings from standardized sets, such as the Boston Naming Test, that have established norms. This task requires activation of essential language cortex, such as *Broca area, arcuate fasciculus, and insular connections* (Duffau et al., 2002) and is essential for language mapping, because anomia is a common feature of postsurgical aphasia.

***e. Word generation.*** Word generation can be tested by requiring the patient to provide a list of words that fit into a specified category or start with a certain letter. Patients may also be required to name an object that is verbally described or visually presented to them by the examiner. *Frontal language areas* are thereby activated with these exercises.

***f. Anomia and dysfunction of the verbal fluency and repetition.*** This is triggered by stimulation of *the arcuate fasciculus and insular connections.*

***g. Dysarthria.*** Dysarthria is triggered by stimulating the *periventricular white matter.*

The last two types of testing are mostly useful for lesions such as gliomas, which also encompass the subcortical white matter adjacent to language centers. Thus, stimulation of the deep resection borders may provide additional information limiting the extent of resection. We believe that mapping both the eloquent cortex and the subcortical white matter is particularly important in maintaining the language function.

*C2. General considerations.* During surgery the patient is required to undertake these tasks during stimulation of various portions of the superior temporal and inferior frontal gyrus. Areas in which baseline normal speech functions (consistently correct responses to the previously described tasks for at least 1 minute) are altered or interrupted can be temporarily marked and subsequently spared from resection. Many centers, including ours, require several such trials of stimulation with speech disruption in order to consider a region of cortex eloquent. We like to have the patient starting the language task at least a couple of seconds before stimulation starts. Sedation, dryness of the mouth, discomfort, or use of drapes can render it difficult to interpret the responses. Thus, we routinely use a Fender cordless amplifier microphone system, which proves to be very helpful.

Any language disruption (e.g., speech hesitancy, trouble with repetition, word generation, naming, or fluency) caused by stimulation, in the absence of triggered epileptiform discharges or electrographic seizures extending beyond the small region underneath the stimulating contacts, can be considered a positive response. In case the epileptiform or epileptic activity is present beyond these contacts, the language deficit may be due to the spread of seizure activity to other regions than the one that was stimulated. It is important to be able to recognize such epileptiform activity. Most of these epileptiform discharges are seen after the stimulation is stopped. Thus, we refer to them as ADs. However, in many situations, there is a rhythmic run leading to these type of discharges, which starts during the stimulation itself (intrastimulation discharges) and which is usually obscured by a dense stimulus artifact. In other cases, these runs triggered by stimulation are not extending beyond the stimulation epoch and are not followed by ADs (Fig. **4-7 b**). In cases of stimulation-triggered language disruption, it is important to identify such intrastimulation runs, in particular if they have a broad field and extend beyond the stimulating contacts. Thus, when we look at the ECoG during epochs of stimulation, with the 60-Hz notch, "on" becomes important. In such cases of broad field intrastimulation runs or poststimulation ADs, we usually disregard the clinical response. In these situations, it is preferable to reduce the length of the epoch of stimulation, if this can be done without jeopardizing the accuracy of testing, while preserving the stimulus intensity and repeating the stimulation. If the stimulation epoch cannot be further reduced, repeating the stimulation at 0.5 mA lower intensity should be attempted.

As in the case of motor mapping, once a positive response is obtained, verifying the reproducibility of the response can be done by having the surgeon occasionally stimulate the identified area. The areas that elicit language disruption are again flagged for guiding the surgeon further (Fig. **4-8 b**). In order to avoid language deficits, we recommend leaving a margin of tissue of approximately 1 cm around regions that demonstrate language deficits on stimulation, even though they may be relatively distant from Wernicke or Broca regions.

A particular situation is mapping the eloquent cortex at the level of the temporo-parietal junction, where the end of the Sylvian fissure is sometimes difficult to recognize visually. The type of responses evoked by the stimulation, sensory symptoms versus language disruption, help localize the position of the stimulated contacts of an implanted grid or the anatomic location of a lesion (supra- or infra-Sylvian). A good example is the case of functional mapping performed for a deep, subcortical cavernous angioma resection. The location of the lesion could not be determined by visual inspection of the cortex. However, according to the MRI images, the lesion was located inferior to the Sylvian fissure. Using the cortical venous system as a guide, the surgeon prepared for defining a path of entry. However, the stimulation of the cortex that was presumably overlying the lesion produced a sensory response, not a language response. Thus, the stimulated region was felt to be superior to the Sylvian fissure . The lesion was indeed located more inferior than the surgeon had initially thought (Fig. **4-8 a–c**).

Mapping threshold is considered to be the smallest intensity at which a positive mapping response is obtained. Of note, because of the variability of stimulus thresholds between different lobes involved in the language function, a separate threshold for frontal and temporo or parietal regions can be considered (see pages 155–57).

Because of time constraints, some surgeons may not wait until a detailed language mapping has been completed. Instead, after establishing a path of entry, they prefer to initiate the resection while the patient is continuing to perform the verbal tasks. We think that this approach can be dangerous, because, once the language deficits are observed, the damage may be irreversible.

(a)

(b)

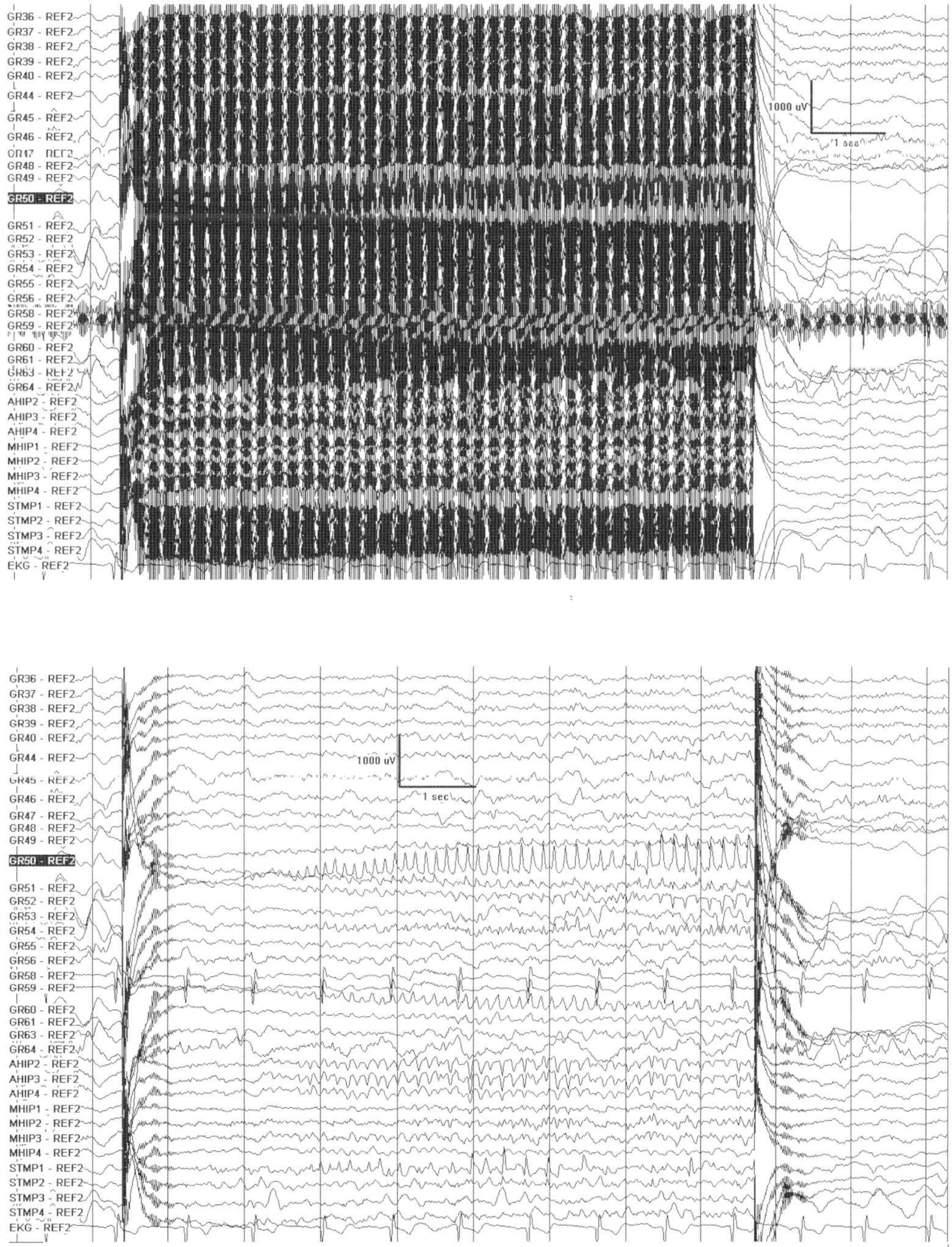


**FIGURE 4-7** ECoG recording during an 8-second epoch of electrical cortical stimulation in language mapping. The recording is done via part of a subdural grid (contacts 36 through 64); two 4-contact depth electrodes, one placed in the anterior part of the hippocampus (AHIP 1 through 4) and the other placed in the middle hippocampus (MHIP 1 through 4); one 4-contact subtemporal strip (STMP 1 through 4). The stimulation is done at contacts 58–59. Details of the grid are shown in Fig. **4-16,** panel **D,** where the customized montage used in Fig. **4-7** is delineated by the bold line. The calibration mark is 1000 $\mu$V and 1 second. **a:** Significant amount of 60-Hz stimulus artifact interfering with the ECoG analysis during the stimulation. **b:** Same stimulation epoch, this time with 60-Hz notch filter "on": notice a gradual buildup of rhythmic, sharp activity, starting at 8 Hz, 800 $\mu$V, and reaching 5 Hz, 1500 $\mu$V, at the end of the stimulation, when it stops abruptly, without following afterdischarges. Notice the broad field of this epileptic run, with best-formed morphology at contact 50, but extending at grid contacts 60 and 51 through 54, the anterior hippocampal depth contacts (AHIP 2–4) and at contact 1 on the subtemporal strip (STMP 1).

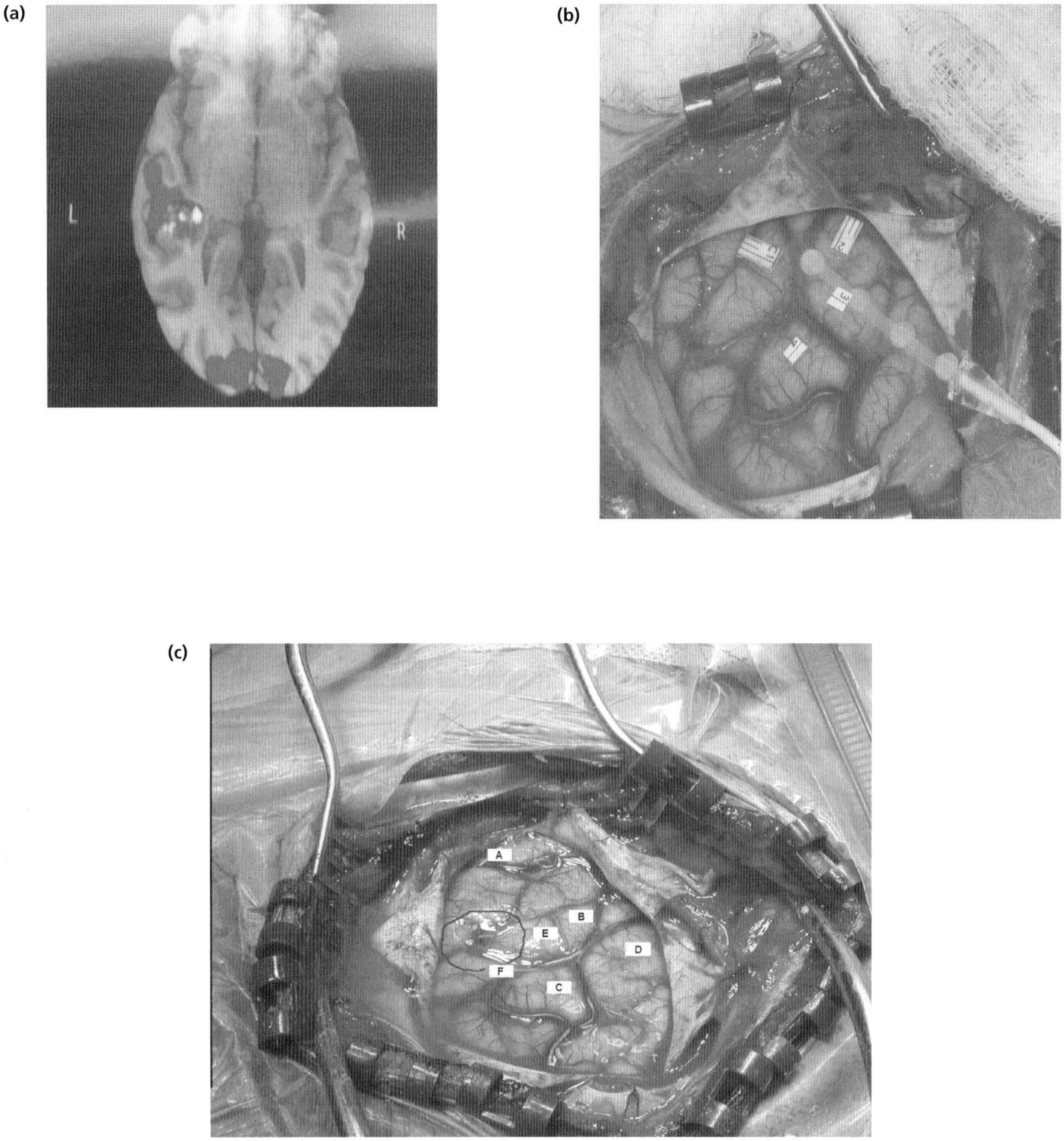


**FIGURE 4-8** Functional language mapping for left temporal cavernous malformation resection. **a:** The image is showing an axial fMRI slide, with areas of language activation located in close proximity to the lesion, anterolateral and superior medial to it. **b:** The image is showing the identified areas of eloquent cortex, flagged for further guidance, as well as placement of the 8-contact strip used for identification of afterdischarges during stimulation. **c:** The topography of surgical field in the same patient: **A:** v. Labbe; **B:** speech arrest; **C** (inconsistently), and **D** (**especially**): right hemisensory complaints; **E:** word choice errors; **F:** limb of v. demarcating superior aspect of the lesion; **Circle:** estimated delineation of lesion. Note that **B** and **E** are located in superior temporal gyrus just above superior temporal sulcus; between **A** and **B** is the middle temporal gyrus; sulcus between **B, E** and **C, D** is the Sylvian fissure.

## V. CORTICAL MAPPING THRESHOLD VARIABILITY

Multiple factors influence the variability of mapping thresholds. Understanding how there factors affect the stimulation threshold, as well as being able to weigh in such effect, becomes essential for an accurate prediction of the individual mapping thresholds. On the other hand, being able to estimate the thresholds is an important element in successful functional mapping and in prevention of devastating neurologic complications.

In this section, we have enumerated several factors that have been reported in the literature to have a direct impact on the functional mapping thresholds . We have also studied the predictive value of these variables in influencing stimulation thresholds, by using a linear regression multivariate analysis. The results of our analysis are summarized as follows.

### 1. Depth of Cortical Suppression

EEG has been used for many years in estimating the depth of anesthesia, by correlating the latter with the average frequencies, amplitudes, and EEG patterns (Esmaeili et al., 2007; Kortelainen et al., 2007; Morimoto et al., 2004; van den Broek et al., 2006). Thus, the depth of cortical suppression could be appreciated by the degree of cortical suppression.

It is generally accepted that an EEG burst-suppression pattern reflects significant suppression of cortical activities. Doyle and his colleague (1999) believe that this suppression is due to the effects of anesthetics on cortical excitability and/or cortical metabolism, oxygen consumption, and cerebral blood flow. In concordance with prior animal studies, which showed that most cortical neurons become electrically silent during flat EEG epochs (Shaul, 1998), we believe that the degree of EEG suppression, appreciated by the length of EEG flats, provides a good estimate of the depth of cortical suppression. We analyzed the impact of EEG suppression on the threshold, and we found that increasing the length of the EEG flats within the EEG burst-suppression pattern by 1 second produces an increase in the mapping threshold by 0.96 mA ($p = 0.002$). This effect was independent from the effect of use of anesthetics, suggesting that it should depend on the blood concentrations of anesthetics and, possibly, on the variability of the degree of cortical suppression at the same drug concentrations, under the direct influence of factors such as age and temperature.

However, it is unlikely that the lengths of the epochs of suppressed EEG also reliably predict the cortical excitability threshold at any given time during stimulation, because the burst-suppression pattern is a succession of both flats and bursts. In their study, Detsch et al. (2002) concluded that the neurons in the primary somatosensory cortex had a significantly higher-amplitude response during bursts than during flats, suggesting that the profound suppression of the sensory information is suspended during EEG bursts. Similarly, studies on rodents suggest that the burst-suppression pattern is a combination of both excitatory and inhibitory mechanisms (Lukatch et al., 2005) and that the anesthetic induced burst suppression EEG activity requires glutamate-mediated excitatory synaptic transmission (Lukatch et al., 2005). In concordance with these studies, we also believe that the burst-suppression pattern is a cyclic succession of excitatory (the bursts) and inhibitory (the flats) epochs, translating into a cyclic succession of higher and lower stimulation threshold.

Thus, our results indicating a relationship between increased mapping thresholds and longer epochs of EEG flats are likely related to an increased probability of stimulating during longer, suppressed epochs.

### 2. Type and Dosage of Anesthetics

Volatile agents have been found to decrease the pyramidal neurons and cortical interneuronal excitability, thus decreasing the amplitudes of D and I waves, decreasing cortical synaptic transmission (Haghighi, 1998; Hicks et al., 1992) and significantly increasing the neocortical GABA (A) receptor-mediated inhibition (Hentschke et al., 2005; Osawa et al., 1994). The depression of cortical activity during anesthesia with propofol has been explained by its effect on the neuronal excitability as a result of selective suppression of Na (+) currents and L-type HVA Ca (2+) conductances in cortical neurons, as well as opening of GABA (A) gated Cl channels (Martella et al., 2005). Other animal and human studies describe a definite effect of propofol on cerebral blood flow, cerebral oxygen consumption, and lactate and glucose metabolism (Vandesteene et al., 1988; Werner et al., 1992). In our study, the effects on mapping thresholds of TIVA and use of inhalational agents decreased by almost one-half (TIVA) and one-third (inhalational agents), respectively, when these effects were adjusted for the induced cortical suppression. After adjusting for the EEG suppression, as a measure as cortical suppression and compared to awake patients, use of TIVA increases the threshold by 2.71 mA

($p$ = 0.011) and the use of inhalational agents by 5.85 ($p$ = 0.0007). The results, however, even when demonstrating a decrease in magnitude of the effect (from 5.23 mA and 8.95 mA, respectively, in the univariate analysis), remained statistically significant and are consistent with prior studies showing that anesthetic agents act not only at the cortical level but also in other regions of neuraxis, involved on the generation of motion. Volatile agents were found to decrease the electrical transmission at the level of the corticospinal axons, as a result of depression of Na (+) currents at the nodes of Ranvier (Burke et al., 2000), and suppress motor pathways at the spinal level, by interfering with the synaptic transmission at the alpha motor neuron (Mavroudakis et al., 1994). Furthermore, others have suggested that the latter is their main effect on generation of movement (Loughnan et al., 1989; Zentner et al., 1992).

## 3. Age

It is well known that a young brain (the first five years of life) shows higher mapping thresholds than those of the adult cortex because of incomplete maturation of the nervous tissue and relative inexcitability of the motor cortex (Goldring et al., 1984; Hines et al., 1940; Jayakar et al., 1992; Sala et al., 2002). Other studies support the assumption that older age is associated with higher mapping thresholds. Because the mean propofol consumption and recovery time are age-dependent, one should expect deeper levels of anesthesia in elderly patients and thus decreased cortical excitability as measured by the burst-suppression pattern (Kreuer et al., 2005). Schwartz et al. (1989) studied the effect of aging on cortical excitability in patients anesthetized with volatile agents and showed that older patients spend a greater proportion of time in electrical silence, because of increased central nervous system sensitivity to such agents, rather than to a decrease in anesthetic clearance. Thus, we believe that the relationship between age and mapping threshold is U-shaped, with higher thresholds expected early and late in life.

The results of our study support the association between aging and higher mapping thresholds. Considering 18 mapping cases where anesthesia consisted of TIVA, the mean length of EEG flats in the older group was significantly longer than the mean length in the younger group (3.7 sec vs. 1.31 sec; $p$ = 0.025). Our results support the hypothesis that the cortical suppression induced by anesthetics is more significant in older patients. In our univariate analysis, patients over the age of 50 had a mapping threshold that was higher by 4.37 mA compared to that of patients of less than 50 years of age ($p$ = 0.0005). However, when adjusted for the confounding effect of cortical suppression, the effect of age became nonsignificant, suggesting that most, if not all, of the effect of age on the mapping threshold is due to the effect of aging on the response of cortical excitability to anesthetics.

## 4. Location and Pathology of the Lesion

Bittar et al. (2000) showed that central lesions are more likely to cause distortion of the somatosensory cortex than noncentral lesions. We agree with his findings. We are of the opinion that, for the purposes of functional mapping, the location and pathology of the lesion need to be considered together.

There are several ways in which the presence of cortical lesions can affect the excitability of the eloquent cortex. Highly invasive pathological type is a reliable indicator of the infiltrative tumoral capability, leading to distorted cortical anatomy and function. Thus, highly infiltrative lesions situated in close proximity of the eloquent cortex could cause an increase in the mapping thresholds. This theory is supported by studies showing an altered interaction among neurons, astrocytes, and blood vessels, when surrounded by neoplastic cells (Hossmann et al., 1992; Linn et al., 1989; Nagano et al., 1993). Similarly, fMRI studies showed a decreased BOLD signal in the sensorimotor areas situated close to glial tumors (Lüdemann et al., 2006; Schreiber et al., 2000). Using fMRI BOLD signal, Krings et al. (2002) investigated the effect of the severity of tumor infiltration in M1 (primary motor cortex) in influencing the BOLD signal of different motor areas and concluded that greater is the infiltration, the lesser is the signal change with motor activation within M1 and the stronger is the signal change in the supplementary motor regions.

Conversely, the presence of well-demarcated tumors is more likely to cause "en bloc" displacement of cortical and subcortical regions, with preservation of the correspondence between somatotopic anatomy and function, in and around the central sulcus. Similarly, the presence of mass effect and edema may, on one hand, directly modify cortical excitability and, on the other hand, cause displacement of the eloquent cortex, with preservation of its relatively normal function (Lehéricy et al., 2000).

Neurodevelopmental lesions, or injuries occurring in the intrauterine period or within the first years of life, can be associated with unexpected locations of the functional relevant regions, as a result of brain plasticity, as well as possible abnormal neural networks, with an imbalance between the excitatory and inhibitory circuits, thus resulting in unexpected location of the eloquent cortex, as well as in unexpected mapping thresholds.

### 5. Stimulation Technique

Mapping threshold variability is also introduced by the variability of the used stimulation technique. Several mapping techniques are described in the literature of specialty, each with their own mapping threshold normative data (Berger et al., 1989; Jayakar et al., 1992; Katayama et al., 1988; Kombos et al., 2000; Neuloh et al., 2002; Sala et al., 2002; Schäffler et al., 1996). A correlation among these normative data is of significant practical importance. From our experience, the mapping thresholds (measured in the current intensity expressed in milliamperes [mA]) established via the Penfield method were overall lower than the mapping thresholds established by the multipulse train technique. This is the reason why the stimulating algorithm used for the two methods uses different caps (25 mA for multipulse train and 15 mA for Penfield). Our data analysis showed that the used of the Penfield method decreased the mapping threshold by 4.50 mA, compared to the use of the multipulse train technique ($p$ = 0.0007). However, when we corrected for the confounding effect of the anesthetics use (given the fact that the multipulse train is almost invariably used in the anesthetized patients, whereas the Penfield method is used in the awake patients), the difference decreased to 0.32 mA and lost its significance.

### 6. Interlobar or Regional Variability of Cortical Excitability

Prior studies have suggested that association cortex, as well as areas with increased number of synapses and connectivity, such as somatosensory and language regions, have an overall higher mapping threshold than primary motor cortex or Broca's area. Along the same lines, Pouratian and colleagues (2004) found a threshold variation of up to 6 mA between cortical sites that were approximately 1 cm apart.

## VI. ECOG DURING FUNCTIONAL MAPPING

### A. Principle

*A1. The negative impacts of stimulation-triggered epileptiform activity.* As previously mentioned, the electrical current applied directly on the cortex during electrophysiologic mapping can trigger epileptiform and epileptic activity. This can have two negative impacts on the mapping procedure. First, the stimulation-induced epileptiform discharges, also known as ADs, can further organize into an intraoperative seizure. This can be a serious threat to the patient's safety, in particular during awake craniotomy. It can also jeopardize further mapping, because of significant sedation introduced by AED medication or confusion, somnolence, and dysphasia in the immediate postictal period (e.g. language mapping). Second, occurrence of ADs can spread to adjacent cortical regions that are different from the stimulated regions, leading to false localization of the eloquent cortex, because the seizure activity itself, rather than the actual stimulation, could be the reason for the positive motor, sensory, or language response.

Thus, continuous ECoG recording should be done during each case of electrophysiologic functional mapping.

*A2. The variability of stimulation-triggered epileptiform activity.* ADs have been widely studied. Electrophysiologic responses to low-intensity stimulation were found to vary between cortical networks and were pathologically enhanced in epileptic tissue such that rhythmic spike and waves were produced (Ajmone Marsan, 1980; Ajmone Marsan et al., 1973; Bates, 1963; Bengzon et al., 1968; Gloor, 1975).

Variability exists in the threshold and duration of ADs in relationship to the type of pathology and structural integrity of the cortical network (Bernier et al., 1990; Cherlow et al., 1977; Jasper, 1954; Walker, 1949). A cortical area of increased excitability (e.g. cortical dysplasia) can translate into baseline epileptiform activity and also result in a facilitated production of large field potentials after the cessation of electrical stimulation and appearance of ADs, particularly when situated in close proximity to the stimulation area. This area could display lower threshold for stimulation-triggered epileptiform activity. Hence, it becomes essential to be able first to assess the baseline, prestimulation cortical excitability, and, then, during the stimulation, to monitor the existing epileptiform activity, if any, as well as to assess for new ADs. This is achieved by the concomitant use of ECoG during the entire mapping procedure. The observed triggered change in the baseline epileptiform activity or the triggered de novo ADs should prompt the clinician to employ caution in further stimulations—especially in the frontal lobe regions, where seizures spread rapidly. Mindful of operative time constraints, the clinician must individualize stimulus intensity at each studied cortical site to avoid production of ADs but also to avoid stimulation below the mapping threshold for that region.

Even though stimulation-induced ADs seem to be more likely to occur in the presence of preexisting, increased cortical excitability, they can be elicited in most patients, despite AED therapy or lesional histopathology.

Variability exists in the location of ADs relative to the location of the lesion to be resected and the site of stimulation. Location of ADs correlates with the variability of the increased cortical excitability and is influenced by the same factors. However, the location of baseline-increased cortical excitability (sometimes translating in baseline epileptiform activity) relative to the location of the brain lesion can also vary significantly. Although cortical dysplasia are notorious for their intrinsic epileptiform potential, other developmental lesions, such as some of the cortical tubers in TS (tuberous sclerosis) or vascular malformations (e.g., meningioma, cavernous angiomas), are rather surrounded by cortical tissue with increased excitability. In other circumstances, the baseline epileptiform activity can be present at a distance from the lesion itself. This may be due to significant tissue damage in the region occupied by the lesion, with decreased local excitability, and it may also be due to the presence of edema or mass effect compressing and irritating the adjacent regions. On the other hand, such activity may also spread at distance and involve more epileptogenic regions through connecting pathways. For example, it is our experience that, in temporal lobe pathology, one of the most commonly seen sites for baseline epileptiform activity remains the mesiotemporal structures.

If stimulation is done at distance from baseline epileptiform activity, it is less likely that the pattern and frequency of such discharges will be changed. (Fig. **4-9 a-1, b**). However, Adam et al. (2004) found that, when stimulating the medial temporal lobe, ADs were triggered at distance, contralaterally. Thus, one could argue that, as a result of connectivity between different regions, a stimulus can induce change in baseline epileptiform activity, which may evolve into a seizure, even at distance from the stimulated area. We have rarely encountered this situation as well. (Fig. **4-9a-2**).

As already mentioned, because preexisting epileptiform activity is a sign of increased baseline excitability, one should expect stimulation in close proximity to such activity to trigger an increase in frequency and organization of such discharges in electrographic seizures (Fig. **4-10 a, c, d**). This observation is supported by data of Bernier et al. (1990), which showed that the location of electrically induced seizures (with or without prior existing epileptiform activity) was a good indicator of the abnormally increased baseline excitability and thus of the cortical resection areas. Some epileptologists have tried to usc cortical stimulation as a tool in depicting the areas of increased cortical excitability. Blume and his colleagues (2004), however, have not found a consistent correlation between the site of stimulus-eliciting ADs and that of spontaneously appearing seizures

Variability also exists in the threshold, morphology, and duration of ADs in relationship to different anatomic locations. Several authors (Lesser et al., 1984; Pouratian et al., 2004) reported a range difference of several milliamperes mA in the such thresholds, observed in cortex less than 10 mm away.

Variability in relationship to timing. Variability exists also in the threshold, morphology, and duration of Ads in relationship to the timing of stimulation (Ajmone Marsan, 1972; Blume et al., 2004; Gloor, 1975; Walker, 1949).

In addition to influencing prior epileptiform activity, stimulation can also trigger de novo ADs, seen mostly after the stimulation stops (ADs) but sometimes preceded by a short intrastimulation run. As previously mentioned on pages 151–54, recognition of such intrastimulation runs becomes important not only for safety reasons but also for validation of the mapping results (see pages 151–54).

Interestingly, on a few occasions, we've seen the reverse happening as well, when cortical stimulation did not trigger ADs, but rather aborted the preexisting baseline discharges (Fig. **4-11**). This finding supports results of prior studies showing that application of a high-frequency stimulation (Lesser et al., 1999) can terminate the ADs. Others have found that application of a low-frequency stimulus (Yamamoto et al., 2002) can have a similar effect. We believe that successful abortion of epileptiform activity depends on the relationship between characteristics of the applied stimulus and those of the epileptiform discharges. For example, Motamedi et al. (2002) found that stimulation, applied early and on the negative peak of the ADs with rhythmic pattern, is more likely to abort them. Thus, the stimulus parameters necessary to abort epileptiform or epileptic activity differ from case to case.

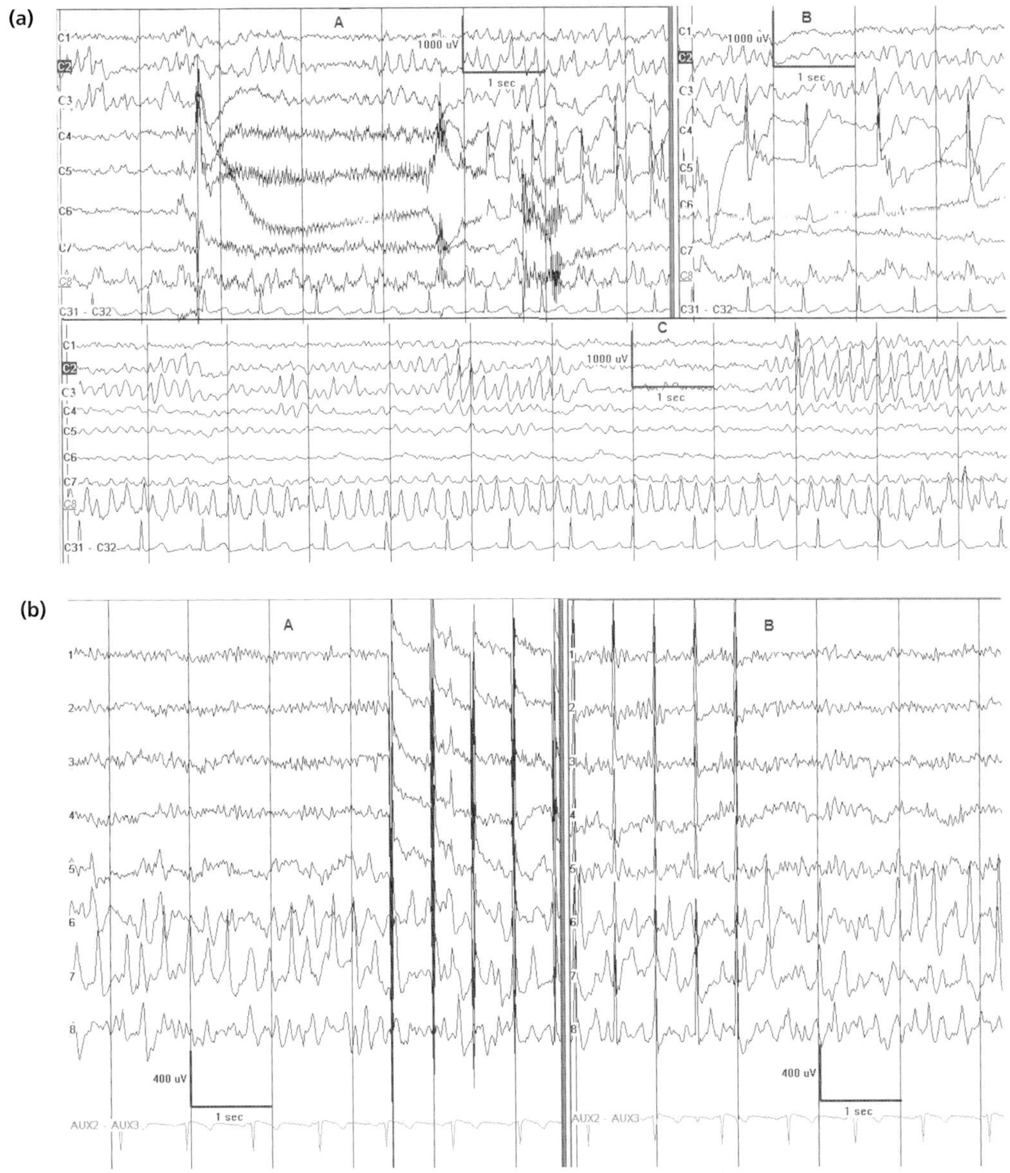


**FIGURE 4-9** The effects of electrical stimulation on cortical excitability. **a:** ECoG recording via an 8-contact (labeled C1, C2, C3, C4, through C8; strip, during and following stimulation with Penfield technique close to contacts C4 and C5. The figure is showing three consecutive epochs: first in panel **A,** followed by the second in panel **B,** and the third in panel **C.** The recording is done in a referential montage, with the reference electrode placed on the contralateral mastoid. C31–C32 channel shows the EKG recording. The calibration mark is 1000 $\mu$V and 1 second. The following epileptiform activity is seen: **1:** C2 and C3 contacts show baseline, prestimulation rhythmic theta activity, intermixed with spikes seen at irregular intervals and with a frequency up to 7 Hz; this activity is not affected by stimulation. **2:** C8 contact shows repetitive sharps, up to 300 $\mu$V and 2–4 Hz, at irregular intervals; several seconds after the stimulation stops, these discharges eventually increase in amplitude and frequency and organize in a rhythmic run of sharp, 500 $\mu$V, 5 Hz epileptic activity. **3:** High-amplitude spike and polyspikes, seen maximal, up to 1000 $\mu$V, at contacts C4, C5, and C6; they are de novo afterdischarges, triggered after stimulation; notice the organization of these discharges and their evolution, with a decrease in frequency from 3–4 Hz toward 1 Hz ( panel **B**). **b:** ECoG recording via an 8-contact (labeled 1 through 8) strip, before, during, and after stimulation with multipulse train technique. Panel **A** shows 4 seconds of prestimulation baseline, followed by 2 seconds of stimulation. Panel **B** shows the last 2 seconds of stimulation, followed by 3 seconds of poststimulation background. The recording is done in a referential montage, with reference electrode placed on the contralateral mastoid. Aux2–Aux3 channel shows EKG recording. The calibration marks in both panels are 400 $\mu$V and 1 second. Notice the baseline, prestimulation spikes maximal at contact 7 and less at 6 and 8, up to 500 $\mu$V and 4 Hz, at irregular intervals. This epileptiform activity was not affected by stimulation, which was done close to contacts 3 and 4.

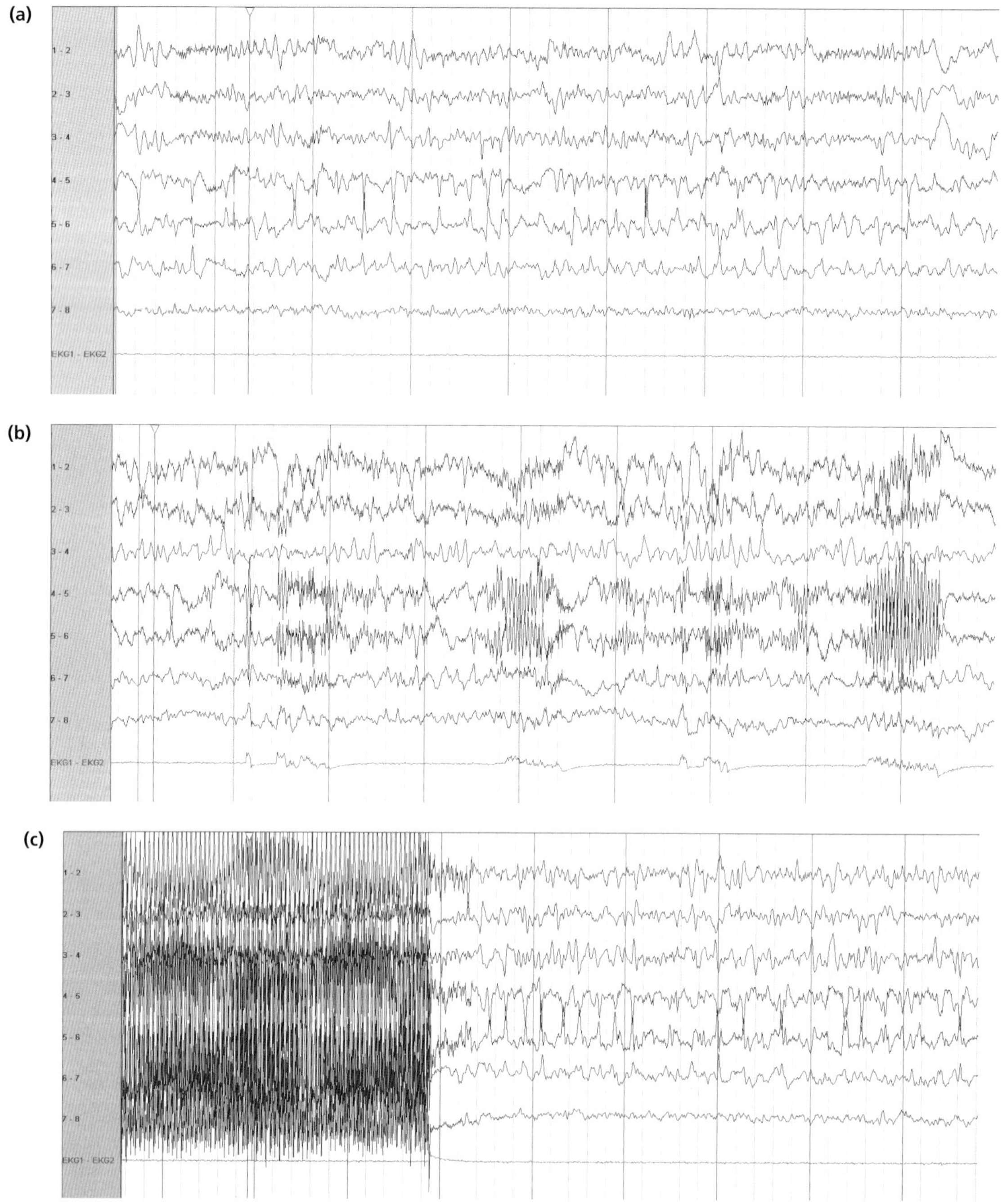


**FIGURE 4-10** ECoG recording during functional language mapping, showing how cortical stimulation via Penfield technique led to organization of baseline epileptiform discharges into an electrographic seizure. The recording is done via an 8-contact (labeled 1 through 8) strip, in a bipolar montage. The time base is 30 mm/sec and the sensitivity is 20 $\mu$V/div. **a:** Baseline epileptiform activity, seen in channels 4–5 and 5–6, prior to cortical stimulation. The spikes show a phase reversal at contact 5 and are seen at irregular intervals, up to 200 msec apart. **b:** Next 9-second epoch, showing brief bursts of 25–30 Hz sharp activity, seen in the same channels (4–5 and 5–6) as the baseline discharges described at **a.** The differential diagnosis includes organized "buzz" of epileptiform activity, as well as electrical artifact. Notice the changes seen in the EKG recording (EKG1–EKG2), synchronous with the fast activity previously described. This activity was considered to be related to stimulus artifact. Of note, filtered muscle artifact can have quite similar morphology, but such artifact has no place in direct cortical recordings (except maybe recording in a referential montage, with a "contaminated" reference electrode, located on the scalp). **c:** After a dense stimulus artifact, seen within the first 3.5 seconds of the next 9-second panel, notice the increase in frequency of the baseline discharges described at **a,** up to 5 Hz or so and occurring at more regular intervals.

**(d)**

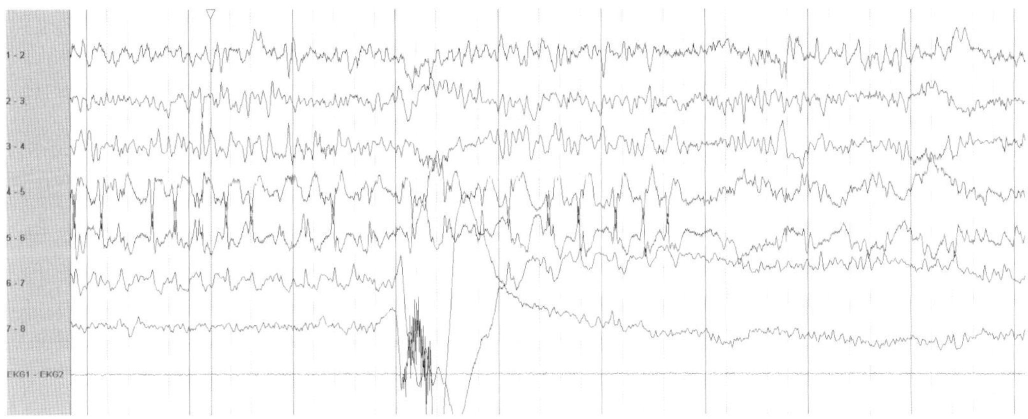


**FIGURE 4-10** *(Continued)* **d:** The panel shows the last 6 seconds of the poststimulation rhythmic run of periodic afterdischarges.

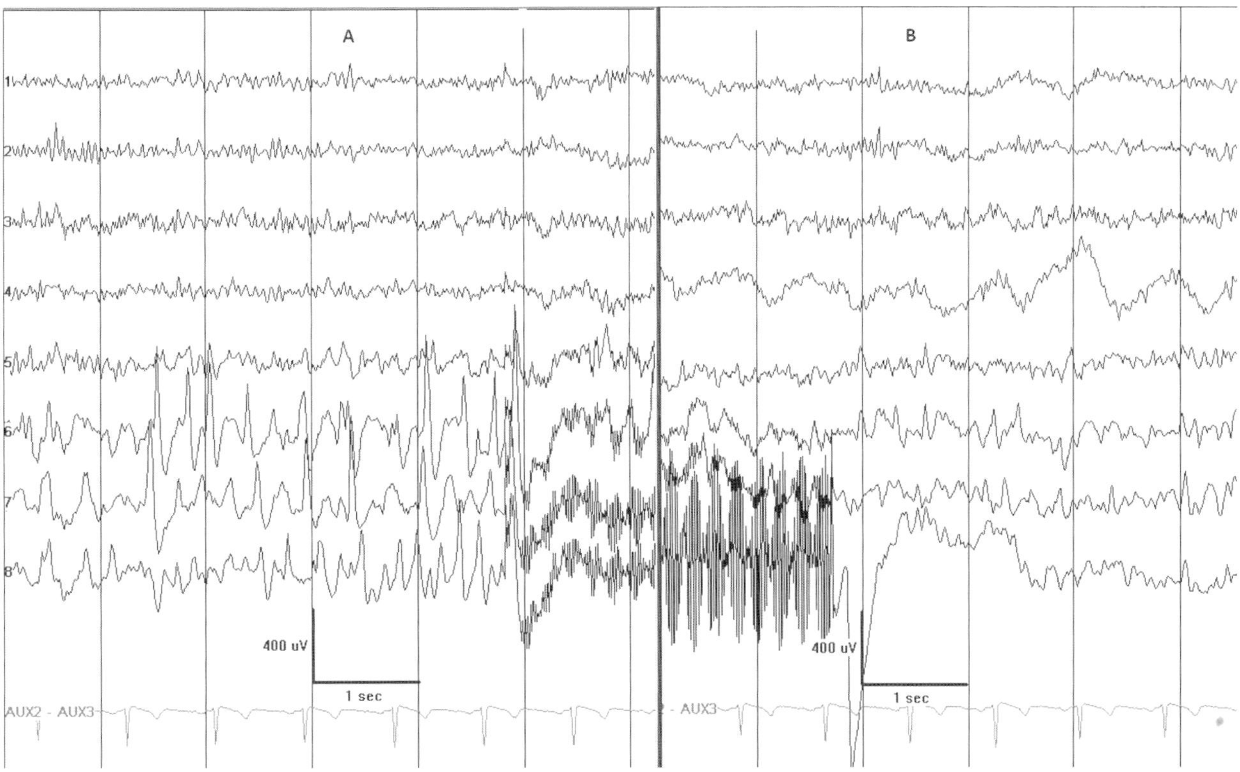


**FIGURE 4-11** ECoG recording during language mapping, showing interruption of baseline epileptiform activity by direct electrical stimulation with Penfield technique within the area of baseline increased excitability. The recording is done via an 8-contact (labeled 1 through 8) strip, in a referential montage, with reference electrode placed on the contralateral mastoid. In panel **A**, notice the baseline, prestimulation spikes, seen at contacts 6 through 8, up to 800 $\mu$V and up to 200 msec apart, at irregular intervals. Stimulus artifact follows, seen in the same channels in the last second of panel A and within the first 2 seconds of panel **B**. Notice the attenuation to disappearance of the baseline epileptiform discharges, following cessation of the stimulation (panel **B**). The calibration marks in both panels are 400 $\mu$V and 1 second.

## B. Methodology

*B1. Technical setup*

***a. The recording technique.*** We commonly use the same 8-contact strip for ECoG recording that was first used for recording of SSEPs during central sulcus localization. The strip, however, may have to be repositioned, close to the regions to be stimulated. The reference can be a needle, placed at the margin of the surgical field or, preferably, at the contralateral mastoid. In awake craniotomies, when attached to the mastoid, this is replaced with a surface stick-on electrode. Even though one may argue that the latter introduces muscle artifact and uneven impedance (compared to the electrodes in direct contact with the cortical surface), this location offers an equidistance between the reference and each of the recording contacts. This becomes important in situations when a precise location of the epileptiform discharges is essential, such as in cases of diagnostic ECoG, where the main goal is delineation of the seizure focus (see pages 98–115).

In cases of functional mapping performed in epilepsy patients undergoing phase II invasive recordings, the ECoG is achieved through the implanted grids, strips, and depth electrodes. In these cases, the reference electrode is one of the four contacts of a strip, placed subdurally, upside down. As in most phase II recordings, the number of channels that need to be included on the review screen is large, and one has to use a very low-sensitivity recording. Such recording makes it very difficult, if not impossible, to identify early onset, fast-frequency epileptiform activities, which usually have a low amplitude. In order to increase the sensitivity of the recording, we like to use a customized montage containing only the necessary channels, involving contacts situated in close proximity to the stimulated contacts, as well as the "active" channels, where the presence of baseline epileptiform activity indicates a decreased threshold for ADs (Fig. **4-7**).

The recording parameters used are low-frequency (LFF) = 1 Hz, high-frequency filter (HFF) = 70 Hz, sensitivity 20–70 $\mu$V/div, time base 30 mm/sec. The notch can be "on" for 60 Hz or "off." Recording with the 60-Hz notch filter "on" becomes important for minimizing the 60-Hz artifact produced by the Penfield stimulation and facilitates the analysis of the ECoG recording during the stimulation. Doing so also enables the detection of any intrastimulation buildup and thus prediction of the imminent appearance of epileptiform discharges, which may, in instances, propagate beyond the stimulation period (see Fig. **4-16**).

*B2. Troubleshooting*

Intraoperative neurophysiologic recordings are notorious for the diversity of artifacts that can obscure the interpretation. ECoG recordings do not make an exception.

***a. Unstable electrode contacts.*** "Loose" contacts can translate in artifacts resembling epileptiform discharges, as well as in a distorted signal-to-noise ratio, when amplitudes of the signal appear disproportionately higher or lower than the signal amplitudes in the adjacent channels. On the other hand, real spikes can be mistaken as electrode artifact, in particular when they have a very brief duration and lack the aftercoming slow wave (Fig. **4-12**). Thus, it becomes important to be able to differentiate electrode artifact from brain activity (see Fig. **4-17 b** and page 124) and, ideally, to eliminate it. We believe that checking the impedances of the strip contacts helps. However, certain manufacturers do not allow impedance checks during cortical recordings, because they consider them unsafe while the recording electrodes are in direct contact with the cortex. We believe that the current density delivered by the machine for such an impedance check is minute and does not pose much risk to the patient. (see pages 115–17).

Irrigation of the unstable electrode with saline may improve the impedance. If all of the impedances are reasonably high but equal, one can still obtain a reliable recording. If one contact introduces significant artifact, it is best to try to reposition the strip. A frequent cause of widespread high impedances remains improper connection of the strip. The connector needs a certain positioning of the contacts. When this position is achieved, securing the connector makes a "snapping" sound. If this is not heard, it is likely that the connection is suboptimal. In many cases, in particular in those with a limited exposure of the surgical field, the surgeon is forced to place the strip blindly, sometimes with few contacts out of sight, under the dura. In such cases, the most distal contacts may not be stable. It is best either to try to reposition the strip or to disable such channels.

***b. Pulsation artifact.*** This is frequently encountered in the ECoG recordings. One should avoid placing the strip contacts on blood vessels: in addition to a poor recording of the cortical electric activity, such location can certainly amplify the pulsation artifact.

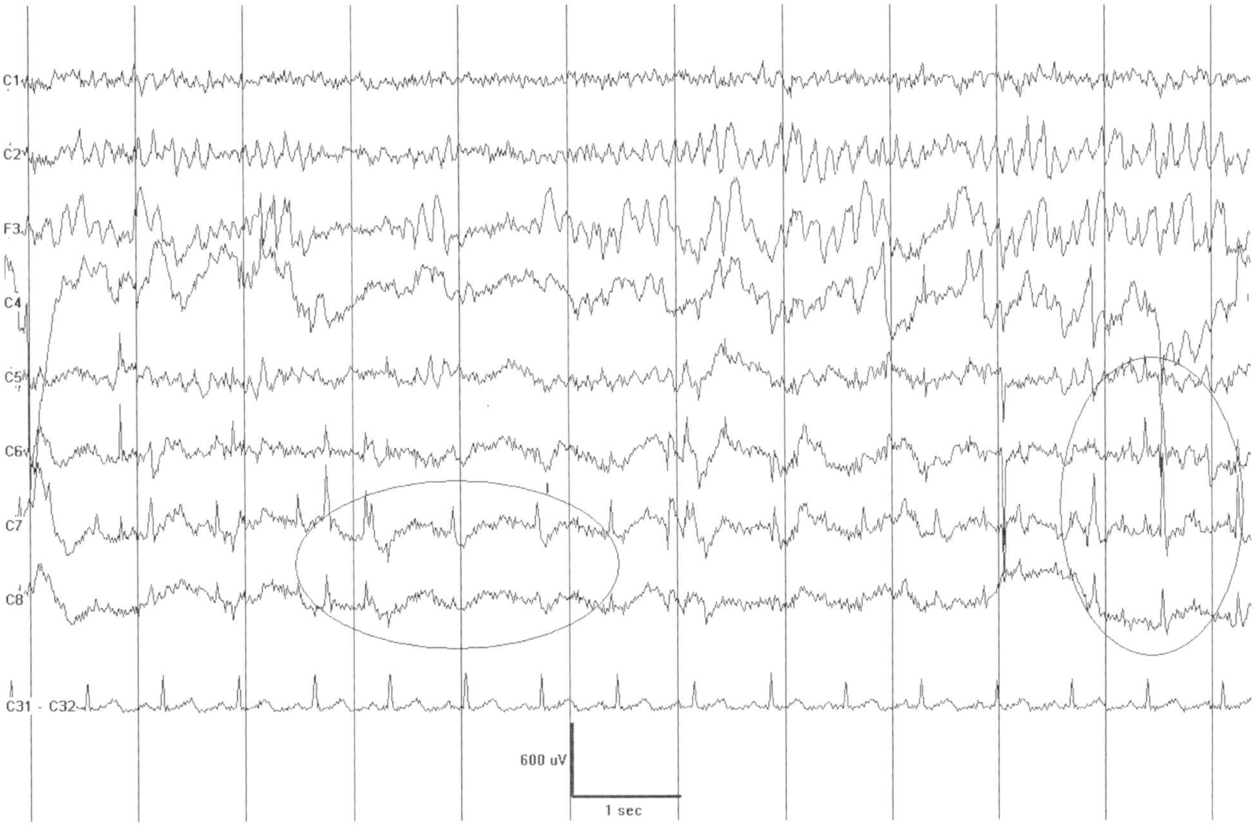


**FIGURE 4-12** ECoG showing baseline brief spikes, not to be mistaken as EKG or electrode artifact. The recording is done via an 8-contact (labeled C1, C2, F3, and C4 through C8) strip in a referential montage, with the reference electrode placed on the contralateral mastoid. The calibration mark is 600 $\mu$V and 1 second. There were very brief spikes, up to 600 $\mu$V and not followed by a slow wave. They are seen in channels C6 through C8, maximal at C7 (circled), at a frequency similar to that of ORS complexes seen in the EKG channel. However, they were not time-locked with the EKG activity, they had an electrical field, and they were aborted by irrigation with ice-cold saline.

***c. 60-Hz artifact.*** The artifact can be decreased by applying the filter 60-Hz notch and also by unplugging electrical equipment (see pages 145–47).

## C. Interpretation

*C1. Background activities.* ECoG analysis allows us to estimate the depth of cortical suppression and thus of the mapping threshold by evaluating the depth of the burst-suppression pattern appreciated in the lengths of the EEG flats (see page 155). The longer that these epochs of EEG flats are, the more significant is the cortical suppression. As previously mentioned, the simple presence of the burst-suppression pattern cannot predict whether successful stimulation, resulting in positive responses, has occurred during the flat epochs.

Additionally, even if burst suppression is generally considered a sign of cortical suppression, there is more than one type of burst-suppression pattern. Although anesthetics within same class display similar patterns (Hoffman et al., 1995), significant variation among different classes of anesthetics has been reported (Akrawi et al., 1996). This is in concordance with our experience. We have seen EEGs with similar lengths of the flats but different length, morphology, and amplitude of the bursts. Significant intraclass, intraindividual variations are seen as well. Even during a constant concentration of anesthetics, relatively fast changes in the lengths of the flats were observed. These changes are most likely due to different states of arousal, when external stimulation can result in decreased duration of suppression and increase in the total duration of the bursts, thus suggesting relative preservation on the cortical reactivity during anesthesia (Hartikainen et al., 1995).

In addition to a good estimate of the mapping threshold, the depth of a burst-suppression pattern also allows appreciation of the imminence of triggered ADs. The deeper the suppression is, the more protective it is against such activity. However, triggered seizures can still occur, even during a reasonably deep EEG burst-suppression pattern.

Bursts of epileptiform activity, in particular polyspikes, may mimic the bursts of a burst-suppression pattern. Their etiology may be easier to determine if their occurrence is related to the stimulation (Fig. **4-13**), if they occur at a high frequency, and/or if their frequency is rapidly changing.

The bursts within this pattern are often characterized by high amplitude, fast, and sharp activity, which needs to be differentiated from epileptiform discharges (Fig. **4-14**). Certain anesthetics (see page 173) determine the presence of a significant amount of sharp beta activity, which can also be mistaken for epileptiform activity. High amplitude, rhythmic mu rhythm is frequently seen over the motor cortex and can also mimic abnormal activity .

*C2. Baseline and triggered epileptiform activity.* ADs are characterized by rhythmic high-voltage spikes and/or polyspike and wave-type complexes that continue after the electrical stimulation and then terminate rather abruptly. Jasper (1954) first depicted the different types of AD morphologies in humans, which can be grouped as (1) sequential spikes (Jasper types A, B, C, H, L), (2) sequential spikes with pauses (K, G), (3) spike-waves (D, F, J), and (4) rhythmic waves (E, I).

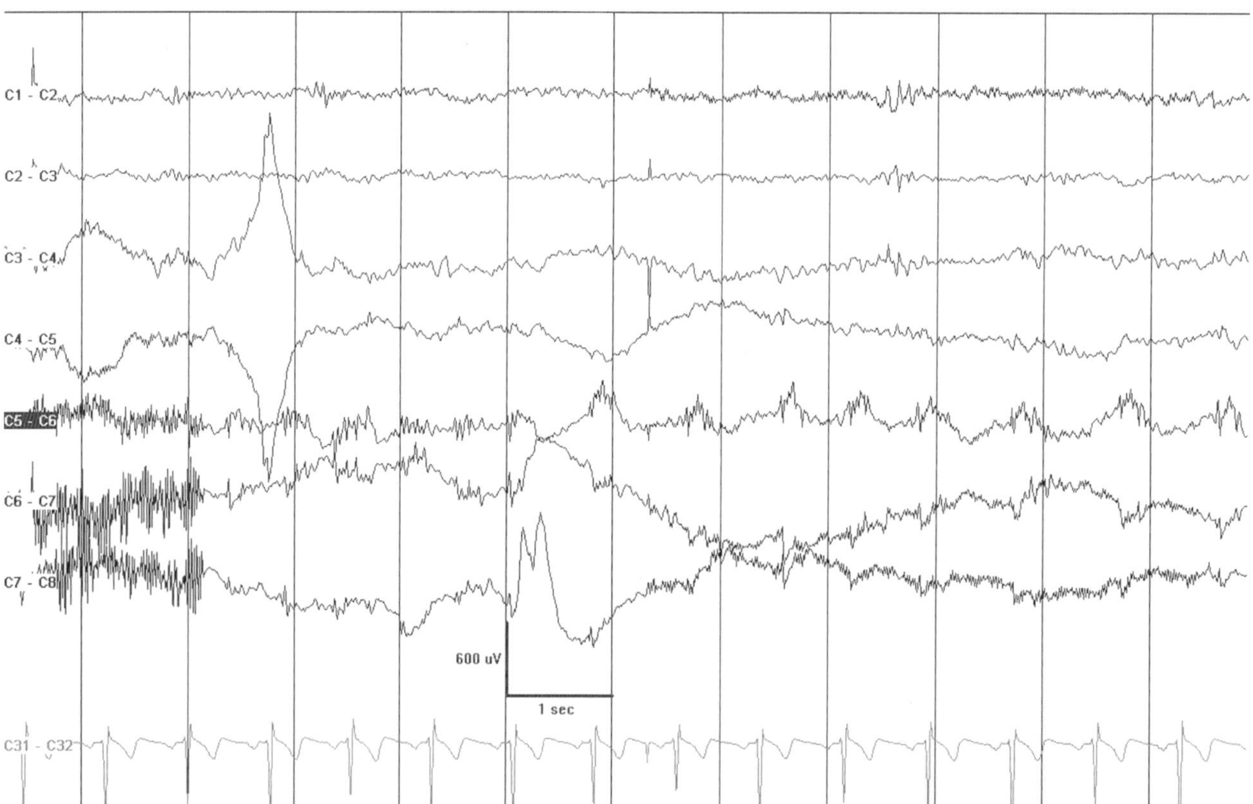


**FIGURE 4-13** ECoG recording during functional mapping with Penfield technique, showing stimulation-triggered polyspikes. The recording is done via an 8-contact strip (labeled C1 through C8), in a bipolar montage. The reference is place on the contralateral mastoid. The calibration mark is 600 $\mu$V and 1 second. Notice the stimulation artifact seen in channels C5–C6, C6–C7, C7–C8, followed by triggered polyspikes, up to 200 $\mu$V, at 1–2 Hz, seen in the same channels, maximal at C5–C6. Also, notice the burstlike morphology of these discharges.

Blume et al. (2004) also found different morphologies for AD: (1) rhythmic waves; (2) bursts of polyspikes; (3) spike-waves; (4) sequential spikes. Our experience is similar to theirs. We also believe that the morphologies and evolution of ADs depend on the morphologies of existing epileptiform activity. Examples of different patterns of baseline and triggered epileptiform activity are shown in Figs. **4-9 a**; **4-10 a, c, d**; **4-12**; **4-13**. The applied stimulation can increase the frequency, periodicity, and organization of preexisting epileptiform discharges (Figs. **10 a, c, d; 4-9 a-2**).

De novo triggered ADs and seizures are seen mostly after stimulation stops, but they are sometimes preceded by a short intrastimulation run. A common evolution of a de novo seizure is from faster, smaller-amplitude activity toward slower, higher-amplitude periodic discharges. Thus, rhythmic theta, alpha, or beta runs are usually present in the beginning of a seizure, followed by a run of high-amplitude rhythmic, periodic spikes, polyspikes, or sharps (Figs. **4-7 b**; **4-14**; **4-15 a**). Of course, a de novo appearance of isolated epileptiform discharges is also possible, with subsequent increase in their frequency (Fig. **4-9 a-3**).

It is easier to see triggered activity when it starts within the multipulse train stimulation epoch. However, during Penfield stimulation, an appropriate filtration of the 60-Hz artifact, by setting the 60-Hz notch "on," can be helpful in identifying an intrastimulus run (see page 162). The length of such a run progressively increases with longer stimulation times and with higher stimulation amplitudes, to the point when it is followed by periodic ADs, seen after the stimulation stops.

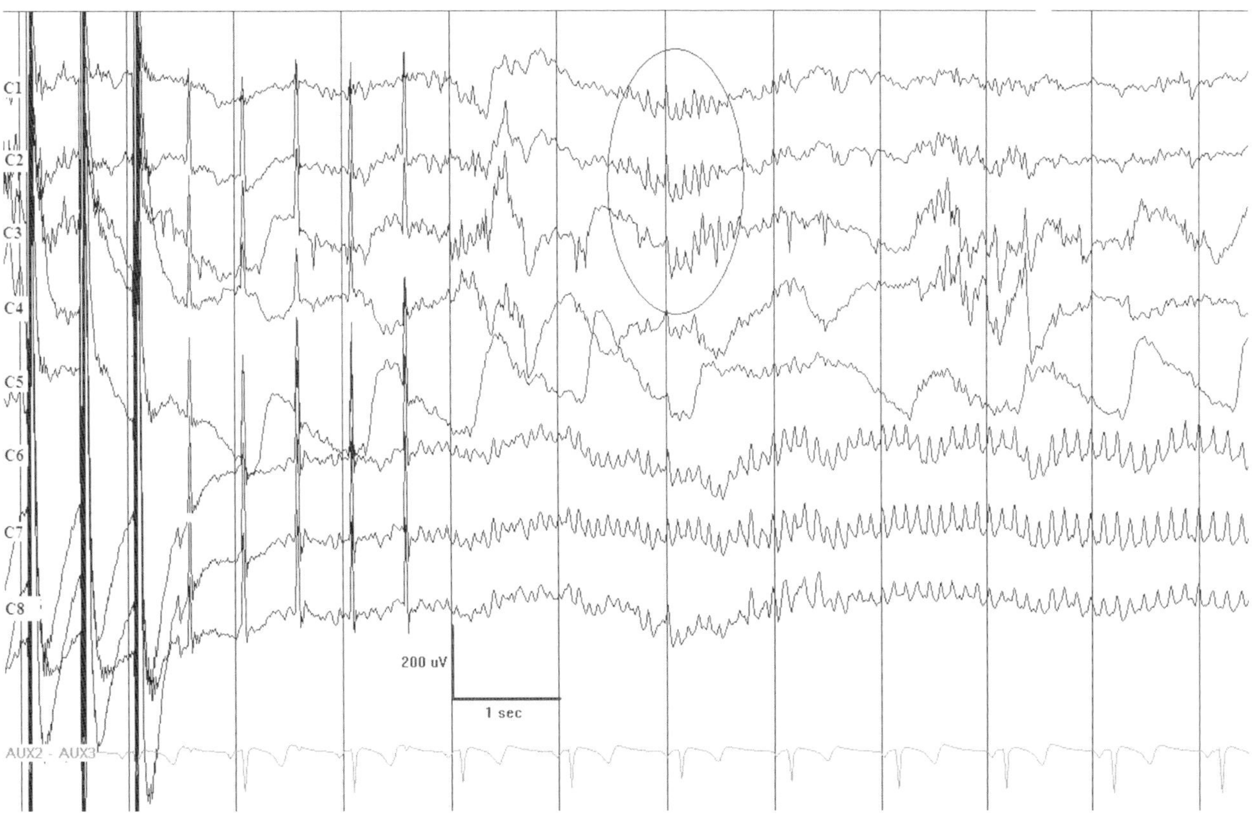


**FIGURE 4-14** Electrographic seizure, triggered by cortical stimulation with multipulse train technique. ECoG recording is done via an 8-contact (labeled C1 through C8) strip, in a referential montage, with the reference electrode placed on the contralateral mastoid. Aux2–Aux3 channel shows EKG recording. The calibration mark is 200 $\mu$V and 1 second. Notice the stimulus artifact present in all channels, within the first 4 seconds of the recording. The stimulation, close to C6 and C7 contacts, triggered a rhythmic run of 70 $\mu$V, 9 Hz sharp activity, seen at contacts C6, C7, and C8, and representing an electrographic seizure. Notice the very similar morphology of the fast activity (circled) present within the bursts of the burst-suppression pattern.

(a)

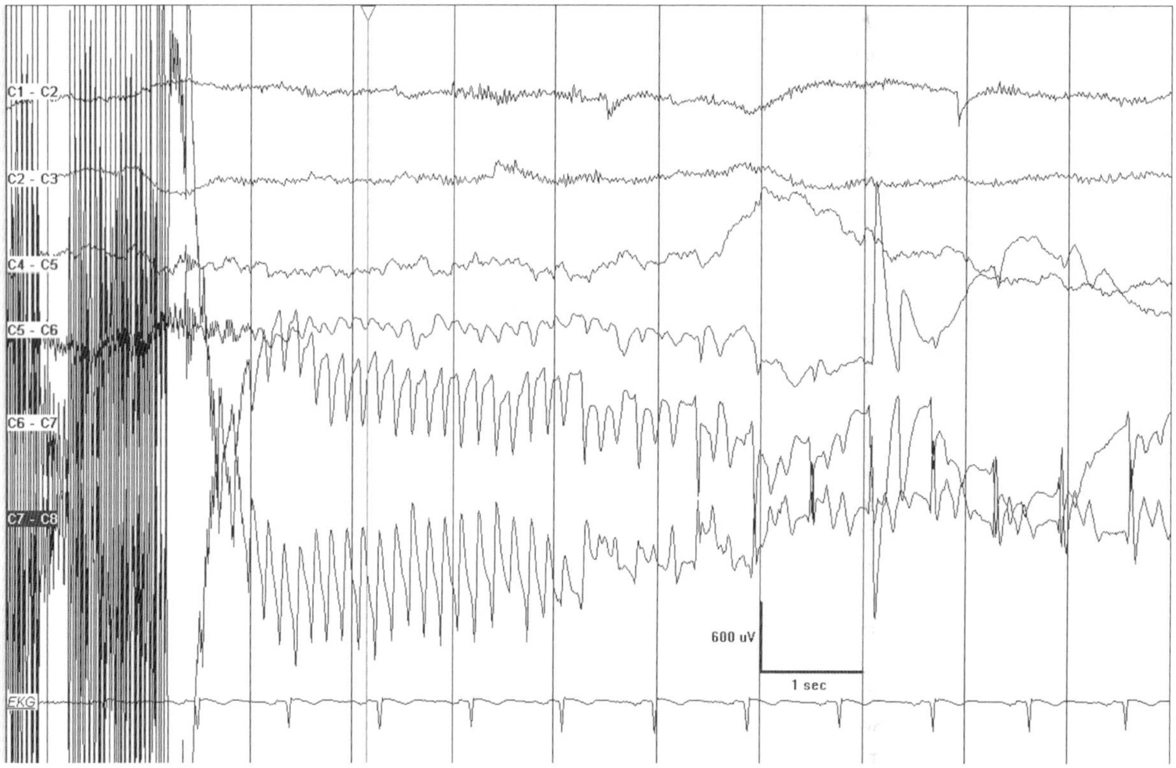


**FIGURE 4-15** Two ECoG recordings showing similar rhythmic theta frequency runs, but of different etiology. **a:** Recording via an 8-contact (labeled C1 through C8) strip, in a bipolar montage. The reference electrode is on the contralateral mastoid. The calibration mark is 600 μV and 1 second. Notice the diffuse and dense stimulus artifact present in the first second of the recording, followed by rhythmic run of up to 1200 μV and 7 Hz sharp activity, present in channels C6–C7 and C7–C8. This activity slows down and is subsequently followed by distinct repetitive spikes at 1–2 Hz and represents a stimulation-triggered electrographic seizure. Notice the phase reversal (contact 7) of the initial rhythmic activity, as well as its morphology.

Thus, one could use the evolution of such runs in predicting the appearance of poststimulation ADs (Fig. **4-16**). However, we've seen many cases when such intrastimulation activity was not followed by the post-stimulation discharges. Thus, without a filtered recording, their presence during stimulation would have been missed (Fig. **4-7 a, b**). In addition to the safety aspect, being aware of such runs becomes important if triggered mMEPs, sensory, motor symptoms, or language disturbance is present. In general, these runs are seen at the stimulating contacts, but sometimes their field extends further, in particular with higher stimulus amplitudes. In such cases, the positive responses may be due to spreading of this epileptic activity to regions other than the one that was stimulated. Thus, a false localization of the eloquent cortex can occur. Our experience is in concordance with the work of Blume and colleagues (2004), who found that ADs may involve more than the stimulation site and thus may cause inaccurate localization cortical function.

*C3. Artifact versus brain activity.* As with any EEG analysis, one has to reliably distinguish artifact from real brain activity.

***a. Unstable contacts.*** Unstable contacts can introduce a "spiky" morphology to the recordings in the channels involved and thus mimic epileptiform activity (Fig. **4-17 b**). Also, an unstable electrode could change the morphology and amplitudes of the recorded brain activity (in comparison with other channels).

***b. Pulsation artifact.*** This is usually due to the pulsatile cortical surface, in particular when the recording contacts are in close proximity to a cortical vessel. This artifact can mimic rhythmic delta slowing as well as epileptic activity

(b)

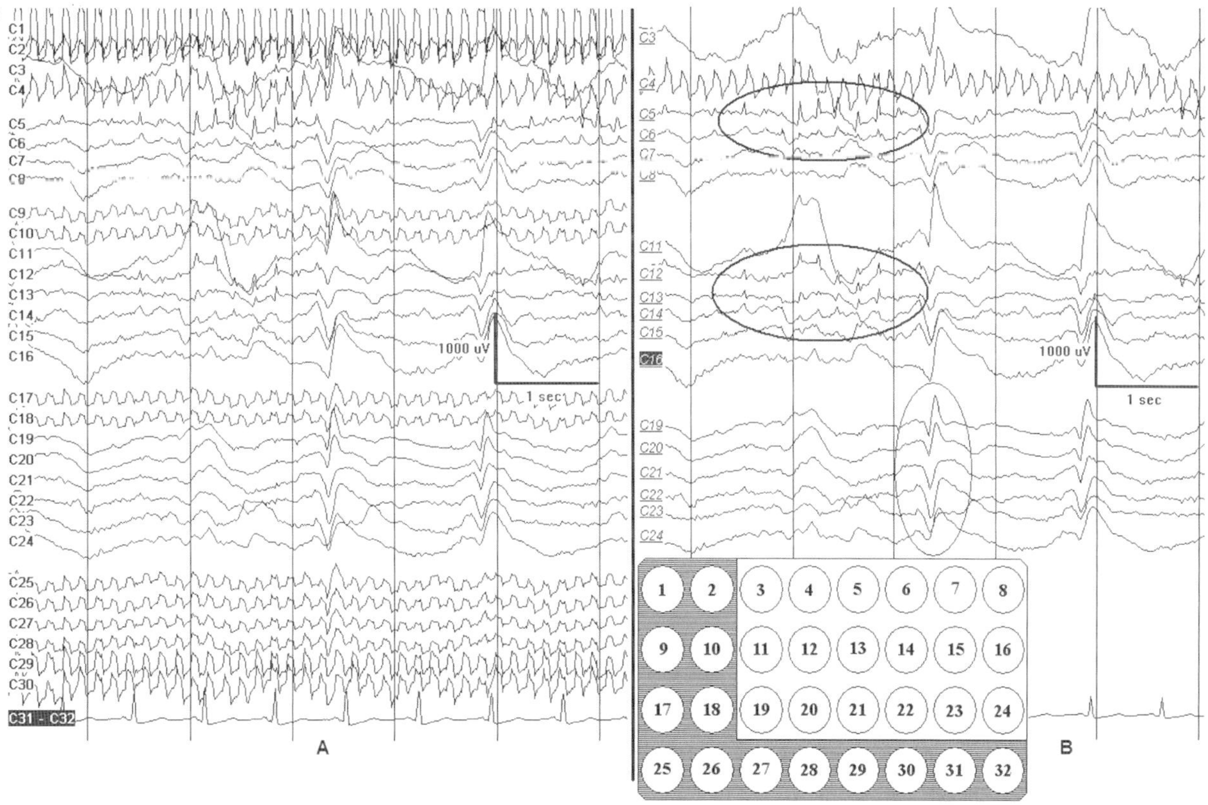


**FIGURE 4-15** *(Continued)* **b:** Recording via a customized 18-contact grid, in a referential montage, with the reference placed on the contralateral mastoid. The surgeon cut row (C25 through C32) as well as columns (1-9-17-25) and (2-10-18-26) Panel **A** shows a display of 30 virtual channels; panel **B** shows a display of the 18 real channels. In panel **A,** notice the rhythmic 6-Hz activity present in channels C1, C2, C4, C9, C10, C17, C18, and C25 through C30. This activity represents electrical artifact and, even without the knowledge of the missing cortical channels, there are several clues for its artifactual nature: (1) there is no reasonable field: the activity is absent channels C3, C5, C11, and C19 through C24; (2) whereas the runs at C1 and C4 have a reasonable morphology for brain activity, the runs at C9, C10, C17, C18, and C25 through C30 have a very unusual morphology; (3) there is a certain monotony and lack of evolution of these theta runs, unlike the run presented at **a;** (4) once the virtual channels are disabled, the lack of field of the rhythmic theta activity becomes evident, because it is seen exclusively in channel C4 (panel **B**). However, the recording shows two types of epileptiform activity: small, brief spikes at contacts C5, C6, C12, and C13 (bold circle) as well as higher-amplitude sharps, with broad field, periodic, up to 1 Hz (e.g., contacts C19 through C24: thin circle). The calibration marks in both panels are 1000 μV and 1 second. Channel C31–C32 shows EKG recording.

(Fig. **4-17 a, b**). We recommend the concomitant use of electrocardiogram (EKG) recording; if the rhythmic activity seen is indeed pulsation artifact, it should be time-locked to the EKG activity (QRS complexes, occasionally T waves): it can be helpful to "drag" the EKG channel right underneath the recording channel with the suspicious activity.

***c. EKG.*** EKG artifact can be misinterpreted as periodic spikes. As in the case of pulsation artifact, EKG recording is essential. The absence of a logical field and the sustained synchrony of such discharges with the QRS complex or T waves help with the differentiation.

***d. 60-Hz artifact.*** The 60-Hz artifact, especially when present in couple of adjacent electrodes, can mimic an epileptic "buzz." If these buzzes are in channels at distance, the lack of a reasonable field can also be an important clue of its artifactual nature. This can become even trickier in the case of presence of real epileptiform activity, in the same channel, displaying the 60-Hz artifact (Fig. **4-10 a–c**).

***e. Electrical artifact.*** This may look very similar to an organized, rhythmic run of epileptic activity. Again, the lack of a field can be a good indicator for artifact. Another important point is that epileptiform or epileptic activity

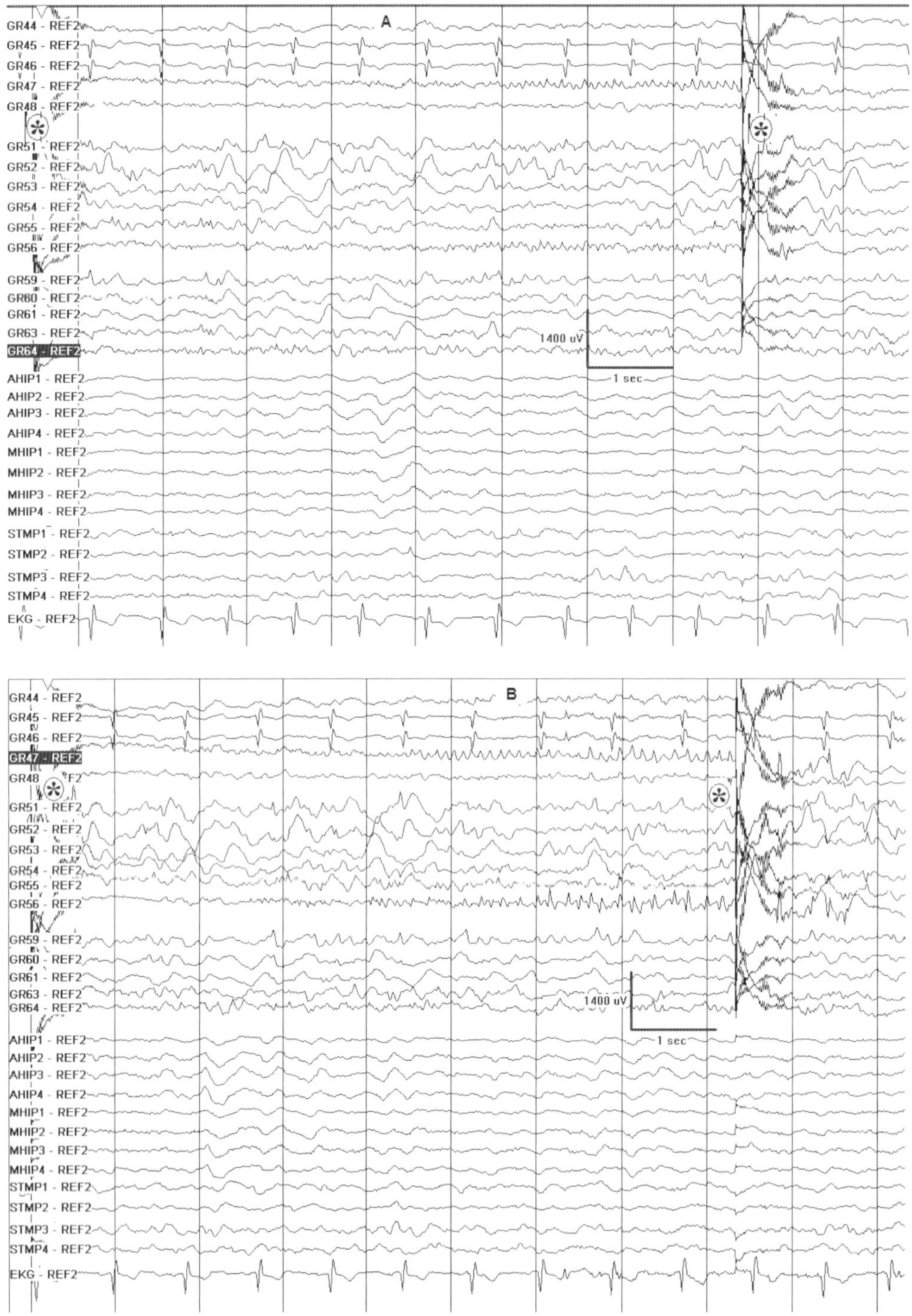

**FIGURE 4-16** Evolution of electrical cortical stimulation-triggered epileptiform discharges. As the stimulus intensity was increased, there was a gradual buildup from a rhythmic intrastimulation run toward poststimulation local and at distance spike-wave discharges (afterdischarges ADs). The ECoG recording was done via a customized grid, two hippocampus-depth electrodes, and a subtemporal strip. For the mapping procedure, however, with stimulation applied at contacts 45 and 46. The reference electrode is one of the 4 contacts of a subdural strip, placed upside down. Each slide shows an 8-second stimulation epoch, using repetitive pulses of 1 msec in width, at 60 Hz and delivered at contacts GR45 and GR46. The recording is done with the 60-Hz notch filter "on": thus, only the beginning and the end of the stimulation can be determined (*). The elimination of most of the artifact allows analysis of the ECoG during the stimulation itself. The stimulus intensity is gradually increased from 3 mA (panel **A**) to 4 mA (panel **B**) and 5 mA (panel **C**).

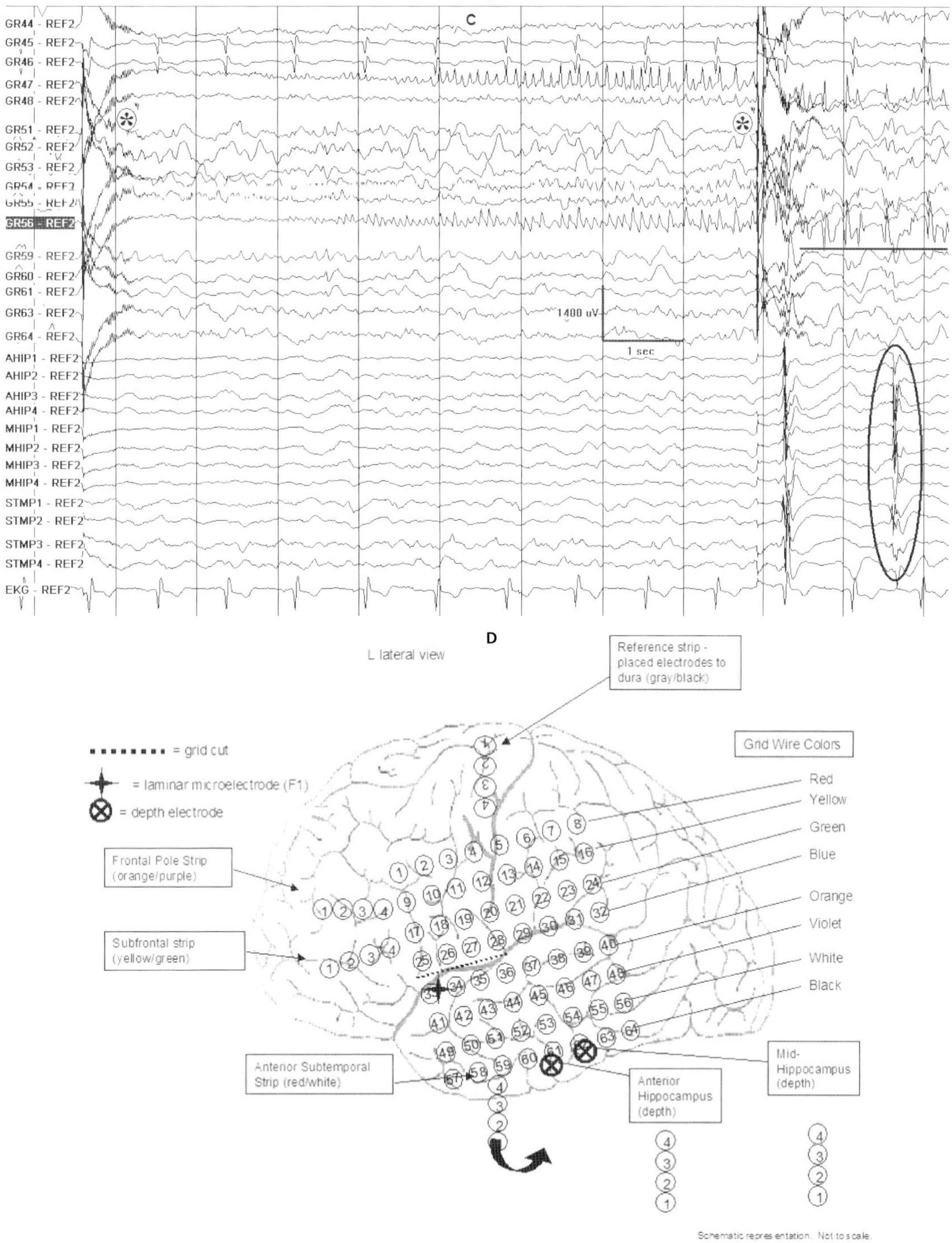


**FIGURE 4-16** *(Continued)* In panel **A**, the recording shows the buildup of a small-amplitude, 8- to 12-Hz rhythmic run at contacts GR47 and GR56, which terminates abruptly once the stimulation is stopped. In panel **B**, a similar run is seen again at the same location, this time with slightly increased amplitude and sharper morphology, in particular at contact GR56. Notice that the frequencies are somewhat slower and that clear spike-wave discharges are present in the second following cessation of the stimulation. In panel **C,** with further increase in stimulus intensity, the run becomes more organized: starting at lower amplitudes (300 $\mu$V) and higher frequencies (12 Hz), it evolves toward higher amplitudes (700 $\mu$V), lower frequencies (7–8 Hz), and sharper morphology. Also notice higher-amplitude and broader-field afterdischarges (underlined and circled), seen after cessation of the stimulation, including high-amplitude spikes and polyspikes present at the subtemporal strip contacts (STMP 1–4) and in the mesiotemporal structures (AHIP 1–4, MHIP 1–4), likely a spread resulting from increased connectivity among different temporal lobe regions. Panel **D** shows the grid orientation and positioning.

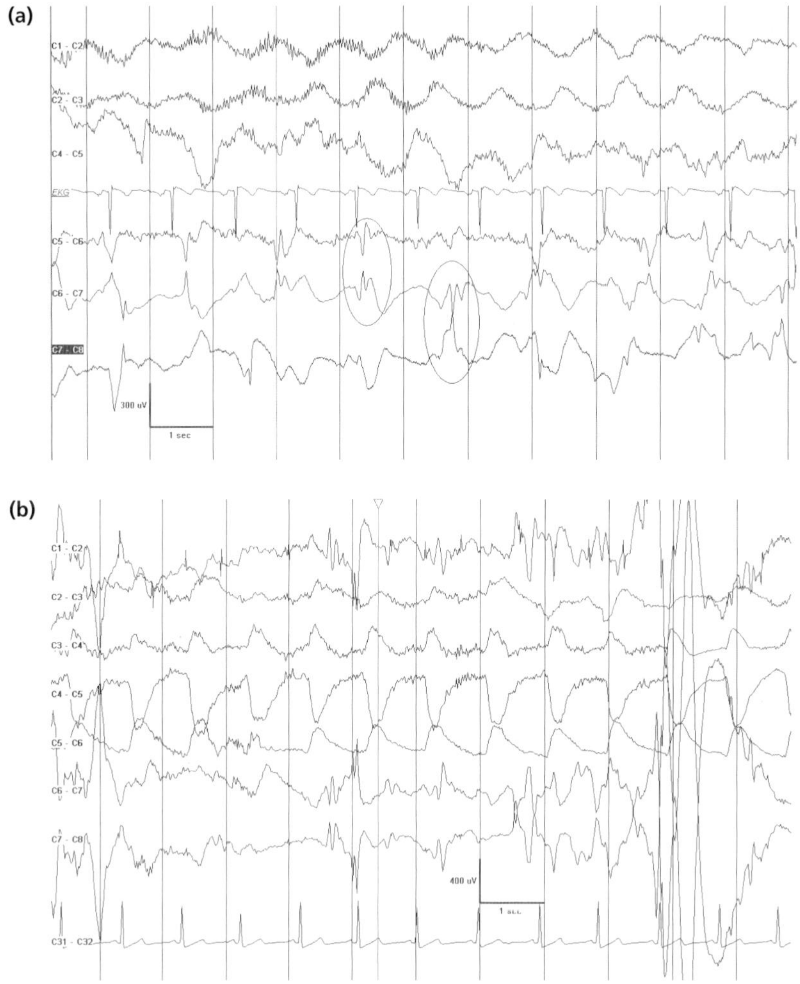


**FIGURE 4-17** ECoGs showing different types of artifact. Both recordings are done using an 8-contact strip (labeled C1 through C8), in a bipolar montage. Channel C31–C32 is showing the EKG recording. The calibration marks are 300 $\mu$V and 1 second (in **a**) and 400 $\mu$V and 1 second (in **b**). **a:** In channels C1–C2 and C2–C3, the recording is showing a rhythmic, 200 $\mu$V, 1 Hz activity, with phase reversal at C2. Notice that the negative peaks in C2–C3 channels are time-locked to the QRS complexes: this activity represents pulsation artifact. Channels C5–C6, C6–C7, and C7–C8 are showing spikes, up to 150 $\mu$V, with phase reversal at C6 or at C7 (circled). These discharges are not time-locked to the QRS complexes. **b:** C1 and C7 electrode artifact translating in high-amplitude, sharp waves, mimicking spike-wave discharges; channels C4–C5, C5–C6, and, to a lesser extent, C3–C4 show rhythmic pulsation artifact.

usually has an evolution or a buildup. A monomorphic appearance, with little or no change in time, can also be a good indicator for its artifactual nature (Fig. **4-15 a, b**).

For troubleshooting some of these artifact, see pages 162–63.

## *D. Management of Stimulation-Triggered Epileptiform Activity*

As mentioned on page 133, ensuring reasonable AED coverage preoperatively is important in patients with prior seizure activity. In case of a triggered seizure or organized epileptiform activity, cortical irrigation with ice-cold saline is very effective in aborting this activity. Many of these patients are already taking propofol. Thus, increasing its infusion rate can also help. Additional versed can be also given in boluses. Intravenous loading of an AED drug, such as Levitiracetam (Keppra) or Fosphenytoin, is also an effective measure. If an electrographic seizure occurs, we recommend continuing the ECoG, in order to ensure complete and lasting cessation of epileptic activity. In the case of mapping during phase

**TABLE 4-1** The effects of commonly used anesthetics on functional cortical mapping

| Type of anesthetics | Effects on central sulcus localization and sensory mapping | Effects on motor mapping via direct cortical stimulation | Effects on cortical excitability, appreciated by ECoG analysis |
|---|---|---|---|
| Benzodiazepines (e.g. diazepam, midazolam) | increase latencies and decrease amplitudes of cortical SSEPs<br>■ central sulcus localization possible, but with caution in correct identification of delayed N19/P22 peaks | significantly reduce amplitudes of mMEPs triggered by cortical stimulation<br>■ recommend avoidance during motor mapping | **low doses:** enhance fast-frequency activity<br>**moderate doses:** cause mild cortical suppression, with increase in theta and delta slowing but do not cause burst-suppression pattern<br>■ caution in distinction between fast sharp activity and real epileptiform or epileptic activity<br>■ protective effect against stimulation-triggered seizures<br>■ increase the mapping threshold |
| Barbiturates (e.g. thiopental) | increase latencies and decrease in amplitudes of cortical SSEPs<br>■ central sulcus localization possible, but with caution in correct identification of delayed N19/P22 peaks | significantly reduce (to disappearance) amplitudes of mMEPs triggered by cortical stimulation<br>■ recommend avoidance during motor mapping | **low doses:** increase in fast-frequency activity<br>**high doses:** cause significant cortical suppression, ranging from burst-suppression pattern to electrical silence<br>■ caution in distinction between sharp activity within bursts and epileptiform activity<br>■ increase the mapping threshold |
| Propofol | mildly increases latencies and decreases amplitudes of cortical SSEPs<br>■ drug of choice for TIVA during central sulcus localization | mildly decreases of amplitudes of mMEPs triggered by cortical stimulation<br>■ drug of choice for TIVA during motor mapping | **low doses:** increases fast-frequency activity<br>**high doses:** causes cortical suppression, manifesting in varied degrees of burst suppression<br>■ caution in distinction between sharp activity within the bursts and epileptiform activity<br>■ increases the mapping threshold |
| Narcotics (e.g. alfentanyl, fentanyl, remifentanyl) | mildly increase latencies and decrease amplitudes of cortical SSEPs<br>■ remifentanyl drug of choice for TIVA during central sulcus localization | minimally decrease amplitudes of mMEPs triggered by cortical stimulation<br>■ remifentanyl drug of choice for TIVA during motor mapping | questionable epileptogenic effect for fentanyl and remifentanyl<br>significant epileptogenic effect for alfentanyl<br>■ strongly recommend avoidance of alfentanyl during electrical cortical stimulation |
| Inhalational agents (e.g. nitrous oxide, desflurane, enflurane, halothane, isoflurane, sevoflurane) | significantly increase latencies and decrease amplitudes of cortical SSEPs (to disappearance)<br>synergistic effect when nitrous oxide combined with a halogenated agent<br>■ strongly recommend avoidance during central sulcus localization, especially when used in combination | significantly decrease in amplitudes of mMEPs triggered by direct cortical stimulation; may affect D waves<br>synergistic effect when nitrous oxide combined with a halogenated agent<br>■ strongly recommend avoidance during motor mapping, especially when used in combination | **low doses:** cause attenuated fast-frequency background<br>**moderate doses** (loss of consciousness):<br>*halogenated agents*- cause high-frequency, rhythmic delta activity with superimposed fast-frequency activity<br>*N20*- causes frontally predominant fast-frequency activity<br>**high doses**(anesthesia):<br>*halogenated agents*- burst suppression<br>*N20*- attenuates the burst suppression produced by halogenated agents<br>*halogenated agents*- questionable epileptogenic<br>■ *halogenated agents*- questionable decrease in threshold for stimulation-triggered seizures, thus use caution during electrical cortical stimulation |

*(Continued)*

**TABLE 4-1** *(Continued)*

| Type of anesthetics | Effects on central sulcus localization and sensory mapping | Effects on motor mapping via direct cortical stimulation | Effects on cortical excitability, appreciated by ECoG analysis |
|---|---|---|---|
| Etomidate | **low sedative/amnestic doses:** increases amplitudes of cortical SSEPs<br>■ useful add-on for TIVA, in particular when poor baseline SSEPs are expected | **low sedative/amnestic doses:** increases amplitudes of mMEPs<br>■ useful add-on for TIVA, in particular when poor baseline SSEPs are expected | **low doses:** increases cortical excitability, thus possible epileptogenic effect<br>**high doses:** causes high-amplitude theta activity and burst-suppression pattern<br>■ may decrease threshold for stimulation-triggered seizures (low doses), thus use caution during electrical cortical stimulation |
| Ketamine | **low doses:** increases amplitudes of cortical SSEPs<br>**high doses:** decreases amplitudes of cortical SSEPs<br>■ useful add-on for TIVA, in particular when poor baseline SSEPs are expected | **low doses:** increases amplitudes of mMEPs triggered by cortical stimulation<br>**high doses:** decreases in their amplitudes, mMEPs remain still easily recordable<br>■ useful add-on for TIVA, in particular when poor baseline SSEPs are expected | high-amplitude theta activity and increase in beta activity<br>increases cortical excitability, in particular in low doses, thus possible epileptogenic effect<br>■ may decrease threshold for stimulation-triggered seizures, thus use caution during electrical cortical stimulation |
| Muscle Relaxants (cisatracurium, vecurorium, succinylcholine) | no effect on the physiology of SSEPs<br>■ may improve the SSEPs recordings by decreasing the muscle artifact in the reference electrode<br>■ may mask the thumb twitch as appreciation of efficient stimulation | dose-dependent effect on amplitudes of mMEPs (to disappearance), in particular after repetitive stimulation<br>■ should be avoided in motor mapping when mMEPs are recorded<br>■ improves the quality of recorded D waves | no effect on cortical excitability<br>decrease muscle artifact introduced in the recording by reference electrode<br>■ "cleaner" ECoG recordings |

II video-EEG recording for epilepsy surgery, we recommend keeping 2–4 mg of Ativan ready in a syringe, as well as saline flush. Special attention needs to be paid to epilepsy patients undergoing the mapping at the end of the monitoring session, when the AED levels are typically low. These patients have typically been tapered off their medication and their cortex is very excitable. One could also attempt to abort electrical stimulation-triggered seizure activity by using high-frequency stimulation, applied at the contacts showing the epileptiform activity. As mentioned on page 158, several authors have confirmed the efficacy of this approach (Lesser et al., 1999; Morrell, 2006). Our experience also is similar to theirs. There has to be a certain match between the type of epileptiform activity and the stimulus parameters, including the stimulus amplitude and frequency that are required to abort such activity.

## VII. ANESTHESIA CONSIDERATIONS

There are three main types of effects of anesthetics that can influence the results of neurophysiologic functional mapping. First is their impact on SSEPs generation, thus on the amplitudes and latencies of the thalamocortical potentials used for identification of the somatosensory cortex, as well as for localization of the central sulcus via phase-reversal technique. Second is their impact on generation of mMEPs. The third is their impact on cortical excitability in general. The latter needs to be evaluated further in terms of (1) their epileptogenic effect, causing a decrease in the stimulation-triggered seizures thresholds; (2) their cortical suppressive effect, causing an increase in the mapping thresholds; (3) their influence on different frequencies activities, introducing challenges in ECoG interpretation.

Whereas the first two effects are appreciated by changes in the amplitudes and latencies of the SSEPs and mMEPs, the third is evaluated by ECoG analysis.

Table **4-1** summarizes these effects for commonly used anesthetics in neurologic surgery requiring functional sensorimotor mapping. For more details regarding their mechanisms of actions and properties, see pages 327–36.

## VIII. CONCLUSIONS

Accurate techniques, with reliable identification of the mapping thresholds, are essential for avoiding permanent neurologic deficits. Language or motor deficits are reported even with judicious use of cortical mapping procedures, but most postsurgical changes are transient rather than permanent. These deficits are likely related to damage of the supplementary regions rather than the primary functional regions (as long as identification of the eloquent cortex was done at the primary cortex mapping threshold). Thus, it is important to understand the intraindividual and interindividual variability of such thresholds and the factors that modify it. ECoG allows the neurophysiologist to monitor for ADs and to assess the baseline cortical excitability, ensuring a safe and reliable mapping procedure. Despite increasing operative times and anesthetic considerations, we believe that the use of electrophysiologic mapping is essential for improving the functional outcomes in patients with lesions near the eloquent cortex. One should, however, adopt a multidisciplinary approach to functional cortical mapping.

## REFERENCES


Adam C, Hasboun D, Clemenceau S, et al. Fast contralateral propagation of afterdischarges induced by stimulation of medial temporal lobe. J Clin Neurophysiol 2004; 21:399–403.

Ajmone Marsan C. Focal electrical stimulation. In: Experimental Models of Epilepsy. Purpura DP, Penry JK, Tower DD, Eds.) 1972; pp 147–172. New York: Raven Press.

Ajmone Marsan C, O'Connor M. Electrocorticography. In: Handbook of Electroencephalography and clinical neurophysiology. (Remond A, Ed.), 1973. Vol 10C, pp 3–49. Amsterdam: Elsevier.

Ajmone Marsan C. Depth electrography and electrocorticography. In: Electrodiagnosis in clinical neurology (Aminoff MJ, Ed.) 1980; pp 167–196. New York: Churchill Livingstone.

Akrawi WP, Drummond JC, Kalkman CJ, Patel PM. A comparison of the electrophysiologic characteristics of EEG burst-suppression as produced by isoflurane, thiopental, etomidate and propofol. J Neurosurg Anesthesiol 1996; 8(1):40–46.

Albright AL, Wisoff JH, Zeltzer P, et al. Prognostic factors in children with supratentorial (nonpineal) primitive neuroectodermal tumors. A neurosurgical perspective from the Children's Cancer Group. Pediatr Neurosurg. 1995;22(1):1–7.

Ammirati M, Vick N, Liao YL, et al. Effect of the extent of surgical resection on survival and quality of life in patients with supratentorial glioblastomas and anaplastic astrocytomas. Neurosurgery. 1987; 21(2):201–206.

Bartholow R. Experimental investigations into functions of the human brain. Am J Med Sci 1874; 67:305–313.

Bates JA. The significance of complex motor patterns in the response to cortical stimulation. Int J Neurol 1963; 4:92–09.

Bengzon AR, Rasmussen T, Gloor P, et al. Prognostic factors in the surgical treatment of temporal lobe epileptics. Neurology 1968; 18(8):717–731.

Berger MS. The impact of technical adjuncts in the management of cerebral hemispheric low-grade gliomas of childhood. J Neurooncol 1996; 28:129–155.

Berger MS, Kincaid J, Ojemann GA, Lettich E. Brain mapping techniques to maximize resection, safety, and seizure control in children with brain tumors. Neurosurgery 1989; 25(5):786–792.

Bernier GP, Richer F, Giard N, et al. Electrical stimulation of the human brain in epilepsy. Epilepsia 1990; 31(5):513–520.

Bittar RG, Olivier A, Sadikot AF, et al. Cortical motor and somatosensory representation: Effect of cerebral lesions. J Neurosurg 2000; 92:242–248.

Blume WT, Jones DC, Pathak P. Properties of after-discharges from cortical electrical stimulation in focal epilepsies. Clin Neurophysiol 2004; 115(4):982–989.

Bonilha L, Kobayashi E, Mattos JP, et al. Value of extent of hippocampal resection in the surgical treatment of temporal lobe epilepsy. Arq Neuropsiquiatr 2004; 62(1):15–20.

Burke D, Bartley K, Woodforth IJ, et al. The effects of volatile anesthetics on the excitability of human corticospinal axons. Brain 2000; (pt 5):992–1000.

Cherlow D, Dymond A, Crandall PH, et al. Evoked responses and after-discharge thresholds to electrical stimulation in temporal lobe epilepsies. Arch Neurol 1977; 34:527–531.

Danks RA, Aglio LS, Gugino LD, Black PM. Craniotomy under local anesthesia and monitored conscious sedation for the resection of tumors involving eloquent cortex. Neurooncol 2000; 49:131–139.

Detsch O, Kochs E, Siemers M, et al. Increased responsiveness of cortical neurons in contrast to thalamic neurons during isoflurane-induced EEG bursts in rats. Neurosci Lett 2002; 317(1):9–12.

Doyle PW, Matta BF. Burst suppression or isoelectric encephalogram for cerebral protection; evidence from metabolic suppression studies. British Journal of Anesthesia 1999; 83(4):580–584.

Duffau H, Capelle L, Sichez N, et al. Intraoperative mapping of the subcortical language pathways using direct stimulation. Brain 2000; 125:199–214.

Duffau H, Denvil D, Lopes M, et al. Intraoperative mapping of the cortical areas involved in multiplication and subtraction: An electrostimulation study in a patient with a left parietal glioma. J Neurol Neurosurg Psych 2002; 73:733–738.

Esmaeili V, Shamsollahi MB, Arefian NM, Assareh A. Classifying depth of anesthesia using EEG features: A comparison. Conf Proc IEEE Eng Med Biol Soc 2007; 1:4106–4109.

Fandino J, Kollias SS, Wieser HG, et al. Intraoperative validation of functional magnetic resonance imaging and cortical reorganization patterns in patients with brain tumors involving the primary motor cortex. J Neurosurg 1999; 91:238–250.

Foerster O. The motor cortex in man in the light of Hughlings Jackson's doctrines. Brain 1936; 59:135–159.

Gloor P. Contributions of electroencephalography and electrocorticography to the neurosurgical treatment of the epilepsies. In: Neurosurgical Management of Epilepsies (Purpura, DP, Penry, JK and Walker, RD, Eds.) 1975; pp 59–105. New York: Raven Press.

Goldring S, Gregorie EM. Surgical management of epilepsy using epidural recordings to localize seizure focus. Review of 100 cases. J Neurosurg 1984; 60:457–466.

Guilburd JN, Lapras C, Guyotat J. Brain tumors in infants. Harefuah. 1989;116(3):133–135.

Haghighi SS. Influence of isoflurane anesthesia on motor evoked potentials elicited by transcortical, brainstem and spinal root stimulation. Neurol Res 1998; 20(6):555–558.

Haglund MM, Berger MS, Shamseldin M, Ojemann GA. Cortical localization of temporal lobe language sites in patients with gliomas. Neurosurgery 1994; 34:567–576.

Hartikainen KM, Rorarius M, Peräkylä JJ, et al. Cortical reactivity during isoflurane burst-suppression anesthesia. Anesth Analg 1995; 81:1223–1228.

Hentschke H, Schwarz C, Antkowiak B. Neocortex is the major target of sedative concentrations of volatile anesthetics: Strong depression of firing rates and increase of GABAA receptor-mediated inhibition. Eur J Neurosci 2005; 21(1):93–102.

Hicks R, Burke D, Stephen J, et al. Corticospinal volleys evoked by electrical stimulation of human motor cortex after withdrawal of volatile anaesthetics. J Physiol 1992; 456:393–404.

Hines M, Boynton EP. The maturation of "excitability" in the precentral gyrus of the young monkey. Contrib Embryol 1940; 178:313–451.

Hoffman WE, Edelman G. Comparison of isoflurane and desflurane anesthetic depth using burst suppression of the electroencephalogram in the neurosurgical patients. Anesth Analg 1995; 81(4):811–816.

Hossmann KA, Linn F, Okada Y. Bioluminescence and fluoroscopic imaging of tissue pH and metabolites in experimental brain tumors of cat. NMR Biomed 1992; 5(5):259–264.

Jasper, HH. Electrocorticography. In: Epilepsy and the Functional Anatomy of the Human Brain (Penfield W, Jasper HH, Eds.) 1954; pp 692–738. Boston: Little Brown.

Jayakar P, Alvarez LA, Duchowny MS, Resnick TJ. A safe and effective paradigm to functionally map the cortex in childhood. J Clin Neurophysiol 1992; 9(2):288–293.

Katayama Y, Tsubokawa T, Maejima S, et al. Corticospinal direct response in humans: identification of the motor cortex during intracranial surgery under general anaesthesia. J Neurol Neurosurg Psychiatry. 1988; 51(1):50–59.

Kim SK, Wang KC, Hwang YS, et al. Epilepsy surgery in children: Outcomes and complications. J Neurosurg Pediatrics 2008; 1(4):277–283.

Kombos T, Suess O, Funk T, et al. Intraoperative mapping of the motor cortex during surgery in and around the motor cortex. Acta Neurochir 2000; 142:263–268.

Kombos T, Suess O, Ciklatekerlio O, et al. Monitoring of intraoperative motor evoked potentials to increase the safety of surgery in and around the motor cortex. J Neurosurg. 2001; 95(4):608–614.

Kortelainen J, Koskinen M, Mustola S, Seppänen T. EEG frequency progression during induction of anesthesia: From start of infusion to onset of burst suppression pattern. Conf Proc IEEE Eng Med Biol Soc 2007; 1:1570–1573.

Korvenoja A, Kirveskari E, Aronen HJ, et al. Sensorimotor cortex localization: Comparison of magnetoencephalography, functional MR imaging and intraoperative cortical mapping. Radiology 2006; 241:213–222.

Kreuer S, Schreiber JU, Bruhn J, Wilhelm W. Impact of patient age on propofol consumption during propofol-remifentanyl anesthesia. Eur J Anaesthesiol 2005; 22(2):123–128.

Krings T, Töpper R, Willmes K, et al. Activation in primary and secondary motor areas in patients with CNS neoplasms and weakness. Neurology 2002; 58:381–390.

Kumabe T, Nakasato N, Nagamatsu K, Tominaga T. Intraoperative localisation of the lip sensory area by somatosensory evoked potentials. J Clin Neurosci 2005;12:66–70.

Lee GP, Loring DW, Smith JR, Lee MR. Predictors of patient inability to cooperate during intraoperative language mapping. Epilepsy & Behavior 2000; 1:327–332.
Lehéricy S, Duffau H, Cornu P, et al. Correspondence between functional magnetic resonance imaging somatotopy and individual brain anatomy of the central region: Comparison with intraoperative stimulation in patients with brain tumors. J Neurosurg 2000; 92:589–598.
Lesser RP, Arroyo S, Crone N, Gordon B. Motor and sensory mapping of the frontal and occipital lobes. Epilepsia 1998; 39:S69–S80.
Lesser RP, Kim SH, Beyderman L, et al. Brief bursts of pulse stimulation terminate afterdischarges caused by cortical stimulation. Neurology 1999; 53(9):2073–2081.
Lesser RP, Lüders H, Klem G, et al. Cortical afterdischarge and functional response thresholds: Results of extraoperative testing. Epilepsia 1984; 25:615–621.
Linn F, Seo K, Hossmann KA. Experimental transplantation gliomas in the adult cat brain. 3. Regional biochemistry. Acta Neurochir (Wien) 1989; 99(1-2):85–93.
Loughnan BA, Anderson SK, Hetreed MA, et al. Effects of halothane on motor evoked potentials recorded in the extradural space. Anaesth 1989; 63:561–564.
Lüdemann L, Förschler A, Grieger W, Zimmer C. Bold signal in the motor cortex shows a correlation with the blood volume of brain tumors. J Magn Reson Imaging 2006; 23:435–443.
Lukatch HS, Kiddoo CE, Maciver MB. Anesthetic-induced burst suppression EEG activity requires glutamate-mediated excitatory synaptic transmission. Cereb Cortex 2005; 15(9):1322–1331.
Martella G, De Persis C, Bonsi P, et al. Inhibition of persistent sodium current fraction and voltage-gated L-type calcium by propofol in cortical neurons: Implications for its antiepileptic activity. Epilepsia 2005; 46(5):624–635.
Mavroudakis N, Vandesteene A, Brunko E, et al. Spinal and brainstem SEPs and H reflex during enflurane anesthesia. Electroencephalogr Clin Neurophysiol 1994; 92:82–85.
Meyer FB, Bates LM, Goerss SJ, et al. Awake craniotomy for aggressive resection of primary gliomas located in eloquent brain. Mayo Clin Proc 2001;76:677–687.
Meyer JR, Roychowdhury S, Russell EJ, et al.. Location of the central sulcus via cortical thickness of the precentral and postcentral gyri on MRI. AJNR Am J Neuroradiol 1996:17:1699–1706.
Morimoto Y, Hagihira S, Koizumi Y, et al. The relationship between the bispectral index and electroencephalographic parameters during isoflurane anesthesia. Anesth Analg 2004; 98 (5):1336–1340.
Morrell M. Brain stimulation for epilepsy: Can scheduled or responsive neurostimulation stop seizures? Curr Opin Neurol 2006; 19(2):164–168.
Motamedi GK, Lesser RP, Miglioretti DL, et al. Optimizing parameters for terminating cortical afterdischarges with pulse stimulation. Epilepsia 2002; 43(8):836–846.
Nagano N, Sasaki H, Aoyagi M, Hirakawa K. Invasion of experimental rat brain tumor: Early morphological changes following microinjection of C6 glioma cells. Acta Neuropathol 1993; 86(2):117–125.
Neuloh G, Schramm J. Mapping and monitoring of supratentorial procedures. In: Neurophysiology in Neurosurgery: A Modern Intraoperative Approach. (Deletis V, et al., Eds.) 2002; pp 339–404. San Diego, CA: Academic Press.
Nitta T, Sato K. Prognostic implications of the extent of surgical resection in patients with intracranial malignant gliomas. Cancer. 1995;75(11):2727–31.
Ojemann JG, Miller JW, Silbergeld DL. Preserved function in brain invaded by tumor. Neurosurgery 1996; 39:253–259.
Osawa M, Shingu K, Murakawa M, et al. Effects of sevoflurane on central nervous system electrical activity in cats. Anesth Analg 1994; 79(1):52–57.
Paolicchi JM, Jayakar P, Dean P, et al. Predictors of outcome in pediatric epilepsy surgery. Neurology 2000; 54(3):642–647.
Penfield W, Boldrey E. Somatic motor and sensory representation in the cerebral cortex of man as studied by electrical stimulation. Brain 1937; (60):389–443.
Penfield W, Kristiansen K (Eds). Epileptic Seizure Patterns. 1951; Springfield, IL: Charles C Thomas.
Penfield W, Jasper H (Eds). Epilepsy and the Functional Anatomy of the Human Brain. 1954; Boston: Little Brown.
Pouratian N, Cannestra AF, Bookheimer SY, et al. Variability of intraoperative electrocortical stimulation mapping parameters across and within individuals. J Neurosurg 2004; 101(3); 458–466.
Prasad A, Worrall BB, Bertram EH, Bleck TP. Propofol and midazolam in the treatment of refractory status epilepticus. Epilepsia 2001;42:380–386.
Sala F, Krzan MJ, Deletis V. Intraoperative neurophysiologic monitoring in pediatric neurosurgery: Why, when, how? Child's Nervous System 2002; 18:264–287.
Sartorius CJ, Berger MS. Rapid termination of intraoperative stimulation-evoked seizures with application of cold Ringer's lactate to the cortex. Technical note. J Neurosurg. 1998; 88(2):349–351.
Scerrati M, Roselli R, Iacoangeli M, et al. Prognostic factors in low-grade (WHO grade II) gliomas of the cerebral hemispheres: the role of surgery. J Neurol Neurosurg Psychiatry. 1996;61(3):291–196.
Schäffler L, Lüders HO, Beck GJ. Quantitative comparison of language deficits produced by extraoperative cortical stimulation of Broca's, Wernicke's, and basal temporal language areas. Epilepsia 1996; 37 (5):463–475.
Schiffbauer H, Ferrari P, Rowley HA, et al. Functional activity within brain tumors: a magnetic source imaging study. Neurosurgery 2001;49(6):1313–1320; discussion 1320–1321
Schreiber A, Hubbe U, Ziyeh S, Hennig J. The influence of gliomas and nonglial space-occupying lesions on blood-oxygen-level-dependent contrast enhancement. AJNR Am J Neuroradiol 2000; 21:1055–1063.
Schwartz AE, Tuttle RH, Poppers PJ. Electroencephalographic burst suppression in elderly and young patients anesthetized with isoflurane. Anesth Analg 1989; 68(1):9–12.
Shaul N. The fundamental neural mechanisms of electroencephalography. Electroencephalography and Clinical Neurophysiology 1998; 106(2):101–107.
Skirboll SS, Ojemann GA, Berger MS, et al. Functional cortex and subcortical white matter located within gliomas. Neurosurgery 1996; 38:678–685.
Stecker MM, Kramer TH, Raps EC, et al. Treatment of refractory status epilepticus with propofol: Clinical and pharmacokinetic findings. Epilepsia 1998; 90:18–26.

Suess O, Suess S, Brock M, Kombos T. Intraoperative electrocortical stimulation of Brodman area 4: A 10-year analysis of 255 cases. Head & Face Medicine 2006; 2:20.

Tan TC, Black PM. Awake craniotomy for excision of brain metastases involving eloquent cortex techniques in neurosurgery. 2001; 7:85–90.

van den Broek PL, van Rijn CM, van Egmond J, et al. An effective correlation dimension and burst suppression ratio of the EEG in rat. Correlation with sevoflurane induced anaesthetic depth. Eur J Anaesthesiol 2006; 23(5):391–402.

Vandesteene A, Trempont V, Engelman E, et al. Effect of propofol on cerebral blood flow and metabolism on man. Anaesthesia 1988; 43 Suppl:42–43.

Vitaz T, Marx W, Victor JD, Gutin PH. Comparison of conscious sedation and general anesthesia for motor mapping and resection of tumors located near motor cortex. Neurosurg Focus 2003; 15:1–5.

Walker AE, Johnson HC, Marshall C. Electrocorticography. Bull Johns Hopkins Hosp. 1949; 84(6):583.

Werner C, Hoffman WE, Kochs E, et al. The effects of propofol on cerebral blood flow in correlation to cerebral blood flow velocity in dogs. J Neurosurg Anesthesiol 1992; 4(1):41–46.

Yang HJ, Nam DH, Wang KC, et al. Supratentorial primitive neuroectodermal tumor in children: clinical features, treatment outcome and prognostic factors. Childs Nerv Syst. 1999;15(8):377–383.

Yamamoto J, Ikeda A, Satow T, et al. Low-frequency electric cortical stimulation has an inhibitory effect on epileptic focus in mesial temporal lobe epilepsy. Epilepsia. 2002; 43(5):491–495.

Yingling CD, Ojemann S, Dodson B, et al. Identification of motor pathways during tumor surgery facilitated by multichannel electromyographic recording. J Neurosurg. 1999; 91(6):922–927.

Zentner J, Albrecht T, Heuser D. Influence of halothane, enflurane, and isoflurane on motor evoked potentials. Neurosurgery 1992; 32:298–305.

# 5

# INTRAOPERATIVE NEUROPHYSIOLOGIC MONITORING AND MAPPING OF THE SPINAL CORD

# 5A

## Intramedullary Spinal Cord Tumor Resection

MIRELA V. SIMON
LAWRENCE BORGES



## I. INTRODUCTION

Intramedullary spinal cord tumors are rare problems, occurring in fewer than one per million of the population. These tumors grow within the substance of the spinal cord and can compress, displace, distort, and engulf the axons and neurons within the spinal cord. These alterations of spinal cord architecture, caused by the tumor, lead to the symptoms of pain, numbness, weakness, ataxia, and bladder and bowel dysfunction that lead the clinician to make a diagnosis of a spinal cord tumor.

Surgery plays a central role in the management and treatment of these tumors. Because many different pathological types of tumors may occur within the spinal cord (Table **5A-1**), a central goal of surgery is to biopsy the tumor and obtain an accurate diagnosis. An accurate diagnosis is central for the proper selection of definitive therapy, which could include complete surgical removal of the tumor, radiation therapy, and chemotherapy. Surgery is the primary definitive therapy for a least 50% of all spinal cord tumors. Most ependymomas, hemangioblastomas, melanocytomas, and metastatic tumors can be removed completely or nearly so during surgery.

**TABLE 5A-1** Common intramedullary spinal cord tumors

1. Ependymoma
2. Astrocytoma
3. Hemangioblastoma
4. Melanocytoma
5. Metastatic Tumors

## II. THE ROLE OF INTRAOPERATIVE NEUROPHYSIOLOGY

Surgical biopsy and definitive surgical removal of tumors from within the spinal cord requires that incisions be made into the spinal cord and planes of dissection be developed around the tumor—separating the tumor from the spinal cord axons and neurons. Despite the fact that these axons and neurons may be adjacent to the tumor, they are most often fully functional components of the spinal cord. Therefore it is imperative that these functional axons and around remain uninjured during tumor dissection, so that the spinal cord function is preserved during surgical removal of intramedullary tumors. Intraoperative dissection of intramedullary tumors is often facilitated by the planes between the tumor and the spinal cord that can be identified by the experienced surgeon using the lighting and optics available in the modern surgical microscope. However, at times these planes can be difficult to determine by visual inspection alone. In addition the close proximity of important functional spinal cord tissue to these planes of dissection means that even the gentlest of surgical dissections within these planes could lead to postoperative neurologic deficit. An important goal of surgery (Table **5A-2**) in many intramedullary tumors (i.e. ependymomas) is complete surgical removal of the tumor. However, no surgeon wants to leave a patient with a new neurologic deficit. Intraoperative neurophysiologic monitoring allows these goals, which can at times be in conflict, to be pursued in greater harmony. Intraoperative neurophysiologic monitoring can provide the surgeon with the courage to pursue a complete tumor resection when the anatomical planes become indistinct. Monitoring can also warn the surgeon that a new neurologic deficit is occurring, even when the anatomical plane appears satisfactory. This ability of intraoperative monitoring to warn the surgeon of an impending or early neurologic deficit can also contribute to the education of the surgeon and potentially speed up the process of acquiring surgical experience. It must always be recognized that—although the goals of spinal cord tumor surgery (Table **5A-2**) can at times seem to conflict—resolution of this conflict is usually mediated by the use of intraoperative neurophysiologic monitoring.

Modern intraoperative neurophysiologic monitoring of spinal cord function consists of monitoring and mapping dorsal column function by somatosensory evoked potentials (SSEPs) and monitoring corticospinal tract function by motor evoked potentials (MEPs).

**TABLE 5A-2** Goals of intramedullary tumor surgery

1. Preservation of Spinal Cord and Nerve Root Function
2. Complete Removal of the Tumor—Where Possible

## III. A HISTORICAL PERSPECTIVE

SSEPs were used initially to monitor spinal cord function during scoliosis surgery. Nash and colleagues (1977) described the technique of monitoring SSEPs during surgery for scoliosis. They described how these potentials could disappear if the vertebral column was distracted or compressed beyond the tolerance of the spinal cord and how these potentials could reappear if these changes were reversed. Working with their neurosurgical colleagues, they also demonstrated the utility of SSEPs for guiding the removal of an intramedullary arteriovenous malformation (Owen et al., 1979). Other surgeons demonstrated how these potentials could predict postoperative dorsal column function after spinal tumor surgery (Macon et al., 1982). Following these and other early investigations, SSEPs were used to monitor many different kinds of spinal surgery. Around this time it also became clear that postoperative motor deficits could occur despite unchanged SSEPs during surgery (Lesser et al., 1986). Therefore it became evident that a direct method of monitoring corticospinal tract function was essential.

Merton and Morton demonstrated in 1980 that the brain could be stimulated directly by applying a high voltage current over the scalp and that this current could activate the corticospinal tract and produce muscle movement (Merton et al., 1980a; 1980b). Deletis and Epstein and their colleagues refined these techniques and used them to monitor spinal cord function during surgery in patients with intramedullary spinal cord tumors (Morota et al., 1997). In this prospective study, they describe their technique of monitoring MEPs and D waves under anesthesia. By correlating clinical outcome with intraoperative observations, this study provides the first guide to interpretation of these potentials in the surgical patient. Other authors have also documented the pattern of changes in MEPS that can be observed during intramedullary spinal cord tumor surgery (Quinones-Hinojosa et al., 2005). Recently Deletis has published a very thoughtful review on this topic and on the specific importance of monitoring both SSEPs and MEPs during spinal cord tumor surgery (Deletis et al., 2008).

The previously cited papers demonstrate clearly that intraoperative monitoring of SSEPs and MEPs can document loss of neurologic function during surgical removal of intramedullary spinal cord tumors. However, the real goal of these monitoring techniques is to improve the clinical outcome of patients with spinal cord tumors who undergo surgery. So, do these monitoring techniques improve clinical outcome? The simple way to answer this question would be to undertake a randomized prospective clinical trial during which some patients had surgery with neurophysiologic monitoring and a control group had surgery without neurophysiologic monitoring. The primary end points of this trial would be the postoperative level of motor and sensory functioning and extent of tumor removal. However, several problems arise when considering such a trial. Spinal cord tumors are rare, they are variable in their longitudinal location within the spinal cord, and they are variable in their degree of adhesion to adjacent spinal cord. Therefore identifying a sufficient number of patients to complete this study would be problematic. In addition, given our present knowledge of this topic, few surgeons would think that such a study was ethical.

Sala et al. (2006) have attempted to answer this question with a retrospective study. Professor Bricolo's neurosurgery unit in Verona, Italy, began using intraoperative neurophysiologic monitoring in September 2000. They studied the surgical outcome in the 50 patients operated on using neurophysiologic monitoring and compared this group to 50 historical control patients who had surgery prior to this time without neurophysiologic monitoring. Both groups were matched by preoperative neurologic status, histologic results, tumor location, and extent of tumor removal. The authors conclude that the clinical outcome was better following surgery with neurophysiologic monitoring for patients with ependymomas, but no different than historical controls in patients with astrocytomas. This observation makes physiologic sense, because the goal of surgery in patients with ependymomas is complete surgical removal of the tumor, whereas the infiltrative nature of astrocytomas usually precludes total removal. Therefore intraoperative neurophysiologic monitoring can be expected to be most valuable in the setting of attempting to remove completely a tumor that abuts the dorsal columns and corticospinal tracts.

One fundamental problem of this study is that the patients in the control and the experimental groups were matched for the extent of tumor removal. Therefore this study could not address the probability that intraoperative neurophysiologic monitoring likely improves the extent of tumor resection while also improving neurologic outcome. Complete surgical resection of ependymomas often includes peeling bits of tumor directly away from the corticospinal tract. Prior to neurophysiologic monitoring, there were situations when the surgeon would become concerned about the indistinct nature of the plane between the tumor and the normal spinal cord at the last stage of tumor dissection. Many times, the surgeon might choose to stop the dissection rather than risk a profound neurologic deficit. Presently, intraoperative neurophysiologic monitoring can provide the surgeon with great insight in this setting and allow a complete tumor resection and good postoperative neurologic function, despite an indistinct anatomical plane. In an era of increasing technology and shrinking health care dollars, one challenge to intraoperative neurophysiologic monitoring is to document the value of this technology in improving surgical and survival outcomes.

## IV. INTRAOPERATIVE NEUROPHYSIOLOGIC MONITORING AND MAPPING: A STEP-BY-STEP APPROACH

It is our practice to monitor both SSEPs and muscle MEPs induced by transcranial electrical stimulation (tcmMEPs) continuously from induction time and before positioning until closure. Additionally, monitoring of neurogenic MEPs (D waves) can be performed once the dura is open and throughout the tumor resection process. All three methods (including principles, stimulation, and recording techniques, neurophysiologic-neuroanatomical correlation and interpretation of the recorded potentials) are detailed on pages 5–26.

In the following, we detail the specifics of neuromonitoring during different stages of the surgical procedure. We offer a step-by-step algorithm for approaching the changes seen in evoked potentials. The latter includes troubleshooting and a plan of action for reversing the neurophysiologic changes.

### 1. Anesthesia Considerations

The neurophysiologic team is an integral part of every surgery to remove or biopsy an intramedullary spinal cord tumor. In fact they are often the first to greet the patient when the patient arrives in the anteroom prior to surgery. Prior to induction of anesthesia, they place all of the surface electrodes for SSEP monitoring. Anesthesia is then induced with propofol or pentothal, and an endotracheal tube is placed, with the assistance of a single dose of succinylcholine. If the patient has significant preoperative muscle weakness or atrophy, a nondepolarizing muscle relaxant is used. Etomidate can also be used for anesthetic induction, especially in patients at risk for hypotension. Etomidate can tran-

(b)

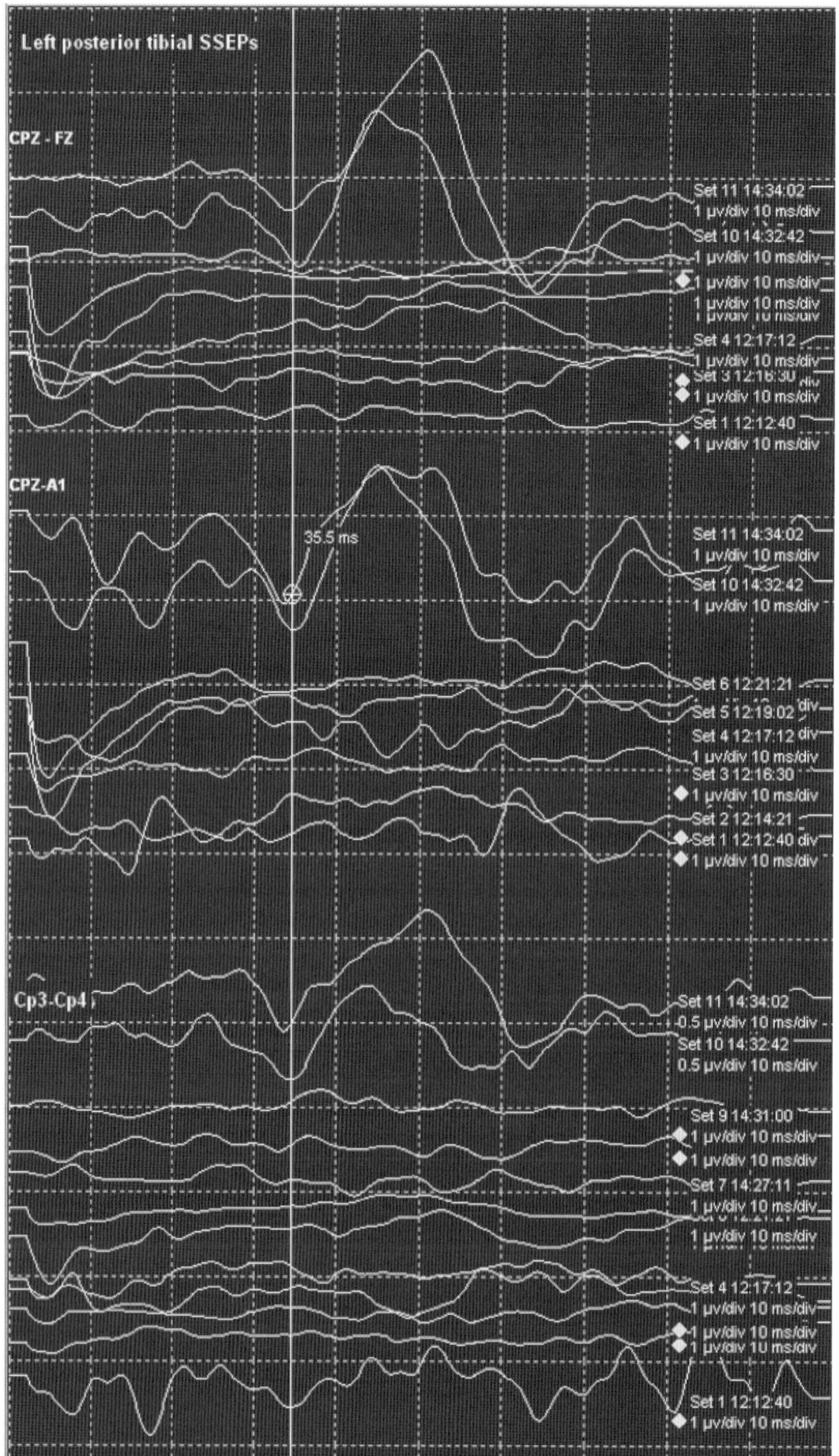


**FIGURE 5A-1** *(Continued)* **b:** Left posterior tibial SSEPs. Sets 1 through 9: after stimulation of the posterior tibial nerve at the ankle, no reliable P37/N45 are obtained. Sets 10 and 11, after stimulation of the left posterior tibial nerve at the popliteal fossa, P37 is obtained with an absolute latency of 35.5 msec.

recordings, especially on the subcortical potentials (N30 and P/N13), recorded from the cervical channel; in general, however, around the time that the SSEPs baselines are obtained, there is a reasonable degree of iatrogenic neuromuscular blockade.

***e. Improving the impedances of the recording electrodes.*** A decreased signal-to-noise ratio can be improved by troubleshooting noisy channels. The raw data window thus allows identification of such channels and the corresponding electrodes. An impedance check should be done, again with the goal of obtaining relatively even impedances, less than 5 kohms.

### B. TcmMEPs

*B1. Absent responses.* One has to ensure that appropriate stimulation is applied. Stimulus artifact signals appropriate stimulation. If no stimulus artifact is present, the hardware connections and the software setup should be checked. For example, if the stimulator labeled "1a" is connected in the software program to the recording muscle group, the cable connecting the stimulating pod to the machine should be plugged into the socket labeled "1a." If no muscle responses are obtained despite correct stimulation and despite maximal stimulation, the most likely cause is iatrogenic neuromuscular block. Several methods (e.g. train of four [TOF]) in assessing the degree of neuromuscular block are described on pages 332–34. Of course, if there is a significant motor baseline deficit, it may not be possible to elicit motor responses. Recently, several authors (Yamamoto et al., 2008; Hayashi et al., 2008) have described a new method to augment the tcmMEPs amplitudes. This method relies in facilitation of tcmMEPs recordings by tetanic stimulation of the peripheral nerves just before the transcranial stimulation. Both posterior tibial nerves are stimulated with a repetitive, high-frequency stimulus (50–100 Hz), at 50 mA for about 5 seconds. The immediate transcranial electrical stimulation elicits a higher-amplitude MEP. These tcmMEPs are known in the literature as the p-MEPs (post-tetanic MEPs). More so, Hayashi et al. (2008) concluded that p-MEPs may be more reliable than the tcmMEPs in detecting postoperative changes in motor function. However, their results were based on a small series of patients, while partial neuromuscular block was used. It is not clear whether their advantage would be maintained in neuromonitoring without the use of muscle relaxants. To this is added the definite inconvenience of this method, because each transcranial stimulation has to be preceded by tetanic stimulation of the peripheral nerves. If done repetitively, over the span of several hours, this may, in fact, cause nerve and muscle fatigue. Moreover, it would interfere with the continuous monitoring of SSEPs, obtained by low-frequency stimulation of the same peripheral nerves. Although we do not recommend replacing the "classical" tcmMEPs monitoring with the newer p-MEPs technique, we believe that the latter may have a more beneficial role in case of significant iatrogenic neuromuscular block. This may be necessary in certain cases of spine surgery, in order to minimize the excessive movement in the field.

*B2. Obtaining some muscle MEPs, but not all of them.* In general, the upper limbs mMEPs are obtained at lower stimulus thresholds than the lower limbs MEPs. This is, in general, due to a more exposed cortical arm area, compared to the leg area, which is located at deeper levels, "hidden" in the interhemispheric fissure. Thus, occasionally, the upper limbs MEPs are obtained but not the lower limbs, despite maximal stimulus amplitude. As mentioned on pages 16–17, we recommend trying different positions for the stimulating corkscrew electrodes: C1/C2 and Cz/6 cm anterior could offer a better stimulation of the leg areas. Another situation is represented by excessive movement with stimulation, usually of the upper body. As previously mentioned, this is generally encountered with C3/C4 positioning. The excessive movement impedes further increase in the stimulus amplitude necessary to trigger MEPs in the lower limbs. Securing the upper limbs diminishes the risk for injury. Changing the electrode position may also help. In some cases, starting a low-rate infusion of muscle relaxant helps reduce the movement and allows further increase in the stimulus amplitude (see pages 179–80).

There are situations when prior motor deficit leads to difficulty in triggering an electrophysiologic response in the corresponding muscle. These MEPs will be (if at all) elicited at a higher stimulus amplitude and will be the first ones to be affected by muscle relaxants, anesthetics, or physiologic factors (e.g. marked systemic hypotension).

The baseline MEPs may be influenced by the presence of a significant iatrogenic neuromuscular blockage as a result of the bolus of muscle relaxant given for intubation. Again, weaker muscles and lower limb muscles (in comparison with upper limb muscles) are the first ones to be affected.

## 3. Positioning the Patient

Once baseline recordings are found to be satisfactory, the patient is positioned into the prone position. We prefer a Jackson Table for all prone surgery, to reduce epidural venous pressure. The head is held fixed in a 3-pin Mayfield headrest for all tumor surgery—regardless of the spinal level being operated. This allows the neck to be in a neutral physiologic position, avoids any pressure on the orbits, and provides a ready point of fixation for the endotracheal tube. Once the patient is in the prone position, post-positioning baseline SSEPs and MEPs are obtained.

## 4. Post-Position Baselines

### A. SSEPs

After the patient is positioned, post-position SSEPs are taken and are compared to the pre-position baseline.

*A1. Increased latencies.* We should start by emphasizing the fact that the absolute latencies of the SSEPs are usually increased in the operating room settings, compared to the outpatient laboratory recordings. This is due to the fact that intraoperative recordings are subject to anesthetic influence. The delay in electrical transmission is encountered mainly at the synaptic level; thus, the greater is the number of synapses in the pathway, the longer is the absolute latency of the evoked response. Because the anesthetics act mainly at the synaptic level, they preferentially affect the polysynaptic pathways, which experience a greater increase in transmission time. Often, the anesthetic-related increase in latency from the baseline SSEPs is first appreciated about one hour postinduction, after opening, when the patient reaches deeper levels of anesthesia. Because positioning of the patient can take as little as 15 minutes, there is usually not enough time for such an increase in latency to develop. However, positioning may take longer, enough for the level of anesthesia (total intravenous anesthesia [TIVA]) to get deeper, causing a latency difference between the pre-position and post-position baselines. If the difference is directly related to TIVA, the morphology of the waveform is not significantly changed, and the amplitudes are only mildly affected. However, when such latency difference (between pre-position and post-position SSEPs baseline) is seen and assumed to be related to propofol administration, we recommend continuing recording the SSEPs for 15–20 minutes or so (before pausing for the opening). This is to ensure that the changes do not worsen dramatically. If they do, the initial assumption that they are related to anesthetics should be reconsidered.

*A2. Worsening of the morphology, decreased amplitudes, and increased latencies.* First, it is possible that, with positioning, the stimulating electrodes have been displaced (Fig. **5A-2**). A particular situation arises when the skin area where these electrodes were applied is moved away from the nerve. SSEPs changes immediately after positioning are unlikely to be caused by inadequate limb positioning resulting in plexus or nerve compressions. These processes require a longer time to develop, and SSEPs changes related to them are described later

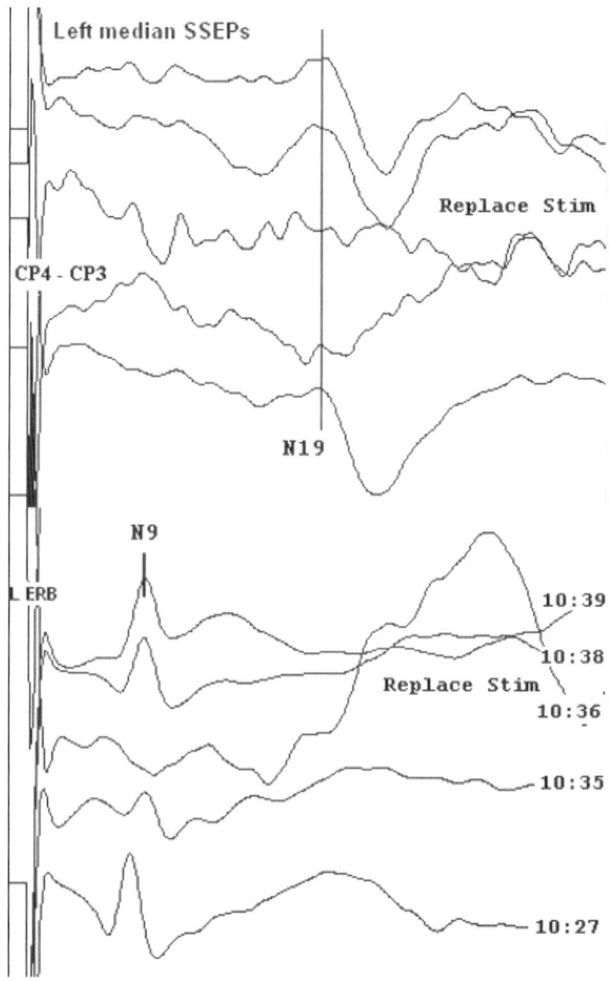


**FIGURE 5A-2** Post-position changes in the left median SSEPs, as a result of displacement of the stimulating electrodes. Notice the loss of the Erb's potential, as well as the cortical N19/P22. After replacement of the stimulator, there is a total recovery of the waveforms.

(see page 189). Second, one should ensure that there has been no administration of inhalational agents that could affect the SSEPs. A third possibility is that real changes related to the patient's repositioning of an unstable spine may have occurred. This situation is more frequently encountered in spine trauma and decompressive surgery and less often in intramedullary spine tumor surgery. This is detailed on pages 210–11.

In the first case, the peripheral potentials (Erb's point or popliteal fossa potential) are also affected, and the abnormality is restricted to the SSEPs of the nerve inappropriately stimulated. In the second case, the peripheral potentials, as well as the cervical subcortical potentials (P/N13, N30, respectively), are normal, but the cortical potentials are not. The changes should also be evident diffusely, but worse in a symptomatic limb. In the third situation, the peripheral potentials are normal, but the subcortical and cortical SSEPs are affected. The location of the changes should correlate with the level of the spine instability.

*A3. Reversal of SSEP polarity* This situation is, in general, related to reversal of the plugged electrodes in the recording headbox (for example, obtaining a P19 instead of a N19 in Cp3-Cp4 channel, after stimulation of the right median nerve, is related to plugging the Cp3 electrode in the Cp4 socket and vice versa).

### *B. TcmMEPs*

*B1. Significant changes in tcmMEPs.* In general, it is difficult to quantify changes in MEPs. A drop in amplitudes of more than 50% at the same stimulus threshold is considered by many authors as significant (Deletis et al., 2008) . However, the changes have been discussed within the context of D-waves monitoring (see pages 201–6). Presence of tcmMEPs, at an amplitude of more than 50% from baseline, has always been seen associated with robust D waves (Kothbauer et al., 1998). In other words, a drop in the amplitudes of tcmMEPs of less than 50% is acceptable. Others follow the all-or-none criteria, considering as significant only the total loss of tcmMEPs (Jones et al., 1996; Zentner, 1989). On the other hand, Calancie and Molano (2008) argued against the "all-or-none" criteria, supporting the use of changes in stimulation thresholds as significant. A change in morphology, from polyphasic to biphasic potentials, has also been interpreted as important (Quinones- Hinojosa et al., 2005). We believe that changes in amplitude, latency, and morphology should always be interpreted in direct relationship with the time span over which they have occurred. This idea is also supported by other authors. Hsu et al. (2008) define a drop in the amplitudes of tcmMEPs by 50% from baseline, occurring within a minute, as significant. We also believe that the changes have to be correlated with any changes of the anesthesia regimen and with the particular surgical step within which they were seen. For example, such changes become important if they occur suddenly. We consider that the time spent with positioning is short enough so that the post-position baselines should be obtained at the same stimulus threshold as the pre-position baselines and without significant changes in the amplitudes of tcmMEPs. A significant drop in the amplitudes of the post-position tcmMEPs, requiring a sudden and significant increase in the stimulus amplitude (more than 20 mA), may be related to additional muscle relaxant given at the positioning. As already specified, the weak (versus normal) and leg (versus arm) muscles are the first to be seen affected. Thus, the changes in tcmMEPs may look asymmetric (Fig. **5A-3**). However, one should carefully analyze and compare the pre-position and post-position amplitudes in all of the muscles. If the changes in tcmMEPs are related to such a muscle relaxant effect, one should see a decline in amplitudes in all mMEPs. For high-amplitude mMEPs, these changes may be difficult to appreciate if the sensitivity of the recording is low. On the other hand, leg MEPs and weak muscles supposedly have smaller amplitudes, and their recording is done at higher sensitivity. Thus, changes in amplitudes become more evident. Moreover, the decline ratio may be bigger in weak muscles. Around positioning, several milligrams of intravenous cisatracurium is often given, and the time necessary to wait for its effect to pass is directly related to the amount administered, but also to the interindividual and intermuscle variability, usually taking longer for the weaker muscles to recover. Unfortunately, the intravenous boluses of muscle relaxants have an immediate, unpredictable, and lingering effect. For example, for a 10-mg bolus, usually a wait of 30 minutes to an hour may be necessary until reliable mMEPs are recorded again. Thus, our practice is to avoid muscle relaxant administration between the pre-position and post-position baselines, particularly in patients in whom possible changes with positioning are expected. Inhalational agents can also affect tcmMEPs recordings. Thus, their use should also be avoided during positioning and, in general, during monitoring. Similar to changes in amplitude, changes in morphology, from polyphasic to biphasic waves, can be seen with muscle relaxants and

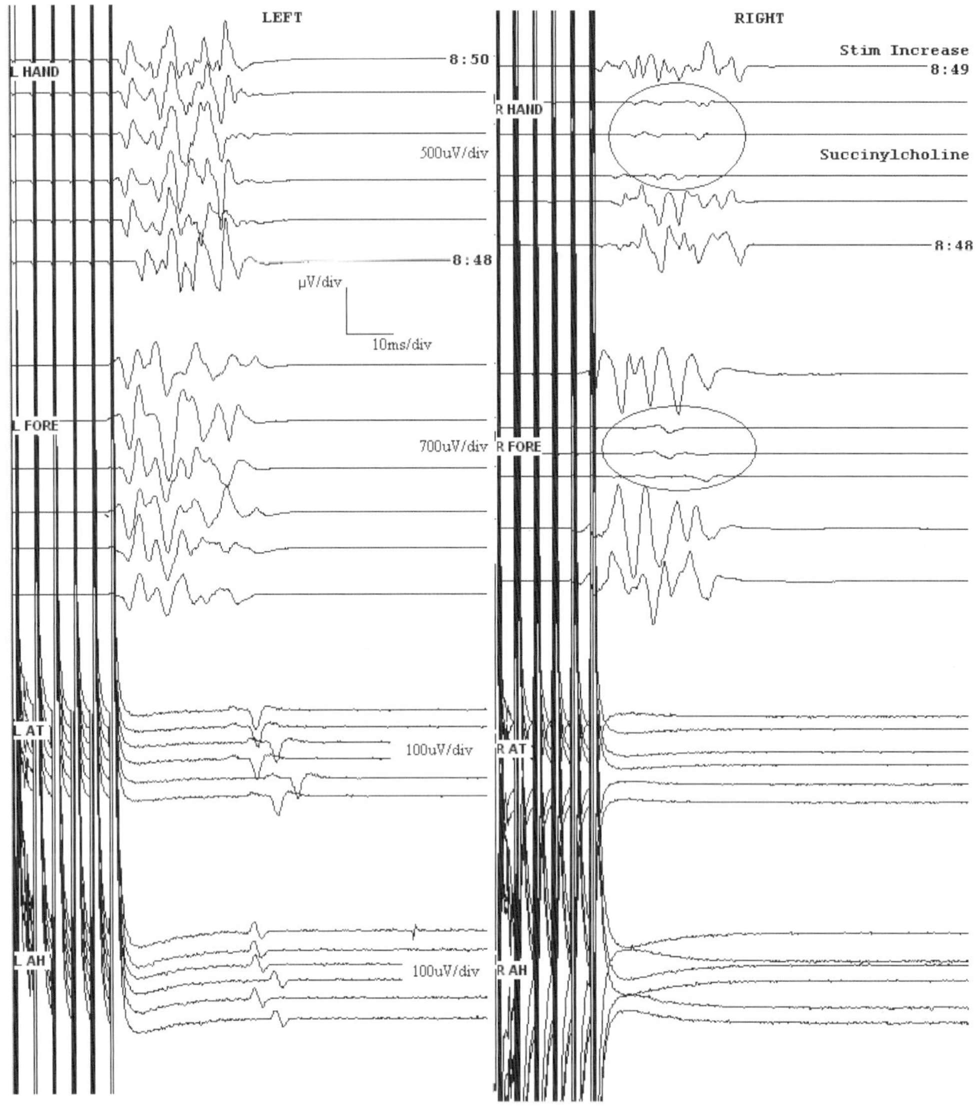


**FIGURE 5A-3** Asymmetric mMEPs changes induced by succinylcholine administration. Symptomatic patient with right arm and bilateral leg weakness. Notice the difference in the baseline mMEPs, minimal in the left leg, absent in the right leg, and smaller in the right arm than in the left arm. Administration of a bolus of succinylcholine markedly decreased the amplitudes in the right abductor pollicis brevis (APB) and brachioradialis MEPs, which required a 40-mA increase in stimulus intensity in order to be obtained.

anesthetics. Because a polyphasic potential is a sign of a robust depolarization of many muscle fibers (see page 25), biphasic potentials may be expected in weak muscles or with impaired innervations. Thus, as with changes in amplitudes, changes in morphology may not only be the consequence of muscle relaxants but also a sign of spine malposition.

If it is inferred that the SSEPs or MEPs changes are directly related to spine positioning, the patient should be turned back supine and repositioned, as shown on pages 210–11.

Both the post-position SSEPs and the tcmMEPs should look nearly identical to the pre-positioning baseline. Any variations between these two sets of potentials must be explained and corrected prior to proceeding with surgery. After post-positioning baseline tcmMEPs are obtained, the extent of the movement with MEPs is reevaluated. The position of the patient's limbs is assessed. Additional padding or restraining straps are used to ensure safe limb position during the movement caused by eliciting the MEPs. Other ways of decreasing the unwanted body movements triggered by transcranial stimulation are described on page 183.

Even though muscle relaxant infusion may help in reducing these movements, we try to refrain from using muscle relaxant during the procedure.

## 5. Opening

Once the post-position SSEPs and tcmMEPs baseline are obtained, we usually pause the monitoring for the opening for a couple of reasons. First, the incision time is characterized by significant electrocautery artifact, which causes either rejection of the trials and inability to average SSEPs or significantly reduced signal-to-noise ratio to the electrical artifact introduced in the record. Second, at least theoretically, during the use of electrocautery, there is an increased risk for ground loops formation (see pages 42–44). However, during this time we ask the surgeon to pause the use of bovie occasionally, so that we can intermittently record several trials of SSEPs and tcmMEPs. During this time a gradual increase in the absolute latencies of the cortical SSEPs can also be noticed as a result of the deepening of the level of anesthesia.

## 6. Laminectomy to Dura Opening

Continuous monitoring of SSEPs and MEPs is started after about 45 minutes or so, at the beginning of the laminectomy, and continued until the tumor is removed, the dura is closed, and the laminoplasty is completed.

### A. SSEPs

The posterior tibial nerve stimulation that produces this amount of sensory evoked potentials causes repetitive dorsiflexion of the feet. This repetitive dorsiflexion of the feet can cause the patient's body and the surgical field to move repetitively if the feet contact the caudal end of the Jackson Table. When the patient is positioned initially into the prone position, the legs are supported on pillows, with the knees flexed and the feet dorsiflexed. The soles of the feet are not permitted to contact the caudal part of the Jackson Table. If the soles of the feet contact the caudal part of the Jackson Table, the repetitive foot dorsiflexion caused by the stimulation of the posterior tibial nerves will cause an annoying repetitive movement at the surgical field. This movement is most apparent and bothersome to the surgeon during microsurgical dissection.

During this phase, some progressive increase in the overall absolute latencies of the cortical SSEPs can be seen because of the effect of propofol (see Fig. **5A-4a**). The main concern becomes a sudden change in latency from one trial to the next or within a couple of trials. The vertical stack display is thus very useful in appreciating the dynamic of the change. Imminent damage to the sensory pathways is usually signaled by a sudden decrease in amplitude, as well as a sudden change in morphology of SSEPs. As previously mentioned, it is our experience that anesthetic changes, although they can affect the latencies and/or amplitudes, cause

(a)

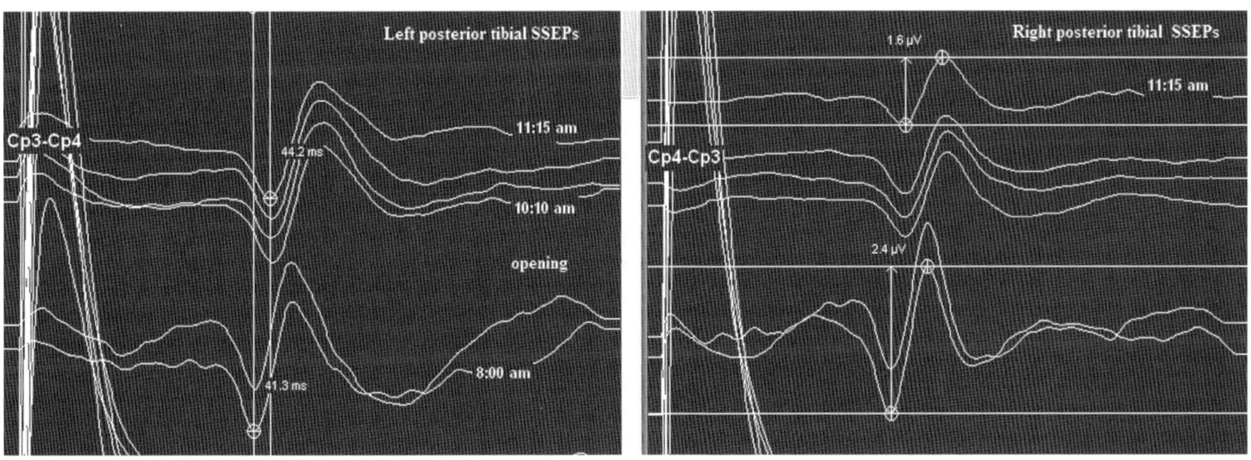


**FIGURE 5A-4** SSEPs and tcmMEPs changes induced by propofol. **a:** SSEPs changes in latencies and amplitudes.

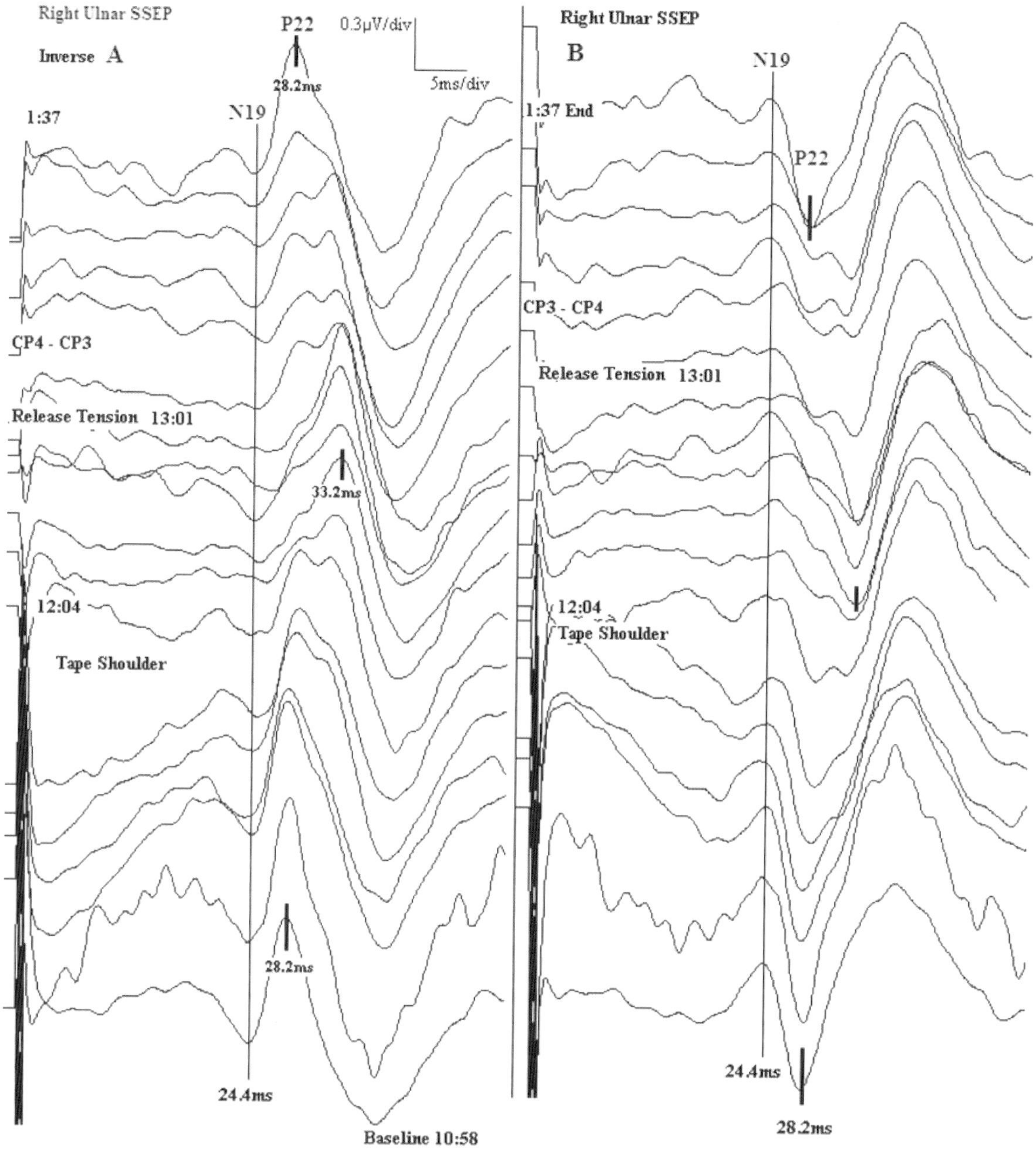


**FIGURE 5A-7** Upper limb SSEPs changes related to arm malpositioning. Right ulnar SSEPs changes, developing within an hour after taping the right shoulder for downward traction of the arm. Panel **A** shows reversal N19/P22 potential; panel **B** shows the correct orientation of the N19/P22 waveforms, according to their polarity (recording is done in the same channel, but with flipped inputs). Because the P22 peak is somewhat higher, it is easier on the eye to follow changes in its absolute latency. Once the tension is released, there is a progressive improvement, with total recovery of N19/P22 absolute latency.

impedances are obtained and if no recognizable D waves are recorded, the electrodes should be repositioned. Warm saline irrigation may also help in improving the impedances. Once the electrodes are appropriately positioned, their position should be secured. Generally, we find the epidural recordings to be more problematic than subdural recordings, resulting in a smaller signal-to-noise ratio.

Another nuisance of such recordings remains the presence of stimulus artifact. This is, in particular, true for

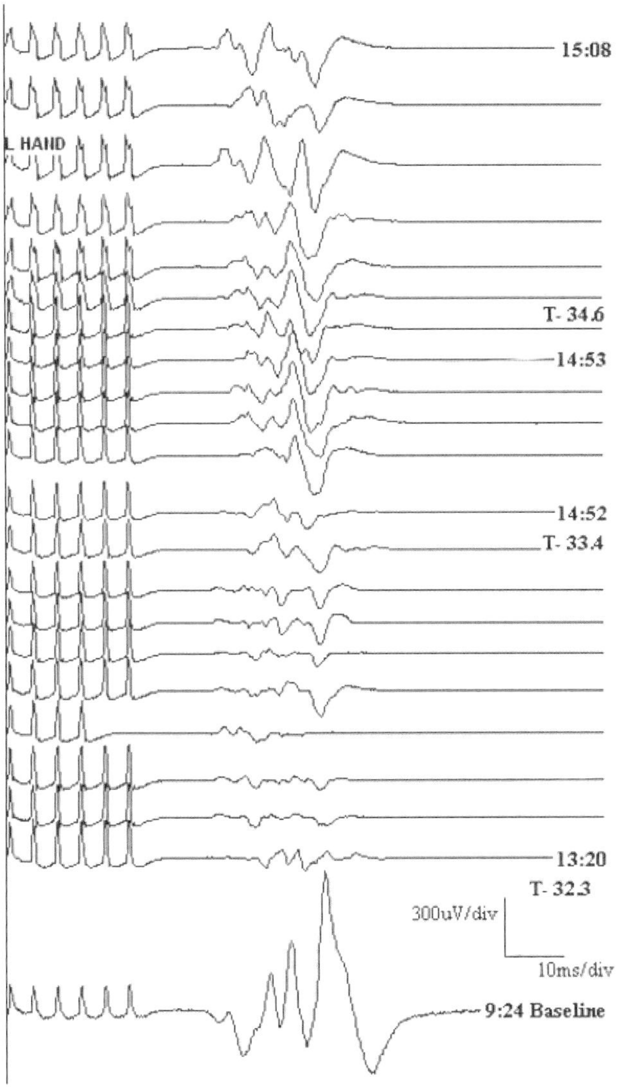


FIGURE 5A-8 Attenuation of mMEPs induced by hypothermia. Warming from 32 to 34 degrees Celsius is associated with improvement in the evoked responses.

obtaining the cervical D waves, because they have a very short latency. Useful information on how to decrease such stimulus artifact is given on pages 288–89.

A frequent situation that can occur is the fluctuation in the amplitude of the D waves due to increased noise it is important to make this specification (e.g. change in position of the epidural electrode). Increasing the number of averaged trials may help. In general, we like to average at least several trials.

In the presence of proximal D waves and the absence of distal D waves, in case the corticospinal tract is affected by the tumor, it is possible that the D waves are not obtainable distal to the tumor because of temporal dispersion of the electrical volleys within the corticospinal tract (even in the presence of mMEPs in the myotomes distal to the spine tumor level).

### *B. Troubleshooting Stimulation*

Because the stimulation method is similar to that of tcmMEPs, its troubleshooting processes are also similar. More details about the stimulating and recording technique can be found on pages 16–23.

Before the actual tumor dissection starts, we do obtain reliable "new" baselines for D waves, tcmMEPs, and SSEPs.

## 8. Starting the Tumor Dissection: Dorsal Column Mapping and Myelotomy

Intramedullary tumors are located within the substance of the spinal cord. Therefore an incision must be made into the spinal cord to identify the tumor and begin to develop a plane of dissection around the tumor. The most common spinal cord incision is a midline myelotomy. An incision is made in the posterior median raphae of the spinal cord, in between the two dorsal columns. The information flow within the dorsal columns is crucial to maintaining the proper balance and perception of lower limb position that is essential for ambulation. Surgical injury to these structures must be minimized. One important way of minimizing injury is to make this incision directly in the median raphae, so that no dorsal column axons are injured. The median raphae can usually be identified by visual inspection along the normal dorsal spinal cord. However, the spinal cord that harbors an intramedullary tumor can be swollen and anatomically distorted. The median raphae may be difficult or impossible to identify accurately by visual inspection alone in this setting.

Intraoperative neurophysiology can play an important role in dorsal column mapping. This can be done with the help of SSEPs, mainly the posterior tibial SSEPs via a few techniques, which are further described as follows.

### *A. Retrograde Recording of Posterior Tibial Nerve Action Potentials*

Quinones et al. (2002) have shown in one patient that a stimulus applied to the dorsal columns can produce a retrograde nerve action potential that can be recorded from the posterior tibial nerves. They have suggested that this stimulus may be used to map the median raphae between the dorsal columns and assist in planning the midline myelotomy. We have not had any experience with this method.

## *B. Anterograde Recording of Spinal SSEPs*

*B1. Principle.* Stimulating the posterior tibial nerves peripherally at the ankles and recording the ascending electrical volleys at the spinal level (spinal SSEPs) have been described by several authors (Krzan et al., 1996; Sala et al., 2008). The principle of this method is based on the somatotopical organization of the large fiber sensory system within the dorsal columns, with the Gracilis tracts situated closest to the midline, one on each side of the median raphae. At their level, the recorded electrical volleys triggered by peripheral stimulation of the ipsilateral posterior tibial nerve will have the highest amplitudes, thus enabling the identification of the neurophysiologic midline (median raphae).

*B2. Stimulation technique.* The method of stimulation, as well as its troubleshooting, is identical to that used for the SSEPs monitoring technique. Details of the technique can be found on pages 5–15, and of the troubleshooting process on pages 181–83.

*B3. Recording technique.* The recording is done via an 8-contact microelectrode (Fig. **5A-9**), placed directly on the dorsal surface of the spinal cord, over the dorsal columns. The row of the 8 contacts should be perpendicular to the spine axis. We use a referential recording, with sterile subdermal needle used as reference and placed by the surgeon in the field. Because the absolute amplitudes of these potentials are important, we prefer that the reference electrode be placed equidistant from all of the contacts of the microelectrode. The low-frequency filter (LFF) of the recording is set at 20 Hz and the high-frequency filter (HFF) at 2 KHz. The HFF is especially important for recording posterior tibial spinal SSEPs.

Based on the absolute latency of the evoked responses (see the following) a time base between 5 and 10 msec/div is appropriate. We use a sensitivity between 1.5–7 $\mu$V/div.

We stimulate the posterior tibial nerve at the ankle, one side at a time, and average at least 200–300 trials for one set. The contact situated the closest to the midline shows the most robust and highest amplitude evoked response, with a gradual loss of amplitudes of the SSEPs because the recording was done at contacts further from midline. Subsequently we stimulate the other posterior tibial nerve, and again we look to see at which contact the highest

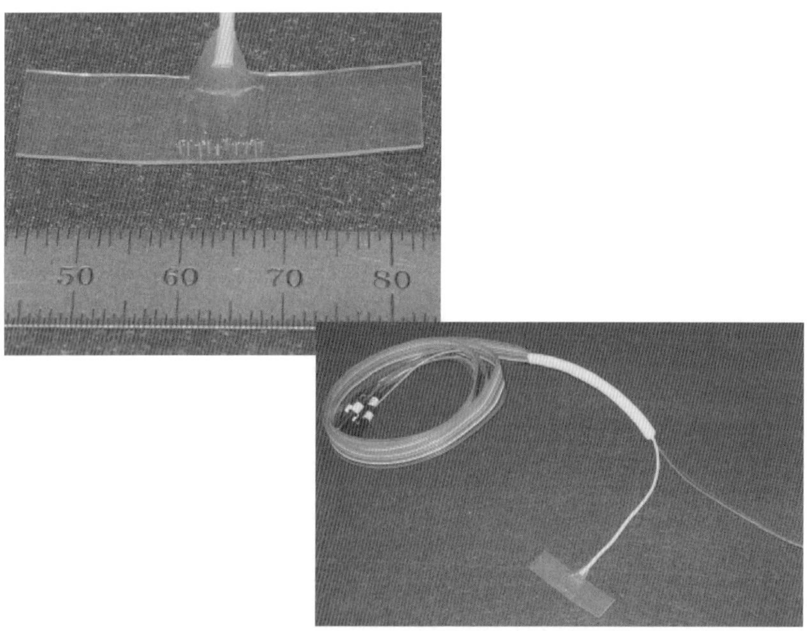


**FIGURE 5A-9** Custom-made microelectrode for recording of spinal SSEPs at the level of dorsal columns.

amplitude response was recorded. The median raphae is thus situated between the two contacts (see Fig. **5A-10**). Similarly, we can stimulate the median or ulnar nerves at the wrist and record the spinal evoked responses in the more lateral parts of the dorsal columns.

*B4. Recorded potentials*

***a. Morphology.*** There is a great variability of the morphology of the spinal SSEPs triggered by stimulation of the peripheral nerves. The potentials evoked by peripheral stimulation of the posterior tibial nerves have a more complex, polyphasic morphology, compared to the spinal responses evoked by peripheral stimulation of the median nerve, with at least several negative, sharp, brief peaks. Their morphology, however, tends to be somewhat simplified at the low thoracic, lumbar levels. Also, because the recording is done from more proximal levels (e.g. cervical), the actual duration of the posterior tibial SSEPs is longer than that of those recorded at the more distal levels, likely because of temporal dispersion.

We usually stimulate the median nerve for the upper limbs SSEPs. As already mentioned, the spinal evoked potentials have a different morphology, usually with a first positive (downward) smaller peak, followed by one or two negative (upward) peaks of higher amplitudes. However, in a simplified way, the recorded ascending volley corresponds thus to N12, as described on page 14.

***b. Amplitudes.*** The amplitudes of the SSEPs vary significantly, according to the spine level where recording takes place and to the peaks (within the entire complex) considered in the measurement. They also have interindividual

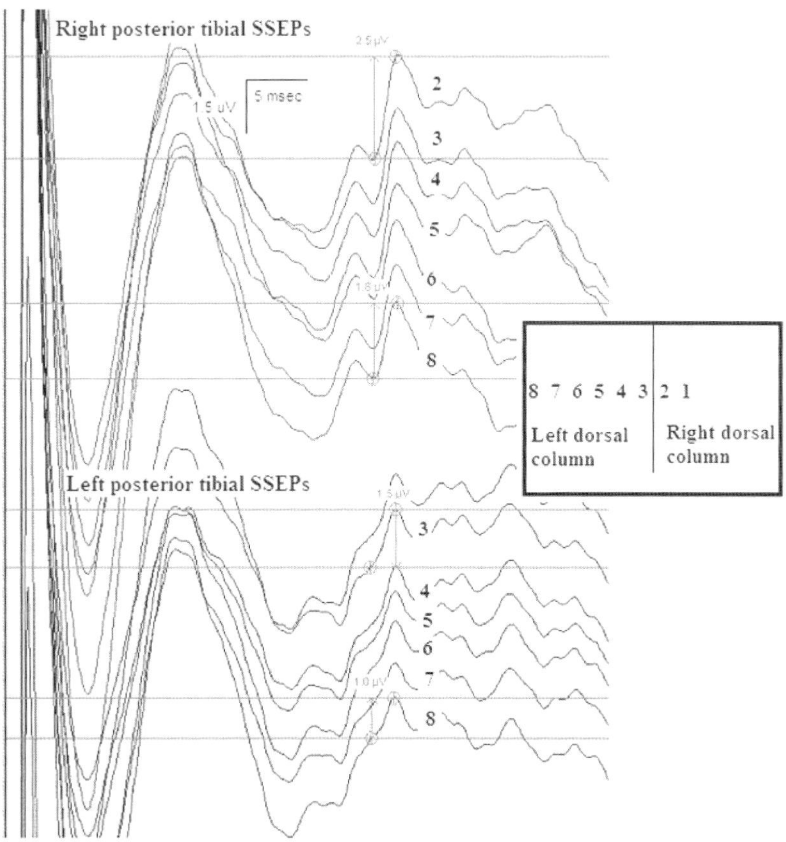


**FIGURE 5A-10** Dorsal column mapping: anterograde recordings of spinal posterior tibial SSEPs. Recordings are done via an 8-contact microelectrode, placed over the dorsal columns, perpendicular on the longitudinal axis of the spinal cord. The posterior tibial nerves are stimulated at the ankles, one at a time. Averaged responses are recorded for each nerve, at the same contacts. The highest amplitude potentials are considered to be recorded at the level of the ipsilateral Gracilis tract, situated close to the median raphae. Thus, the location of the latter is determined, in between the two contacts recording from the Gracilis tracts, right and left, respectively. The present recording was done at the C5 level. The morphology of the posterior tibial spinal SSEPs is expected to be more complex than shown: the simplified morphology is likely related to a low HFF setup (500 Hz).

variability. Tables **5A-3** and **5A-4** show the amplitudes of the highest negative peak, at different levels, for both posterior tibial and median nerve stimulation. In general, the posterior tibial SSEPs evoked by posterior tibial nerve stimulation seem to have the highest amplitude at the lumbosacral level, if recorded with an epidural electrode. Comparing the amplitudes of these recorded potentials at the thoracic and cervical levels, these amplitudes are greater at the thoracic level. The amplitudes of the spinal median SSEPs recorded at different cervical levels are higher, up to several times higher than the amplitudes of the posterior tibial SSEPs recorded at the same level. The recordings obtained directly from the spinal cord surface, which were recorded epidurally, have significantly higher amplitudes (usually 2–3 times) than some of those presented in Tables **5A-3** and **5A-4**.

***c. Latencies.*** As expected, the absolute latency of these evoked responses depends upon the distance between the stimulation and recording sites and varies with the spinal level at which the recording is obtained. In fact, if one places the microelectrode parallel to the spine axis rather than perpendicular to it, a lag between the peaks can be seen, because the traveling volley is recorded at different spinal levels; the peak recorded by the most proximal contact has the longest latency. By measuring the interpeak latency between the most proximal and most distal peak and knowing the distance between the recording contacts, one could calculate the individual central conduction velocity.

Table **5A-3** shows the expected absolute latencies and amplitudes of the spinal evoked responses recorded at different levels, after stimulation of the posterior tibial nerve at the ankle. Some of the values were calculated by adding 9 msec (the very approximate expected latency of the popliteal fossa potential) to the absolute latencies obtained by stimulation of the posterior tibial nerve at the knee (in Table **5A-3** for subjects EF and MR, both 14-year-old males). Of course, this latency also depends on the patient's height, as well as on existing comorbidities affecting the peripheral nerve conduction (e.g. peripheral neuropathy). In summary, the spinal posterior

**TABLE 5A-3*** Absolute latencies and amplitudes of the SSEPs recorded at different spine levels after stimulation of the posterior tibial nerve at the ankle

| Spine (vertebral) level of recording | Absolute latency of first positive peak (msec) | Absolute latency of most negative peak (msec) | Amplitude (μV) |
|---|---|---|---|
| C4 (21 yo) | 31.2 | 35.4 | 2.7 (subdural recording) |
| C7 (EF, 14 yo) | 23.7 | 25.5 | 0.6 (epidural) |
| T2 (9 yo) | 13.5 | 14.1 | 0.55 (epidural) |
| T4 (EF, 14 yo) | 23 | 24 | 0.6 (epidural) |
| T5 (EF, 14 yo) | 22.7 | 23.5 | 1.2 (epidural) |
| T6 (EF, 14 yo) | 22.6 | 23.3 | 2.1 (epidural) |
| T7 (EF, 14 yo) | 22.3 | 22.9 | 2.7 (epidural) |
| T8 (EF, 14 yo) | 21.9 | 22.5 | 4 (epidural) |
| T9 (EF, 14 yo) | 21.1 | 22.1 | 3.5 (epidural) |
| T10 (MR, 14 yo) | 20.4 | 23.7 | 1.8 (epidural) |
| T11 (MR, 14 yo) | 20 | 22.5 | 7 (epidural) |

**TABLE 5A-4*** Absolute latencies and amplitudes of the SSEPs recorded at different cervical levels after stimulation of the median nerve at the wrist

| Spine (vertebral) level of recording | Absolute latency of the first positive peak (msec) | Absolute latency of the most negative peak (msec) | Amplitude (μV) |
|---|---|---|---|
| C1 | 10.6 | 12.5 | 8.2 (subdural recording) |
| C2 | 10.4 | 11.7 | 19.4 (subdural); 5 (epidural) |
| C3 | 10 | 12 | 10.3 (epidural) |
| C4 | 9.8 | 11.9 | 17.6 (epidural) |
| C6 | 9.6 | 11.8 | 17 (subdural) |

*Tables **5A-3** and **5A-4** are after Halter JA, 1995.

tibial spinal evoked responses should be expected within the 25–33 msec range for SSEPs recordings at the cervical level and within the 25–20 msec range for recordings at different thoracic levels. For the median SSEPs, the expected latency for the highest negative peak falls within the range of 11.5–14. For more details regarding the recording and interpretation of spinal evoked potentials, we recommend the *Atlas of Human Spinal Cord Evoked Responses* by Halter (1995).

*B5. Troubleshooting.* If uninterpretable responses are obtained, several questions should be asked.

1. *Is the stimulation appropriate?*
   First, one has to ensure that stimulation connections and setup are appropriate and that no significant stimulus artifact impedes identification of our responses.
2. *Is the recording microelectrode appropriately placed?*
   In order to be able to appreciate the amplitude gradient of these responses, as a function of the distance from the midline, the recording electrode has to be placed perpendicular on the spine axis. A different orientation may confuse the picture. For example, no differences among the amplitudes or latencies recorded at the 8 contacts may be found if the microelectrode is placed parallel to the long axis of the spine. The recording surfaces need to make good contact with the spine surface. We use impedance checks and expect the impedances to be equal, even if slightly higher than 5 kohms (usually 5–10 kohms is okay). Irrigation with warm saline helps, in many instances, in establishing a good contact with the spine surface. Once good contact is obtained, the electrode should be secured in place—a sterile compress soaked in saline placed over it may help keep it in place.
3. *Are the recording parameters appropriate ?*
   Inappropriate time base, sensitivity, and filter settings can hinder the identification of the waveforms. A too narrow recording frequency band may significantly change the morphology of the recorded SSEPs. For example, decreasing the HFF simplifies and distorts the morphology of the posterior tibial SSEP, recorded at the thoracic level, by smoothing the waveform appearance (see Fig. **5A-10**).
4. *Is there appropriate averaging?*
   Because of the significant amount of electrical interference, especially 60 Hz, that is usually encountered with such recordings, a good averaging can be of great help. If the signal-to-noise ratio is small, our practice is to double or triple the amount of averaged trials. One should also make an effort to identify the source and eliminate the artifact. Knowledge of the expected spinal evoked response (at which latency) helps focus one's attention on a certain time epoch, while "ignoring" others, containing artifactual peaks.
5. *Is there peripheral neuropathy?*
   If the patient is known to have peripheral neuropathy, stimulating at the knee level can be tried—of course, the expected latencies for the recorded potentials shorten, usually by 7–10 msec.
6. *Is there mechanical compression of the tumor?*
   Excessive temporal dispersion of the traveling electrical volleys can be caused by mechanical compression of the tumor at or distal to the spinal level where recording is done, resulting in conduction block. This may result in unreliable responses.

## C. Anterograde recording of cortical SSEPs.

*C1. Principle, technique, and interpretation of evoked responses.* Stimulation of the spinal cord with scalp recording of the evoked cortical somatosensory responses has been reported by several authors (Berić et al., 1986; North et al., 1991). In an attempt to neurophysiologically identify the median raphae, we have also stimulated the dorsal columns directly, with a handheld bipolar stimulating electrode, and recorded the somatosensory evoked response from the scalp electrodes. We have found these recordings useful for identifying the functional axons of the dorsal columns, so that they can be avoided when making a midline myelotomy spinal cord incision.

The method relies on assessing the changes in polarity of the scalp SSEPs as the surgeon stimulates either the left or the right dorsal columns. Stimulating the medial part of the right dorsal column depolarizes the right Gracilis tract and thus generates cortical SSEPs with the same polarity as stimulating the right posterior tibial SSEPs. This translates in the Cp4-Cp3 channel in a positive downward peak, equivalent to the P37 potential but occurring earlier (because the stimulation is done at the spinal level). When the surgeon is moving the stimulator to the left and crosses the median raphae, stimulation of the now medial part of the left dorsal column depolarizes the left Gracilis tract; thus the polarity in the Cp4-Cp3 channel reverses, showing a negative, upward peak. As a control, the evoked potential recorded in the Cpz-Fz channel should always show a positive peak, regardless of the site (right or

without interfering with microsurgical dissection. The MEPs that produce patient movement are typically obtained every 5 minutes. However, the surgeon often requests more frequent MEPs during dissection of parts of the spinal cord directly adjacent to the corticospinal tracts.

### A. Continuous Monitoring of SSEPs During and After Myelotomy.

Some authors (Whittle et al., 1986) have suggested that monitoring the SSEP is of limited value to the surgeon, because this response often disappears after making a midline myelotomy. We disagree with this assertion. Occasionally, the SSEPs, particularly the posterior tibial SSEPs, are indeed injured by myelotomy. When this happens, the changes are most often asymmetric, and thus the unaffected side is still reliable for further monitoring (Fig. **5A-13**). These situations occur mostly when the neurophysiologic dorsal column mapping was limited. On other occasions, however, particularly after successful mapping, SSEPs are minimally or not at all affected by myelotomy. In such circumstances, reliable SSEPs present after the midline myelotomy are very useful in subsequently monitoring the extent of spinal cord retraction. After making the midline myelotomy and beginning the tumor dissection, the midline pial edges of the dissected spinal cord are gently retracted with 7-0 prolene sutures to facilitate further intramedullary tumor dissection. Overdistraction of the spinal cord and adjacent dorsal columns can cause a change in the SSEPs, alerting the surgeon of the need to relax the retracted edges of the spinal cord.

Thus, having preserved, reliable SSEPs postmyelotomy is important for continuing the prevention of iatrogenic injury during future spinal cord manipulation and during the actual tumor resection. On a more positive note, we have also seen improvement in the SSEPs, after decompression of a swollen spinal cord by removal of an intramedullary tumor.

### B. Continuous Monitoring of Combined mMEPs and D Waves

The purpose in monitoring MEPs is to preserve function of the corticospinal tract axons and distal motor function. Therefore the neurophysiology team must quickly communicate any changes in MEP to the surgeon, and subsequent steps must be taken to address the problem. Deletis, Epstein, and their colleagues have identified the changes in MEP that occur with injury to the corticospinal tract, and they have used these changes to offer guidance to enable the neurophysiologic/surgical team to recognize early corticospinal tract injury (Deletis et al., 2008; Kothbauer et al., 1998; Morota et al., 1997; Sala et al., 2006). Their recommendations combine the information from mMEP and D-wave monitoring and are weighted heavily toward analysis of the D-wave amplitude. We also believe that monitoring the D-wave amplitude is very important, and it should be used in conjunction with mMEPs whenever possible.

Several authors have stated that patients who experience a complete loss of MEP, accompanied by D-wave amplitude preserved to no less than 50% of baseline, sustain only a transient new motor deficit (Kothbauer et al., 1998; Morota et al., 1997; Sala et al., 2007). This is explained by the fact that mMEPs are the result of depolarization of the alpha motorneuron, not only by the corticospinal tract but also by other descending tracts, such as the propriospinal tract, which could be injured during myelotomy and resection. Their injury causes immediate and significant changes to disappearance of the mMEPs. Meanwhile, absence of significant changes in D waves indicates the relative integrity of the corticospinal tract and thus guarantees future progressive and complete recovery of the postoperative motor function. We have, however, seen one patient who met these criteria but sustained a new and permanent motor deficit (see the following). Therefore, it must be emphasized that the current recommendations will undoubtedly become more precise as these techniques evolve and as further experience is gained.

The surgeon performing microsurgical intramedullary spinal cord dissection can be reassured as long as the tcm-MEP and D-wave recordings remain unchanged from baseline. We have not encountered any patients with new postoperative motor deficits in this setting (Fig. **5A-14**; Fig. **1-9**, page 24).

However, the surgical and anesthesia team must be prepared to respond when the MEP and D-wave recordings deviate from their baseline. If significant MEPs changes, as described on pages 185–87, occur, the recommended response of the surgical and anesthesia team to these changes can be best remembered using the acronym TIP: Time, Irrigation, Papaverine/Pressure (Sala et al., 2007). The best initial surgical response is to stop immediately performing the surgical maneuver that led to the change. In large tumors, the surgeon may simply shift the microsurgical dissection to another part of the surgical field or opposite side of the spinal cord. Sometimes it is best to stop operating altogether for between 5 and 30 minutes,

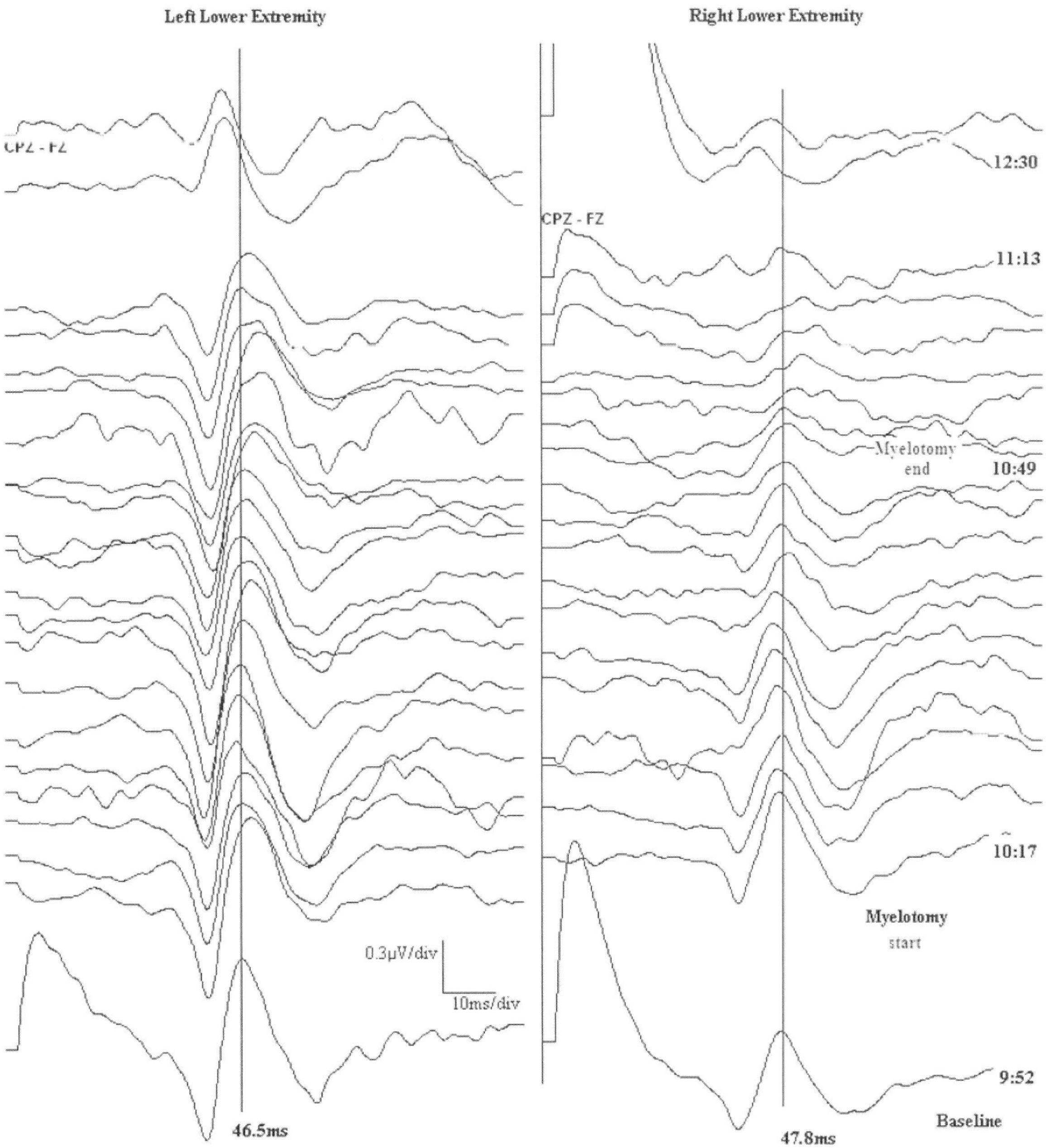


**FIGURE 5A-13** Asymmetric right posterior tibial SSEPs changes as a result of possible injury or stunning of the electrical transmission within the right dorsal column during myelotomy.

to give the delicate spinal cord tissue time to recover. During this cessation of surgery, irrigation of the surgical field with warm saline helps wash away any deleterious metabolites or potassium (Fig. **5A-15**). Including papaverine in the irrigant also causes local arteriolar dilatation and improves local spinal cord blood flow; this can be particularly useful when resection and dissection cause local vasospasm. The anesthesia team can simultaneously elevate mean arterial blood pressure, as another means of improving local spinal cord blood flow. If these

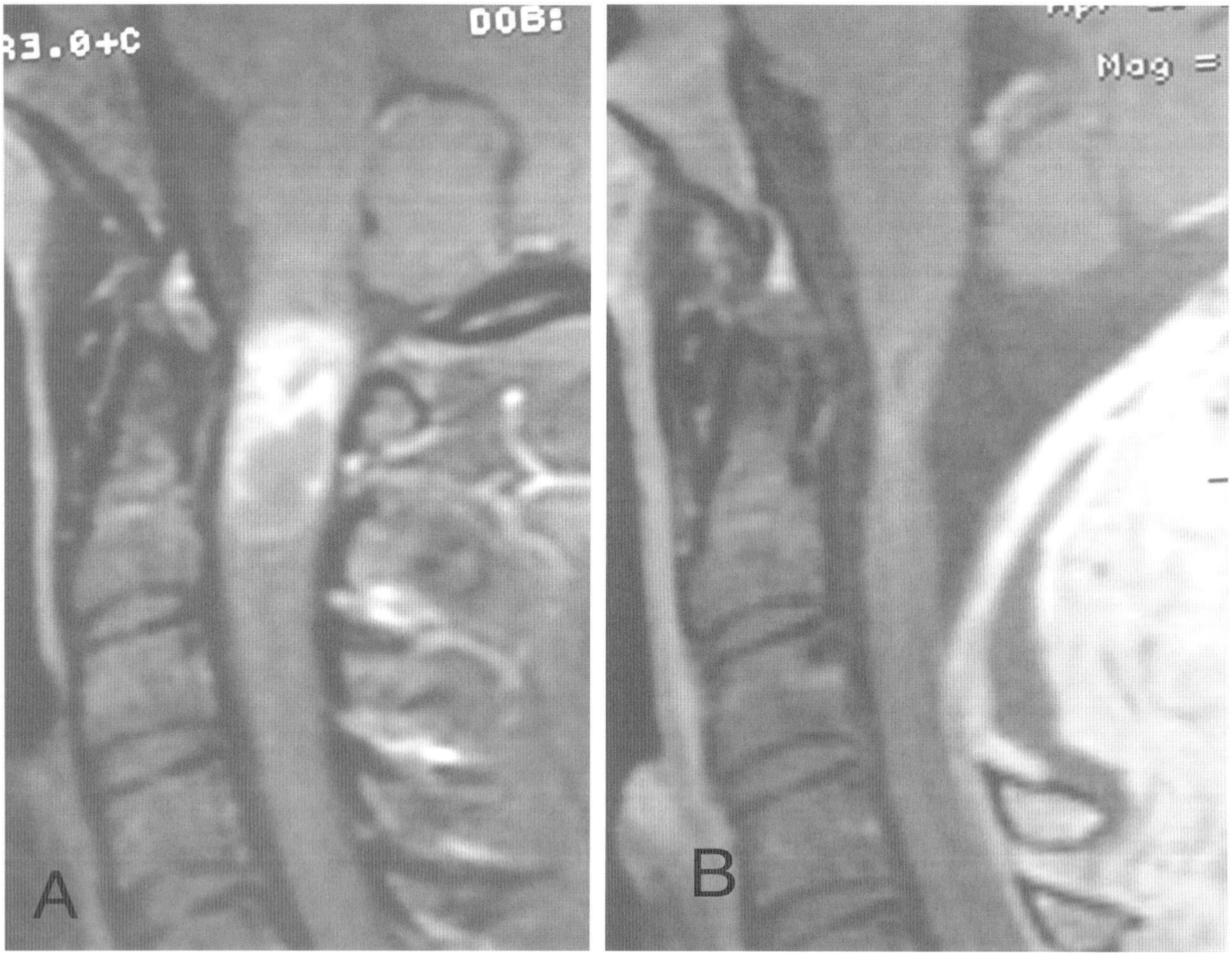


**FIGURE 5A-14** Cervical spinal cord ependymoma. No change in SSEP and tcmMEP after tumor resection. Postoperative motor function was initially unchanged from preoperative motor function and improved further several months after surgery. Complete tumor resection. **A**. Preoperative MRI scan. **B**. Postoperative MRI scan. The intraoperative recording is shown in Fig. 1-9 on page 24.

interventions result in improvement in the MEP, surgery may usually be safely continued.

In addition to vasospasm, vascular insults can result from electrocautery used for hemostasis, which could potentially cause thrombosis of vessels important for the spinal cord blood supply. Inappropriate use of cusa could also potentially damage the cord, by aspirating normal nervous tissue. Thus, special care should be given in case of removal of infiltrative and highly vascular tumors, and the cusa settings should be chosen appropriately.

If the maneuvers previously described do not lead to improvement in the MEP, the surgeon must consider whether or not to continue the tumor resection. This can be a difficult decision. Certainly, no surgeon wants to leave the patient with a new and possibly permanent neurologic deficit. However, in the case of an intramedullary ependymoma, the best method of controlling the tumor long term is complete surgical resection of the tumor at the initial operation. To help decide the correct response to this uncertainty, most surgeons turn to analysis of the D-wave amplitude (Table **5A-5**). Surgery may usually be safely continued if the D-wave amplitude has not changed despite alterations in morphology, threshold, or amplitude of the MEP. Many authors also suggest that surgery may also be continued in the face of complete loss of the MEP, as long as the D-wave amplitude remains greater than 50% of its

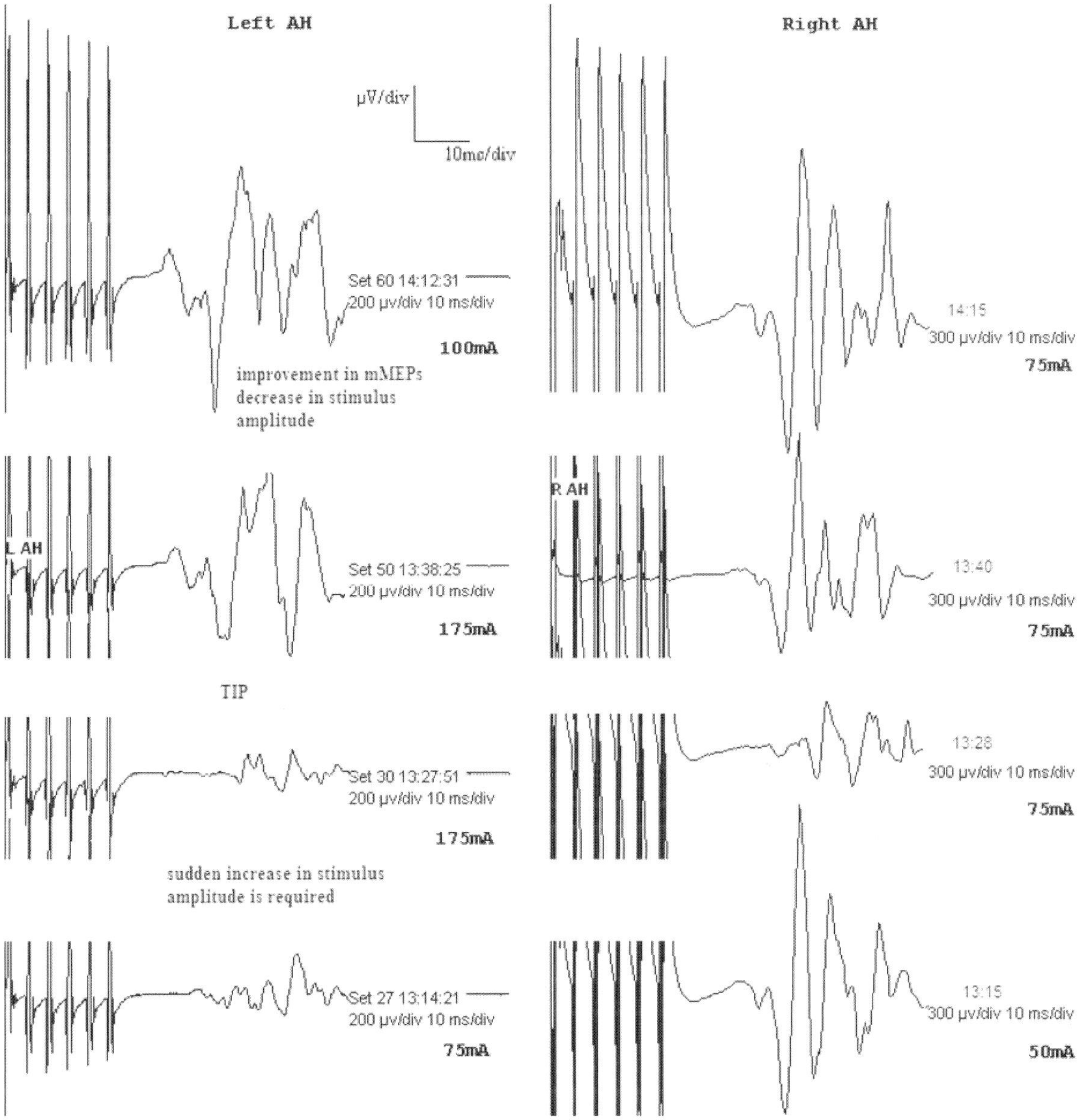


**FIGURE 5A-15** Improvement in tcmMEPs with warm saline irrigation, papaverine, increase in the blood pressure, and pausing the resection (TIP). At 13:27, the left AH tcmMEPs trial required a stimulus intensity higher by 100 mA, compared to the previous trial. A milder change was seen as well in the right AH tcmMEPs, which required only a 25-mA increase in the stimulus amplitude. These changes occurred relatively fast, during the tumor resection. The surgeon stopped the resection and irrigated the surgical field with warm saline and papaverine. The anesthesiologist increased the systemic blood pressure. Within 10 minutes after initiation of TIP, the tcmMEPs showed a definite improvement in their amplitudes. This improvement continued, and, at 50 minutes after initiation of TIP, the left AH MEPs were back to baseline. Notice the decrease in the required stimulus amplitude, from 175 to 100 mA.

normal baseline, without producing a new permanent neurologic deficit. As stated previously, we urge caution in this setting, because we have seen a case of permanent new motor deficit after losing the tcmMEP, despite a D-wave potential preserved at 50% of baseline (Figs. **5A-16** and **5A-17**). However, the immediate postoperative paraplegia improved slowly over time, yet the motor function never returned to the preoperative baseline. As in the case of other evoked potentials (SSEPs and tcmMEPs), using a strict numeric safety threshold (i.e.

**TABLE 5A-5** Using the tcmMEP and D-wave data to guide surgery and predict outcome

| tcmMEPs | D-wave Amplitude | Surgeon Response | Postoperative Motor Function |
|---|---|---|---|
| No change | No change | None | Same as preop |
| ↓ Amplitude<br>↑ Threshold<br>Change in Morphology | No change or > 50% of baseline | TIP* | Same as preop or transient deficit |
| Lost | No change or > 50% of baseline | TIP | New deficit that often improves |
| Lost | Below 50% of baseline | STOP | Permanent new deficit; possible plegia |

* TIP: Time, Irrigation, Papaverine/Elevate Blood Pressure (Sala et al., 2007).

50% from baseline) can be considered arbitrary. In our case, when the mMEPs could no longer be obtained, the amplitudes of the D waves were decreased, but they were still greater than 50% from their baseline. Based on these results, it was decided to continue tumor resection. However, when the resection was stopped, D waves were close to 50% of baseline. Even though the patient's immediate postoperative deficit improved over time, he never recovered his muscle strength completely. Thus, we recommend a more conservative approach: in case of total loss of tcmMEPs, the resection should be continued only if D-waves remain at or close to baseline. Particularly, if one notices a pattern of constant decline in D-wave amplitudes over time, even while remaining above 50% from baseline, we suggest that the resection should be stopped. Our observation is in accordance with what other authors have reported. Burke et al. (1998) advise that a drop in the averaged D-wave recordings of more than 20% from baseline should be considered a warning sign. If the D-wave amplitude declines to less than 50% of its preoperative baseline, there is a general consensus that surgery should be stopped, and a significant new postoperative motor deficit can be expected.

As mentioned on pages 193–95 and 16–26, in some instances it is not possible to obtain D waves. In the absence of recordable D waves, most surgeons stop surgery if a significant decline occurs in the tcmMEP.

## *C. Mapping the Corticospinal Tract*

A promising method for mapping the corticospinal tract during intramedullary tumor resection is based on the recently described collision technique (Deletis et al., 2001; Deletis, 2006). This technique consists of activating the corticospinal tract by simultaneous transcranial electrical stimulation and direct spinal cord stimulation. Continuous transcranial electrical stimulation causes a descending volley, which is recorded via epidural/subdural spinal electrodes as D waves. Intermittently, the surgeon also stimulates the exposed spinal cord, via a handheld stimulator. When the probe is in close proximity to the corticospinal tract, it depolarizes it and causes an antidromic volley, which is expected to collide with and annihilate the D wave. Thus, the disappearance of the D wave helps identify the location of the corticospinal tract.

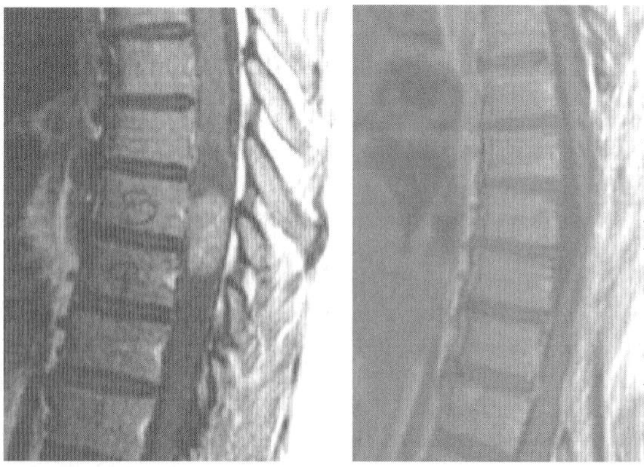

**FIGURE 5A-16** Thoracic spinal cord ependymoma complete removal. As the last piece of tumor was being removed, the MEP was lost bilaterally. The D-wave amplitude remained better than 50% of its baseline. SSEP remained unchanged. The patient awoke paraplegic; his leg strength improved slowly and only partially. Two years after surgery, he is able to walk several steps but has significant spasticity. A. Preoperative MRI scan. B. Postoperative MRI scan. Fig. 5A-17 shows the intraoperative recordings.

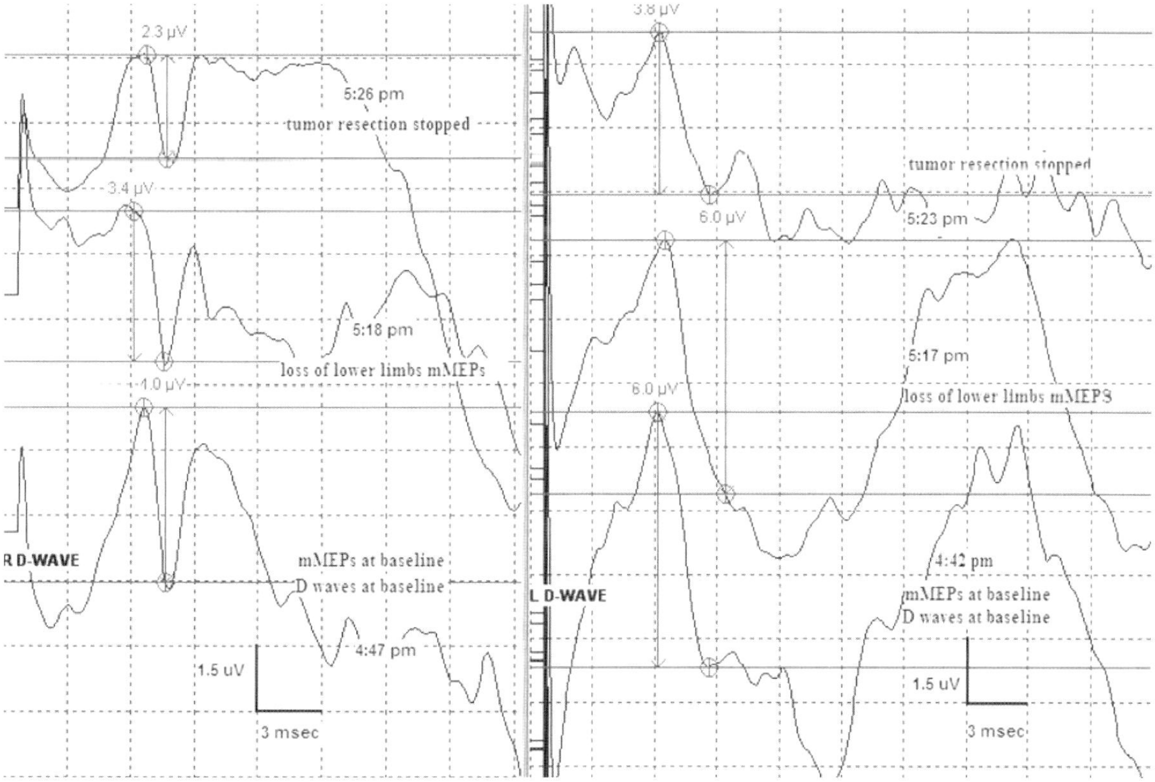


**FIGURE 5A-17** Loss of lower limbs mMEPs (not shown) and preservation of D waves at more than 50% from baseline (trials at 5:18 pm on the left and 5:17 pm on the right) encouraged the surgeon to continue the tumor resection. However, the amplitudes of the D waves continued to decrease. At the end of the tumor resection, their amplitudes were close to 50% from baseline. The patient woke up paraplegic, and, even though his motor strength improved, it never recovered to preoperative baselines.

## 10. Closing

Once the resection is completed, the epidural electrodes used for D-wave recordings are taken out, and the surgeon proceeds with dura closure and laminoplasty reconstruction. We continue SSEPs and mMEPs monitoring during this last phase, because we have encountered neurophysiologic changes during it. When they occur, they are most likely related to "closure in tension," with additional pressure exerted on the spinal cord. Luckily, this is a relatively rare encounter.

Intraoperative neurophysiologic monitoring has been an important technological advance in the surgical management of patients with intramedullary spinal cord tumors. It has allowed the surgery to be performed more safely and offers the possibility of better tumor resection, with improved postoperative neurologic function. It is hoped that more neurosurgical centers will report their experiences with monitoring so that the technique and its predictive value may be improved.

## REFERENCES

Berić A, Dimitrijević MR, Sharkey PC, Sherwood AM. Cortical potentials evoked by epidural stimulation of the cervical and thoracic spinal cord in man. Electroencephalogr Clin Neurophysiol 1986;65(2):102–110.

Burke D, Hicks RG. Surgical monitoring of motor pathways. J Clin Neurophysiol 1998;15(3):194–205.

Calancie B, Molano MR. Alarm criteria for motor-evoked potentials: what's wrong with the "presence-or-absence" approach? Spine 2008;33(4):406–414.

Deletis V, Bueno De Camargo A. Interventional neurophysiological mapping during spinal cord procedures. Stereotact Funct Neurosurg 2001;77(1–4):25–28.

Deletis V. Intraoperative neurophysiology of the corticospinal tract of the spinal cord. Suppl Clin Neurophysiol. 2006;59:107–112.

Deletis V, Sala F. Intraoperative neurophysiological monitoring of the spinal cord during spinal cord and spine surgery: A review focus on the corticospinal tracts. Clin. Neurophysiol 2008;119:248–264.

Halter JA. Spinal cord evoked potentials recorded at different vertebral levels. In: Atlas of Human Spinal Cord Evoked

**TABLE 5B-2** Pathological conditions where decompressive spine surgery may be required

1. Spine deformities (e.g. scoliosis, kyphosis).
2. Spondylosis. Degenerative disease of the spine. Acute disc herniation.
3. Trauma. Unstable vertebral fractures. Spinal column misalignment.
4. Pathological vertebral fractures (e.g. multiple myeloma, metastatic carcinoma such as breast, lung cancer).
5. Intradural extramedullary tumors (e.g. schwannomas, lipomas).
6. Chiari malformation.
7. Epidural abcess.
8. Epidural hematoma.

The specific role played by intraoperative neurophysiologic monitoring in scoliosis surgery and repair of aortic aneurysms is covered in subsequent sections of this chapter and is not further discussed here.

## II. INTRAOPERATIVE NEUROPHYSIOLOGIC MONITORING: A STEP-BY-STEP APPROACH

The two main neurophysiologic tests used for neuromonitoring during decompressive surgery are SSEPs and tcmMEPs. Additional protection of the nerve roots during spine manipulation may be achieved by adding free-run and triggered EMG recordings involving the myotomes of interest. The use of this latter test for root monitoring and mapping during decompressive surgery is further detailed on pages 278–79.

Details of the methodologies of SSEPs and tcmMEPs testing, as well as interpretation of the evoked responses, can found on pages 5–33.

Similar to the approach to neuromonitoring for intramedullary tumor resection surgery, we describe the following step-by-step approach to neuromonitoring in decompressive surgery. The emphasis is on the particularities resulting from the special pathology and surgical circumstances of the latter. Descriptions of the neuromonitoring steps common to other spine surgeries (e.g. anesthesia considerations, obtaining SSEPs and tcmMEPs baselines, troubleshooting stimulation and recording) can be found in Subchapter **5A**.

### 1. Positioning a Mechanically Unstable Spine

Traumatic injury to the vertebral column and spinal cord is a common problem that can cause permanent quadriplegia or paraplegia. In many instances the individual sustaining these injuries is rendered quadriplegic or paraplegic at the moment of impact. In this setting there is little that can be done to reverse the neurologic deficit. However, an individual who has sustained a significant vertebral column injury but is neurologically intact remains at risk for developing a spinal cord injury until the vertebral column injury has healed. Many vertebral column injuries require surgery to stabilize and fuse the bony spine, to facilitate subsequent healing without a spinal deformity. Surgery might require that the patient is turned into the lateral or prone position or that the neck be flexed or extended. A mechanically unstable spine may not tolerate these position changes without causing spinal cord compression and loss of neurologic function, despite the utmost caution on the part of the surgery and anesthesia team in positioning the patient. Intraoperative neurophysiologic monitoring can play an invaluable role in alerting the surgery and anesthesia team to a new impairment in spinal cord function, brought on by positioning, so that it can be quickly corrected. Commonly, prompt repositioning of the patient allows the spinal cord function to return to normal and allows the surgery and anesthesia team to devise a different management strategy.

Ofiram et al. (2006) have called attention to the special problems posed by individuals with skeletal dysplasia during positioning for spinal surgery. Many of these developmental spine pathologies affect the craniovertebral junction and can be associated with instability at this location. They report on three patients who had a loss of evoked potentials following prone positioning for spinal surgery. In one individual, repositioning of the head in flexion allowed the potentials to return to normal. In the two other individuals, the surgery was stopped, and alternative management options were pursued. In all three instances, it is likely that the intraoperative neurophysiologic monitoring prevented severe postoperative neurologic deficits. They advocate that post-positioning SSEPs and MEPs be used in all patients with skeletal dysplasia.

Our practice is to obtain baseline pre-positioning tcmMEPs and SSEPs recordings, before the patient is positioned for the surgery. As soon as the positioning is completed, post-positioning tcmMEPs and SSEPs recordings are obtained. Any variance in amplitude, wave morphology, latency, or stimulation threshold between the pre-positioning and post-positioning studies must be investigated and explained. Often the explanation for the change is straightforward, such as a disconnection in a recording or stimulating electrode. Sometimes the anesthesiologist has administered a bolus of an anesthetic agent in preparation for the positioning. Pages 179–86 give a detailed description of anesthesia considerations and troubleshooting techniques during the pre- and post-position SSEPs and tcmMEPs baseline recordings.

However, if no clear explanation is found, the patient is returned to the original supine position, and new baseline SSEPs and tcmMEPs studies are obtained. Additional investigations in this setting, if these new studies remain abnormal, may include radiographic study of the vertebral column or even waking the patient to assess neurologic function. If the SSEPs and tcmMEPs recordings fail to improve, it may be necessary to abandon the surgery.

Fig. **5B-1** gives an example of one situation where intraoperative neurophysiologic monitoring detected post-position tcmMEPs changes and prevented a neurologic catastrophe.

## 2. Changes Related to Malposition of the Limbs

At times the post-positioning baseline has alerted us to a completely unpredicted problem. This patient was positioned into the prone position on the Jackson Table in anticipation of decompressing the extensive spinal stenosis that was present from T9 to the sacrum. The position is known as "superman" and is employed in posterior approaches for lumbosacral surgeries. The head was in a neutral position, and the shoulders were placed in approximately 45 degrees of flexion and 90% abduction. The arms were positioned on arm boards in this abducted position, so that the anesthesia team had access to the hands. The arms appeared to be positioned satisfactorily. However, after 15 minutes in the prone position, attenuation of SSEPs was observed, with an increased latency in the left arm. Although the tcmMEPs remained intact in the left arm, the left arm was repositioned straight along the patient's body, with the left hand pointing to the patient's feet. Following this repositioning of the left arm, the SSEPs responses from the left-hand stimulation promptly returned to normal. The SSEPs and tcmMEPs remained normal throughout the remainder of the operation, and the patient awoke from surgery with normal neurologic function (Fig. **5B-2**). It seems likely that an unusual muscle insertion may have been present at the thoracic inlet, leading to compression of the brachial plexus with the left arm abducted. In the absence of the intraoperative monitoring, this patient would likely have woken up with at least numbness and paresthesias of the left hand. This problem was prevented by the use of intraoperative neurophysiologic monitoring and by adhering to the protocol of obtaining pre-positioning and post-positioning baseline recordings.

In a prospective study that included 230 patients undergoing lumbosacral surgery, Chung et al. (2009) showed that ulnar SSEPs monitoring is a reliable method to detect possible compressive neuropathy resulting from

(a)

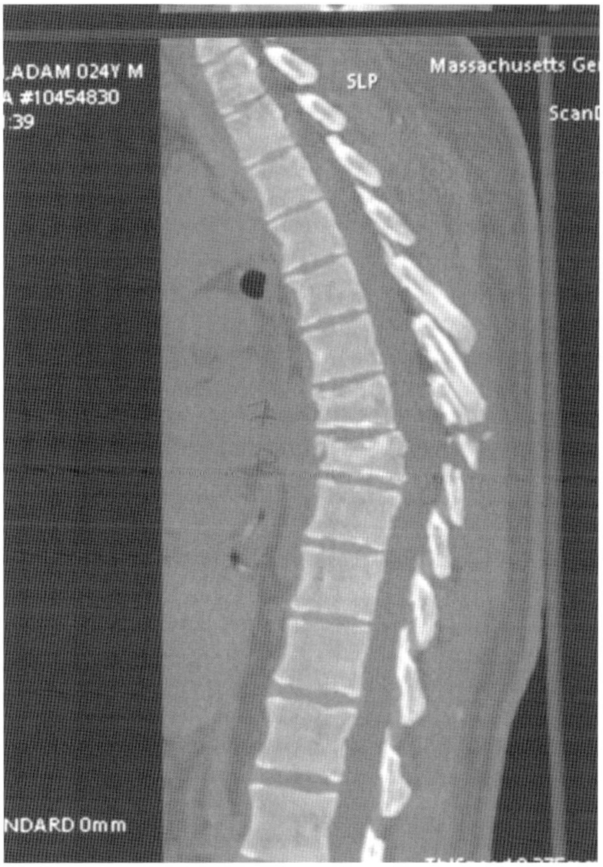


**FIGURE 5B-1** Loss of muscle MEPs with positioning of an unstable spine. This individual sustained a flexion-distraction fracture at T7 and T8 following a motorcycle crash. Early kyphotic deformity can be seen in the sagittal CT scan that was performed following this injury. **a:** Despite this significant vertebral column injury, the patient had normal neurological function. It was initially believed that the best management strategy would be a posterior instrumented stabilization using multiple pedicle screws above and below the fracture. However, when the patient was turned into the prone position for this operation, the lower limbs muscle motor evoked potentials (mMEPs) significantly diminished (right more than left) within several minutes on placing the patient into the prone position on a Jackson Table.

(b)

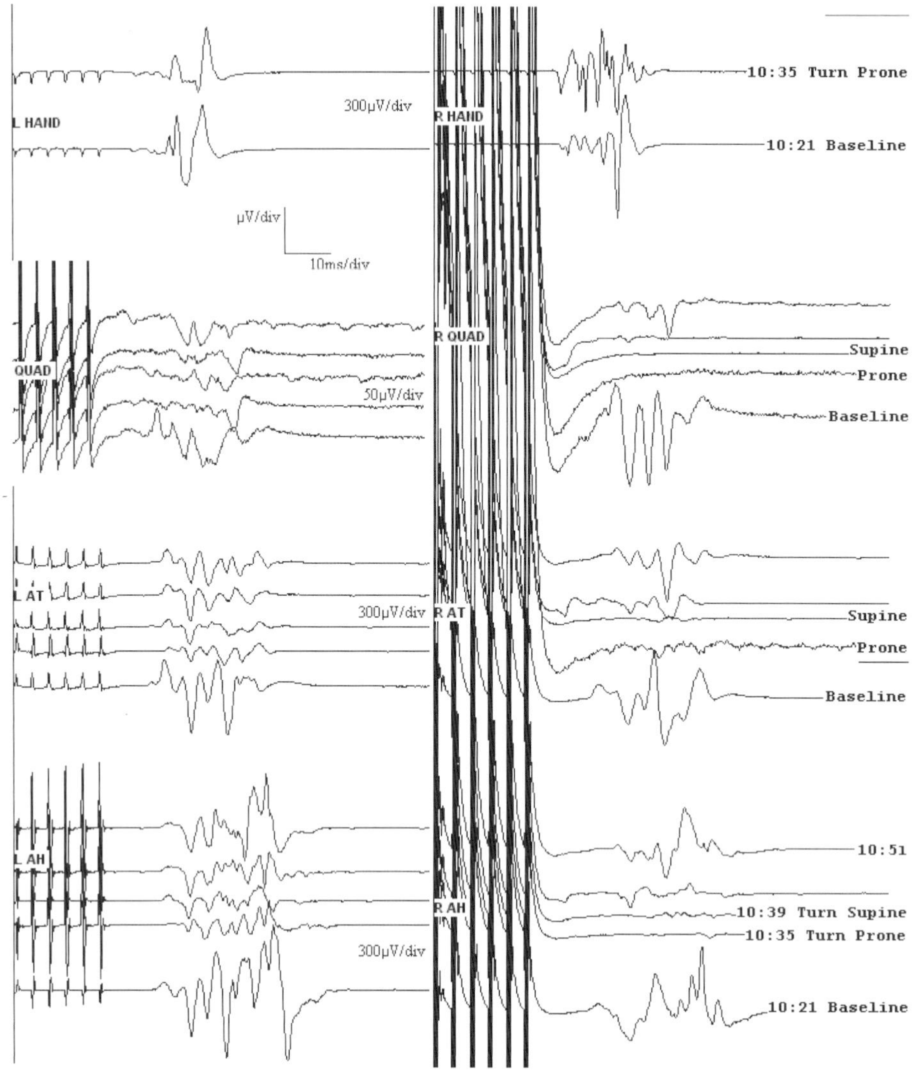


**FIGURE 5B-1** *(Continued)* **b**: The patient was immediately repositioned into the supine position and fortunately the motor evoked potentials returned. Surgery was deferred at that time. The patient was returned to the operating room several days later and underwent an anterior decompression at T7-T8 through a right thoracotomy. This was followed several days later by a posterior stabilization using the pedicle screw construct that had been planned initially.

arm malpositioning. We also agree with other authors (Bartels et al., 2007; Baumann et al., 2000) that certain conditions, such as obesity and diabetes, may be predisposing factors to development of such compressive neuropathies.

In addition to the potential arm malposition in the "superman" position employed in posterior approaches of lumbosacral surgeries, another situation is that of downward traction on the shoulders by taping, in anterior cervical spine surgery, which can also result in a significant brachial plexopathy (see Fig. **5A**-7, page 194). An impressive review in 3,806 patients who underwent anterior cervical surgery (Schwartz et al., 2006) showed the benefit of ulnar SSEPs and hand tcm-MEPs monitoring in detection and thus prevention of such problems. Along the same lines, Labrom and colleagues (2005) studied retrospectively about 434 cases of pediatric patients who underwent scoliosis surgery, and they concluded that monitoring the ulnar nerve SSEPs is a useful method in prevention of brachial plexopathy

(c)

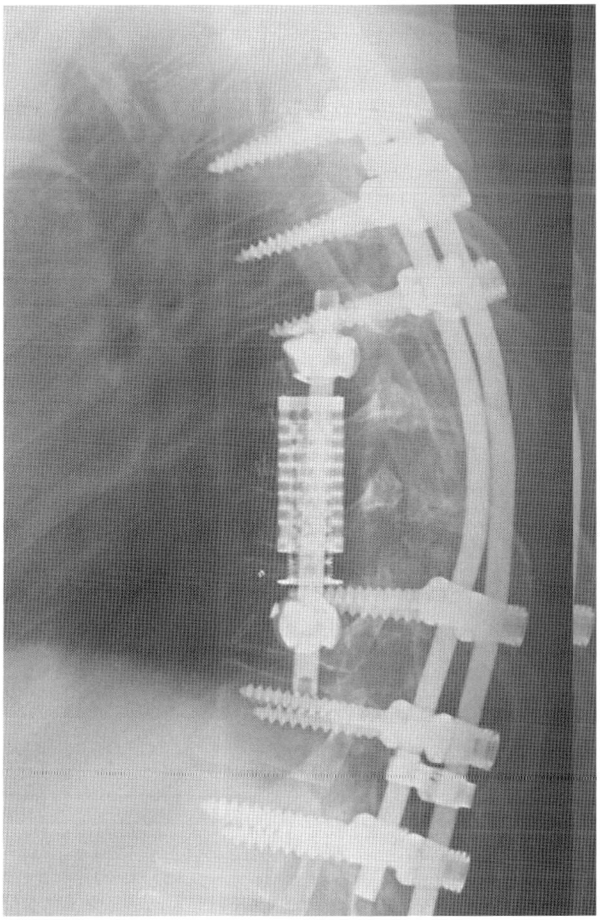

**FIGURE 5B-1** *(Continued)* **c**: The mMEPs and SSEPs remained within normal limits during the second procedure. The patient's injuries healed well allowing a return to a full and normal life.

resulting from arm positioning. Also, in this series, the posterior approach with prone positioning had significantly higher incidence of such complication than the anterior approach.

## 3. Changes Related to Spine Exposure

During this time, there may be a progressive increase in the SSEPs absolute latencies, as well as a progressive decrease in their amplitudes. A progressive increase in the stimulus amplitude threshold for tcmMEPs may occur as well. All of these changes are generally related to systemic factors, particularly effects of anesthetics, and are discussed in detail on pages 179–87. Continuous recording of SSEPs and tcmMEPs is difficult, if not impossible, during this time, because of the almost constant use of bovie. However, it is important to ask the surgeon to pause the opening occasionally so that SSEPs and tcmMEPs are recorded intermittently. The goal is to avoid long gaps between two consecutive trials and to obtain new baselines before a possible injurious step.

There are specific situations when the maneuvers used for gaining access to the spine could compromise nervous tissue. The resulting neurophysiologic dysfunction can be thus detected by intraoperative monitoring. The temporal relationship to the actual injurious maneuver, as well as the type of SSEPs/MEPs changes, can also help in determining the mechanism of injury.

### *A. Manipulation of Abdominal Vasculature*

A rare problem that has been identified by the use of intraoperative neurophysiologic monitoring has been the problem of thrombotic occlusion of the iliac artery during anterior lumbar spine fusion. Surgical exposure of the lower lumbar levels requires careful retraction of the abdominal vessels. During this step, progressive vascular occlusion of these vessels could result from vessel manipulation. Isley et al. (2007) reported a case of a 46-year-old man who underwent uncomplicated anterior lumbar fusion at L4-L5 and L5-S1. During the abdominal closure, after a relative delay from the injurious event, the neurophysiology team observed a complete loss of SSEPs in the left leg. In such cases, loss of all of the evoked potentials, including the peripheral popliteal fossa response, indicates limb ischemia. Investigation of this new and unexpected finding by the surgical team led them to uncover loss of palpable pulses in the left lower extremity. This individual was found to have a thrombotic occlusion of the left iliac artery, likely related to surgical retraction, and was treated emergently with thrombectomies of the internal and external iliac arteries, with patch angioplasty of the left common and proximal external iliac arteries. The prompt identification of this unexpected complication by the neurophysiologic monitoring team resulted in an immediate intervention that allowed this patient to have an excellent vascular and neurologic outcome.

### *B. Test Occlusion of Segmental Arteries*

Gaining access to different spinal levels and facilitating spine instrumentation, during anterior deformity and/or decompressive surgery, sometimes requires ligation

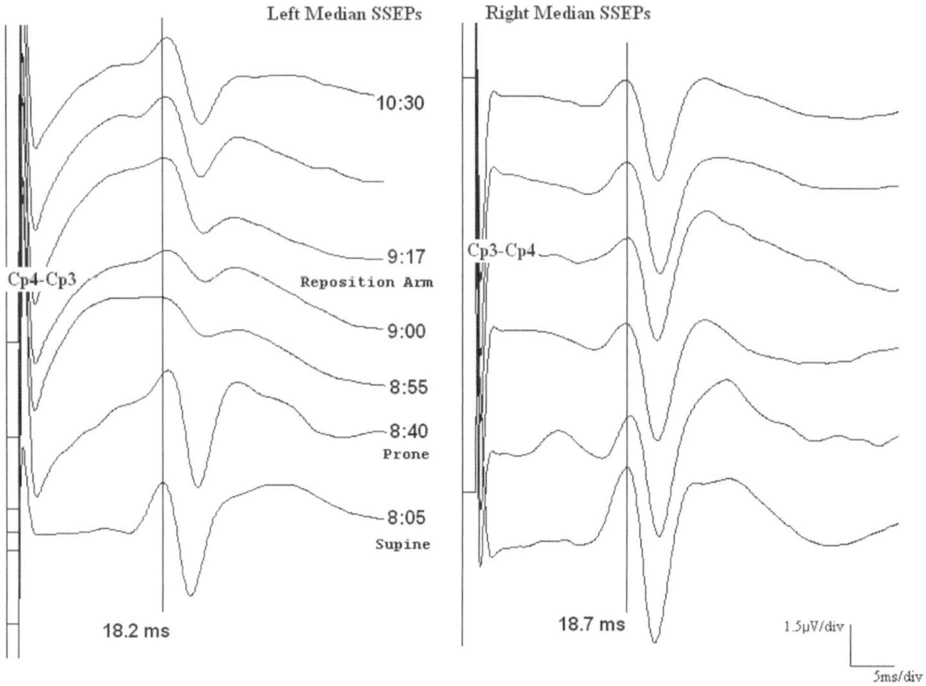


**FIGURE 5B-2** Intraoperative SSEPs recordings showing attenuation of left median SSEPs after positioning into the prone position with the left arm abducted on an arm board. This SSEP attenuation corrected with repositioning the left arm into a straight position along the patient's body with the left hand pointed at the feet.

of segmental arteries, which could have potentially catastrophic consequences on spinal cord vascular supply. Thus, the vessels considered for ligation are first sequentially test clamped. The role of neurophysiologic testing is to detect spinal cord ischemia. Thus, during test clamping, the emphasis should be on tcmMEPs, rather than SSEPs monitoring. Moreover, unlike the changes seen with vascular thrombosis, spinal ischemia causes acute changes in mMEPs. Consequently, tcmMEPs recording is performed continuously, for about 7–10 minutes, with each clamp. If mMEPs changes are detected, the clamped vessel is not sacrificed. More details and examples on positive test clamping are given in Fig. **5C-4** (page 228).

## 4. Changes Related to Spine Manipulation

Although we have found intraoperative neurophysiologic monitoring to be most important when positioning patients for decompressive spinal surgery, we also advocate its continuation throughout the surgical procedure. It is reassuring for the surgical team to know that the spinal cord function is intact during the surgical exposure needed to access and decompress a spinal cord that looks frighteningly compressed on the preoperative MRI scan. Table **5B-1** gives a list of surgical steps during spine manipulation when the spinal cord could be significantly compromised. Use of intraoperative neurophysiologic testing during these steps ensures early identification of spinal cord dysfunction and intervention by reversing the injurious surgical maneuver before irreversible damage occurs.

As mentioned previously, a special situation occurs in the anterior approach for cervical spine surgery in patients with myelopathy, when there may be changes in head and neck position during the repositioning of retractors to gain exposure to the vertebral column. The persistence of normal SSEPs and tcmMEPs during these maneuvers tells the surgical team that the spinal cord is tolerating these changes.

### *A. Pedicle Screws Placement*

Monitoring can also provide significant reassurance during placement of pedicle screws. Although most surgeons use some radiographic means of determining the correct trajectory for pedicle screw placement, it is always comforting to know that the SSEPs and tcmMEPs are normal. Additionally, free-run and triggered EMG can be used for ensuring an intact pedicle screw. More details regarding the use of intraoperative neurophysiology during pedicle screw placement are found on pages 277–78.

## B. Accidental Slips of Instruments

Simple use of surgical instruments, some applied under pressure and tension (e.g. during pedicle screw and rod placement), could potentially cause accidental blunt trauma of the spinal cord, especially after laminectomy, when the cord is exposed. This could result in spinal cord edema, hematoma, or vascular spasm, depending upon the severity of the injury. Even though it is a rare occurrence, we have encountered a few such situations (see Fig. **5B-3**).

## C. Other Specific Manipulation Techniques

More information regarding causal relationship between specific spine manipulation techniques (e.g. passing sublaminar wires, rods tightening), mechanisms of spinal cord injury, and SSEPs/tcmMEPs changes can be found on pages 228–34.

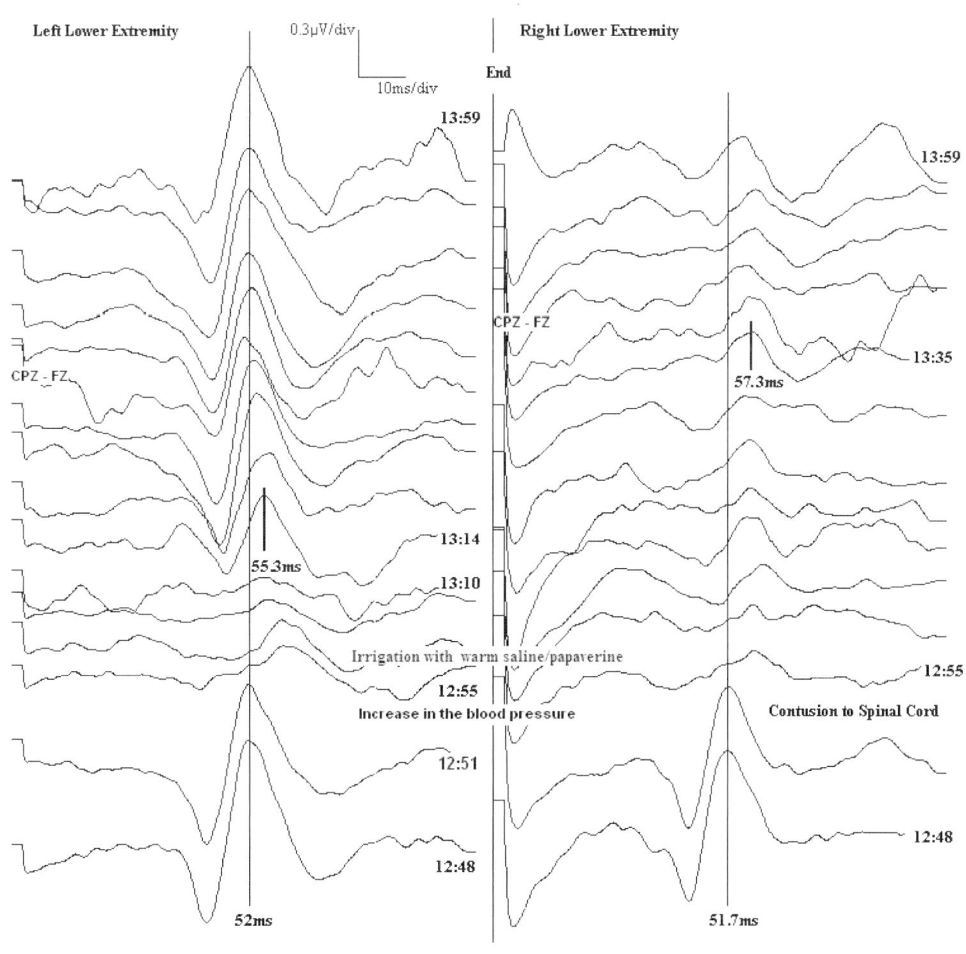


**FIGURE 5B-3** Spinal cord contusion caused by accidental slipping of spinal cord instrumentation during decompressive surgery. An almost sudden attenuation in the posterior tibial SSEPs was seen, likely induced by "stunning" and possibly vascular spasm. Thus, the systemic blood pressure was increased, and irrigation with warm saline and papaverine was performed along with an intravenous bolus of solumedrol. About 25 minutes after the iatrogenic injury, a definite improvement of the left posterior tibial SSEPs, followed by some improvement in the right posterior tibial SSEPs, was seen about 45 minutes after the injury.

(a)

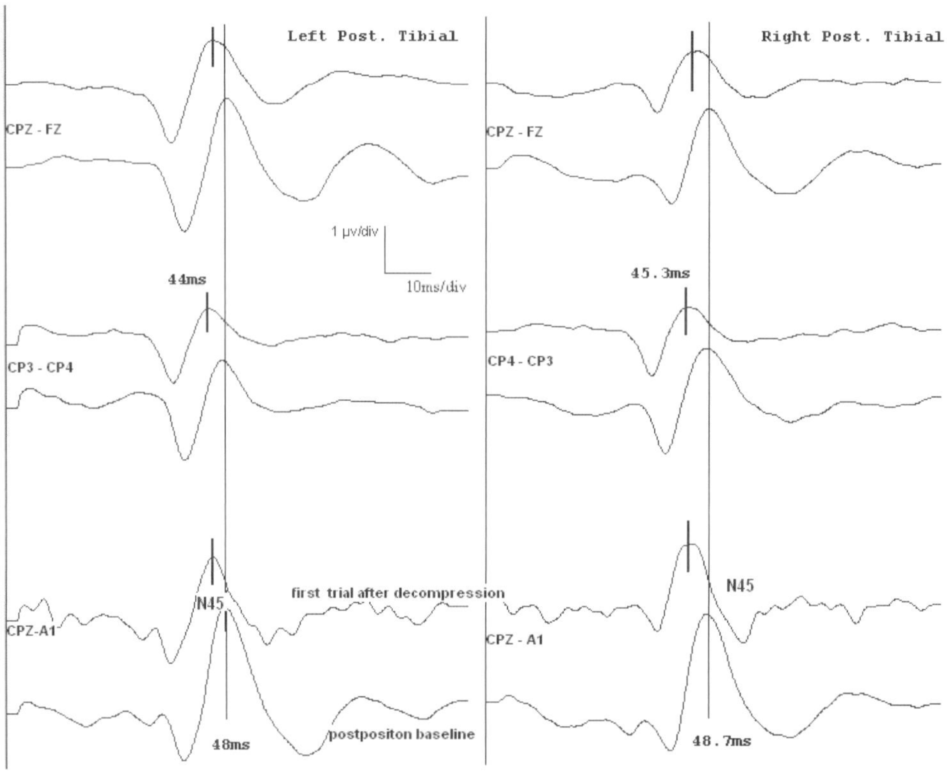


**FIGURE 5B-4** Improvement of SSEPs with decompression is shown. **a:** Decompressive surgery, with resection of a low thoracic, intradural, extramedullary tumor. With decompression, there is an improvement in the absolute latencies of the posterior tibial SSEPs (P37/N45), about 3.4 msec on the right and 4 msec on the left. Notice that the amplitudes continue to be smaller than at baseline, likely as a result of an anesthetic effect.

### *D. Improvement of the Evoked Responses*

Occasionally, a significant improvement in the SSEPs or tcmMEPs is seen after decompression. These improvements are mainly seen in the absolute latencies. They should be differentiated from the decrease in latencies associated with a decrease in the level of anesthesia, which is to be expected at the end of the surgery. When such improvement in latencies is related to decompression, it should immediately follow it. On the other hand, an anesthetic-related change is usually progressive. We also had more dramatic improvements of SSEPs, from no recordable baseline posterior tibial SSEPs to reproducible waveforms after decompression (see Figs. **5B-4 a,b**). Similarly, we had cases when the tcmMEPs amplitudes and latencies significantly improved with decompression (Fig. **5B-5**). Neurophysiologic improvement tends to be seen after surgery for acute or subacute cord compression.

## 5. Changes Related to Surgical Closure

Intraoperative neurophysiologic monitoring has also been valuable during closure of the surgical wound. It has alerted us to a laminoplasty reconstruction that caused compression of the thoracic spinal cord and also to an incipient postoperative hematoma. The prompt identification of these problems led to their being corrected during the initial surgery without any neurologic consequence.

## 6. Steps to be Taken in Case of Significant Changes

### *A. Determining Whether the Change is Real*

Before calling any significant neurophysiologic changes directly related to the surgical procedure, a thorough and efficient troubleshooting of stimulation and recording should be done. Anesthesia and the effects of other

(b)

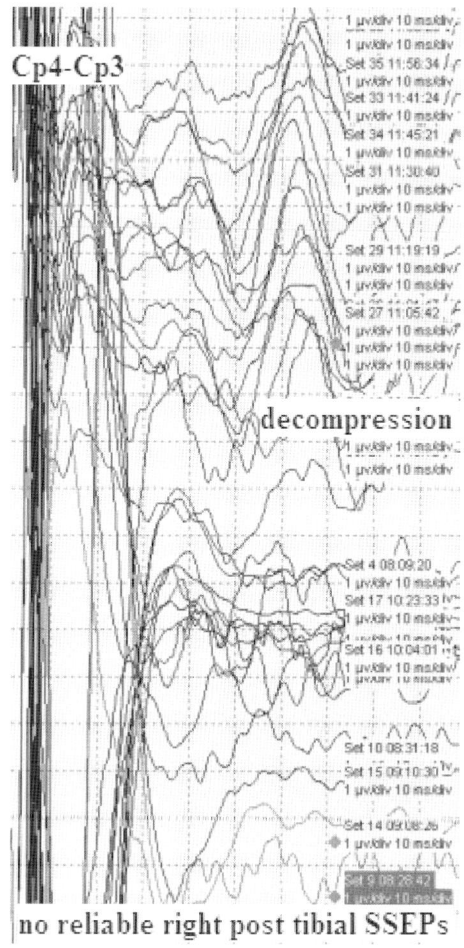


**FIGURE 5B-4** *(Continued)* **b**: Acute spinal cord trauma with T11 fracture. The patient has significant sensory deficit in the lower extremities, particularly in the right foot. No reliable right posterior tibial SSEPs are obtained at baseline. After decompression, delayed but reproducible right posterior tibial SSEPs are seen.

systemic factors should be ruled out. Details about these processes are found on pages 179–207.

A situation frequently encountered in decompressive surgery is represented by unexpected administration of muscle relaxants, sometimes in boluses. This is usually done during critical moments during the surgery, many times during instrumentation, in an attempt to control unacceptable movement in the surgical field. Unfortunately, this may hinder a reliable tcmMEPs recording. As mentioned in the previous Subchapter, Yamamoto et al. (2008) has recently described the post-tetanic facilitation as a method to enhance the amplitudes of the motor evoked responses in such conditions and thus to continue MEPs monitoring while eliminating the unwanted movement (see page 183 for details).

### *B. Establishing the Cause, the Mechanism, and the Pathology of Injury*

Once a significant change is detected, several factors are important in establishing the cause, mechanism, and pathology of injury. These factors are (1) type of change (SSEPs vs. tcmMEPs); (2) extent (cortical, subcortical, peripheral SSEPs; upper versus lower limb, unilateral versus bilateral); (3) timing (e.g. relationship with certain maneuvers); (4) pace (abrupt versus progressive).

*B1. Type of changes.* SSEPs versus tcmMEPs changes point toward the location of the injury: anterior-lateral or posterior cord. SSEPs changes tend to be more severe in peripheral limb ischemia than tcmMEPs changes, whereas tcmMEPs changes are more significant than SSEPs changes in spinal cord ischemia. Both SSEPs and tcmMEPs changes can result from a significant spinal cord injury below a certain level, in significant brachial plexopathy affecting the SSEPs and tcmMEPs involving the respective peripheral nerves (e.g. ulnar SSEPs and hand tcmMEPs changes).

*B2. Extent of changes.* Loss of cortical and subcortical upper limbs SSEPs, with preservation of brachial plexus potential, can signal a cervical spinal cord injury. Similarly, preservation of upper limbs SSEPs, with loss of lower limbs SSEPs in a thoracic spine surgery, can signal a thoracic spinal cord injury. Unilateral or bilateral loss of all waveforms in upper limbs SSEPs, including that of N9, can point toward arm malpositioning, with development of brachial plexopathy. Loss of all waveforms in posterior tibial SSEPs, including that of popliteal fossa potential, may signal a lower limb ischemia.

*B3. Timing.* Observation of a cause and effect relationship between a surgical maneuver and the neurophysiologic change is essential. In order to achieve this, one has to be familiar with the more injurious circumstances within the surgical procedure and thus be able to monitor these situations closely.

This familiarity is also important in determining the resulting pathology. Along these lines, the surgical steps specified in Table **5B-1** may have different

(a)

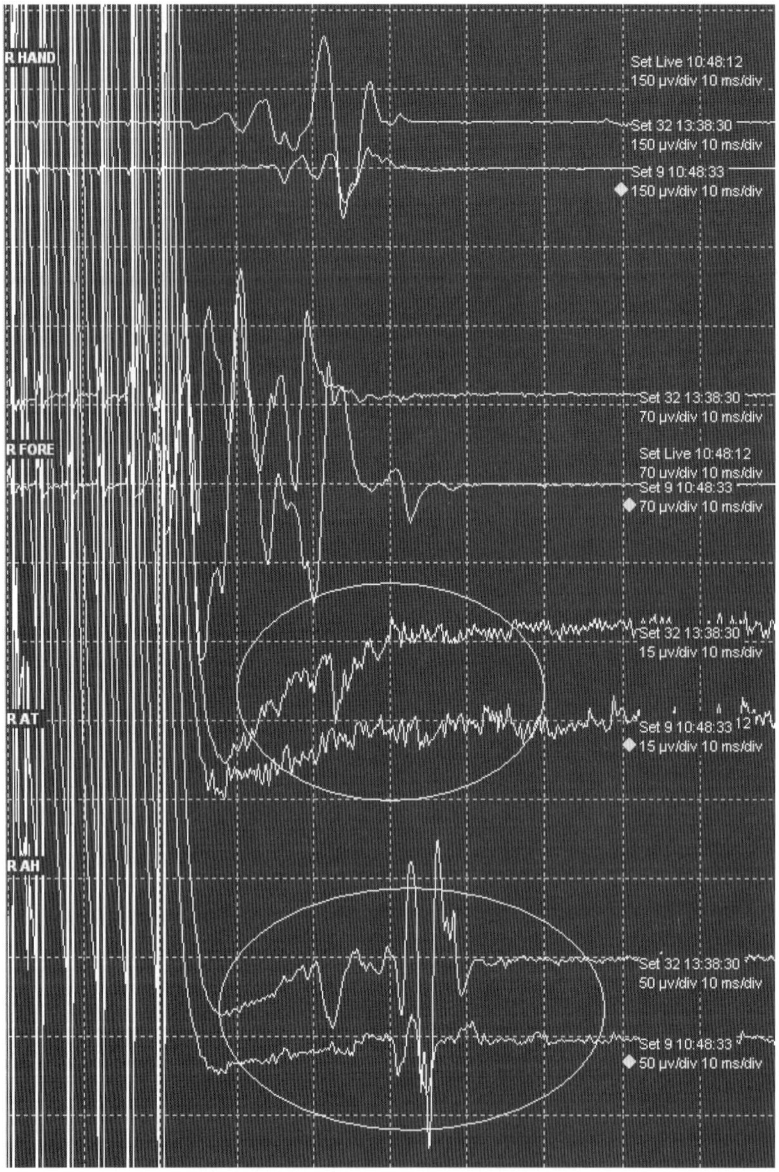


**FIGURE 5B-5** Improvement of tcmMEPs with decompression. **a:** C5-C6 disc herniation with acute cord compression. The patient is symptomatic, with right hand and lower limb weakness. Anterior cervical discectomy and fusion (ACDF) is performed. Following decompression, there is a definite improvement of the right hand (APB-ADM), AT, and AH tcmMEPs. This correlated with postoperative clinical improvement.

consequences, depending upon the mechanism of injury. For example, blunt, sudden trauma or sustained compression may cause spinal cord contusion, edema, or bleeding with hematoma formation. Furthermore, ischemia may develop as a result of direct compression of the vessels, vascular spasm, vascular thrombosis, vessel interruption, or as a consequence of progressive development of edema or hematoma with indirect vessel compression. Neurophysiologic changes seen with distraction and derotation of the spinal column are usually a result of vascular compromise. Manipulation of vessels may result in progressive (because of thrombosis) or sudden (vascular spasm or rupture) vascular compromise. Clamping or ligation of an important segmental artery causes sudden vascular compromise as a result of spinal cord ischemia.

(b)

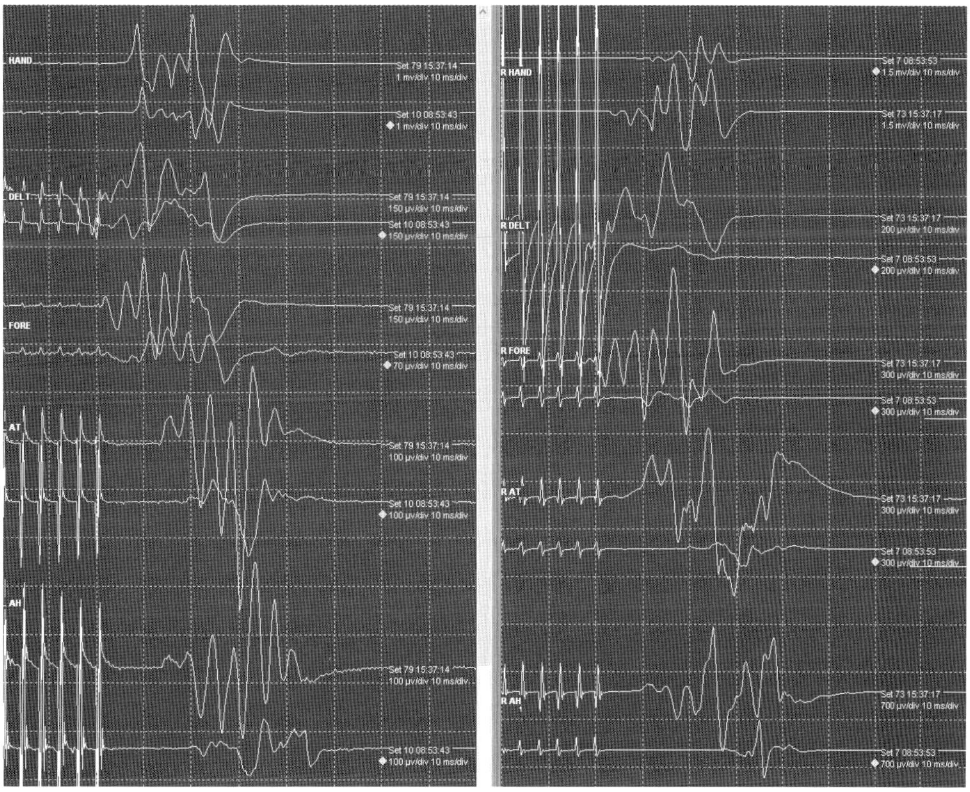


**FIGURE 5B-5** *(Continued)* **b:** Decompressive cervical spine surgery, with diffuse improvement of the tcmMEPs. No muscle relaxant was used.

As a consequence, obtaining new SSEPs and tcmMEPs baselines just before each of such steps is essential, as is constant maintenance of systemic factors, such as anesthetic levels, blood pressure, temperature, etc. Thus, continuous communication between the neurophysiology, surgery, and anesthesia teams is very important.

*B4. Pace.* The rapidity with which a change develops is significant in determining the mechanism and pathology of injury. For example, a slow progressive (usually more than 40 minutes) loss of all peripheral and central lower limb SSEPs, followed by less severe yet progressive changes in the tcmMEPs of all of the muscles of that lower limb, usually points toward lower limb ischemia. A sudden loss of lower limbs tcmMEPs, either unilateral or bilateral, which develops within several minutes, followed by usually less pronounced SSEPs changes, should point toward spinal cord ischemia.

Thus, in order to be able to determine the pace at which a change occurs, one has to ensure continuous monitoring of both SSEPs and tcmMEPs, without long gaps between subsequent trials. Periods during surgery when the surgeon is constantly using electrocautery or instrumenting may render it difficult, if not impossible, to stimulate constantly. In such situations, if the gap is getting longer than several minutes, our practice is to ask the surgeon to pause the cautery or the manipulation so as to enable averaging a new SSEPs set or obtaining new tcmMEPs trials.

It is also important to use a display of the recording that allows a better appreciation of the evolution of these evoked responses over time: vertical stack for SSEPs monitoring and the history log for tcmMEPs monitoring ( see Fig. **1-4bA**, page 13, and Fig. **5D-9**, page 250).

## C. Management

*C1. Reverse the injurious maneuver.* Eliminating the cause of the neurophysiologic dysfunction is the first and the most important step to take, regardless of the pathology of injury. Taking the instrumentation out, releasing the clamp, and decreasing the correction may automatically reverse the changes. Depending upon the severity of the insult and its mechanism, the reversal of neurophysiologic changes may take minutes to hours.

*C2. Treat the resulting pathology*

***a. Spinal cord ischemia.*** In these situations, increasing the systemic blood pressure or, in cases of significant blood loss, starting blood transfusion can be helpful. If vascular spasm is the cause, irrigation with warm saline or administration of papaverine could be effective. Spinal cord ischemia may also result from mass effect with direct compression on the vessels. Thus, for an immediate release, evacuation of the compressive mass (e.g. hematoma) or administration of steroids (in case of spinal cord edema) should be considered. However, there is significant controversy regarding the use of the latter in acute ischemic injury (Adams et al., 2007).

***b. Spinal cord edema.*** Acute trauma as a result of spine instrumentation may result in significant edema, with mass effect causing neurophysiologic dysfunction. In such cases, immediate administration of an intravenous bolus of methylprednisolone, followed by a constant infusion for the following 24–48 hours, may help in ameliorating the mass effect and improving the neurologic outcome (Bracken, 2002). However, recent studies showed that the benefit of steroids in acute trauma of the spine remains controversial (Hurlbert, 2006). This may be due to the fact that, immediately after the injury, a complex cascade of events (e.g. ischemia, hypoxia, cellular ionic shifts, free-radical production, programmed apoptosis) develops. Thus, in addition to the immediate "mass effect" factor, these events have a delayed impact, which is seen days to weeks after the injury and which may not get a definite benefit from the steroid treatment. Moreover, it may be that the increase in its adverse events (e.g. wound infection, pulmonary embolism or infections) overcomes its neuroprotective benefit (Hadley, 2002).

Our practice is to administer intravenous steroids, within the first 24 hours after the injury, if we expect development of acute edema with immediate and significant mass effect. Depending upon the severity of the injury, we consider a high-dose intravenous bolus (e.g. 10 mg/kg) and may continue with an infusion of 1–3 mg/hour for the first 24 hours. It is our experience that physicians are generally reluctant to give an appropriate high dose of steroid bolus. On the other hand, we refrain from administering steroids if there is an increased risk for infection (e.g. immunosuppression, lung disease).

Other neuroprotective agents have been investigated. Tirilazad mesylate is thought to act as a membrane stabilizer. However, when 48 hours of tirilazad administration was compared with 24 hours as well as 48 hours of steroids administration, there was no difference in the neurologic outcome at 1 year among the three groups (Bracken et al., 1997, 1998). Another agent is GM-1 ganglioside, which is thought to prevent apoptosis and increase the neural sprouting. Even though patients receiving this drug showed better functional recovery than the patients in the placebo group (Geisler et al., 1991), no significant benefit from the drug was seen in a long-term follow-up (Geisler et al., 2001). These and other neuroprotective agents need to be further investigated, and they are not routinely used in clinical practice.

Intraoperative neurophysiologic monitoring is technically demanding. It requires close coordination between the neurophysiology, surgery, and anesthesia teams. At times it can give false-positive alerts, suggesting that the spinal cord is being injured when in fact the patient awakens with normal neurologic function (Kim et al., 2007). It has been our observation that having a dedicated neurophysiologic monitoring team that is integrated seamlessly into all complex and significant spinal surgeries greatly reduces the possibility of false-positive alerts. We believe that broader adoption of intraoperative neurophysiologic monitoring by all teams performing complex and significant spinal surgery will improve its reliability and predictive value and ultimately improve the outcome of complex spinal surgery.

## REFERENCES

Adams HP Jr, Del Zoppo G, Alberts MJ, et al. Guidelines for the early management of adults with ischemic stroke: A guideline from the American Heart Association/American Stroke Association Stroke Interventional Council, and the Atherosclerotic Peripheral Vascular Disease and Quality of Care Outcomes in Research Interdisciplinary Working Groups: The American Academy of Neurology affirms the value of this guideline as an educational tool for neurologists. Stroke 2007;38:1655–1711.

Bartels RH, Verbeek AL. Risk factors for ulnar nerve compression at the elbow: A case control study. Acta Neurochir (Wien) 2007;149(7):669–674; discussion 674.

Baumann SB, Welch WC, Bloom MJ. Intraoperative SSEP detection of ulnar nerve compression or ischemia in an obese patient: A unique complication associated with a specialized spinal retraction system. Arch Phys Med Rehabil 2000;81(1):130–132.

Bracken MB, Shepard MJ, Holford TR, et al. Administration of methylprednisolone for 24 or 48 hours or tirilazad mesylate for 48 hours in the treatment of acute spinal cord injury. Results of the Third National Acute Spinal Cord Injury Randomized

Controlled Trial. National Acute Spinal Cord Injury Study. JAMA 1997 May 28;277(20):1597–1604.

Bracken MB, Shepard MJ, Holford TR, et al. Methylprednisolone or tirilazad mesylate administration after acute spinal cord injury: 1-year follow up. Results of the third National Acute Spinal Cord Injury randomized controlled trial. J Neurosurg 1998;89(5):699–706.

Bracken MB. Steroids for acute spinal cord injury. Cochrane Database Syst Rev 2002;(3):CD001046. Review.

Chung I, Glow JA, Dimopoulos V, et al. Upper-limb somatosensory evoked potential monitoring in lumbosacral spine surgery: A prognostic marker for position-related ulnar nerve injury. Spine J 2009;9(4):287–295.

Geisler FH, Coleman WP, Grieco G, et al. The Sygen multicenter acute spinal cord injury study. Spine 2001 Dec;26(24 Suppl):S87–98.

Geisler FH, Dorsey FC, Coleman WP. Recovery of motor function after spinal-cord injury—a randomized, placebo-controlled trial with GM-1 ganglioside. N Engl J Med 1991 Jun 27;324(26):1829–1838. Erratum in: N Engl J Med 1991;325(23):1659–1660.

Hadley HM. Pharmacological therapy after acute cervical spinal cord injury. Neurosurgery 2002 Mar;50(3 Suppl):S63–72.

Hurlbert RJ. Strategies of medical intervention in the management of acute spinal cord injury. Spine 2006 May 15;31 (11 Suppl):S16-21; discussion S36.

Isley MR, Zhang XF, Smith RC, Cohen MJ. Intraoperative neuromonitoring detects thrombotic occlusion of the left common iliac arterial bifurcation after anterior lumbar interbody fusion. J Spinal Disord Tech 2007;20:104–108.

Kim DH, Zaremski J, Kwon B, et al. Risk factors for false-positive transcranial motor evoked potential monitoring alerts during surgical treatment of cervical myelopathy. Spine 2007;32:3041–3046.

Labrom RD, Hoskins M, Reilly CW, et al. Clinical usefulness of somatosensory evoked potentials for detection of brachial plexopathy secondary to malpositioning in scoliosis surgery. Spine 2005;30(18):2089–2093.

Lesser RP, Raudzens P, Luders H, et al. Postoperative neurological deficits may occur despite unchanged intraoperative somatosensory evoked potentials. Ann Neurol 1986;19:22–25.

Machida M, Weinstein SL, Yamada T, et al. Dissociation of muscle action potentials and spinal somatosensory evoked potentials after ischemic damage of spinal cord. Spine 1988;13:1119–1124.

Morota N, Deletis V, Constantini S, et al. The role of motor evoked potentials during surgery for intramedullary spinal cord tumors. Neurosurgery 1997;41:1327–1336.

Nash CL Jr., Lorig RA, Schatzinger MA, Brown RH. Spinal cord monitoring during operative treatment of the spine. Clinical Orthopedics and Related Research 1977;126:100–105.

Ofiram E, Lonstein JE, Skinner S, Perra JH. "The disappearing evoked potentials": A special problem of positioning patients with skeletal dysplasia. Spine 2006:31:E464–E470.

Schwartz DM, Sestokas AK, Hilibrand AS, et al. Neurophysiologic identification of position-induced neurologic injury during anterior cervical spine surgery. J Clin Monit Comput 2006;20(6):437–444.

Vauzelle C, Stagnara P, Jouvinroux P. Functional monitoring of spinal cord activity during spinal surgery. Clinical Orthopedics and Related Research 1973;93:173–178.

Yamamoto Y, Kawaguchi M, Hayashi H, et al. The effects of the neuromuscular blockade levels on amplitudes of posttetanic motor-evoked potentials and movement in response to transcranial stimulation in patients receiving propofol and fentanyl anesthesia. Anesth Analg 2008;106(3):930–934.

should be weighted on an individual basis. Successful monitoring can be achieved, even in the pediatric group with cerebral palsy and associated epilepsy disorder (Master et al., 2002).

Details about the methodology and interpretation of recorded SSEPs, tcmMEPs, and EMG activity are found on pages 5–33. In the following, we provide the specifics of intraoperative neurophysiology in relationship to various stages of this surgical procedure.

## V. INTRAOPERATIVE NEUROPHYSIOLOGIC MONITORING: A STEP-BY-STEP APPROACH

### 1. Induction of Anesthesia and Pre-Position SSEPs and tcmMEPs Baselines

As with all other types of surgical procedures, neurophysiologic monitoring starts immediately after induction. The setup for SSEPs is usually done before induction, whereas the setup for tcmMEPs is done after induction, because the latter requires insertion of subdermal needle recording electrodes and corkscrew stimulating electrodes. Obtaining pre-position baseline is important in further assessing appropriate positioning of the patient. Anesthesia used is TIVA with propofol and remifentanyl. In an attempt to decrease the risks associated with high doses of propofol administration, in particular in children, new drugs have been tried, such as dexmedetomidine. Studies of the latter during pediatric spine deformity surgery showed that, although it does not seem to have a significant impact on SSEPs recordings, it may, in fact, cause a decrease in the tcmMEPs amplitudes. Thus, we recommend caution during its use in tcmMEPs monitoring, until more extensive research is done. Similarly, avoidance of inhalational agents and, if possible, of muscle relaxants (in particular, intravenous boluses) is recommended.

It is the experience of several authors (Langeloo et al., 2003), including ours, that preoperative neurologic deficits and conditions such as neuromuscular scoliosis have a higher proportion of abnormal tcmMEPs baselines. Also, one should expect a higher incidence of tcmMEPs changes in these patients.

Further anesthesia considerations and troubleshooting pre-position SSEPs and MEPs baselines can be found on pages 179–83.

### 2. Positioning the Patient: Post-Position SSEPs and tcmMEPs Baselines

The classical surgical approach to address either a scoliotic or a kyphotic deformity is the posterior approach. Positioning for surgery is extremely important to avoid peripheral neurologic injury and is the responsibility of all members of the operating room team. The patient is placed prone on an appropriate operating room table. Care is taken to pad all bony prominences, including the region of the ulnar nerves at the elbows and over the iliac crests. Additionally, pressure is avoided over the eyes, and the neck is carefully placed to avoid the extremes of flexion, extension, and lateral bending.

Of note, anterior approaches can also be employed for isolated thoracolumbar curves, or in combination with a posterior approach, for curves that are stiff and difficult to correct without "loosening" them up, by releasing the contracted anterior longitudinal ligament and annulus/disc spaces over the apex of the curvature.

Once the patient is positioned, post-position SSEPs and tcmMEPs baselines are obtained, and, in case of changes from pre-position baselines, troubleshooting is employed. Further details are given on pages 184–87, as well as a description of criteria used for determining the significance of SSEPs and tcmMEPs changes.

A special consideration should be given to arm positioning, in particular in the "superman" prone positioning for posterior approach (Labrom et al., 2005). This may result in compressive brachial plexopathy, followed by significant yet reversible immediate postoperative neurologic deficits. Prevention of such problem is done by appropriately positioning the arms, as well as by continuously monitoring the ulnar nerve SSEPs and the tcmMEPs in the hand muscles. Detailed description of this situation is given on pages 211–13.

### 3. Opening and Exposure of the Spine

With the posterior approach, the skin is incised, and the dissection is carried down to the deep fascia, which is incised over the spinous processes. The paraspinal muscles are dissected off the posterior elements, including the spinous processes, laminae, over the facet joints, to the tips of the transverse processes. The facet joints are then removed to increase flexibility of the spine and increase the likelihood of arthrodesis, and the instrumentation is then placed. Occasionally, osteotomies need to be carried out over the apex of the curvature.

As already mentioned, there are situations when anterior approach is preferred, for isolated thoracolumbar curves or, in combination with a posterior approach, for curves that are stiff and difficult to correct without "loosening" them up, by releasing the contracted anterior longitudinal ligament and annulus/disc spaces over the apex of the curvature. Thus, gaining quick and good access to these structures is important. However, depending upon the spinal segments of interest, these may require additional caution. For example, gaining access to the lumbar segments requires traction applied on the abdominal wall and other structures, such as the pelvic vessels.

### A. Abdominal muscle contraction—Iatrogenic muscle relaxation

During anterior approach, resistance encountered because of contraction of the abdominal muscles is a common complaint from the surgeon during this opening phase. In many such circumstances, more iatrogenic muscle relaxation is required. Given the significant and unpredictable effect of intravenous muscle relaxants boluses, we prefer an increase in the infusion of muscle relaxant, which may have less unpredictable consequences on tcmMEPs. All of these details are discussed with the surgeon and with the anesthesiologist before the surgery, and any changes to the anesthetic regimen are communicated to the monitoring team. It is also important to acknowledge that the opening for the anterior approaches may be done by a general surgeon, who may not be aware of the impact that radical changes in anesthetics may have on the neurophysiologic testing. If administration of more muscle relaxant is necessary, the amount given and the time are documented. Also, we suggest that a new tcmMEPs trial is obtained before the muscle relaxant is given. The effect of an intravenous bolus may take up to 1 hour to wear off. It may also be or look asymmetrical or disproportionate. This, of course, depends on the baseline muscle strength, as well as on the patient's weight and muscle bulk. Further details regarding troubleshooting the effects of muscle relaxants on tcmMEPs can be found on pages 179–80, 183, and 185–87. It is important that, at the end at the opening phase and before the instrumentation begins, a reliable new pre-instrumentation tcmMEPs baseline is obtained.

### B. Intermittent monitoring

During opening, the electrocautery is frequently used, and spine manipulation does not occur. Thus, it is difficult and not essential to monitor the SSEPs and MEPs continuously. Moreover, safety concerns exist regarding continuous monitoring and concomitant use of the monopolar electrocautery (see pages 42–44). Thus, significant gaps between two consecutive trials can be seen. However, asking the surgeon to pause occasionally is important in order for new SSEPs and MEPs trials to be recorded; they provide an idea about the trend of evoked responses under deepening of anesthesia.

Situations can arise during the surgery when real SSEPs and MEPs changes may occur. Thus, it is important to discuss with the surgeon and anesthesiologist before the surgery, in order to anticipate the timing of these occurrences and to tailor the use of necessary anesthetics and muscle relaxants so that reliable neurophysiologic responses can be obtained, when needed.

Along the same lines, Yamamoto and colleagues (2008) used post-tetanic facilitation to increase significantly the amplitudes of the motor evoked responses, thus allowing reliable monitoring of the motor pathways, even in the case of significant iatrogenic neuromuscular block. A detailed description of the method can be found on page 183.

### C. Traction on pelvic vessels: Vascular compromise

This situation is rare but can occur during anterior approach. An example is given on page 213, when, in the process of exposing the lumbar spine, traction on the iliac artery was applied. This caused vascular thrombosis, with compromise of the blood supply to the ipsilateral leg. Diffuse SSEPs changes developed over time, signaling peripheral limb ischemia and prompting vascular treatment.

### D. Ligation of Segmental Arteries

The blood supply to the spinal cord arises primarily from the segmental arteries that branch from the aorta. The blood supply to the anterior aspect of the cord is more vulnerable than the blood supply to the posterior cord. During anterior deformity surgery, the segmental vessels are frequently purposefully ligated at operative levels to facilitate spinal release, decompression, osteotomy, or instrumentation. These vessels are often test clamped initially, to judge the implications of permanent ligation on spinal cord function via intraoperative neurologic monitoring. As also mentioned on pages 213–14, if the clamped vessel has a significant contribution to the spinal cord blood supply, sudden and significant changes in the tcmMEPs occur within the first 10 minutes or so after the clamping, signaling presence of spinal cord ischemia. If the test is positive, the surgeon will not permanently ligate the respective segmental artery (see Fig. **5C-4**). Releasing the clamp should be promptly followed by a reversal of the changes seen in the tcmMEPs. A lag in tcmMEPs recovery may signify arterial vasospasm. In such case, increasing the blood pressure and local irrigation with warm saline should help.

Boachie-Adjei et al. (1991) have demonstrated that it is safe to ligate segmental vessels unilaterally at multiple levels. Thus, care should be taken to avoid ligation of segmental vessels bilaterally at any given spinal cord level, to avoid compromise of the blood supply at that level.

We have, however, encountered situations when unilateral ligation of one or several segmental arteries resulted in ipsilateral spinal cord ischemic changes (see Fig. **5C-4**). Along the same lines, the Artery of Adamkiewicz (arteria radicular magna) represents a special consideration in anterior deformity surgery, because it is felt to represent a major supply of blood to the lower thoracic portion of the cord. Its damage or ligation has been postulated to produce catastrophic neurologic sequellae in anterior deformity surgery.

## 4. Instrumentation of the Spine

The cross-sectional anatomy of the spinal canal can be visualized as a tunnel. The roof of the tunnel is the lamina, the floor is the vertebral body and the separating discs, the pedicles are the supporting pillars, and the

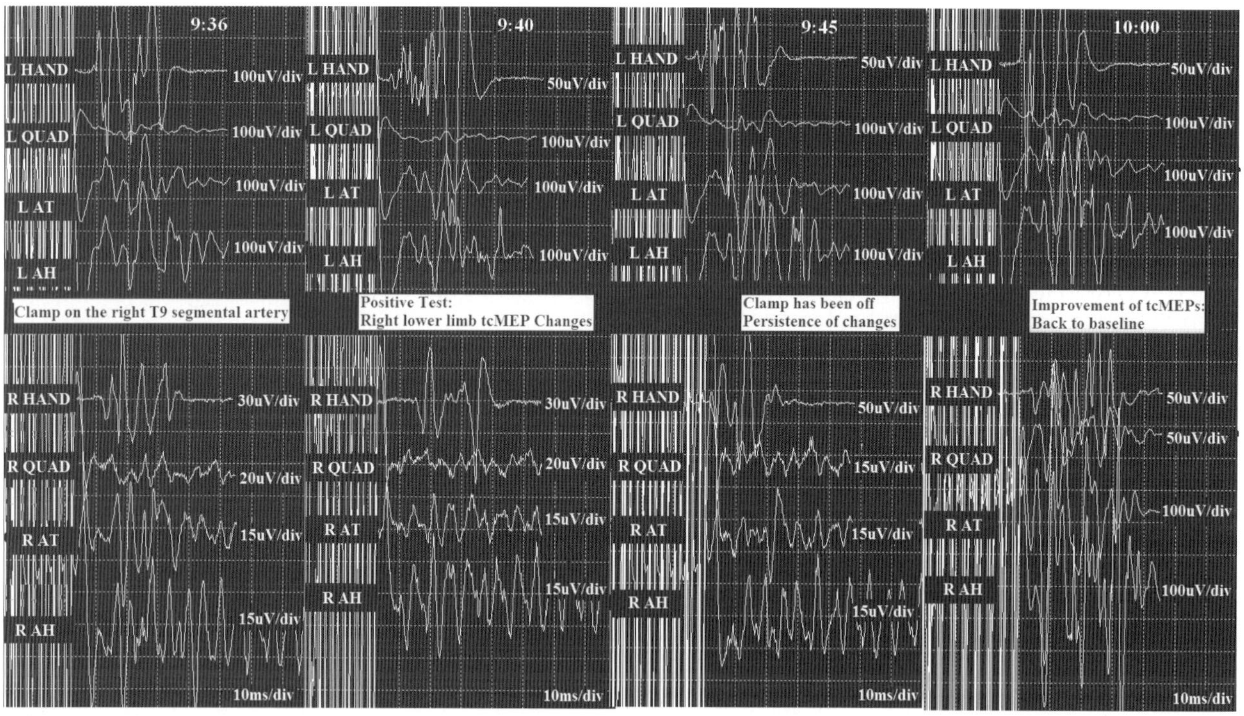


**FIGURE 5C-4** Test clamping of the segmental arteries. The procedure is T9-L1 thoracolumbar fusion for idiopathic scoliosis, anterior approach. In an attempt to gain better access to the intervertebral spaces, T9, T10, and T11 segmental arteries were individually and sequentially clamped, one side at a time. When the right T9 segmental artery was clamped at around 9:36 am, there was a sudden decrease in the amplitudes of the right lower limb tcmMEPs and a very subtle decrease on the left as well. Although no further worsening was seen on the left, the right-sided changes worsened within the next 5 minutes (trial at 9:40 am). The test was considered positive, and the clamp was promptly removed. However, the changes persisted beyond the clamp time (see set at 9:45 am). It was considered that this may be a result of vasospasm of the previously clamped segmental artery. The anesthesiologist increased the systemic blood pressure to 190 mmHg. The surgeon irrigated the artery with warm saline and papaverine. At 10:00 am, a prompt increase in the lower limb tcMEPs was seen, with return to the preclamp baselines. As a consequence, the surgeon did not ligate the right T9 segmental artery.

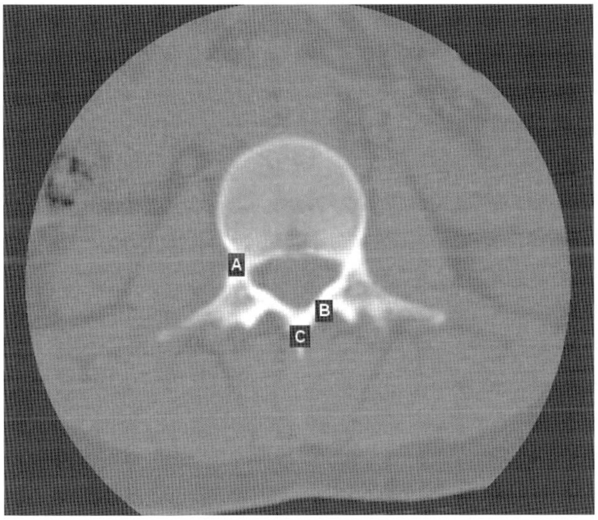


**FIGURE 5C-5** CT scan with axial view, showing cross-sectional anatomy. **A:** Pedicle; **B:** Laminae; **C:** Spinous Process.

foramen, through which the nerve roots pass, are the exits (Fig. **5C-5**).

## A. Anterior Approach

When annulectomies and discectomies are carried out in deformity surgery with anterior approach, the cartilaginous endplates or ring apophyses are separated from the bony endplates of the vertebral bodies. The disc's annulus and nucleus pulposis are removed back to—and sometimes including—the posterior longitudinal ligament (Fig. **5C-6**). This allows for an increased bony surface area for arthrodesis and diminishes the likelihood of disc material being displaced into the spinal canal with corrective maneuvers.

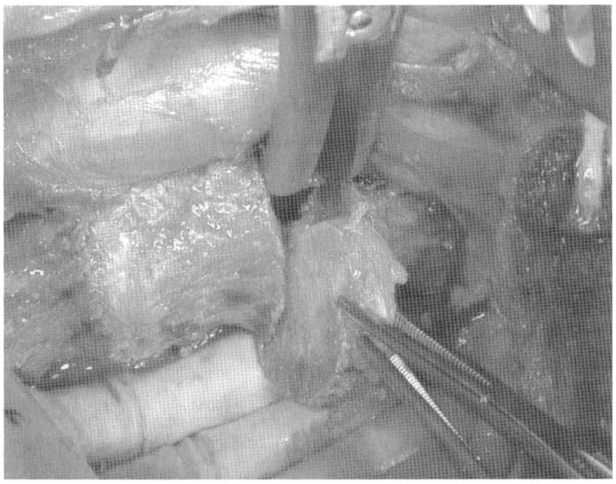

**FIGURE 5C-6** Anterior spinal arthodesis s/p excision of nucleus pulposis/disc.

## B. Pedicle Screw Placement

Pedicle screws can be placed in the cervical spine, as well as throughout the thoracic, lumbar, and sacral spines. Pedicles are placed from posterior to anterior and have the potential for injuring the spinal cord or nerve root, if placed too medially, or impinging on a nerve root, if placed too superiorly or inferiorly (see Fig. **5C-7**). Special considerations exist in the cervical spine, where the pedicle diameters are quite small at C2 and C7. At these levels, not only are the spinal cord and nerve roots at risk, but so is the vertebral artery. Likewise, lateral mass screws that are placed in the cervical spine place the nerve root and vertebral artery at risk. Intraoperative neurophysiology can be used to check the appropriate placement of the pedicle screws, which, if malpositioned, could cause a breech in the pedicles and produce nerve root damage. The technique involves free-run and triggered EMG and is detailed on pages 277–78**.** Concomitantly, SSEP and MEPs monitoring can add further help prevent spinal cord or root damage, as well as compression of vertebral artery. When the vertebral artery is compressed, it produces brainstem ischemia, resulting in changes of both cortical SSEPs and MEPs, as well as anterior cervical spine ischemia, with corresponding changes in tcmMEPs and less so in SSEPs.

## C. Instrumentation after Laminectomy

A special consideration must be given to the open canal during spinal instrumentation in deformity surgery. The lamina or "roof" of the spinal canal is often completely or partially removed during deformity surgery—to facilitate hook or wire placement, identify pedicles, perform osteotomies, or excise an abnormal vertebral element. Once the laminectomy has been performed, there is an increased risk for mechanical injury to the spinal cord by the surgical team. For example, sublaminar hooks and sublaminar wires that are frequently utilized in the cervical and thoracic spine have the potential for impinging on the spinal cord during passage or corrupting the blood supply to the spinal cord.

We have seen several cases of changes in SSEPs and MEPs with placement of the sublaminar wires, as a result of compression at the spinal level, where the

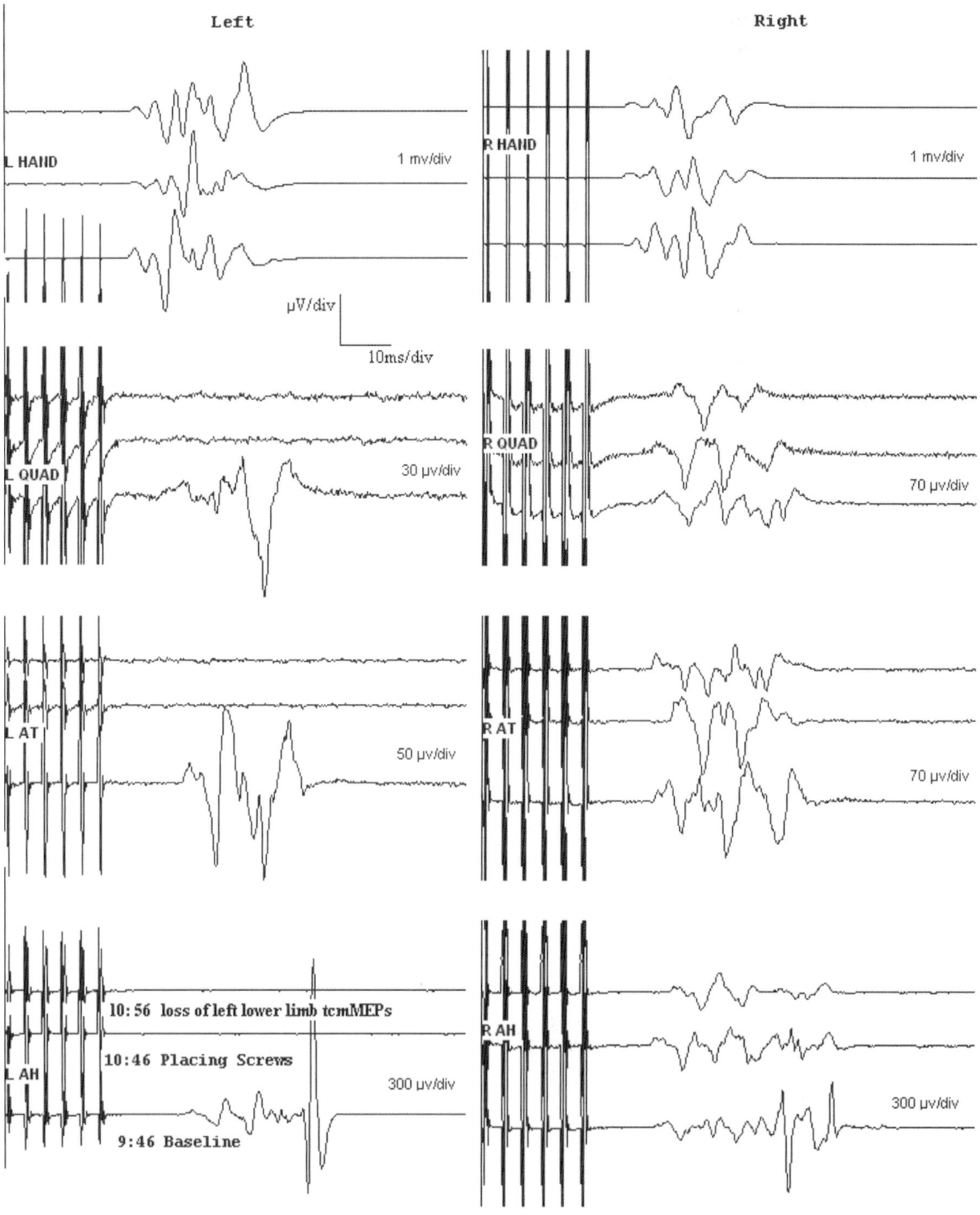


**FIGURE 5C-7** Posterior spine fusion T4–T11 for complete spine fracture at T8, in a patient with ankylosing spondylitis. Instumentation of the pedicle at T8–T9 resulted in total loss of left lower limbs tcmMEPs, without recovery. This correlated with left lower limb postoperative monoplegia. Immediately upon defining a path of entry (hole) at T8, T9 pedicles, the surgeon noticed cerebrospinal fluid (CSF) leak; thus no further screws were placed at this level. Unfortunately, 10 minutes later, there was total loss of left lower limbs tcmMEPs, without future recovery, despite an increase in the blood pressure and administration of intravenous solumedrol.

instrumentation takes place. These changes can occur strictly in the myotome innervated from the corresponding spinal level, if the alphamotor neurons and the ventral roots are involved, but also below it, if the mechanical compression involves the long tracts. Most of these changes are relatively abrupt and signal mostly vascular ischemia as a result of vascular compression. Thus, during passing the sublaminar wires, we pay special attention to tcmMEPs monitoring, which is more sensitive to spinal cord ischemia (see pages 16–26 and 237–66). This also enables immediate feedback to the surgeon, because tcmMEPs do not require averaging, and tcmMEPs changes, if any, can be expected to develop within minutes. Fig. **5C-8** shows an example of SSEPs and tcmMEPs changes as a result of sublaminar wires passed. Prompt removal of the wires reversed the changes.

## *D. Corrective Maneuvers*

The spinal cord is probably most vulnerable to injury and resultant dysfunction during the maneuvers employed by surgeons to correct deformities and spinal balance. These manipulations include cantilever, rotation, distraction, in situ bending, compression, and translation. These

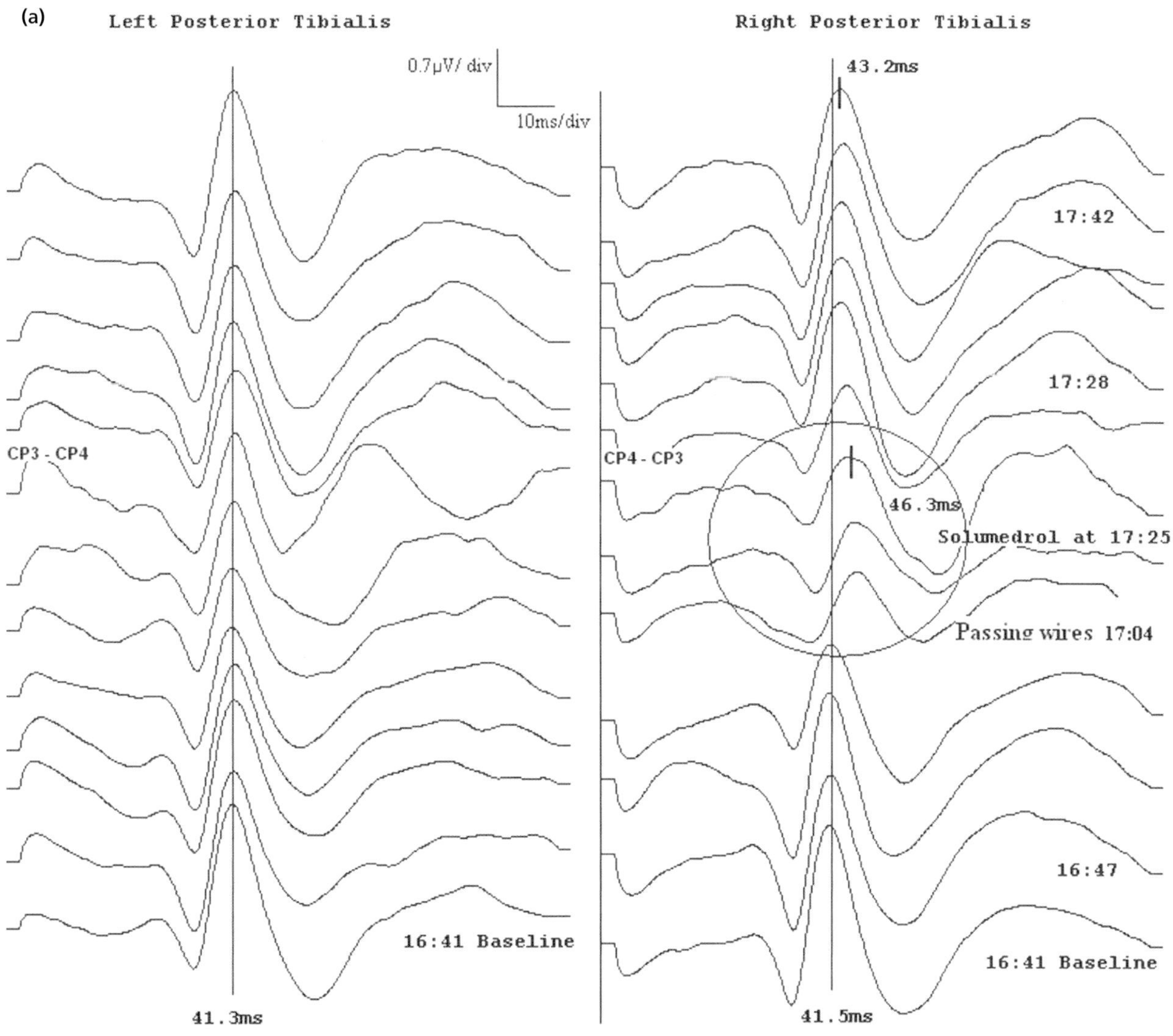


**FIGURE 5C-8** SSEPs **(a)** and tcmMEPs **(b)** changes as a result of spinal cord compression by sublaminar wires. The changes occurred within minutes after the wires were passed. The wires were immediately removed; the surgeon irrigated the field with warm saline, and the blood pressure was increased. Definite improvement was seen within 20 minutes.

(b)

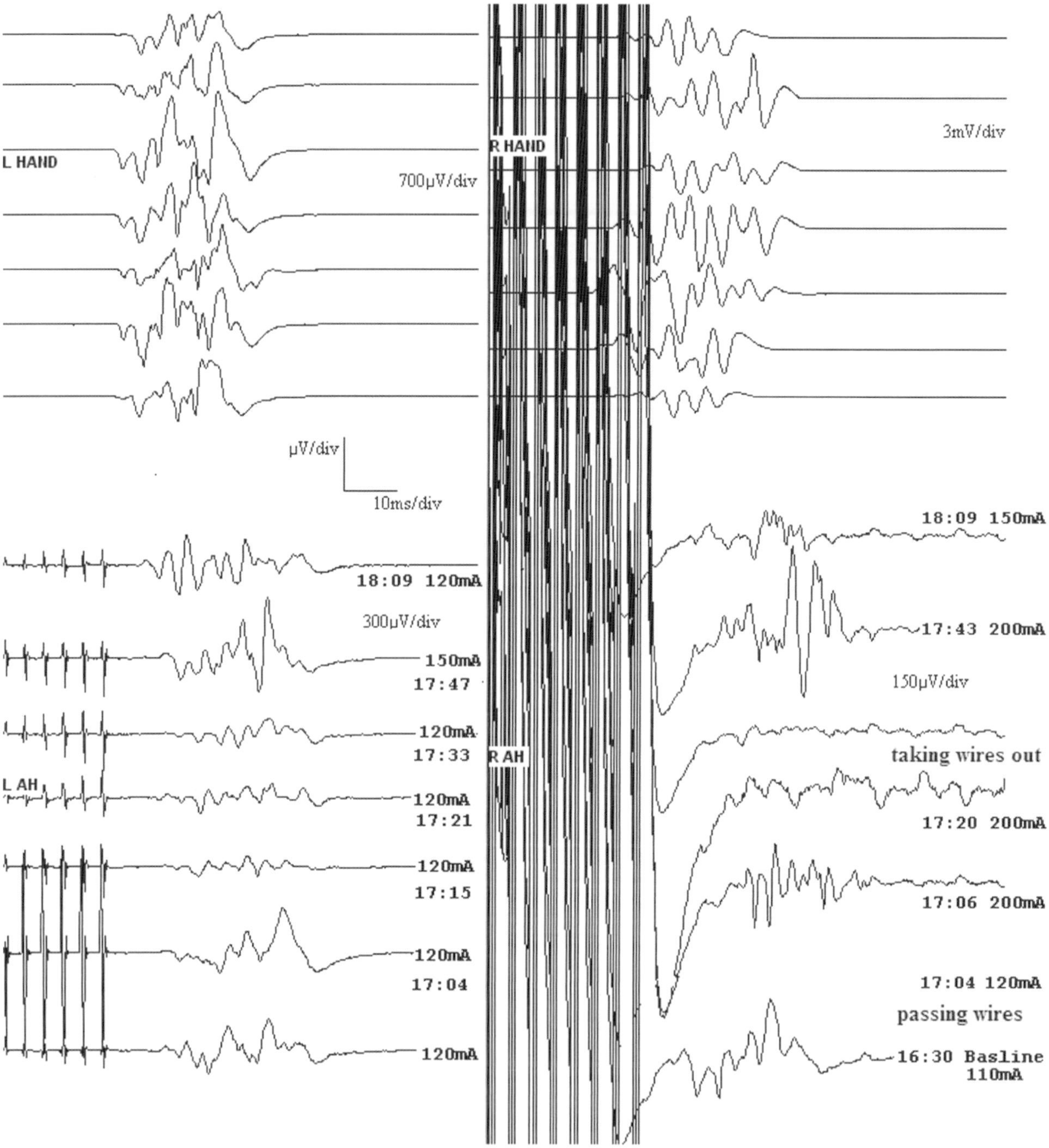


FIGURE 5C-8 b *(Continued).*

may be performed in combination and often occur simultaneously.

A cantilever maneuver is employed in both kyphosis correction (see the following) and scoliosis correction. It involves connecting the spinal rods to the spine, via screws, hooks, or wires, at either the cephalad or caudad extremes of the spinal fixation, and gradually manipulating the rod(s) down to the contralateral extreme, thereby correcting the deformity. When utilized in kyphosis, this shortens the posterior bony aspect of the spinal column but has the potential for lengthening the cord. Cord monitoring is critical during this maneuver to uncover any subtle spinal cord embarrassment. Generally, signal changes during these manipulative maneuvers result from reversible interruptions in blood supply and can be readily undone with timely and effective communication to the surgeon.

Translation and rotation maneuvers are designed to correct coronal plane deformities and rotational malalignment of the scoliotic spine. Signal abnormalities can occur during these manipulations, which may be reversible with timely communication and appropriate interventions. As cord monitoring technology has become more sophisticated, many surgeons believe that changes in the monitoring signal can occur with almost any deformity manipulation and that cord monitoring is useful in determining which of these are likely to be transient and which are likely to be permanent if no intervention is undertaken. As such, the rapidity of resolution needs to be communicated during these critical maneuvers.

Finally, distraction along the concavity of a deformity and compression along the convexity are universally employed during deformity surgery. These maneuvers have the potential to stretch or impinge upon the neural elements, resulting in spinal cord injury. Any manipulation of the spinal column has the same potential to affect the function of the cord or isolated nerve roots adversely. Additionally, this type of manipulation, during posterior correction of scoliosis, can result in acute spinal cord ischemia as a result of development of an anterior spinal artery syndrome (Stöckl et al., 2005). Thus, the most critical time for close spinal cord monitoring and timely communication among the surgeon, anesthesiologist, and spinal cord monitoring professional is at this point in the operation. The advantage of tcmMEPs and SSEPs monitoring over the Stagnara wake-up test is the ability to intervene earlier in the cascade of events. The avoidance of hypotension, close spinal cord monitoring, and timely communication at this point in the surgical procedure are all critical pieces to maximizing the positive outcome for the patient.

We have occasionally seen significant tcmMEPs changes during rod tightening. Immediate release of the tension, which sometimes requires taking the rod out, can reverse the changes. As previously mentioned, these changes are the result of vascular compromise, and, with timely intervention, irreversible damage can easily be avoided.

It is important to recognize the changes that signal neurophysiologic dysfunction and need to be acted upon. Thus, ruling out other causes for changes in SSEPs and MEPs, such as technical causes, effects of anesthetics, hypotension, or low hypothermia, is essential. Detailed technical troubleshooting and differential diagnosis aids are given in Subchapter **5A**. As with any type of surgical procedure requiring intraoperative monitoring, good and constant communication among surgical, anesthesia, and neurophysiologic teams is important.

If neurophysiologic changes are present, the first step is to reverse the maneuver that caused these changes. Second, because many of these changes result from vascular compromise, with or without vasospasm, avoiding hypotension and irrigation with warm saline can be helpful.

### *E. Kyphosis Correction with Spinal Osteotomy*

Smith Petersen osteotomy (SPO), performed at multiple spinal thoracic levels, was found to be one of the highest risk procedures related to loss of evoked responses in surgical correction of Sheuermann's kyphosis (Cheh et al., 2008). The first mechanism of injury is considered to be spinal cord hypoperfusion. These data are supported by the findings of Bradford and colleagues (1980), showing that a simple increase in the mean arterial blood pressure reversed the neurophysiologic changes associated with this procedure, whereas the correction itself was not adjusted. Also, the type and pace of changes also support the vascular theory (i.e. the sudden loss of tcmMEPs with concomitant preservation of the SSEPs). The second mechanism is probably an overcorrection of the kyphotic malalignment. We agree with Cheh and colleagues (2008) that a combination of both mechanisms may have a more dramatic effect. In the following, we briefly describe a few approaches and the mechanisms by which they may result in spinal cord injury.

*E1. Posterior approach.* Lee et al. (2006) showed that a posterior-only approach does not entirely eliminate the risk of damage of the anterior spine artery. The anterior column may become unstable during the posterior column release. After multiple SPOs resulting in posterior compression, associated with correction, the anterior vertebral chain may be lengthened, thus stretching the anterior spinal artery; also, the length of the spinal canal may shorten (Bridwell et al., 2001).

*E2. Vertebral column resection.* This maneuver increases the instability of the spinal column, thus causing abrupt subluxation, which may result in spinal cord injury. This can be avoided by placing a temporary rod to stabilize the column. Also, during this phase, posterior cord compression by an insufficient laminectomy may occur. Third, because the spinal column is shortened, spinal cord buckling may thus occur. In order to avoid the latter complication, an anterior cage should be placed; this increases the anterior height of the canal.

### 5. Closure

Delayed neurologic deficits, occurring after the actual injurious surgical steps took place, have been described in several types of surgery. Laminoplasty in tension is one example (see page 216). Delayed paraplegia after thoraco-abdominal aneurysm (TAA) repair is most likely the result of inappropriate spinal cord perfusion in the postoperative period (see pages 261–62). Delayed onset deficits have also been described in spinal deformity surgery (Cheh et al., 2008). One of the mechanisms may be related to a borderline appropriate spinal cord perfusion, which becomes inappropriate in certain circumstances (e.g. with fluctuations of blood pressure). Thus, we recommend continuing neuromonitoring throughout the wound closure.

### 6. Management Strategies in Case of Neurophysiologic Changes

If neurophysiologic changes are present, the first step is to reverse the maneuver that caused these changes. For example, releasing the correction should be done promptly. It may also be necessary that the rod(s) be removed and replaced with others that are less bent. Second, because many of these changes result from vascular compromise, with or without vasospasm, avoiding hypotension and irrigation with warm saline can be helpful. Thus, when hypoperfusion is suspected, a prompt increase in the mean arterial blood pressure should be attempted. Cheh and colleagues recommend keeping the mean blood pressure above 80 mmHg; for this purpose, they also suggest adding a dopamine drip, if necessary. In this case, releasing the correction should be done promptly. It may also be necessary that the rod(s) be removed and replaced with others that are less bent. Significant SSEPs and tcmMEPS changes are more likely to occur when there is significant blood loss, systemic hypotension, or anemia. Intravenous administration of steroids is somewhat controversial, yet it may help in certain cases of definite blunt trauma of the spine with mass effect caused by vasogenic edema (see page 220). Pages 216–20 give a detailed algorithm to follow in case of neurophysiologic changes related to spine manipulation.

## VI. CONCLUSION

Intraoperative neuromonitoring is now a routine part of many surgical procedures. It has allowed insight into the nervous system during spinal deformity surgery and provides critical information to the surgeon, allowing for reversal or avoidance of neural insults. Neuromonitoring success requires good communication and good cooperation between the surgeon, the anesthesiologist, the neurophysiologist, the technician, and nursing staff. Although it is still a young and evolving field, intraoperative neuromonitoring has already become an invaluable tool in spinal deformity surgery.

## REFERENCES

Beals RK, Robbins JR, Rofe B. Anomalies associated with vertebral malformations. Spine 1993; 18:1329–1338.

Boachie-Adjei O, Dendrinos GK, Ogilvie JW, et al. Management of adult spinal deformity with combined anterior-posterior arthrodesis and Luque-Galveston instrumentation. J Spinal Disord. 1991;4(2):131–141.

Bradford DS, Ahmed KB, Moe JH, et al. The surgical management of patients with Scheuermann's disease: A review of twenty-four cases managed by combined anterior and posterior spine fusion. J Bone Joint Surg Am 1980;62(5):705–712.

Bridwell KH, Kuklo TR, Lewis SJ, et al. String test measurement to assess the effect of spinal deformity correction on spinal canal length. Spine 2001;26(18):2013–2019.

Cheh G, Lenke LG, Padberg AM, et al. Loss of spinal cord monitoring signals in children during thoracic kyphosis correction with spinal osteotomy: Why does it occur and what should you do? Spine 2008;33(10):1093–1099.

Herring JA. Developmental dysplasia of the hip. In: Tachjian's Pediatric Orthopaedics 2002;pp 213–315. Philadelphia: W. B. Saunders Co.

Hsu B, Cree AK, Lagopoulos J, Cummine JL. Transcranial motor-evoked potentials combined with response recording through compound muscle action potential as the sole modality of spinal cord monitoring in spinal deformity surgery. Spine 2008;33(10):1100–1106.

Labrom RD, Hoskins M, Reilly CW, et al. Clinical usefulness of somatosensory evoked potentials for detection of brachial plexopathy secondary to malpositioning in scoliosis surgery. Spine 2005;30(18):2089–2093.

Langeloo DD, Lelivelt A, Louis Journée H, et al. Transcranial electrical motor-evoked potential monitoring during surgery for spinal deformity: A study of 145 patients. Spine 2003;28(10):1043–1050.

Lee SS, Lenke LG, Kuklo TR, et al. Comparison of Scheuermann kyphosis correction by posterior-only thoracic pedicle screw fixation versus combined anterior/posterior fusion. Spine 2006;1(20):2316–2321.

Legatt A. Current practice of motor evoked potential monitoring: Results of a survey. Journal of Clin Neurophysiology 2002;19:454–460.

Master DL, Thompson GH, Poe-Kochert C, Biro C. Spinal cord monitoring for scoliosis surgery in Rett syndrome: Can these patients be accurately monitored? J Pediatr Orthop 2008;28(3):342–346.

Noonan, KJ, Walker T, Feinberg, JR, et al. Factors related to false-versus true-positive neuromonitoring changes in adolescent idiopathic scoliosis surgery. Spine 2001;27(8):825–830.

Nuwer MR, Dawson EG, Carlson LG. Somatosensory evoked potential spinal cord monitoring reduced neurologic deficits after scoliosis surgery: Results of a large multicenter survey. Electroencephalogr Clin Neurophysiol 1995;96:6–11.

Pelosi L, Lamb J, Grevitt M, et al. Combined monitoring of motor and somatosensory evoked potentials in orthopaedic spinal surgery. Clin Neurophysiol 2002;113(7):1082–1091.

Stöckl B, Wimmer C, Innerhofer P, et al. Delayed anterior spinal artery syndrome following posterior scoliosis correction. Eur Spine J 2005;14(9):906–909.

Ulkatan S, Neuwirth M, Bitan F, et al. Monitoring of scoliosis surgery with epidurally recorded motor evoked potentials (D wave) revealed false results. Clin Neurophysiol 2006;117(9):2093–2101.

Wever DJ, Tønseth KA, Veldhuizen AG, et al. Curve progression and spinal growth in brace treated idiopathic scoliosis. Clin Orthop Relat Res 2000;(377):169–179.

Yamamoto Y, Kawaguchi M, Hayashi H, et al. The effects of the neuromuscular blockade levels on amplitudes of posttetanic motor-evoked potentials and movement in response to transcranial stimulation in patients receiving propofol and fentanyl anesthesia. Anesth Analg 2008;106 (3):930–934.

# 5D

# Thoracoabdominal Aneurysm Repair

MICHAEL A. PECK, MIRELA V. SIMON,
AND MARK F. CONRAD



## I. INTRODUCTION

Spinal cord ischemia (SCI) remains a potentially devastating complication of thoracic aneurysm (TA) and thoracoabdominal aneurysm (TAA) repair, with an overall incidence of 16% in Crawford's benchmark series (Crawford et al., 1986).

SCI occurs from aortic cross-clamping of the descending thoracic aorta or from the positioning of an endovascular graft. Normally, paired segmental arteries throughout the thoracoabdominal region provide a rich collateral network that supplies the spinal cord. Degenerative aneurysms can cause occlusion of the majority of these segmental intercostal and lumbar arteries from mural thrombus and aortic plaque formation (Jacobs et al., 2002). Cross-clamping can further compromise the remaining patent segmental arteries by preventing them from supplying blood to the arterial network that perfuses the cord. Similarly, placement of an endovascular graft can cause SCI by obstructing the patent intercostal arteries.

Several techniques have been used with variable success to evaluate, monitor, and maintain the spinal cord perfusion before, during, or after the surgery. These include preoperative imaging of the spinal cord vascularization, use of atriofemoral bypass and shunts, prevention of the steal phenomena, SSEPs and tcmMEPs monitoring to guide selective intercostal reimplantation, and cerebrospinal fluid (CSF) drainage. In addition, neuroprotective measures, such as preconditioning, epidural cooling (EC), and pharmacologic agents can have a good impact. Despite these advances, SCI is estimated to complicate central aortic surgery in at least 5–10% of patients; even in the hands of experts (Coselli et al., 1999; Safi et al., 1996; Conrad et al., 2007). Along these lines, the beneficial role of intraoperative neurophysiology monitoring in improving the postoperative neurologic outcome is now widely recognized. Combining the use of intraoperative neurophysiologic testing with other techniques and adjuncts has further lowered paraparesis and paraplegia rates (Jacobs et al., 2006; Achouh et al., 2007; Ogino et al., 2006).

In this chapter we review the contemporary management of TAA (both open and endovascular repair), with emphasis of the role of intraoperative neurophysiology to detect and correct SCI before permanent neurologic deficit occurs. Additional benefits of SSEPs, tcmMEPs, and EEG monitoring (e.g. detection of cerebral and limb ischemia) are described as well. In addition to intraoperative neuromonitoring, we also address the aforementioned techniques that can be employed to maintain adequate spinal cord perfusion, along with known neuroprotective adjuncts used during these procedures.

## II. THE VASCULARIZATION OF THE SPINAL CORD

Studies by Adamkiewicz, dating to more than 130 years, ago mark the beginning of recorded interest in the blood supply of the spinal cord (Adamkiewicz, 1882). This work described the anterior and posterior spinal arteries, their anastomoses, and the major inputs into the blood supply of the spinal cord. (Fig. **5D-1**) Specifically, blood reaches the upper spinal cord (~C1 through T3 level) via subclavian branches, such as vertebral (~C3 level), deep cervical, and ascending cervical (~C6-C7 levels) arteries (Turnbull et al., 1966; Lazorthes et al., 1966). The central region (~T3 through T7 level) of the cord is supplied from multiple segmental branches from the aorta, namely intercostal arteries, with one branch at T7 level that is present more consistently (Sandson et al., 1989). The distal region of the cord extends from T8 to the cone and receives vascularization via radicular arteries from intercostal and lumbar arteries. In particular, a large segmental branch in the lower thoracic or upper abdominal aorta, which enters the spinal canal between T9 and T12 most of the time, has been designated as the artery of Adamkiewicz (or radicular artery). The most distal part of the spinal cord, or conus, is supplied primarily from the hypogastric artery and its branches (Griepp et al., 2007). Specifically, a cone artery (Desporges-Gotteron artery), originating from the internal iliac artery, can be present at L2 or L5 levels (Garcin et al., 1962). There is an entire network of small arteries in the spinal canal, perivertebral tissues, and paraspinous muscles that form multiple anastomotic pathways outside the spinal canal, as described by Lazorthes (Lazorthes et al., 1971). It is generally accepted that the cervical spinal cord has a more solid vascularization, relatively evenly distributed bilaterally, whereas vascularization of the thoracic and lumbar regions is better represented on the

(a)

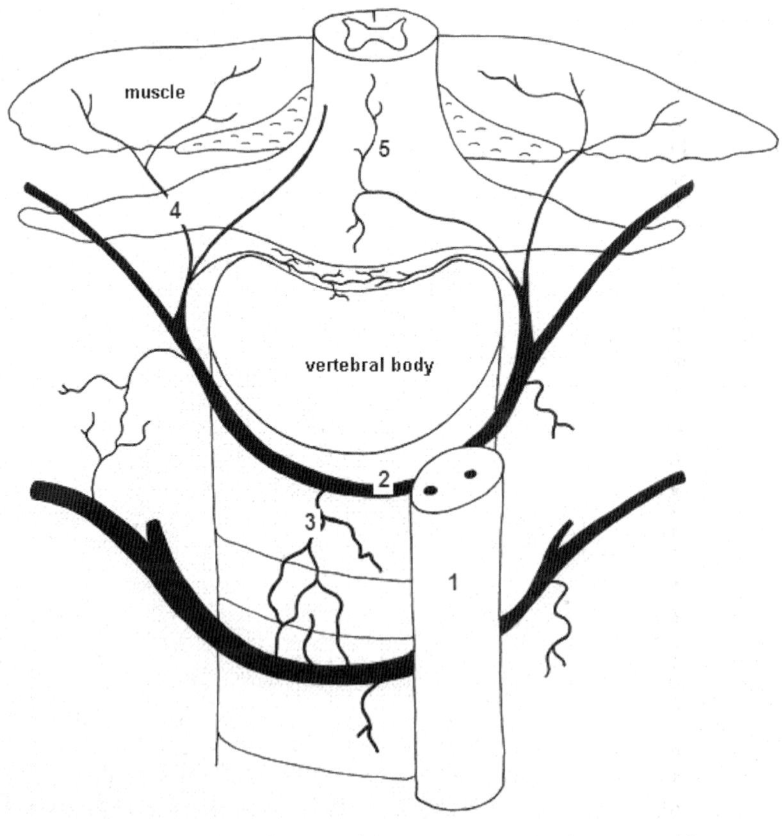


**FIGURE 5D-1** Vascularization of the spinal cord. **a:** The aorta **(1)** gives rise to paired segmental (intercostal or lumbar) arteries **(2)** and a rich collateral network forms, including perivertebral **(3)**, paraspinous muscle **(4)**, and other branches that together supply the anterior spinal artery **(5)**.

(b)

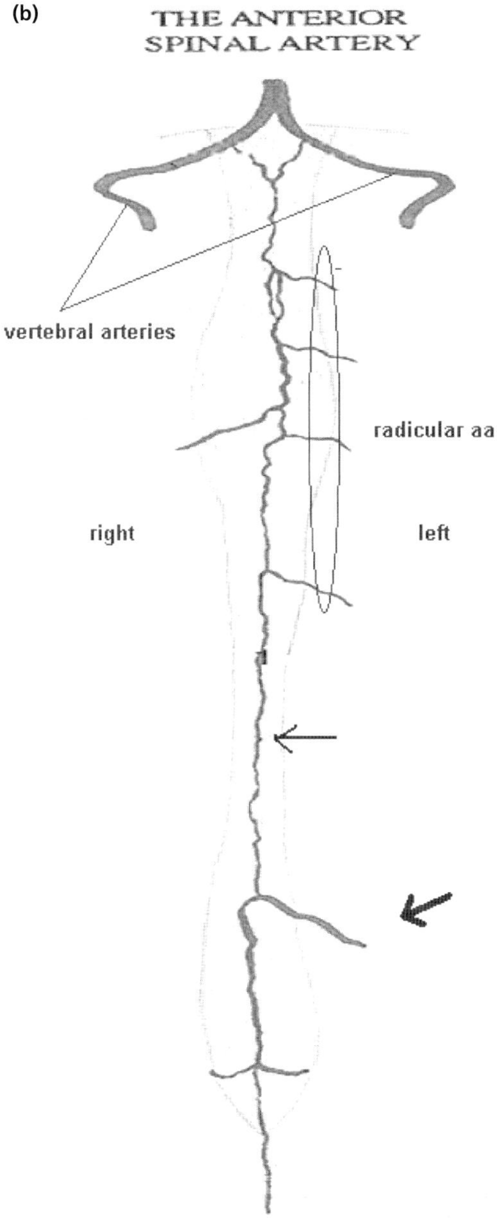


(c)

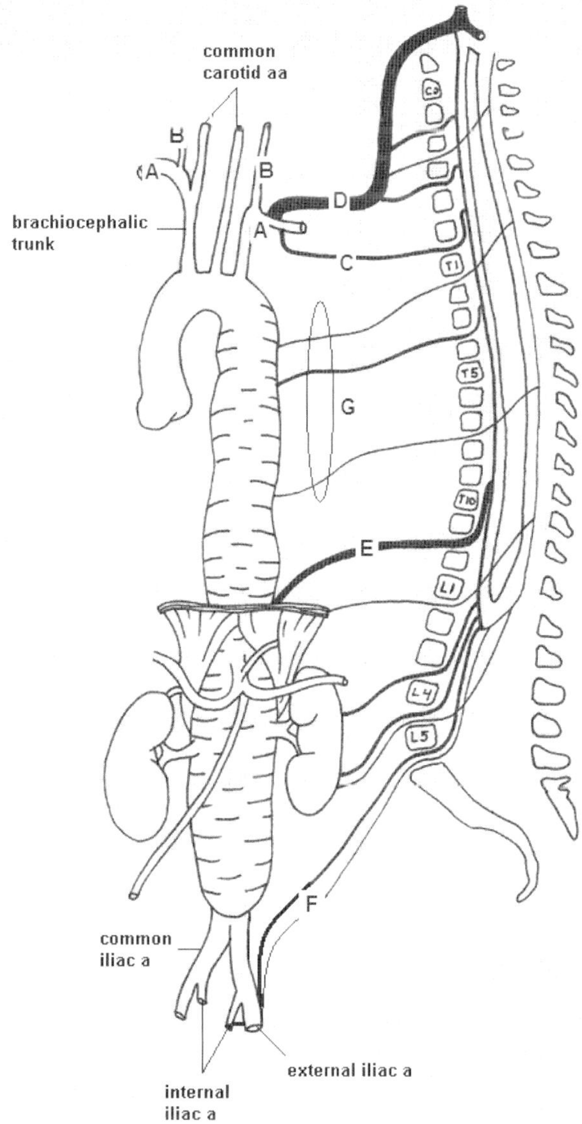


**FIGURE 5D-1** *(Continued)* **b:** Anterior spinal artery (thin arrow) arising from anastomosis of radicular arteries, of which the great radicular artery (artery of Adamkiewicz, thick arrow), and branches of vertebral arteries. Notice the richer vascularization of the left side of the spinal cord. **c:** Major vessels contributing to the spinal cord blood supply. The spinal cord vascularization has three main sources: (1) the proximal collateral flow, arising from the vertebral (**D**), ascending cervical (**B**), and deep cervical (**C**) arteries, all branches of subclavian (**A**) arteries; (2) radicular arteries, arising from thoracic intercostal (i.e. segmental) arteries (**G**)—notice the important blood supply given by the great radicular artery (**E**); (3) distal collateral flow, arising from the lumbar and hypogastric (internal iliac) arteries (**F**).

left (Carpenter, 1991). Recent laboratory and clinical observations have supported the concept of a collateral arterial network that supplies the cord in the absence of patent intercostal arteries. Inputs to this network include the segmental vessels, subclavian arteries, and hypogastric arteries and their branches. This network can increase flow from one source when another is reduced. Alternatively, nutrient flow can be reduced if another low-resistance pathway is opened, such as an open excluded aortic segment, iliac, or visceral vessel as a result of cross-clamping during aneurysm resection, resulting in hypoperfusion and SCI (Griepp et al., 2007).

The importance of extrasegmental vessels for inflow into the spinal cord has been demonstrated in a number of laboratory studies. Christiansson et al. (2001) found that oxygen tension in the intrathecal space could be maintained in a viable range, without the development of acidosis, after all segmental arteries were ligated in a porcine model. When the subclavian arteries were clamped in addition, the $PO_2$ decreased to zero and acidosis in the intrathecal space ensued. Strauch et al. used MEPs to monitor spinal cord function in a porcine model to show that the number of segmental vessels that could safely be ligated was decreased substantially if either the subclavian or median sacral (equivalent of hypogastric) artery was ligated (Strauch et al., 2003). These findings have been supported clinically, because patients with prior aortic surgery or coverage of the subclavian artery are at higher risk of SCI than those who have not had interruption of these collateral pathways (Baril et al., 2006; Lombardi et al., 2003). Similarly, prior aortic abdominal surgery, disrupting the collateral flow from the hypogastric and lumbar arteries, can significantly increase the risk of SCI during occlusion of the segmental arteries, as further detailed on page 261.

## III. OPERATIVE TECHNIQUES. PREVENTING ACUTE SPINAL CORD ISCHEMIA

### 1. Determining the Operative Approach: Open TAA Repair Versus TEVAR

A successful surgical strategy begins with preoperative planning. Defining the type of aortic pathology (TA or TAA) and the specific anatomic classification (Fig. **5D-2**) has direct implications for both the technical conduct of the operation and the incidence of operative complications, particularly SCI. A type II TAA requires the most complex form of reconstructions and is also associated with the highest risk of SCI, whereas a type I TAA has a similar incidence of SCI as a TA (Conrad et al., 2008). Aneurysm extent is currently the main determinant of operative approach, because an entirely endovascular repair is presently only possible in patients with true TA, whereas those with TAA are eligible for thoracic endovascular aortic repair (TEVAR) only in the setting of a hybrid procedure involving a combined open and endovascular approach.

In contemporary practice, a dynamic, fine-cut, contrast-enhanced CT scan, often with helical reconstruction, provides the surgeon with the following: (1) an accurate assessment of aneurysm size and extent; (2) a baseline study with which future images may be

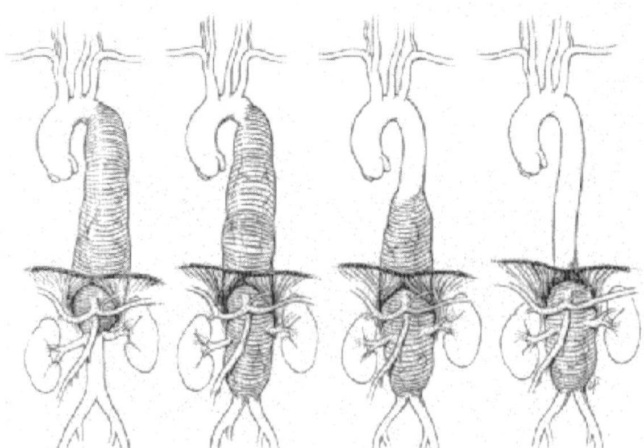

| TYPES OF TAA | I | II | III | IV | Total |
|---|---|---|---|---|---|
| Patients | n=121 | n=69 | n=164 | n=101 | n=455 |
| Major Paraplegia | 11 (9.1%) | 10 (14.5%) | 19 (11.6%) | 3 (2.9%) | 43 (9.5%) |

**FIGURE 5D-2** Spinal cord ischemia (SCI) incidence stratified by Crawford type in 455 TAA repairs.

compared; (3) determination of anatomic suitability for endovascular repair; (4) the anatomic extent of resection for open repair; and (5) a sense of the risk of subsequent SCI based upon the arterial anatomy. In our practice, all patients with anatomically suitable TA are treated with TEVAR, and those with TAA who are appropriate risk surgical candidates undergo operative repair.

## 2. Open TAA Repair

Our preference is the selective use of atriofemoral bypass with sequential cross-clamp application (Fig. **5D-3**) and the use of a clamp-and-sew technique with adjuncts (Fig. **5D-4**) in type III and IV TAA. Some key aspects of this approach are aimed at reducing complications of paraplegia, renal insufficiency, and postoperative respiratory failure. In an effort to minimize the risk of paraplegia, we use selective segmental arteries reimplantation, under the direct guidance of SSEPs/tcmMEPs monitoring. In order to ensure reliable intraoperative neurophysiology monitoring, we stopped using regional hypothermic protection to the segment of the spinal cord typically at risk for ischemic injury during TA/TAA repair, as described previously (Fig. **5D-5**). Renal protection is achieved through the direct installation of renal preservation fluid (4°C lactated Ringer's solution with 25 g of mannitol per liter) into the renal artery ostia after opening the aorta.

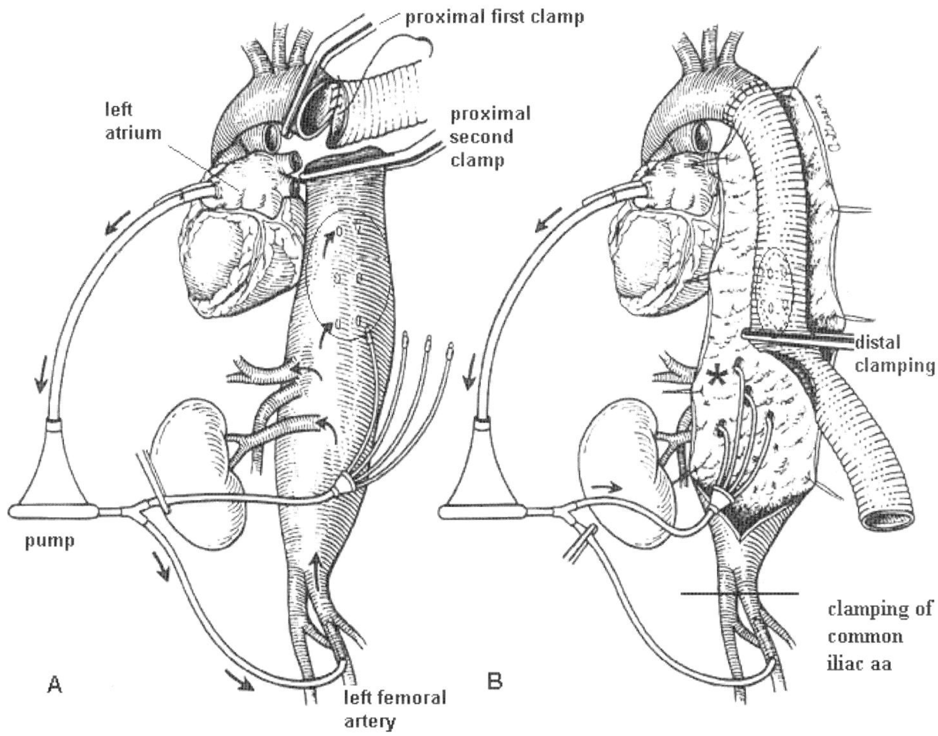


**FIGURE 5D-3** Operative approach to TAA repair, using distal aortic perfusion (left atriofemoral bypass). Using the heparin-impregnated Bio-Medicus pump, perfusion distal to the proximal cross-clamp is initially maintained via the femoral artery (**A**) and then by multiple perfusion catheters to the visceral and renal vessels, once reconstruction proceeds distally (**B**). In panel **A**, notice the first and second proximal clamps, positioned close to each other, in order to perform the first graft anastomosis. Also, notice the origin of paired segmental arteries (circled). In panel **B**, the clamps were moved distally, below their origin. The aorta is now open. Notice the origin of visceral vessels vascularized now by multiple perfusion catheters (*). Also, notice the sewed (attached to the graft) button of segmental arteries (circled), considered critical for maintenance of adequate spinal cord blood supply. In both panels, the bypass maintains retrograde perfusion. Oxygenated blood is pumped from the left atrium to the left femoral artery, below the most distal clamp. Once the distal clamp is moved below the origin of the visceral vessels, the femoral catheter may or may not be clamped, and the retrograde flow is redirected via the perfusion catheters, to ensure the appropriate blood supply to the pelvic organs.

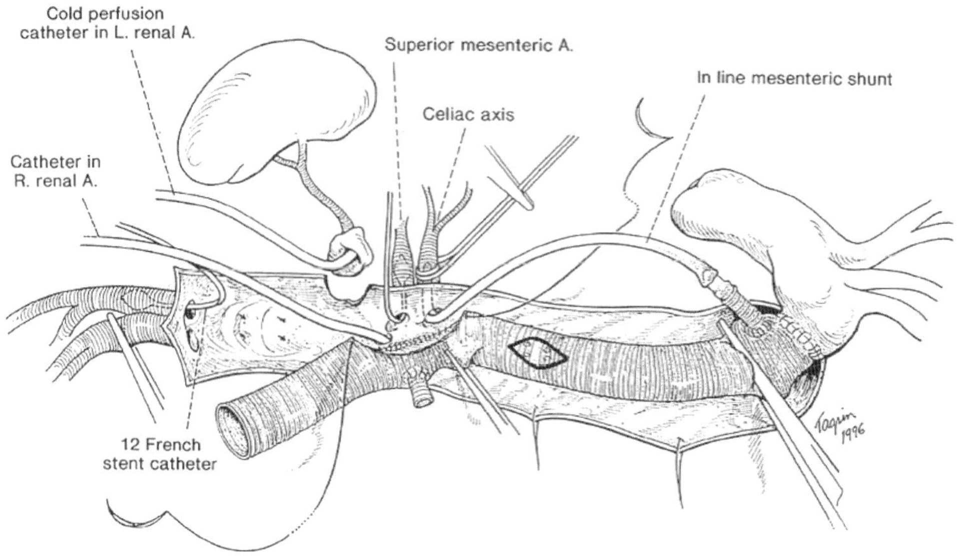


**FIGURE 5D-4** Operative approach for repair of type II thoracoabdominal aneurysm, with a clamp-and-sew technique, supplemented with regional hypothermic adjuncts for renal and spinal cord protection. The entire aneurysm sac is continuously exposed after proximal cross-clamp application. After completion of the proximal anastomosis, pulsatile arterial perfusion is established into the mesenteric circulation through in-line mesenteric shunting into either the celiac axis (as depicted) or the superior mesenteric artery. Thereafter, critical intercostal vessels are reconstructed (thick elipse), and a single inclusion button anastomosis for reconstruction of celiac, superior mesenteric, and right renal arteries is possible in the majority of cases. The left renal artery is reconstructed with a side-arm graft.

## *A. Physiologic Changes with Aortic Cross-Clamping*

Aortic cross-clamping, for open TAA repair using atriofemoral bypass, involves the placement of two clamps: one proximal to the aneurysm and a second just distal to the first, such that only a short segment of aorta is excluded via the bypass circuit. For further reference, we name these proximal clamps one and two, respectively. This allows for the creation of a proximal anastomosis while still preserving flow to the majority of the spinal cord. This flow is achieved in a retrograde fashion via the atriofemoral bypass (see Fig. **5D-3**). After completion

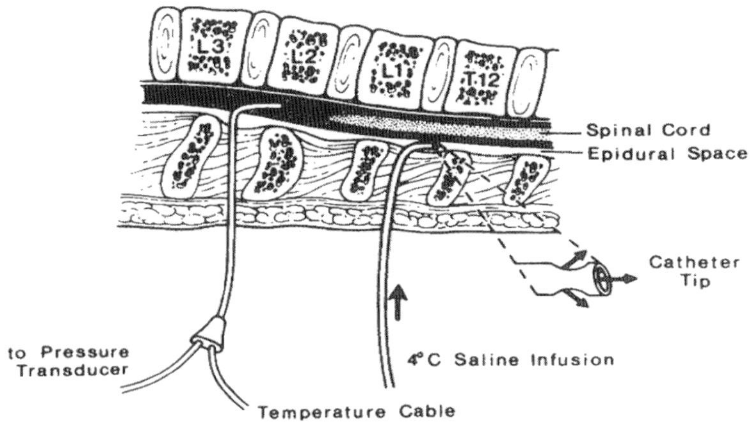


**FIGURE 5D-5** Regional spinal cord hypothermic protection is achieved via epidural cooling. A 4°C epidural saline infusion (arrow) is begun in anticipation of cross-clamping. Cerebrospinal fluid temperature and pressure are simultaneously monitored with a separate intrathecal catheter.

of the proximal anastomosis, the clamps are moved distally in a stepwise fashion, to create ultimately the distal anastomosis. Despite this careful stepwise approach and continuous perfusion, aortic cross-clamping can result in physiologic alterations that render the spinal cord ischemic. This results from an inadequate number of patent intercostals and lumbar vessels as a result of mural thrombus and calcifications of the aortic wall occluding these segmental arteries. On the other hand, it is also our experience that the slow development of occlusion of the segmental arteries, as in the case of atherosclerotic lesions, may have allowed enough time for development of the collateral flow from elsewhere thus, in case clamping takes out of the circulation these nonfunctional vessels, the spinal cord perfusion is not affected. On the contrary, in the case of sudden development of an aortic aneurysm, as in the case of trauma, many of the segmental arteries may still have an important role in spinal cord blood supply; thus taking these vessels out of the circulation by aortic clamping may cause SCI.

The SCI that may result from aortic cross-clamping, can be exacerbated by a "steal" phenomenon within the clamped aortic segment, as a result of backbleeding through patent segmental vessels. This reversal of flow draws blood away from the spinal cord (Griepp et al., 2007).

Christiansson et al. (2000) sought to better understand the pathophysiology of SCI using an intrathecal $PO_2$ detection catheter to investigate the functional vascular anatomy of the spinal cord during various modes of aortic clamping and selective interruption of segmental arteries in a porcine model. Single thoracic aortic cross-clamping caused a rapid decrease of intrathecal $PO_2$ to undetectable levels. The free runoff of blood to the distal aorta and away from the spinal cord with this technique rendered the collateral circulation insufficient and rendered the cord ischemic. The application of a second clamp to the distal aorta reduced intrathecal $PO_2$ by only 50%. This difference, in comparison to the proximal clamping, is explained by the prevention of steal through the intercostal arteries. Another variation, the addition of a subclavian artery clamp, excluded all proximal collateral circulation and resulted in severe SCI (Christiansson et al., 2001). Another double clamp model by Ishizaki et al. (1999) successfully identified spinal cord feeding arteries by using intrathecal oxygen monitoring. By allowing perfusion of these arteries with oxygenated blood, intrathecal $PO_2$ and spinal-evoked potentials were improved. Finally, Meylaerts et al. (2000) were able to prevent paraplegia, resulting from 60 minutes of aortic clamping, with the addition of selective segmental artery perfusion in a porcine model.

## B. The Overall Goals of the Therapy

The ultimate goal of the surgical treatment of TAA is prevention of aneurysm-related death caused by rupture, while minimizing the risk of perioperative complications such as SCI. We recently reported our results of 455 TAA repairs performed over a 20-year period and found that SCI differed by aneurysm extent, with type II and III degenerative TAA having the greatest incidence of SCI (Fig. **5D-2**) (Conrad et al., 2007). Various surgical and medical adjuncts have been used with varying degrees of success, leading to a significant reduction in the incidence of SCI (Coselli et al., 1999; Safi et al., 1996; Cambria et al., 2000). Such strategies, aimed at decreasing the incidence and severity of end-organ injury and SCI after TAA repair, can be stratified into two categories. The first category includes those procedures aimed at maintaining appropriate visceral and spinal cord blood supply during the operation: (1) preoperative identification of critical intercostal (segmental) arteries; (2) the use of shunts and bypass circuits for maintenance of distal aortic perfusion during placement of the cross-clamps; (3) prevention of a steal phenomena; (4) CSF drainage; and (5) reconstruction of critical segmental arteries identified intraoperatively, with the help of intraoperative neurophysiology. The second general category involves a wide variety of neuroprotective adjuncts, such as preconditioning, hypothermic adjuncts, and pharmacologic agents, meant to attenuate excitatory and inflammatory mechanisms that contribute to the development of ischemic injury. Adjuncts from both categories are continued into the postoperative period, as passive CSF drainage is instituted for 48 hours, as is maintenance of systolic blood pressure to aid spinal cord perfusion.

### *B1. Maintaining appropriate spinal cord blood supply: the role of intraoperative neurophysiology*

***a. Preoperative imaging.*** Preoperative imaging is used to identify the radicular artery and any relevant collateral vessels. With degenerative aneurysms, any number of the usually paired segmental lumbar and intercostal arteries are occluded by thrombus within the aneurysm (Jacobs et al., 2002). It thus becomes essential to evaluate the spinal cord circulation, both preoperatively and perioperatively. In an effort to apply an anatomic approach to the prevention of spinal cord injury, some surgeons have used magnetic resonance angiography (MRA) or

patient. Thus, we believe it is essential to identify those segmental arteries that have a significant role in the spinal cord blood supply and to reimplant only those arteries. This way, precious time can be saved—by reducing the overall clamp time.

We are currently using transcronically evoked muscle motor evoked potential (tcmMEPs) in types I to III TAAs to guide reimplantation of intercostals when changes occur. This approach has been supported in a porcine model where the use of MEPs allowed a stepwise sacrifice of nearly all segmental arteries, with preservation of spinal cord function. The authors further concluded that the unusual demands on the collateral circulation after segmental artery sacrifice support the use of distal aortic perfusion and maintenance of a constant mean arterial pressure in the early postoperative period (first 48 hours) (Halstead et al., 2007).

A novel technique for maximal reimplantation of intercostal arteries during TAA was reported by Woo et al. (2007). In a small series (n = 8) 8-mm grafts were extended from the proximal to the distal aortic graft, with a spatulation of the graft allowing a side-to-side anastomosis of the intercostal arteries (Fig. **5D-7**). This technique allowed an average of 5 of 7 sets of paired intercostals arteries to be reimplanted without complications or resulting neurologic deficits. The purported benefit of this technique is the continuous flow loop that is created, which improves graft patency and avoids the need for selective reimplantation of intercostal arteries. However, in the absence of a direct comparison between intercostal reimplantation and monitoring with selective reimplantation, the latter technique is currently favored by most surgeons.

***e. Monitoring tcmMEPs and SSEPs.*** One adjunct that should reduce SCI is the ability to assess the adequacy of spinal cord perfusion and spinal cord function during cross-clamping. Monitoring of evoked potentials, MEPs (i.e. tcmMEPs) or SSEPs, allows continuous and dynamic determination of spinal cord integrity, as well as the adequacy of interventions initiated to restore perfusion.

Detailed description of the methodology of SSEPs and tcmMEPs and interpretation of the recorded evoked responses for both methods are given on pages 5–26.

Monitoring tcmMEPs has been found to be a highly reliable technique to assess spinal cord function during TAA repair (van Dongen et al., 2001). In a series of 112 consecutive TAA repairs, Jacobs et al. (2006) demonstrated that all patients who had reliable MEP responses at the end of the procedure were able to move their legs normally, even if the amplitude was less than 10% compared with the initial values. In addition, complete absence of MEPs was associated with paraplegia that did not resolve. The overall rate of paraplegia in this study was 4.5%. and all were in type II TAAs.

Several animal studies have addressed the association between MEPs amplitude and neurologic impairment. Murakami et al. (2004) established the threshold of a decrease of 75% in MEPs amplitude from baseline to be indicative of neuronal loss and neurologic deficits in rabbits. However, Kakinohana et al. (2005) reported a high failure rate of MEPs to predict correctly postsurgical paraplegia, in a rat model using intra-aortic balloon occlusion, that resulted in paraplegia despite preservation of MEPs amplitude. To date, no human data are

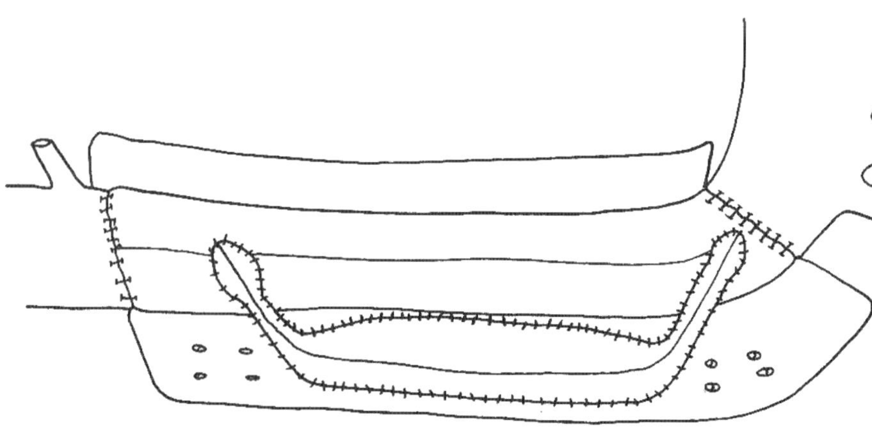

**FIGURE 5D-7** An intercostal patch graft, with spatulation of the graft allowing a side-to-side anastomosis of the graft to the posterior aortic wall incorporating multiple pairs of intercostal arteries.

available that relate specific MEP amplitude changes to neurologic symptoms. Schepens et al. (2004) used an amplitude decline of at least 50% as an intraoperative indication to intervene, whereas Jacobs et al. (2006) used the threshold of a decrease greater than 75% in MEP amplitude as a sign of critical SCI, even in the face of no identifiable functional deficits. They postulated that these patients have a tenuous blood supply to the cord that, with postoperative hypotension, could result in delayed paraplegia.

Despite the previously mentioned variability in interpreting changes in the tcmMEPs, we believe that significant changes may signal a reduction in the spinal cord blood supply and warrant immediate maneuvers to improve spinal cord perfusion. We also believe that tcmMEPs changes can be determined as significant only within the surgical and anesthesia context and only after correlation with factors such as the preoperative motor strength.

For example, we consider as significant changes sudden drops in amplitudes, occurring within 2–10 minutes, or sustained progressive drops, within 10–40 minutes. of more than 75% from baseline, or a necessity to increase the stimulus amplitude by more than 30 mA, in order to maintain a certain amplitude. A total loss of tcmMEPs, despite a maximum stimulus intensity of 200 mA, is considered a definite sign of pending postoperative deficit. However, these cutoff thresholds need to be interpreted within the surgical and anesthesia context and correlated with the patient's preoperative motor strength. For example, low-amplitude MEPs (e.g. 30 μV) were obtained at baseline at 180 mA, but several hours later no MEPs were obtained in the same muscles, despite an intensity of 200 mA. This situation is much less worrisome than a baseline MEPs amplitude at 3 mV, obtained at 100 mA, with loss of MEPs despite maximal stimulus. Moreover, because of fluctuations from one trial to another, measurements are done in several trials at a given time, and an averaged amplitude is compared between the results obtained at different times.

As mentioned on pages 16–26, mMEPs are known to be more sensitive to SCI than the SSEPs. This is because the alpha motor neuron interposed in the motor pathway is gray matter and more sensitive to ischemia than the large, myelinated fibers of the sensory system, traveling within the dorsal columns and tested by SSEPs. Thus, with SCI, one should expect changes first in the tcmMEPs, followed by changes in SSEPs. Changes in tcmMEPs caused by significant SCI are expected to occur fast, within few minutes (2–10 minutes usually). In addition to SCI, limb (peripheral) ischemia also causes changes in tcmMEPs. This is as a result of ischemia of the peripheral nerves and muscles. These changes occur at a slower pace, usually within 10–40 minutes or so and may be seen first in the most distal muscles (e.g. foot or hand). It is our experience that, in these cases, tcmMEPs changes usually lag behind changes in SSEPs, which tend to occur first. We believe this may be related to the difference in the size and myelin content of motor versus sensory fibers within the peripheral nerves.

Monitoring of the spine for detection of spinal cord injury can also be achieved by using SSEPs that offer the advantage over tcmMEPs of not being affected by paralytics and not causing body movements that could be quite bothersome for the surgeon. Earlier studies by Crawford et al. (1988) found a high incidence of false-positive SSEPs studies leading to the conclusion that SSEPs are unreliable for monitoring the spinal cord. However, more recently, Achouh et al. 2007) used three-channel SSEPs during TAA repair that excluded false-positive changes as a result of peripheral nerves or cerebral events, thus making SSEPs more reliable. Despite a low sensitivity (62.5%), SSEPs had a very high negative predictive value (99.2%), such that there was a low event probability rate of spinal cord injury in the absence of SSEPs changes. In the 20% of patients who experienced SSEPs changes, corrective measures aimed at restoring and optimizing oxygen delivery to the spine were performed, including reimplantation of additional patent intercostal arteries, lowering CSF pressure with drainage, increasing hemoglobin level, and increasing distal aortic perfusion pressure with volume or vasoconstrictors. Overall, SSEPs monitoring remains less encouraging than MEPs in predicting perioperative SCI. On the other hand, SSEPs do have an important role in signaling peripheral limb ischemia; thus, recording peripheral potentials, such as popliteal fossa and Erb's point potentials, is important. In cases of peripheral limb ischemia, all peripheral, subcortical, and cortical potentials are affected. As mentioned previously, the SSEPs changes related to limb ischemia may occur before the MEPs changes: a significant increase in the latencies is usually seen, followed closely by significant drop in amplitudes and loss of morphology. SSEPs may be useful in detecting cerebral ischemia as well, either in the middle cerebral artery (MCA) territory (ulnar or median SSEPs) or within the anterior cerebral artery (ACA) territory (posterior tibial SSEPs). However, when the subcortical structures are affected, both upper and lower SSEPs may be affected. In general, the cortical SSEPs show changes, whereas the cervical potential and the peripheral potentials are unchanged.

*Important steps in SSEPs/tcmMEPs monitoring*

1. *Pre-position and post-position baselines.*

As already mentioned in the previous sections, the SSEPs setup is done with the patient awake, whereas the set up for tcmMEPs is done after the induction takes place because of the need to place subdermal needle recording electrodes in the corresponding muscles, as well as corkscrew scalp electrodes for the transcranial electrical stimulation. Details about the setup technique can be found on pages 5–26. As a special note, for this surgical procedure, it is important that the SSEPs recording channels include peripheral popliteal fossa and Erb's point recordings, as well as subcortical (Cs5-Fz) and cortical channels (Cpc-Cpi, Cpc-A1, for upper limbs, and Cpz-Fz, Cpi-Cpc, for lower limbs SSEPs). For the tcmMEPs, we use hand (APB-ADM muscles) channel as reference and three lower limb muscle channels for active recordings (quadriceps, AT and aH muscles). Details about troubleshooting obtaining baseline recordings are found on pages 179–87.

2. *Opening.*

We usually do not continuously monitor during the opening time. However, we perform several trials whenever possible, in order to avoid big gaps between the trials.

3. *New baselines before starting the atriofemoral bypass.*

It is important to obtain new SSEPs and MEPs baselines, because SSEPs and tcmMEPs changes from the initial baselines, but strictly related to anesthesia, may have occurred. Insertion of the left femoral artery cannula can contribute to development of left lower limb ischemia, during the retrograde perfusion of the left lower limb. On the other hand, right lower limb ischemia may become a problem if an arterial line is inserted in the right femoral artery. Thus, following all of the evoked responses is important (Fig. **5D-8**). In such cases of lower limb ischemia, the posterior tibial SSEPs changes are progressive and occur relatively slowly, over 40 minutes or more. The changes are diffuse and involve all evoked responses, the popliteal fossa (PF), and subcortical and cortical potentials. The landmark is represented by marked increased in the absolute latencies of P37/N45, followed by significant changes in amplitudes and morphology. It is possible that the stimulation threshold necessary to obtain robust tcmMEPs has increased. From this moment on, no muscle relaxants or inhalational agents should be given, and the anesthesia should consist exclusively of TIVA (propofol and remifentanyl). Also, EC should be avoided. Our experience suggests that a temperature below 32°C may not be compatible with obtaining reliable tcmMEPs and SSEPs recordings.

4. *New baselines on atriofemoral bypass and proximal clamping.*

Once the proximal clamp is in place and the atriofemoral bypass functioning (see pages 242–43), we start using the tcmMEPs continuously for about 10 minutes. It is important to bear in mind that the retrograde perfusion of the spinal cord depends upon the mean distal blood pressure in the pump. An adequate mean distal blood pressure lies anywhere between 50 and 80 mmHg, with a mean around 60 mmHg, although there is some interindividual variability. Obtaining tcmMEPs helps us determine whether 50, 60, or 70 mmHg is the necessary value and thus guides the anesthesiologist's aims in maintaining for such a value. Again, what is a significant change varies from author to author. We mostly look at sudden decreases in amplitudes, which require sudden (within minutes) increases of 10–20 mA in the stimulus amplitudes. It is important to remember that changes of SCI can occur rapidly, sometimes within a couple of minutes. Thus, rapid successive stimulations, every half minute, are recommended. Inadequate distal mean blood pressure results in SCI and changes in tcmMEPs that occur quickly, within minutes (Fig. **5D-9**). Once it becomes clear that the distal mean blood pressure is adequate, decrease in the stimulation frequency every 5 minutes or so can be done. Sometimes, stimulation is limited by significant movements, present in the surgical field, that may disturb the surgeon. In this situations one has to rely on the judgment and discretion of the surgeon to resume stimulation. As a general rule, every time the distal mean blood pressure falls below the safe level, we obtain another tcmMEPs trial. Immediate placement of a second clamp, close to the first clamp, should not affect the evoked responses, because the aortic segment between the two clamps is very small. The first anastomosis is located in between the two clamps (see pages 242–43).

5. *Clamping great vessels at the aortic arch.*

In the case when the aortic arch is also involved in the aneurysm, clamping the subclavian arteries, while taking out of the circuit the segmental arteries, can expedite the occurrence of SCI because of interruption of the proximal collateral flow—in this case, the tcmMEPs changes are likely seen before the SSEPs changes and occur abruptly.

Clamping of the subclavian artery can also cause upper limb ischemia; from our experience, in this

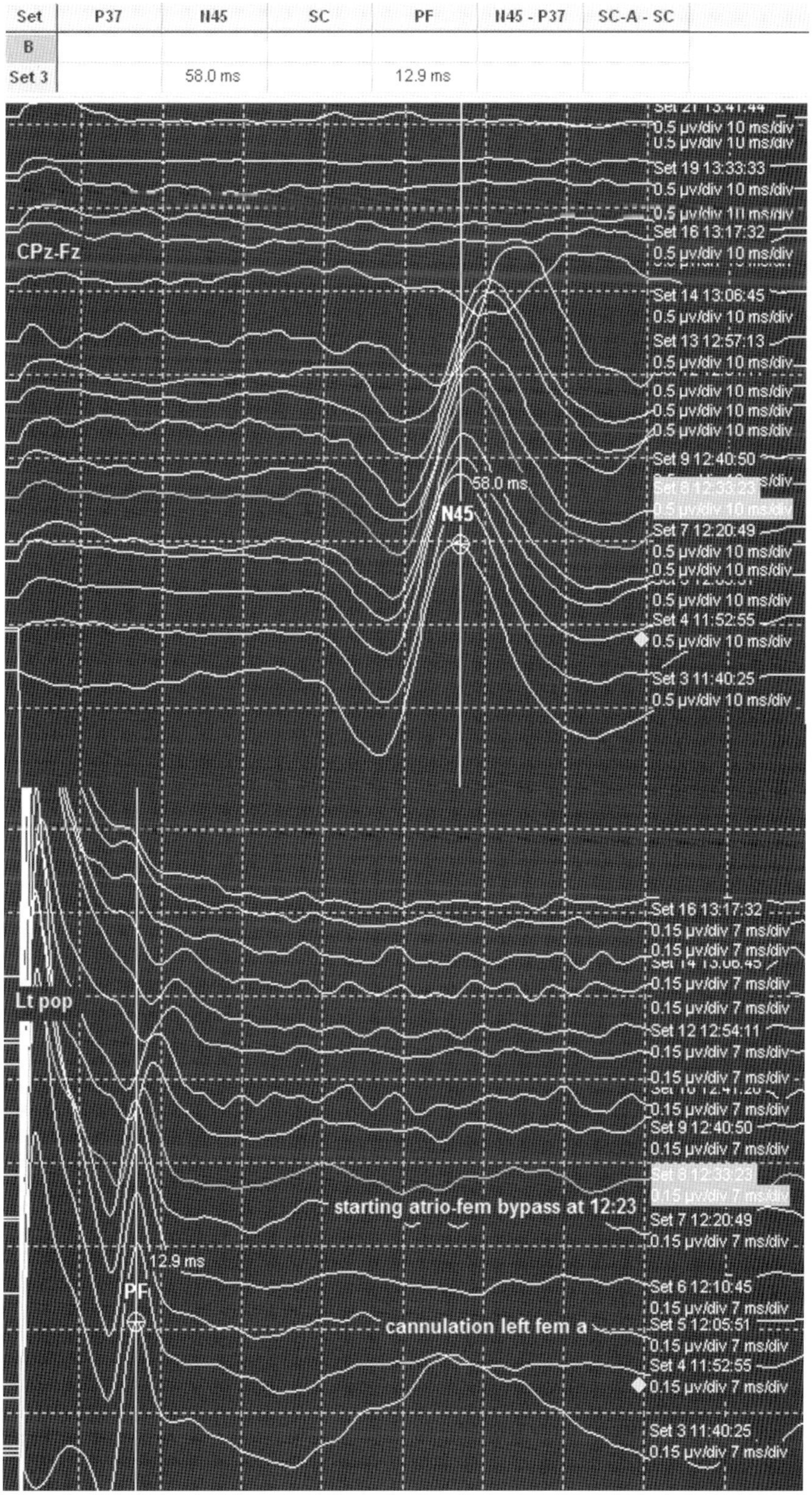


**FIGURE 5D-8** Left lower limb ischemia, developing on retrograde flow maintained via left atriofemoral bypass. Notice the progressive changes in the left posterior tibial SSEPs: the attenuation to disappearance of popliteal fossa potential (PF) and the significant increase in absolute latency, followed by decrease in amplitudes and change in morphology of P37/N45. After the bypass is started, the left posterior tibial SSEPs are lost within 40 minutes.

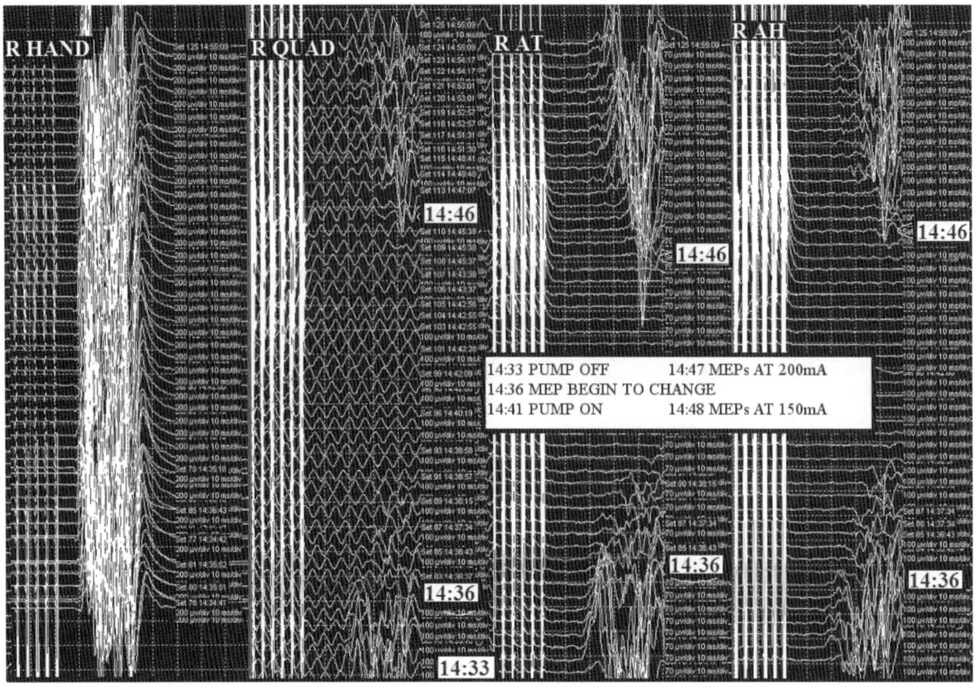


**FIGURE 5D-9** SCI when the bypass pump was turned off for a few minutes. Notice the sudden loss of right lower limb tcmMEPs (quadriceps, AT and AH), while the hand MEPs are unchanged. The changes occurred within 2–3 minutes after the pump was turned off. Definite recovery started to be seen within several minutes after the pump was turned back on.

case, the SSEPs changes precede the tcmMEPs changes and occur at a slower pace than in the case of SCI (see Fig. **5D-10**).

There are situations when changes in hemodynamics that occur with TAA surgery, particularly involving the aortic arch, may cause significant cortical ischemia (e.g. sudden hypotensive episodes, clamping of carotid system, resulting in low cerebral flow or thromboembolic events). The occurrence of significant cortical SSEPs changes (see Fig. **5D-11**), in the presence of robust cervico-medullary potential and PF, may signal such adverse events. These changes must be correlated with the particular surgical step in which they occurred. If cortical ischemia is suspected, prompt increase in the systemic proximal blood pressure is important. Thus, adding EEG monitoring to cases involving the aortic arch may be helpful. However, in order to assess EEG changes reliably, one has to ensure that no burst-suppression pattern is present.

6. *Clamping distal to the aortic portion thought to contain the segmental arteries.*

This is a critical step during surgery, and the surgeon often requests frequent tcmMEPs. Again, just before moving the clamp down in this position, we often run tcmMEPs as our new baseline. Once the third clamp is on, we stimulate every half minute or so for about 10 minutes. If no sudden drops (more than 50%) are seen from one set to another, we clear the region, but we continue monitoring the tcmMEPs. If sudden changes have occurred, the surgeon is informed (Fig. **5D-12**). The anesthesiologist further increases the pressure in the bypass, and the surgeon proceeds to identify the opening of segmental arteries in the aortic wall and reimplants them in the graft. Some of these openings are easily recognized because of the presence of backbleeding (see pages 242–43). It becomes clear that these arteries are open, and the surgeon attempts to reimplant them in a stepwise fashion. Sometimes it is not necessary to reimplant all segmental vessels. We have had cases when reconnecting only some of them reversed the tcmMEPs changes. However, in other situations, the opening of the segmental artery that is important and that needs to be reimplanted is not easily identified, for example as in the case of an obstructing intramural thrombus. In such cases, especially in the presence of EC, it would have not been possible to know whether SCI is still present

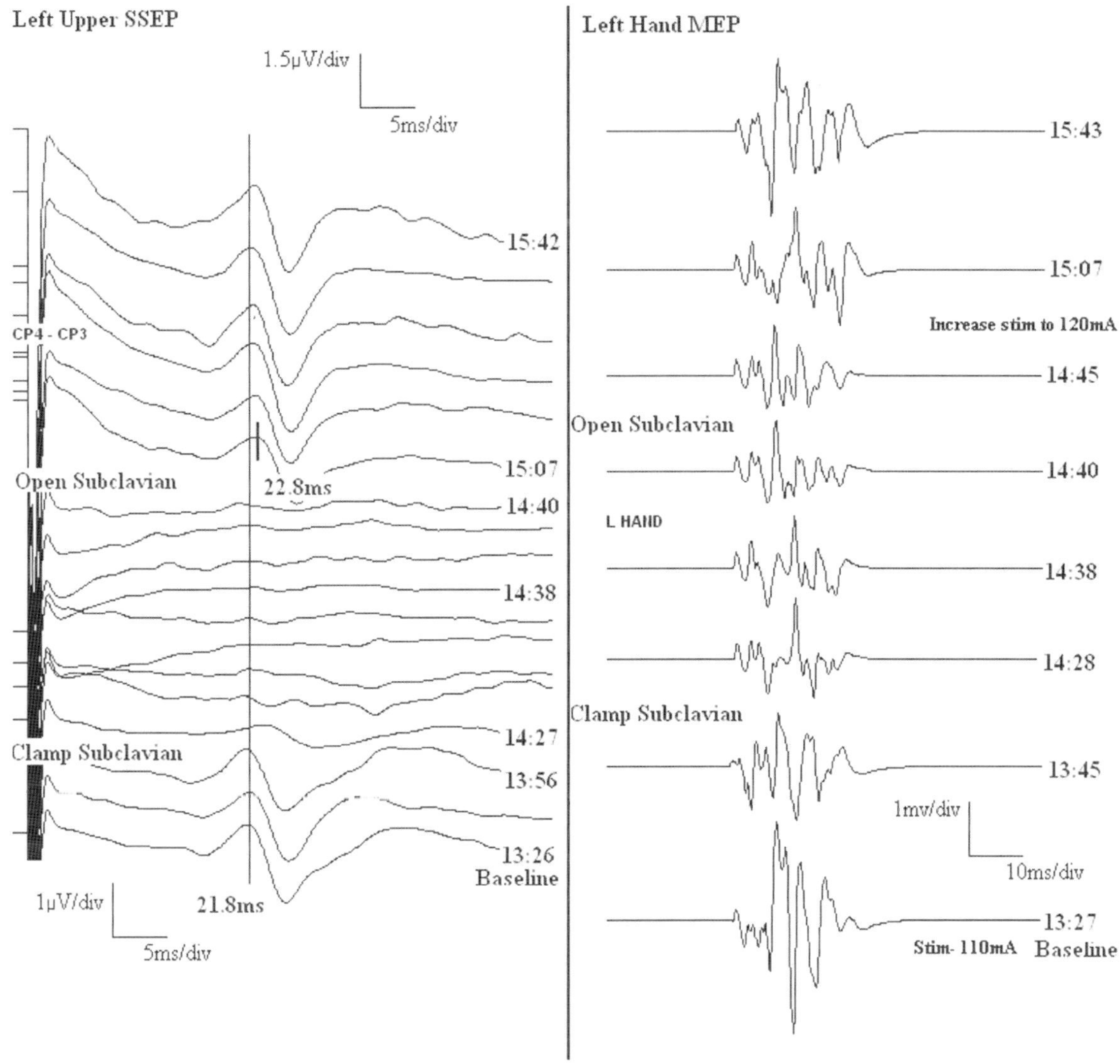


**FIGURE 5D-10** Left upper limb ischemia as a result of clamping of the left subclavian artery. Notice the dramatic changes in the left median SSEPs; the total disappearance of N19/P22 occurs at about half an hour into the clamp; similarly, after the artery is unclamped, it takes about half an hour for the changes to reverse. Also notice the less dramatic changes in the left hand MEPs.

without continuous tcmMEPs. Persistence of tcmMEPs suggests to the surgeon that there are still segmental arteries to look for.

*7. Clamping distal to the origin of the pelvic visceral arteries.*

For reconstruction of the portion of the aorta containing the origin of the pelvic visceral vessels, the clamp is moved distal to it. In order to maintain the perfusion of the pelvic organs, the retrograde flow is redirected toward them via the perfusion catheters; with or without clamping of the femoral catheter (see Fig. **5D-3B**). Opening the femoral catheter ensures some retrograde perfusion to the legs.

This is the moment when lower limb ischemia can frequently occur because of interruption of appropriate blood supply to the lower limbs. This causes diffuse ischemia of both muscles and peripheral nerves. The degree of limb ischemia depends on the position of the distal clamp, above or at the level of the common iliac arteries, with or without open femoral catheter (see previous discussion). The vascular injury can be acute

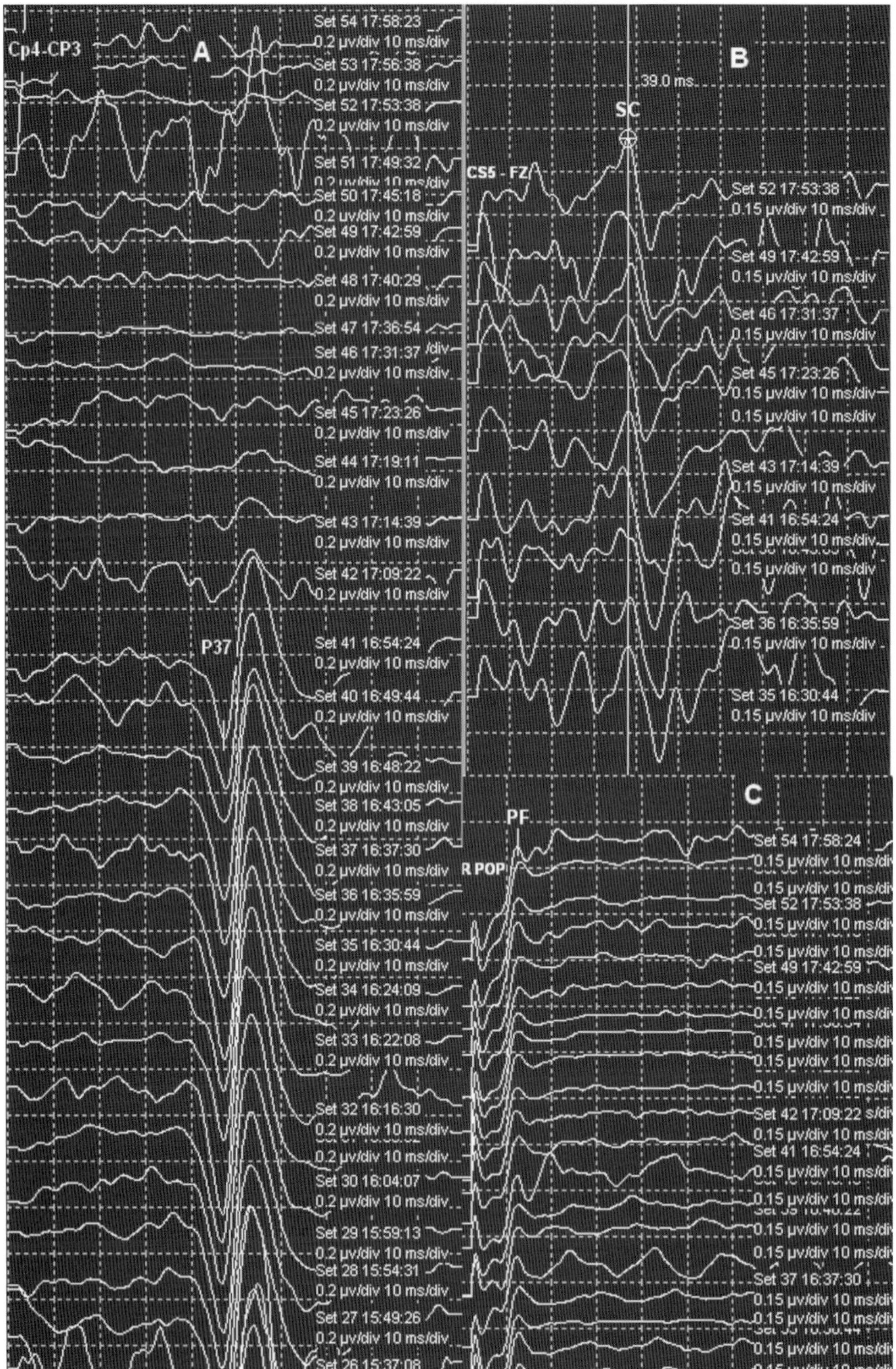


**FIGURE 5D-11** Left hemispheric cortical ischemia during aortic arch aneurysm repair, likely the result of a thromboembolic event. Panel **A** shows a sudden change to total loss of the cortical right posterior tibial SSEPs (P37/N45), which occurred within minutes. Panel **B** shows relatively preserved subcortical posterior tibial SSEPs (N30). There is a decrease in the signal-to-noise ratio in this channel, likely a result of significant muscle tension artifact (the use of muscle relaxant is avoided, in order to obtain reliable tcmMEPs). However, the subcortical potentials are present and reproducible. Panel **C** shows unchanged peripheral PF potentials. The change in cortical SSEPs, with preservation in the subcortical and peripheral potentials, supports the diagnosis of cortical ischemia.

(a)

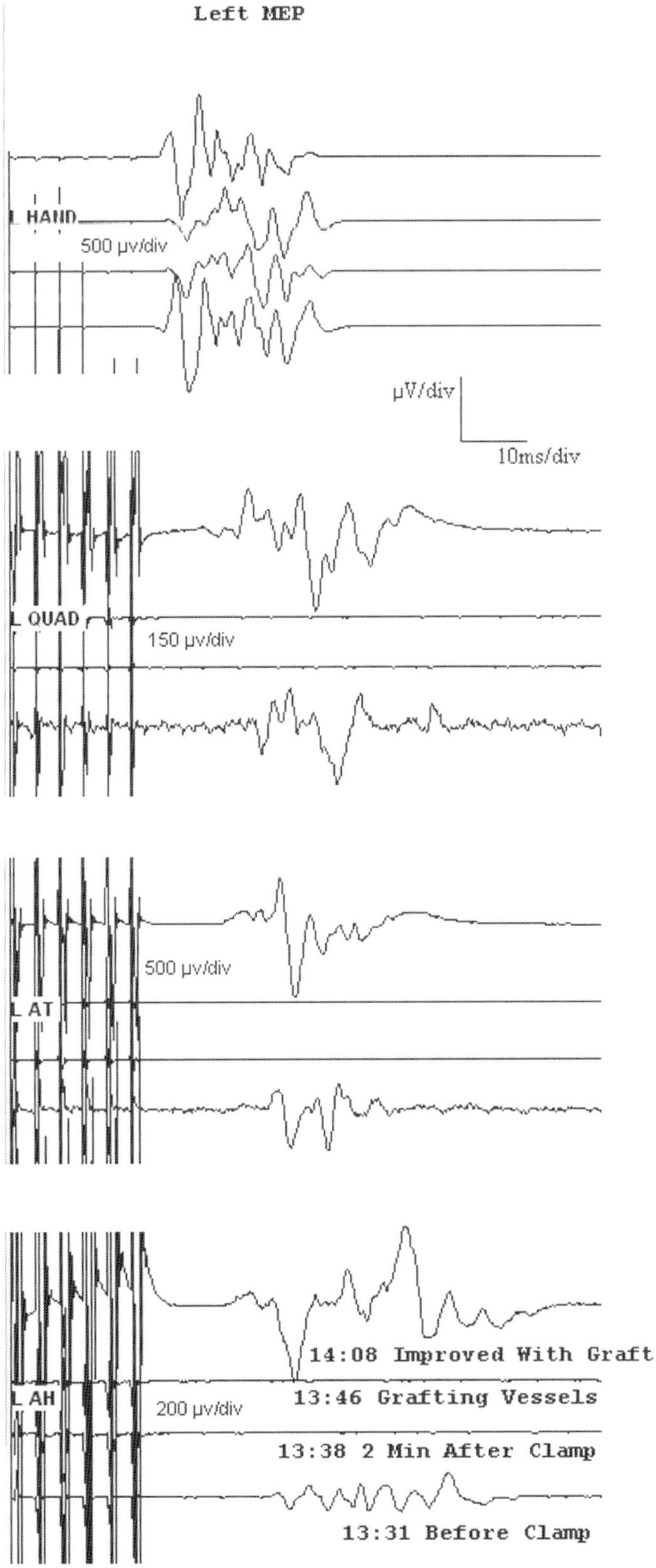


**FIGURE 5D-12** Two examples of SCI resulting from cross-clamping of the aorta distal to the origin of critical segmental arteries. SCI is detected by sudden changes in the lower limbs tcmMEPs (**a**, left; **b**, right). In such situations, the mean distal blood pressure is increased with the hope of improving the distal collateral flow to the spinal cord.

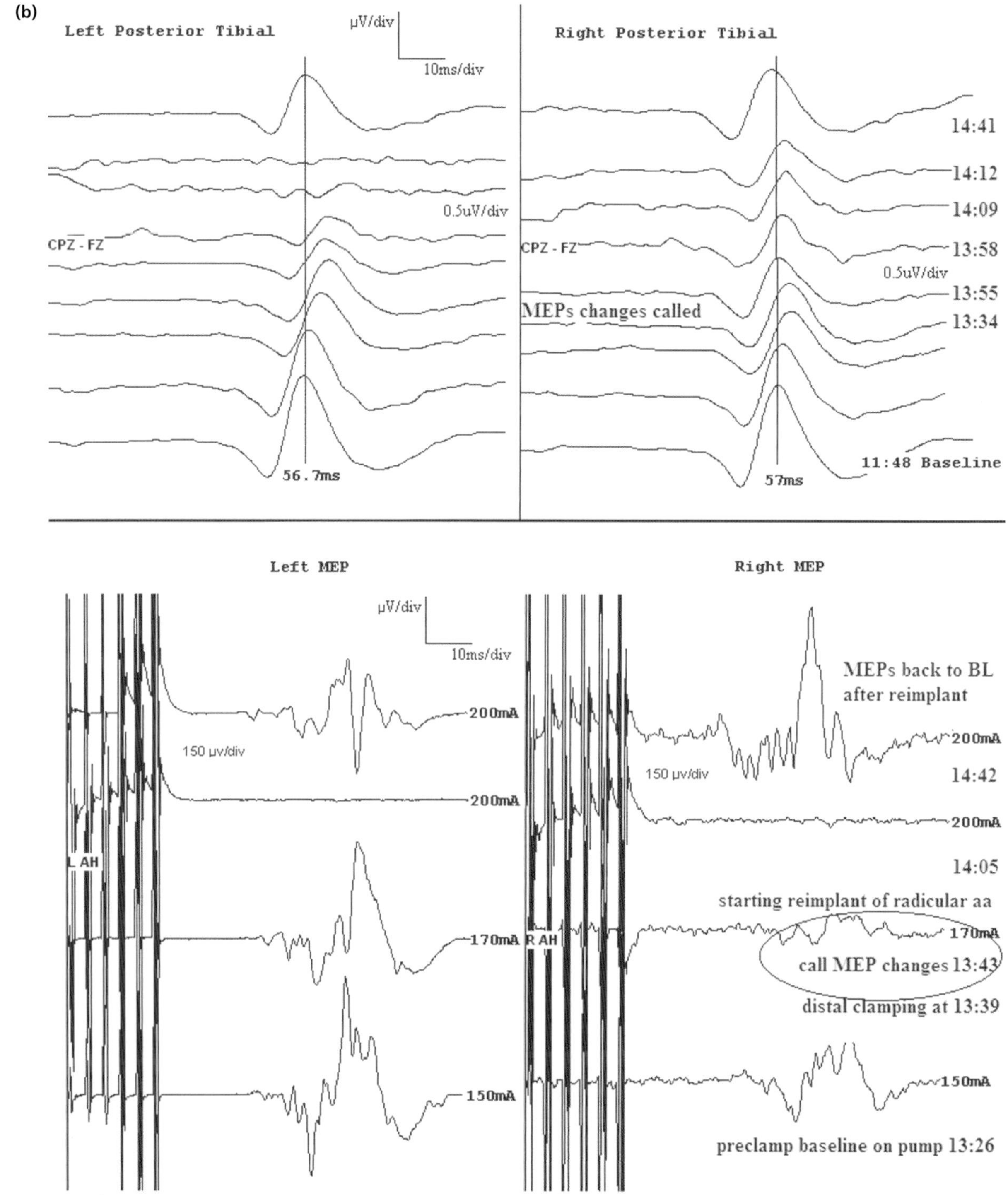


**FIGURE 5D-12** *(Continued)* Guided by the presence of backbleeding, the surgeon searches for patent segmental arteries (where backbleeding is present), with the goal of reattaching them to the graft. If after doing this, no improvement in the MEPs is seen, the surgeon continues to look for more segmental arteries, some of which may be masked by a thrombus or atheroma. Most of such vessels are already obstructed. However, others may retain function. In **b**, notice the disproportionate changes seen between right lower limb tcmMEPs and SSEPs.

and relatively serious. The neurophysiologic changes can occur quickly, faster than in the slower developing ischemia because of cannulation of the femoral artery. As with the case of limb ischemia caused by arterial cannulation, SSEPs changes tend to occur first—with tcmMEPs changes lagging behind—are diffuse, and are remarkable for early and significant increase in the absolute latencies of the cortical potentials (see Fig. **5D-13**). The degree of limb ischemia and the thus of the neurophysiologic changes can be asymmetric between the two legs, depending on preexisting hypoperfusion and on the exact location of the clamp.

The tcmMEPs changes also involve all of the lower limbs muscles, and the AH MEPs changes may be more significant than in MEPs changes corresponding to more proximal muscles. These changes usually resolve once the clamps are removed.

However, there are cases when the hypogastric (internal iliac) arteries contribute to the spinal cord vascularization. In such case, clamping of the iliac arteries may, in fact, cause SCI. Thus, tcmMEPs rather than SSEPs changes occur first and develop more rapidly than in the case of limb ischemia.

8. *Closing.*

Once all of the clamps and the atriofemoral bypass are removed, the most critical part of the surgery being over, one may be tempted to stop the neuromonitoring as well. However, we like to continue monitoring during closure to ensure that there is no pattern of decrease in MEPs and to alert the anesthesiologist if we notice alarming decreases in tcmMEPs associated with smaller systemic blood pressure. This type of information can be very valuable in the immediate postoperative period, when one wants to ensure that the systemic blood pressure offers adequate spinal blood flow supply, particularly when the collateral flow of the spinal cord may be challenged in new ways.

Table **5D-1** gives examples of different types of neurologic insults detected by SSEPs, tcmMEPs, and EEG monitoring during TAA surgery and their mechanisms of occurrence.

***f. Cerebrospinal fluid drainage.*** The rationale for the use of CSF drainage is based upon animal evidence indicating that a decrease in CSF pressure to less than 10 mm Hg during clamping of the thoracic aorta enhances spinal cord perfusion and maintains medullary perfusion pressure, which is protective against ischemic injury (Berendes et al., 1982; Bower et al., 1989; Griffiths et al., 1978; Miyamoto et al., 1960). A spinal drain is placed preoperatively, which is then used preemptively to drain CSF intraoperatively and can be continued during the initial 48-hour postoperative period, as needed, to maintain a spinal pressure of 10 mm Hg or less. Although the use of CSF drainage to prevent paraplegia has been suggested on the basis of physiologic aspects of aortic cross-clamping as well as animal evidence, the effectiveness of this intervention remains controversial (Ling et al., 2000; Cina et al., 2004). Wherein a randomized trial and two cohort studies did not find CSF drainage to be effective in preventing paraplegia, the highest volume of CSF drained was 50 mL; a significant number of patients had CSF pressure greater than 10 mm Hg during the study, and CSF drainage was continued only during surgery and not in the postoperative period (Murray et al., 1993; Crawford et al., 1991; Coselli et al., 2002). Alternatively, a recent meta-analysis determined that the number of patients needed to treat with CSF drainage to prevent one case of paraplegia was 10 (Cina et al., 2004). Coselli et al. (2002) demonstrated the usefulness of CSF drainage in a prospective randomized study, and Safi et al. (1994) reported that delayed paraplegia was reversed with aggressive CSF drainage after recognition of neurologic deterioration.

*B2. Neuroprotective techniques and adjuncts*

***a. Regional spinal cord hypothermia.*** Based upon favorable experimental evidence and the concept of regional spinal cord hypothermia, Davison et al. (1994) devised a method to provide moderate hypothermia directly to the region of the spinal cord that is at risk for ischemic injury during TAA repair, and this group has been using EC for types I to III TAA repairs since 1993 (Conrad et al., 2007). Their initial experience included only a small number of patients at greatest risk for paraplegia (i.e. type I and II TAAs) (Davison et al., 1994), but a subsequent study included 170 patients treated with EC, which represented 77% of all cases (Cambria et al., 2000). The technique involves the placement of a spinal/epidural catheter at the T10 to T12 level and advanced cephalad, which permits the CSF temperature to be cooled, in anticipation of aortic cross-clamping, to a target of 25–27°C with the continuous infusion of iced (4°C) saline. After reperfusion of the lower extremities, EC is stopped, and continuous passive CSF drainage is initiated and continued for 48 hours. CSF temperature and pressure are simultaneously monitored with a separate intrathecal catheter in the subarachnoid space at the L3–4 interspace, and a gradient of 30 to 40 mm Hg was

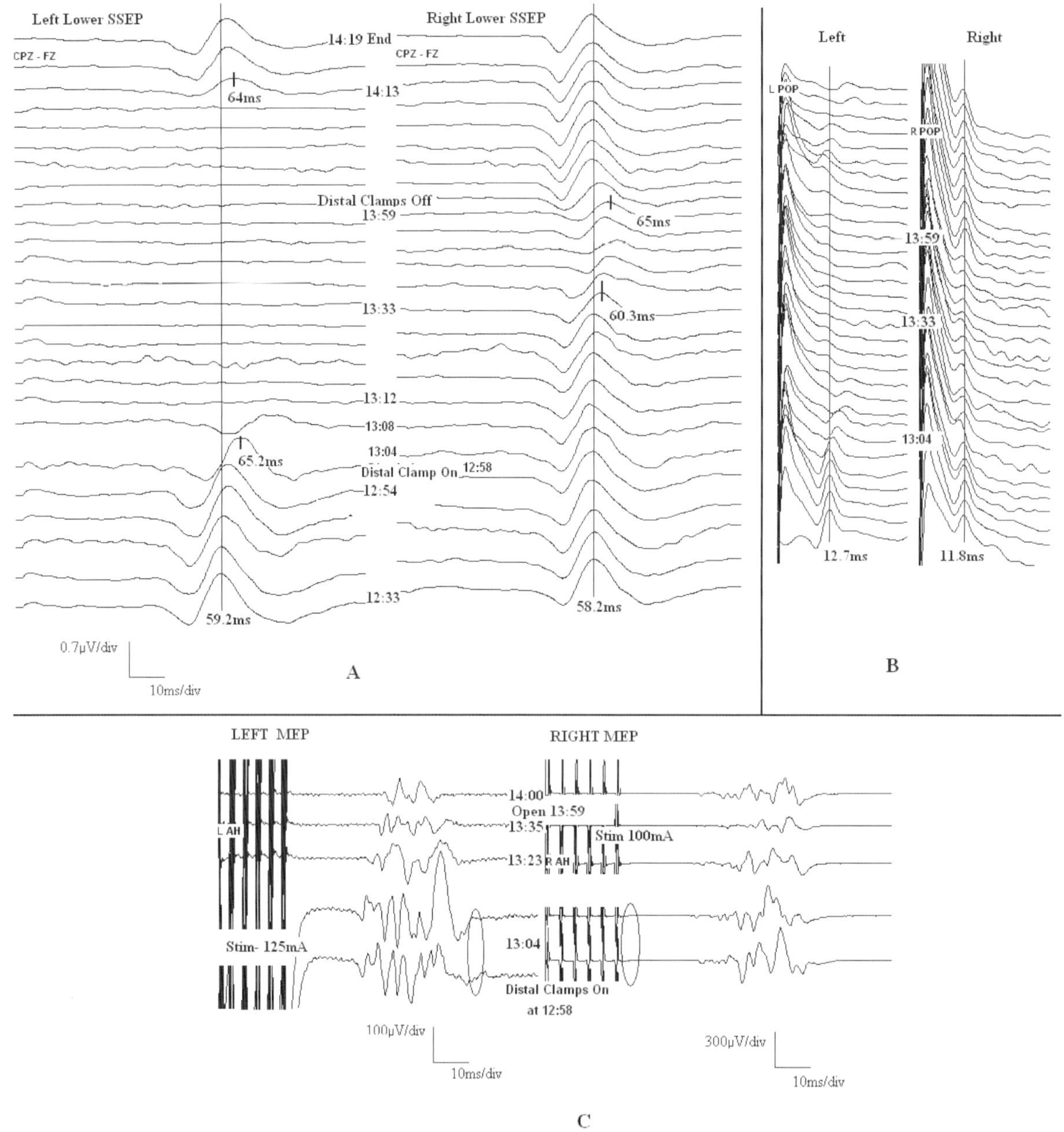


**FIGURE 5D-13** Lower limb ischemia resulting from the clamping of the common iliac arteries. Notice significant diffuse changes, left more than right, in the posterior tibial SSEPs, including attenuation to disappearance of the cortical P37/N45 potentials (panel **A**), as well as of the popliteal fossa potential (panel **B**). The changes are similar to those occurring during cannulation of the left femoral artery, with retrograde blood flow to the left leg. However, they tend to occur more rapidly, seen as early as 5 minutes after the clamping and being quite dramatic within 15 minutes. The changes in lower limbs mMEPs are less significant and somewhat lag behind the changes in the posterior tibial SSEPs (panel **C**).

**TABLE 5D-1.** Neurophysiologic changes expected with different mechanisms of neurologic insults during TAA surgery

| Type of insult | Mechanism | Neurophysiologic changes | Action |
|---|---|---|---|
| Limb ischemia, less severe | Cannulation left femoral a. for bypass (esp. when on pump)<br>Cannulation right femoral a. for arterial line (esp. when on pump)<br>Wire insertion in femoral a. for endovascular graft | Diffuse posterior tibial SSEPs changes: PF, N30, and P37/N45<br>Early increase in the absolute latencies, followed by decrease in amplitude and change in morphology to total loss of SSEPs ~40–45 min, changes as early as 10 min<br>SSEPs changes more dramatic and earlier than MEPs changes in the muscles in the same limb<br>Asymmetric | Reposition cannula, wire<br>Increase distal mean BP in the pump |
| Limb ischemia, more severe | Clamping of common iliac aa. and femoral cath.<br>Clamping of subclavian a. in aneurismal repairs involving the aortic arch | Same changes as above in posterior tibial SSEPs (or in median/ulnar SSEPs: Erb's, P/N13, and N19/P22)<br>Changes occurring faster; total loss of SSEPs can occur in several minutes<br>Diffuse or asymmetric<br>SSEPs changes>MEPs changes | Try to reposition clamp<br>Increase distal mean pressure in the pump |
| Spinal cord ischemia | Aortic cross-clamping w/ inadequate distal mean BP in the pump<br>Aortic cross clamping, eliminating critical segmental arteries<br>Additional cross clamping of subclavian a. or hypogastric a.<br>Endovascular graft obstructing the segmental aa | Sudden changes, with loss of lower limb MEPs, within minutes (as early as 2 min)<br>Can have changes in the upper limbs MEPs( cervical spine ischemia—clamping of the subclavian a.)<br>Can be asymmetrical<br>Invariably, MEPs changes precede and are more dramatic than SSEPs changes | Increase distal mean BP in the pump<br>Proceed with identification and reimplantation of critical segmental aa<br>Reposition clamp or graft |
| Cortical ischemia | Clamp great vessels of aortic arch<br>Thromboembolic event because of wire manipulation in TEVAR<br>Significant hypotensive episode | Sudden (minutes) attenuation to disappearance of cortical SSEPs (N19/P22, P37/N45, or both)<br>Preservation of subcortical and peripheral potentials<br>Sudden diffuse loss of tcmMEPs, especially in cortical strokes<br>Mostly asymmetric<br>EEG monitoring very useful: ipsi > bilateral attenuation of the faster frequencies and slowing | Increase in the proximal mean BP<br>Change clamp position<br>Consider shunting<br>Deep burst suppression as neuroprotective measure (however, not compatible with EEG monitoring) |

maintained between the mean arterial and CSF pressures (Fig. **5D-5**). With this technique there was an overall incidence of SCI of 7% (12 of 170), with the vast majority of these occurring in type I and II TAAs. Two-thirds of the deficits occurred in a delayed fashion, and only three cases resulted in devastating permanent paraplegia (Cambria et al., 2000). In a comparative study of TAA repair with and without EC, in cases performed between 1987 and 2001, over half of the 337 patients had EC, which reduced the risk of spinal cord injury significantly compared to no EC (10.6% vs. 19.8%) (Cambria et al., 2002). A potential limitation of EC is that CSF pressure rises during cooling. However, the cord is cooled at the critical anatomic level during cross-clamping, and the neuroprotective effect of this appears to outweigh the potential disadvantages of increased CSF pressure. The neuroprotective effect from moderate hypothermia is preferred, because, despite the exact operative technique, some degree of obligatory ischemia interval of the critical intercostal segment (T9–L1) occurs with resection of type I and II and most type III TAAs.

Unfortunately, using EC renders neurophysiologic spinal cord monitoring impossible. This is based on the fact that, at low temperatures, usually below 32°C, the SSEPs and tcmMEPs are significantly affected and become unreliable.

We advise against the use of EC in favor of SSEPs/tcmMEPs monitoring. Even though it is considered a neuroprotective measure, EC may not prevent a neurologic postoperative deficit from occurring. In fact, it renders impossible the appreciation of the effect of selective clamping on the spinal cord neurophysiology. Although allowing en bloc reimplantation of the intercostal arteries, it makes it impossible to distinguish between selective roles of different intercostal arteries and thus the need for selective reimplantation.

***b. Preconditioning.*** Preconditioning as a means of protecting the spinal cord has been described (Toumpoulis et al., 2003, 2004). The spinal cord of a pig was initially exposed to a brief period of ischemia, to increase spinal cord tissue tolerance to subsequent episodes of ischemia. Experimentally they compared early ischemic preconditioning to late ischemic preconditioning (80-minute reperfusion interval versus 48 hours) and discovered that early preconditioning was superior both to late and to no exposure to mild ischemia. The benefit of early ischemic preconditioning has been confirmed by others in rabbit and rat models (Zvara et al., 1999; Fan et al., 1999).

***c. Neuroprotective agents.*** Pharmacologic agents have been administered, either systemically or via the intrathecal route, to both humans and in animal models in an attempt to prevent neurologic injury in thoracoabdominal aneurysm repair. Svensson et al. (1998) demonstrated a combined benefit of CSF drainage and dilation of vessels with intrathecal papaverine, with a resulting reduction in the incidence and severity of neurologic injury. Archer et al. (1994) described similar results with a combination of CSF drainage and intrathecal naloxone. With regard to animal models, Cassada et al. (2001) found that an adenosine $A_{2A}$ receptor agonist (ATL-146e) with known anti-inflammatory properties resulted in spinal cord functional protection from ischemia in rabbits. This suggests that at least one component of spinal ischemic injury results from a leukocyte-mediated reperfusion response during the critical post-ischemic time interval. Pharmacologic protection has also been confirmed in a porcine model, with either retrograde or systemic administration of this same adenosine analogue (Reece et al., 2006). Despite the success of numerous pharmacologic agents in attenuating ischemic spinal cord injury, their use has not yet become widespread in clinical practice. We do not recommend the use of intravenous steroids, such as methyl-prednisolone, in acute ischemic injury of the spinal cord. In fact, according to some authors, this may be harmful (Adams et al., 2007).

## 3. TEVAR

Endoluminal treatment of isolated TA with stent grafting was introduced in 1994 by Dake et al. (1994). Since then, comparative (versus open repair) trial data are emerging for the spectrum of thoracic aortic pathology (Bavaria et al., 2007; Cho et al., 2006; Greenberg et al., 2005; Leurs et al., 2004); Makaroun et al., 2005). Despite a lack of late results from clinical trials and a paucity of level I data from randomized controlled trials, the commercial availability of an endovascular stent graft has caused the treatment paradigm for a variety of thoracic aortic pathologies to evolve toward stent-grafting strategies. TEVAR offers the benefit of aneurysm exclusion without the physiologic insult associated with thoracotomy and cross-clamping of the proximal aorta. The procedure is conceptually simple, involving the insertion of a presized, covered, stent graft that is deployed under fluoroscopic guidance (Fig. **5D-14**). Thus, the graft is passed over guide wires, through the femoral artery, accessed at the groin, and up to the region of the aorta that needs to be treated. Once the graft is in place, ballooning the areas situated proximal and distal to the region of the aorta

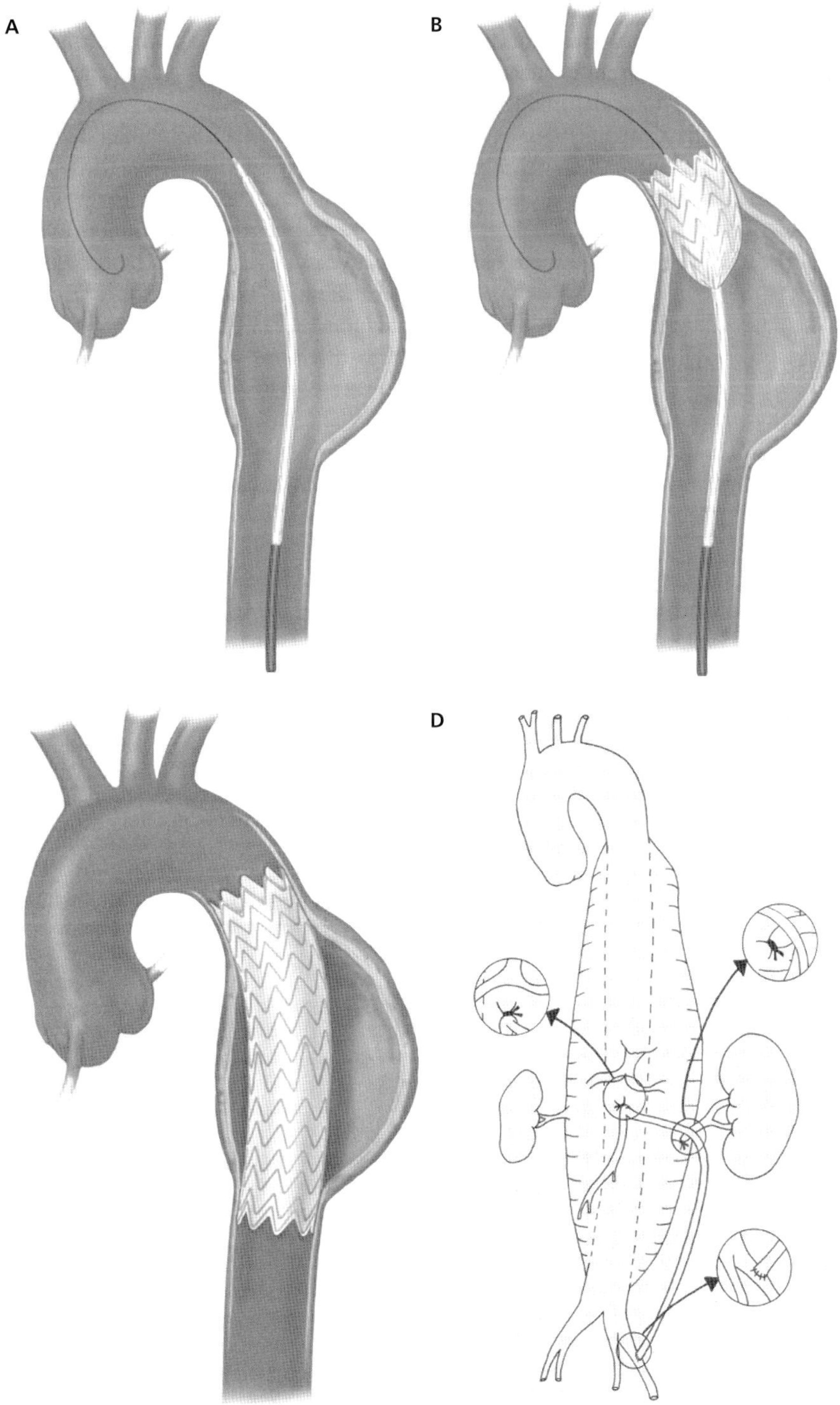


**FIGURE 5D-14** Panels **A** through **C**: cartoon depiction of endovascular repair of descending TA. **A**: Descending TA, with adequate proximal and distal seal zones. A stiff wire has been advanced to the level of the aortic valve, and the device is positioned under fluoroscopic guidance; **B**: Initiation of device deployment; **C**: The completed endovascular repair; **D**: Because of anatomic constraints of a 2-cm proximal and distal seal zone, TEVAR is precluded for most TAA. Extension of stent graft repair to more extensive aneurysms, using a hybrid operation, involves surgical bypasses to the renal and visceral vessels and coverage of the aneurysm, including the origin of these vessels with a stent graft.

**TABLE 5D-3** SCI and mortality in TEVAR clinical trials

| | | | 30-day Mortality (%) | | | Paraplegia (%) | | |
|---|---|---|---|---|---|---|---|---|
| Reference | N (Endo) | N (Open) | Endo | Open | p-value | Endo | Open | p-value |
| Makaroun: Gore TAG (2005)[81] | 140 | 94 | 2.1 | 11.7 | 0.004 | 3.0* | 14.0* | 0.01 |
| Kwolek: Medtronic Valor (2006)[83] | 195 | – | 2.1 | – | – | 1.5 | – | – |
| Matsumura: Zenith TX2 (2008)[84] | 160 | 70 | 1.9 | 5.7 | <0.01 | 1.3 | 5.7 | 0.01 |

*Includes total SCI rate.

including the most recent publications, indicate that the mortality of TA/TAA repair remains in the 7% to 10% range (Table **5D-3**) (Conrad et al,. 2007; Jacobs et al., 2006; Coselli et al., 2004; Estrera et al., 2001; Safi et al., 2003; Svensson et al., 1993; Estrera et al., 2003; Galloway et al., 1996; Verdant et al., 1995; Yamauchi et al., 2006; Coselli et al., 2007; Grabitz et al., 1996). The overall incidence of SCI has dropped dramatically during this last decade with the use of these adjuncts (distal aortic perfusion, CSF drainage, intercostal artery reimplantation, and EC) (3.3%–12%) but remains a major issue, especially in patients with more extensive type I and II lesions (Conrad et al., 2007; Jacobs et al., 2002; Coselli et al., 2004; Estrera et al., 2001; Safi et al., 2003; Svensson et al., 1993; Estrera et al., 2001; Galloway et al., 1996; Verdant et al., 1995; Yamauchi et al., 2006; Coselli et al., 2007; Grabitz et al., 1996). The results summarized in Table **5D-3** represent best-case scenarios from centers of excellence, and the real-world experience has not been as optimistic. Rigberg et al. (2006) evaluated over 1,000 cases from a California medical database and found a 30-day mortality of 19% that increased to 31% at 1 year. Cowan et al. (2003) reviewed over 1,500 patients from the National Inpatient Sample (NIS) and found an overall mortality for TAA repair of 22.3%, with higher-volume centers having better outcomes. In contrast, a recent update of our experience over a period of 20 years had an overall operative mortality of 8.2% (6.8% elective and 12.9% emergent) and a paraplegia rate of 9.5% (Conrad et al., 2007). EC has reduced the overall incidence of SCI and also appears to minimize the overall severity of the injury, but the problem has hardly been eliminated. Jacobs et al. (2006) showed that the use of intraoperative MEPs, as a guide to adequacy of spinal cord perfusion, could reduce the actual paraplegia rate to 4.2% in types I and II TAA; however, the early mortality rate of 13.4%, possibly secondary to increased operative time and complexity of revascularizing threatened portions of the spinal cord, is a huge price to pay.

Late neurologic events can occur from hours to weeks after TAA repair. Factors commonly associated with the delayed onset of a neurologic deficit include an unstable arterial blood pressure (mean <60 mm Hg), hypoxemia, low hemoglobin, and an increased CSF pressure (Estrera et al., 2003; Azizzadeh et al., 2003). These conditions overcome the ability of the collateral flow to maintain a reasonable and safe blood supply to the spine. Estrera et al. (2003) found late events in 2.7% of patients in their series. This group also reported that additional preoperative risk factors, responsible for delayed neurologic events by multivariate analysis, included acute dissection, type II TAA, and renal insufficiency.

As already mentioned, maintaining some of the neuroprotective measures beyond the clamp and in the immediate postoperative period has become essential. Thus, postoperative management hinges upon maximizing spinal cord perfusion, by maintaining the mean arterial pressure at approximately 90 mm Hg and continuously draining CSF for 48 hours, to maintain the pressure at 10 mm Hg. Patients are usually extubated on the first postoperative day and spend three to four days in the intensive care unit.

## V. CONCLUSION

Although spinal cord injury remains a devastating complication of open repair of the thoracic aorta, the rate of paraplegia has decreased to the 5–10% range with the use of current adjuncts and surgical techniques. Most SCI occurs in patients with type II TAAs, and this must be factored into operative planning and preoperative discussions with patients. In summary, an efficient surgical approach to the thoracic aorta should give special attention to both minimizing aortic clamp time and performing adjunctive measures to protect the spinal cord, cerebral hemispheres, and limbs from ischemic injury. Over the past several years, the use of intraoperative neurophysiology, SSEPs and, in particular, tcmMEPs monitoring gained a significant role in reducing the total clamp time by (1) expediting the detection of the segmental arteries in need for reimplantation; (2) alerting the anesthesiologist when the mean distal blood pressure in the bypass is below the level necessary to maintain good spinal cord perfusion; (3) identifying limb ischemia, resulting from direct clamping (e.g. iliac artery or subclavian artery clamping) or endovascular obstruction of the origin of the corresponding arteries (in TEVAR) by the endovascular graft; (4) identifying limb ischemia, resulting from cannulation of the left femoral artery for atriofemoral bypass placement or graft insertion in TEVAR; (5) identifying limb ischemia from insertion of a right femoral artery arterial line; (6) identifying cerebral ischemia caused by clamping vessels at the aortic arch, by obstruction of their origin by the endovascular graft (TEVAR), or by an embolic event (e.g. dislocation of a plaque in the aortic arch by the wire in TEVAR).

There is real risk of delayed paraplegia; thus protective efforts must continue into the postoperative period. We believe that intraoperative neurophysiology can play a significant role in minimizing the postoperative deficits—not only in open TAA repair but also in the endovascular repair of the thoracic aneurysm; thus we encourage its use in both surgical techniques.

Despite considerable progress, no single operative strategy to date has been totally effective in eliminating the problem of SCI and the resultant substantial morbidity that follows TAA repair; we recommend the use of a combination of techniques and methods to assess and maintain appropriate spinal cord perfusion and to increase the resistance of spinal cord tissue to ischemic injury.

## REFERENCES

Achouh PE, Estrera AL, Miller III CC, et al. Role of somatosensory evoked potentials in predicting outcome during thoracoabdominal aortic repair. Ann Thorac Surg 2007;84:782–788.

Adamkiewicz A. Die Blutegefasse Des Menschlichen Ruckenmarkes. Sitzungsberichte Akademie der Wissen Schaften in Wein—Mathematische-Naturwissen Schaftliche Klasse—Abteilun 1882;84:101–130.

Adams HP Jr, Del Zoppo G, Alberts MJ, et al. Guidelines for the early management of adults with ischemic stroke: A guideline from the American Heart Association/American Stroke Association Stroke Interventional Council, and the Atherosclerotic Peripheral Vascular Disease and Quality of Care Outcomes in Research Interdisciplinary Working Groups: The American Academy of Neurology affirms the value of this guideline as an educational tool for neurologists. Stroke 2007;38:1655–1711.

Archer CW, Wynn MM. How we do it. Cardiovasc Surg 1999;7:593–596.

Archer CW, Wynn MM, Hoch JR, et al. Combined use of spinal fluid drainage and naloxone reduces risk of neurologic deficit in the repair of thoracoabdominal aneurysms. J Vasc Surg 1994;19:236–248.

Azizzadeh A, Huynh TT, Miller CC III, et al. Postoperative risk factors for delayed neurologic deficit after thoracic and thoracoabdominal aortic aneurysm repair: A case-control study. J Vasc Surg 2003;37:750–754.

Baril DT, Carroccio A,Ellozy SH, et al. Endovascular thoracic aortic repair and previous or concomitant abdominal aortic repair: Is increased risk of spinal cord ischemia real? Ann Vasc Surg 2006;20(2):188–194.

Bavaria JE, Appoo JJ, Makaroun MS, et al. Endovascular stent grafting versus open surgical repair of descending thoracic aortic aneurysms in low-risk patients: A multicenter comparative trial. J Thorac Cardiovasc Surg 2007;133:369–377.

Berendes JN, Bredee JJ, Schipperheyn JJ, Mashhour YAS. Mechanism of spinal cord injury after cross clamping of the descending thoracic aorta. Circulation 1982;66(2Pt2):I112–115.

Black JH, Davison JK, Cambria RP. Regional hypothermia with epidural cooling for prevention of spinal cord ischemic complications after thoracoabdominal aortic surgery. Semin Thorac Cardiovasc Surg 2003;15:345–352.

Borst HG, Jurmann M, Buehner B, Laas J. Risk of replacement of the descending aorta with a standardized left heart bypass technique. J Thorac Cardiovasc Surg 1994;107:126–133.

Bower TC, Murray MJ, Gloviczki P, et al. Effects of thoracic aortic occlusion and cerebrospinal fluid drainage on regional spinal cord blood flow in dogs: Correlation with neurologic outcome. J Vasc Surg 1989;9:135–144.

Cambria RP, Clouse WD, Davison JK, et al. Thoracoabdominal aneurysm repair: Results with 337 operations performed over a 15-year interval. Ann Surg 2002;236(4):471–479.

Cambria RP, Davison JK, Carter C, et al. Epidural cooling for spinal cord protection during thoracoabdominal aneurysm repair: A five-year experience. J Vasc Surg 2000;31:1093–1102.

Carpenter MB. Blood supply of the central nervous system. In: Core Text of Neuroanatomy (Carpenter MB, Ed.) 1991;pp 434–462. Baltimore: William & Wilkins.

Cassada DC, Tribble CG, Kaza AD, et al. Adenosine analogue reduces spinal cord reperfusion injury in a time-dependent fashion. Surgery 2001;130:230–235.

Cheung AT, Pochettino A, McGarvey ML, et al. Strategies to manage paraplegia risk after endovascular stent repair of descending thoracic aortic aneurysms. Ann Thorac Surg 2005;80:1288–1289.

Cho JS, Haider SE, Makaroun MS. Endovascular therapy of thoracic aneurysms: Gore TAG trial results. Semin Vasc Surg 2006;19:18–24.
Christiansson L, Hellberg A, Koga I, et al. A new method of intrathecal $PO_2$, $PCO_2$, and pH measurements for continuous monitoring of spinal cord ischemia during thoracic aortic clamping in pigs. Surgery 2000;127:571–576.
Christiansson L, Ulus AT, Hellberg A, et al. Aspects of the spinal cord circulation as assessed by intrathecal oxygen tension monitoring during various arterial interruptions in the pig. J Thorac Cardiovasc Surg 2001;121:762–772.
Cina CS, Abouzahr L, Arena GO, et al. Cerebrospinal fluid drainage to prevent paraplegia during thoracic and thoracoabdominal aortic aneurysm surgery: A systematic review and meta-analysis. J Vasc Surg 2004;40:36–44.
Conrad MF, Cambria RP. Contemporary management of descending thoracic and thoracoabdominal aortic aneurysms: Endovascular versus open. Circulation 2008;117:841–852.
Conrad MF, Crawford RS, Davison JK, Cambria RP. Thoracoabdominal aneurysm repair: A 20-year perspective. Ann Thorac Surg 2007;83:S856–S861.
Coselli JS, Bozinovski J, LeMaire SA. Open surgical repair of 2286 thoracoabdominal aortic aneurysms. Ann Thorac Surg 2007;83:S862–S864.
Coselli JS, LeMaire SA. Left heart bypass reduces paraplegia rates after thoracoabdominal aortic aneurysm repair. Ann Thorac Surg 1999;67:1931–1934.
Coselli JS, LeMaire SA, Conklin LD, Adams GJ. Left heart bypass during descending thoracic aortic aneurysm repair does not reduce the incidence of paraplegia. Ann Thorac Surg 2004;77:1298–1303;discussion 1303.
Coselli JS, Lemaire SA, Koksoy C, et al. Cerebrospinal fluid drainage reduces paraplegia after thoracoabdominal aortic aneurysm repair: Results of a randomized clinical trial. J Vasc Surg 2002;35:631–639.
Cowan JA Jr, Dimick JB, Henke PK, et al. Surgical treatment of intact thoracoabdominal aortic aneurysms in the United States: Hospital and surgeon volume-related outcomes. J Vasc Surg 2003;37:1169–1174.
Crawford ES, Crawford JL, Safi HJ, et al. Thoracoabdominal aortic aneurysms: Preoperative and intraoperative factors determining immediate and long-term results of operations in 605 patients. J Vasc Surg 1986;3:389–404.
Crawford ES, Mizrahi EM, Hess KR, et al. The impact of distal aortic perfusion and somatosensory evoked potential monitoring on prevention of paraplegia after aortic aneurysm operation. J Thorac Cardiovasc Surg 1988;95:357–367.
Crawford ES, Svensson LG, Hess KR, et al. A prospective randomized study of cerebrospinal fluid drainage to prevent paraplegia after high risk surgery on the thoracoabdominal aorta. J Vasc Surg 1991;13:36–46.
Dake MD, Miller DC, Semba CP, et al. Transluminal placement of endovascular stent-grafts for the treatment of descending thoracic aortic aneurysms. N Engl J Med 1994;331:1729–1734.
Davison J, Cambria R, Vierra D, et al. Epidural cooling for regional spinal cord hypothermia during thoracoabdominal aneurysm repair. J Vasc Surg 1994;20:304–10.
Estrera AL, Miller CC III, Huynh TT, et al. Preoperative and operative predictors of delayed neurologic deficit following repair of thoracoabdominal aortic aneurysm. J Thorac and Cardiovasc Surg 2003;126:1288–1294.
Estrera AL, Rubenstein FS, Miller CC III, et al. Descending thoracic aortic aneurysm: Surgical approach and treatment using the adjuncts cerebrospinal fluid drainage and distal aortic perfusion. Ann Thorac Surg 2001;72:481–486.
Etz CD, Halstead JC, Spielvogel D, et al. Thoracic and thoracoabdominal aneurysm repair: Is reimplantation of spinal cord arteries a waste of time? Ann Thorac Surg 2006;82:1670–1677.
Fan T, Wang CC, Wang FM, et al. Experimental study of the protection of ischemic preconditioning to spinal cord ischemia. Surg Neurol 1999;52:299–305.
Galloway AC, Schwartz DS, Culliford AT, et al. Selective approach to descending thoracic aortic aneurysm repair: A ten-year experience. Ann Thorac Surg 1996;62:1152–1157.
Garcin R, Godlewski S, Rondot P. Clinical study of vascular myelopathies (in French). Rev Neurol (Paris) 1962;106(6):558–591.
Grabitz K, Sandmann W, Stuhmeier K, et al. The risk of ischemic spinal cord injury in patients undergoing graft replacement for thoracoabdominal aortic aneurysms. J Vasc Surg 1996;23:230–240.
Greenberg RK, O'Neill S, Walker E, et al. Endovascular repair of thoracic aortic lesions with the Zenith TX1 and TX2 thoracic grafts: Intermediate-term results. J Vasc Surg 2005;41:589–596.
Griepp RB, Griepp EB. Spinal cord perfusion and protection during descending thoracic and thoracoabdominal aortic surgery: The collateral network concept. Ann Thorac Surg 2007;83:S86–869.
Griffiths IR, Pitts LH, Crawford RA, Trench JG. Spinal cord compression and blood flow. I: The effect of raised cerebrospinal fluid pressure on spinal cord blood flow. Neurology 1978;28:1145–1151.
Gutsche JT, Cheung AT, McGarvey ML, et al. Ann Thorac Surg 2007;84(4):1195–2000.
Halstead JC, Wurm M, Etz C, et al. Preservation of spinal cord function after extensive segmental artery sacrifice: Regional variations in perfusion. Ann Thorac Surg 2007;84:789–794.
Heinemann MK, Brassel F, Herzog T, et al. The role of spinal angiography in operations on the thoracic aorta: Myth or reality? Ann Thorac Surg 1998;65:346–351.
Husain AM, Swaminathan M, McCann RL, et al. Neurophysiologic intraoperative monitoring during endovascular stent graft repair of the descending thoracic aorta. J Clin Neurophysiol. 2007;24(4):328–335.
Ishizaki M, Sugiyama S, Uchida H, et al. Identification and selective perfusion of the spinal cord-feeding arteries by intrathecal $PO_2$ monitoring for spinal cord protection. Eur J Vasc Endovasc Surg 1999;18:17–24.
Jacobs MJ, de Mol BA, Elenbaas T, et al. Spinal cord blood supply in patients with thoracoabdominal aortic aneurysms. J Vasc Surg. 2002;35:30–37.
Jacobs MJ, Mess W, Mochtar B, et al. The value of motor evoked potentials in reducing paraplegia during thoracoabdominal aneurysm repair. J Vasc Surg 2006;43:239–246.
Kakinohana M, Kawabata T, Miyata Y, Sugahara K. Myogenic transcranial motor evoked potentials monitoring cannot always predict neurologic outcome after spinal cord ischemia in rats. J Thorac Cardiovasc Surg 2005;129:46–52.

Kawaharada N, Morishita K, Abe T, et al. Thoracoabdominal or descending aortic aneurysm repair after preoperative demonstration of the Adamkiewicz artery by magnetic resonance angiography. Eur J Cardiothorac Surg 2002;21:970–974.

Kieffer E, Fukui S, Chiras J, et al. Spinal cord arteriography: A safe adjunct before descending thoracic or thoracoabdominal aortic aneurysmectomy. J Vasc Surg 2002;35:262–268.

Kwolek CJ, Fairman R. Update on thoracic aortic endovascular grafting using the Medtronic Talent device. Semin Vasc Surg 2006;19:25–31.

Lazorthes G, Gouaze A, Bastide G, et al. Cervical spinal cord arterial vascularizations: Study of substitutions anastomoses (in French). Rev Neurol (Paris) 1966;115:1055–1068.

Lazorthes G, Gouaze A, Zadeh JO, et al. Arterial vascularization of the spinal cord: Recent studies of the anatomic substitution pathways. J Neurosurg 1971;35:253–262.

Leurs LJ, Bell R, Degrieck Y, et al. Endovascular treatment of thoracic aortic diseases: Combined experience from the EUROSTAR and United Kingdom Thoracic Endograft registries. J Vasc Surg 2004;40:670–679.

Ling E, Arellano R. Systematic overview of the evidence supporting the use of cerebrospinal fluid drainage in thoracoabdominal aneurysm surgery for prevention of paraplegia. Anesthesiology 2000;93:1115–1122.

Lombardi, JV, Carpenter JP, Pochettino A, et al. Thoracoabdominal aortic aneurysm repair after prior aortic surgery. J Vasc Surg 2003;38:1185–1190.

Makaroun MS, Dillavou ED, Kee ST, et al. Endovascular treatment of thoracic aortic aneurysms: Results of the phase II multicenter trial of the Gore TAG thoracic endoprosthesis. J Vasc Surg 2005;41:1–9.

Matsumura JS, Cambria RP, Dake MD, et al. International controlled clinical trial of thoracic endovascular aneurysm repair with the Zenith TX2 endovascular graft: 1-year results. J Vasc Surg 2008;47:247–257.

McGarvey ML, Cheung AT, Szeto W, Messe SR. Management of neurologic complications of thoracic aortic surgery. J Clin Neurophysiol 2007;336–343.

Meylaerts SA, de Haan P, Kalkman CJ, et al. Prevention of paraplegia in pigs by selective segmental artery perfusion during aortic cross-clamping. J Vasc Surg 2000;32:160–170.

Minatoya K, Karck M, Haverich A, et al. The impact of spinal angiography on the neurological outcome after surgery on the descending thoracic and thoracoabdominal aorta. Ann Thorac Surg 2002;74(Suppl):1870–1872;discussion 1892–1898.

Miyamoto K, Ueno A, Wada T, Kimoto S. A new and simple method of preventing spinal cord damage following temporary occlusion of the thoracic aorta by draining the cerebrospinal fluid. J Cardiovasc Surg 1960;1:188–197.

Murakami H, Tsukube T, Kawanishi Y, Okita Y. Transcranial myogenic motor-evoked potentials after transient spinal cord ischemia predicts neurologic outcome in rabbits. J Vasc Surg 2004;39:207–213.

Murray MJ, Bower TC, Oliver WCJ, et al. Effects of cerebrospinal fluid drainage in patients undergoing thoracic and thoracoabdominal aortic surgery. J Cardiothorac Vasc Anesth 1993;7:266–272.

Nijenhuis RJ, Jacobs MJ, Jaspers K, et al. Comparison of magnetic resonance with computed tomography angiography for preoperative localization of the Adamkiewicz artery in thoracoabdominal aortic aneurysm patients. J Vasc Surg 2007;45:67–85.

Ogino H, Sasaki H, Minatoya K, et al. Combined use of Adamkiewicz artery demonstration and motor-evoked potentials in descending and thoracoabdominal repair. Ann Thorac Surg 2006;82;592–596.

Reece TB, Okonkwo DO, Ellman PI, et al. Comparison of systemic and retrograde delivery of adenosine A2A agonist for attenuation of spinal cord injury after thoracic aortic cross-clamping. Ann Thorac Surg 2006;81:902–909.

Rigberg DA, McGory ML, Zingmond DS, et al. Thirty-day mortality statistics underestimate the risk of repair of thoracoabdominal aortic aneurysms: A statewide experience. J Vasc Surg 2006;43:217–222.

Safi HJ, Bartoli S, Hess KR, et al. Neurologic deficit in patients at high risk with thoracoabdominal aortic aneurysms: The role of cerebral spinal fluid drainage and distal aortic perfusion. J Vasc Surg. 1994;20:434–444.

Safi HJ, Harlin SA, Miller CC, et al. Predictive factors for acute renal failure in thoracic and thoracoabdominal aortic aneurysm surgery. J Vasc Surg 1996;24:338–344.

Safi HJ, Miller CC III, Huynh TT, et al. Distal aortic perfusion and cerebrospinal fluid drainage for thoracoabdominal and descending thoracic aortic repair: Ten years of organ protection. Ann Surg 2003;238:372–380.

Sandson TA, Friedman JH. Spinal cord infarction: Report of 8 cases and review of the literature. Medicine 1989;68:282–292.

Schepens M, Dossche K, Morshuis W, et al. Introduction of adjuncts and their influence on changing results in 402 consecutive thoracoabdominal aortic aneurysm repair. Eur J Cardiothorac Surg 2004;25:701–707.

Stone DH, Brewster DC, Kwolek CJ, et al. Stent-graft versus open-surgical repair of the thoracic aorta: mid-term results. J Vasc Surg 2006;44:1188–1197.

Strauch JT, Spielvogel D, Lauten A, et al. Importance of extrasegmental vessels for spinal cord blood supply in a chronic porcine model. Eur J Cardiothoracic Surg 2003;817–824.

Svensson LG, Crawford ES, Hess KR, et al. Experience with 1509 patients undergoing thoracoabdominal aortic operations. J Vasc Surg 1993;17:357–370.

Svensson LG, Hess KR, D'Agostino RS, et al. Reduction of neurologic injury after high-risk thoracoabdominal aortic operation. Ann Thorac Surg 1998;66:132–138.

Toumpoulis IK, Anagnostopoulos CE, Drossos GE, et al. Early ischemic preconditioning without hypotension prevents spinal cord injury caused by descending thoracic aortic occlusion. J Thorac Cardiovasc Surg 2003;125(5):1030–1036.

Toumpoulis IK, Papakostas JC, Matsagas MI, et al. Superiority of early relative to late ischemic preconditioning in spinal cord protection after descending thoracic aortic occlusion. J Thorac Cardiovasc Surg 2004;128(5):724–730.

Turnbull IM, Brieg A, Hassler O. Blood supply of cervical spinal cord in man: A microangiographic cadaver study. J Neurosurg 1966;24:951–965.

van Dongen EP, Schepens MA, Morshuis WJ, et al. Thoracic and thoracoabdominal aortic aneurysm repair: Use of evoked potential monitoring in 118 patients. J Vasc Surg 2001;34:1035–1040.

Verdant A, Cossette R, Page A, et al. Aneurysms of the descending thoracic aorta: Three hundred sixty-six

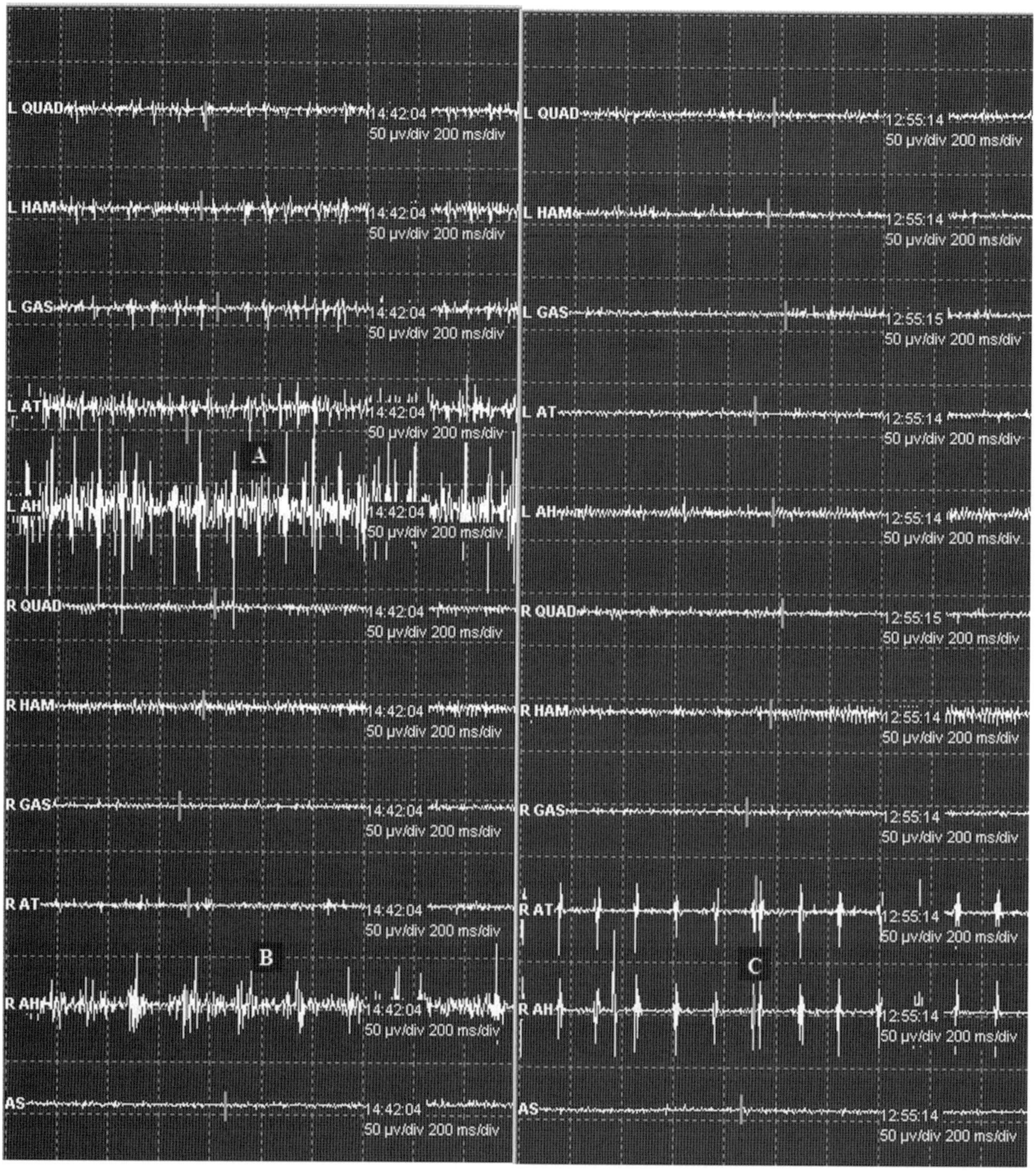


**FIGURE 6-4** Abnormal free-run EMG activity, suggesting mechanical irritation of the peripheral nerves. **Left** panel shows isolated neurotonic (**A**) and myokimic (**B**) discharges present in left AH and right AH muscle, respectively, triggered by mechanical irritation of the motor fibers. Of note, the discharges present in the right AH muscle are of smaller amplitude than, but of similar morphology with, the classical myokimic discharges. **Right** panel shows periodic motor unit potentials MUP firing (**C**), present in the right AH and AT muscles, as a result of irritation of the sensory fibers. Again, the amplitudes of the MUPs are smaller than usually described in such situations (less than 100 μV).

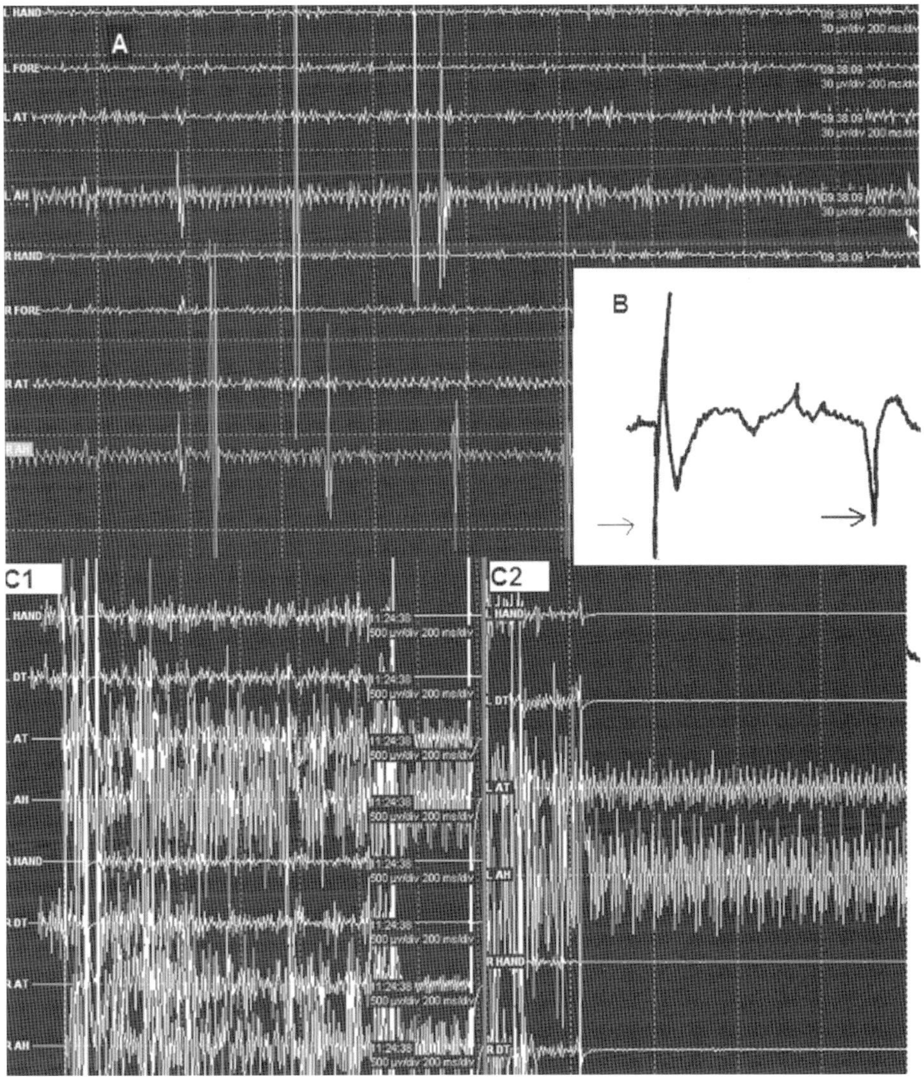


**FIGURE 6-5** Other types of free-run EMG activity. Panel **A:** Examples of spontaneous motor unit potentials that are due to "light anesthesia." Panel **B:** Abnormal spontaneous muscle activity, fibrillation potentials (thin arrow), and positive sharp waves (thick arrow): sign of muscle deinnervation. Panel **C:** High-frequency electrical artifact: **C1:** diffuse, **C2:** focal.

occurrence in certain surgeries and need not always signal postoperative nerve root damage. These are concerning when they are directly related to manipulation and are persistent, especially when they linger after the manipulation has stopped. Irrigation with cold water can also "irritate" and cause spontaneous muscle responses, in one or more myotomes, depending on the location of the irrigation.

***d. False-negative results.*** Free-run EMG activity induced by mechanical irritation is known to have a good sensitivity. However, in case of vascular injury or sharp sudden transection of the nerve, we may not get the typical EMG response to signal iatrogenic injury. Moreover, mechanical activation of the remaining distal stump of a transected nerve can also result in EMG activity in the corresponding muscle, thus giving the false impression that the injured nerve is still in continuity.

In order to avoid false-negative results, Crum et al. (2008) recommend using subcutaneous needle recording electrodes rather than using intramuscular (hook wire) electrodes, because the latter offer a very localized sampling. The presence of neurotonic discharges elsewhere in the muscle can therefore be missed. On the other hand, Skinner et al. (2008) used a porcine model to compare the sensitivity of surface electrode with intramuscular electrode recordings. These investigators concluded that false negatives can occur when surface electrodes are used, because the detection of neurotonic discharges requires near-field recordings. Crum and colleagues, however, have posited that using of a pair of intramuscular electrode results in limited muscle sampling; along the same lines, Skinner et al. (2008) recommend doing multiple intramuscular recordings from the same muscle. This type of setup is difficult to follow routinely in practice. We tend to agree with Crum and colleagues that using subcutaneous needle recordings is a practical approach in these circumstances and enhances the chances of detecting neurotonic discharges.

As in triggered EMG recordings, we recommend against the use of muscle relaxants in these procedures, in order to avoid false-negative responses.

*B2. TcmMEPs recordings.* Transcranial electrical stimulation and continuous recording of the triggered muscle MEPs (mMEPs) in the myotomes of interest is also used for continuous, simultaneous monitoring of multiple nerve roots. Mok et al. (2008) have shown that, in an animal model, tcmMEPs are sensitive in picking up nerve root injury. We routinely use this method in surgeries with high risk of lumbosacral roots injury, such as of the cauda equina surgery or untethering of the cord (Fig. **6-6**).

*B3. Recording of the bulbocavernous reflex.* The bulbocavernous reflex can also be followed intraoperatively by stimulating the pudendal nerve and recording over the anal sphincter (Kothbauer and Novak, 2004).

*B4. SSEPs recordings* The sensory pathways can be monitored with stimulation of the tibial nerve and recording over the dorsal column or scalp (Kothbauer and Novak, 2004). However, even though this technique has a high specificity, its sensitivity is low. Here we

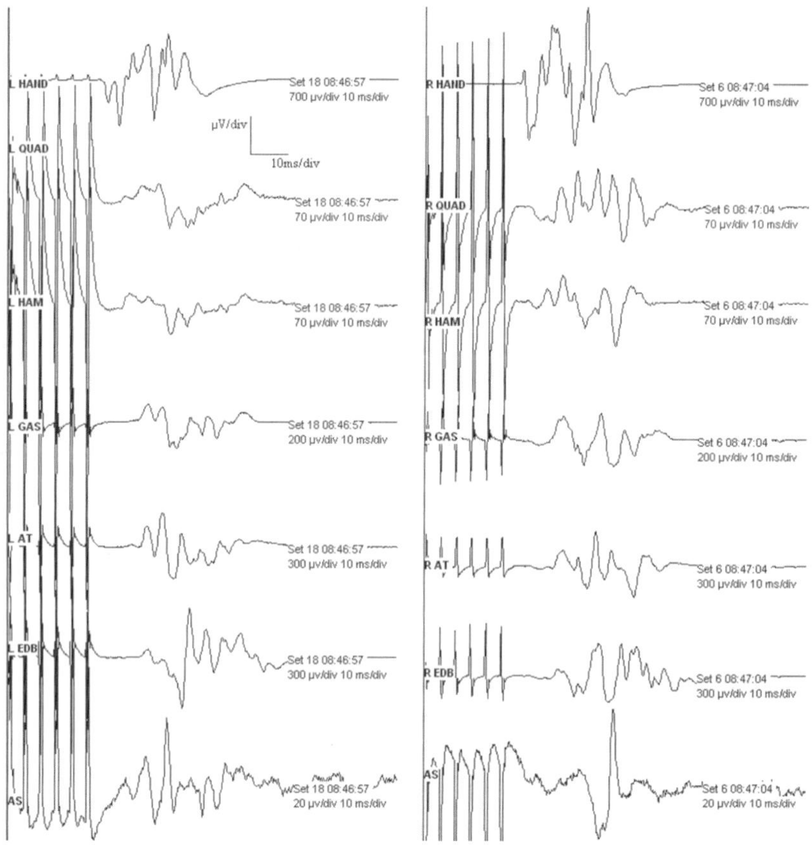


**FIGURE 6-6** Continuous monitoring of the lumbosacral nerve roots via tcmMEPs. Notice the wide sampling of the myotomes. The hand mMEPs are taken as reference.

concur with other authors (Gunnarsson et al., 2004; Krassioukov et al., 2005) and suggest using both free-run EMG (known to have a high sensitivity) and SSEPs monitoring (offering higher specificity) because they play a complementary role.

Once the filum or the tethering mass is identified, it is tested for the absence of a motor response and severed. After the spinal cord has been untethered and sufficient amount of lipoma resected to permit closure of the thecal sac, the closure may begin. In cases of a tethering mass such as a lipoma, the goal of the operation is to untether the cord, not to achieve a complete resection of the lipoma. The dura is reconstructed with lumbar fascia or a dural substitute. The safe performance of spinal cord untethering depends on the identification of functional neural elements and cannot be optimally performed without nerve root monitoring.

## 2. Lumbar Pedicle Screw Insertion

Instrumentation of the spine thoracolumbar spine is done as an adjunct to fusion and, in most instances, involves insertion of pedicle screws. One complication of hardware insertion is malposition of the screw, resulting in compression of a nerve root. This usually occurs typically with a medial or inferior pedicle wall breach, and the accuracy of screw placement is 90 to 95%, depending on the method utilized (Kosmopoulos and Schizas, 2007). Several methods have been proposed to reduce the incidence of this complication. Intraoperative fluoroscopy, image guidance (Kosmopoulos and Schizas, 2007; Rajasekaram et al., 2007), with either free-hand or robotically assisted screw placement (Togawa et al., 2007), and laminotomy with direct visualization of the pedicle (Di Silvestre et al., 2007) have all been evaluated. Intraoperative monitoring can be used to detect a medial pedicle wall breach and to reposition a screw intraoperatively (Raynor et al., 2007). It does not exclude the use of other means of improving the accuracy of the placement and can therefore be combined with imaging techniques to improve safety.

The technique involves stimulation of a pedicle finder instrument that is used to create an opening down the pedicle shaft and thus of the hole created in the pedicle. Stimulation of the pedicle screw itself, which is typically made out of titanium, is also done routinely.

For several reasons, we believe that it is necessary to stimulate the hole as well as the pedicle screw. First, if stimulation is performed after the initial pedicle finder instrument is inserted or following the application of the drill bit (Moed et al., 1998), a breach may be detected before a screw of larger diameter is inserted. Because some screws have high electrical resistance, testing them alone, and not the holes as well, can give false-negative results. Second, if a breach is not detected by stimulating the hole, inserting the screw can itself enlarge and perforate a previously intact hole. Thus, stimulating the screw becomes important.

EMG is recorded over the lower extremity musculature, and the threshold needed to elicit a CMAP is assessed. A large review of this technique has been published recently (Raynor et al., 2007). The authors showed that the lower the threshold needed to elicit a response was, the more likely was the occurrence of a breach. With a threshold response of less than 2.8 milliamperes (mA), a positive response always indicates a breach in the pedicle wall. If the response is greater than 8 mA, the probability of a breach is 0.3%. Intermediate responses demonstrate a graded likelihood of a breach. If a breach is detected, the screw is repositioned or removed. Although gross malposition causing injury to a nerve may not be avoided, this technique avoids, at the very least, a return to the operating room to revise a malpositioned screw causing radiculopathy. In addition to lumbar screw placement, pedicle screw stimulation has been used for thoracic pedicle screw insertion as well, with recoding over the intercostals muscles or the appropriate segment of the rectus abdominis muscle (Shi et al., 2003). Electrical stimulation of spinal instrumentation may be especially useful when the anatomy is distorted by the presence of a spinal deformity or with difficult screw insertion, such as iliac screws (Moed et al., 1998). It serves as an adjunct to radiographic means of safe screw insertions, with the goal of reducing the incidence of a neurologic deficit.

However, there is a wide variability of the stimulation thresholds used to signal a breach in the pedicle. This results from several factors. First, the pedicles themselves are of different thicknesses at different spinal levels. Within the same spinal level, the thickness can also present an intraindividual variability. Second, presence of a chronic nerve injury, such as chronic radiculopathy, causes increase in the stimulation threshold that is necessary for triggering a CMAP, which is sometimes difficult to predict. Third, no universally accepted normative data are available—published data from different authors presents a variety of thresholds, of either stimulus intensity or voltage, in direct relationship with a variety of stimulus parameters. It has been suggested that an accurate interpretation of the neurophysiologic data would require that each nerve root at risk should be initially stimulated before the actual pedicle manipulation. This allows testing the stimulation technique, assessing the degree of neuromuscular block, as well as establishing an individual stimulation threshold that serves as reference, when testing the screw and hole placement. However, this approach is quite impractical: it requires additional time, and the nerve roots are not always accessible for stimulation. For all of these reasons and because damage to the nerve roots may have already occurred by the time the positioning of the holes and screws is checked, we do not use this procedure routinely for pedicle screws placement.

## 3. Nerve Root Decompression

Surgery for the treatment of nerve root compression is common. Nerve roots can be compressed by disc herniations, hypertrophied ligamentum flavum, osteophytes, or trapped within a narrowed foramen. According to the Agency for Healthcare Research and Quality (AHRQ), lumbar diskectomy alone is the third most performed operation in the United States. Intraoperative injury to a nerve root can occur during a decompressive procedure, related to traction or compression of the nerve. Again, intraoperative neurophysiology plays a role in both mapping and monitoring of the nerve roots.

### A. Mapping with Triggered EMG

Triggered EMG recordings (CMAPs) are also used to identify ventral nerve roots by observing the appearance of the triggered responses in the corresponding myotomes. Likewise, one can assess the success of the decompression by comparing the stimulation thresholds necessary to trigger a CMAP response before and after decompression, respectively, especially in patients with acute compression. If the decompression was successful, the postdecompression stimulation threshold may decrease compared to the predecompression value. Further injury of the nerve root should result in an increase in the stimulation threshold. Matsui and colleagues (1997) have demonstrated a significant correlation between the severity of preoperative sensory and motor radiculopathic symptoms and the stimulation threshold of the corresponding nerve root compared to an intact nerve root. Furthermore, decompression was followed by a decrease in the stimulation threshold of the related nerve root.

### B. Monitoring

*B1. Free-run EMG and tcmMEPs recordings.* In an effort to improve the safety of nerve root decompression procedure, continuous monitoring of the electrical transmission in the nerve roots is done by recording the muscle activity in the corresponding myotomes. Nerve roots are normally insensitive to gentle manipulation intraoperatively. However, when they have been injured or compressed, even slight mechanical distortion can trigger a motor response. This can manifest itself intraoperatively with movement of an extremity in patients who are not fully paralyzed pharmacologically. In an effort to minimize iatrogenic injury to the nerve during decompression, some surgeons avoid neuromuscular blockade following the initial induction dose, to facilitate the observation of movement related to excess nerve manipulation. A more reliable way of assessing intraoperative activation of a nerve root than looking for gross movement is to perform free-run EMG during the decompression and to look for abnormal responses triggered by mechanical irritation, as discussed on pages 273–76. In this instance, the recording of muscle activity is done via recording electrodes placed over the appropriate muscles, which may differ from the standard electrode placement used in monitoring of the spinal cord. Although it may seem obvious, it has been shown that conventional placement of upper extremity transcranial motor-evoked potential electrodes over the hand fails to predict more proximal nerve root injury (Fan et al., 2002). Thus, a preoperative review of the myotomes of interest and communication between the surgeon and neurophysiologist is important (see Table **6-1**).

One of the complications of cervical decompressive procedures is postoperative palsy of the C5-innervated muscles (Sakaura et al., 2003). Although the mechanism is unknown and more than one cause may exist, a number of authors have attempted to limit this complication with the use of intraoperative monitoring. Jimenez and colleagues (2005) found that the occurrence of C5 palsy was reduced from 7.3% before the institution of intraoperative monitoring compared to 0.9% after its adoption. Free-run intraoperative EMG has been used successfully to identify patients in whom a deficit is likely to occur postoperatively, presumably averting a more serious injury (Fan et al., 2002). In this study, the authors combined free-running EMG with transcranial electrical motor-evoked potential while recording over the deltoid and biceps muscles. In general, tcmMEPs recordings can be done from the same muscles as free-run EMG recordings, but it is uncertain whether this procedure is helpful in preventing nerve root injury. Some authors have found it to be insensitive when used alone (Tanaka et al., 2006). However, in another recent study, the combining of transcranial electrical stimulation muscle MEPs with free-run EMG was found to increase the sensitivity for detecting a postoperative deficit (Bose et al., 2007). Thus, whenever possible, we use combined tcmMEPs and free-run EMG recordings for monitoring the activity in the nerve roots.

*B2. SSEPs recordings.* Monitoring can also be done via SSEPs. However, because of multiple root supply of the stimulated peripheral nerves (median, ulnar, and posterior tibial, respectively), this method is not very sensitive in picking up isolated nerve root injuries. Thus, we do not use it alone for these purposes. However, as mentioned previously, changes in SSEPs are more specific than free-run EMG abnormal activity alone, and several authors recommend the combination of the two methods in complex surgical procedures involving conus medullaris and cauda equina (Krassinoukov et al., 2005).

Dermatomal SSEPs have also been used in detecting sensory nerve root injuries. However, they present significant technical challenges. They have small amplitudes, require increased averaging, and can be done only one at a time—thus, a thorough sampling of all of the nerve roots of interest results in a slow turnaround time. Moreover, interindividual variability of the dermatomal supply can be misleading in choosing the appropriate site of stimulation for monitoring individual nerve roots.

Finally, as in the case of mapping with triggered EMG, monitoring during nerve root decompression surgery may have a role in assessing the adequacy of the decompression. Although this requires further study, the intraoperative resolution of abnormal electrophysiologic activity has been used to evaluate the decompression of a nerve root demonstrating abnormal baseline activity (Chappuis and Johnson, 2007).

## III. NERVE TUMOR SURGERY

### 1. The Roles of Intraoperative Neurophysiology

As with any surgical procedure, and especially with peripheral nerve surgeries, it is essential to have a good understanding of the local anatomy around the pathological region. For example, detailed knowledge of anatomy of the brachial plexus (and of the lumbar plexus, respectively), helps the surgeon and the neurophysiologist in choosing which nervous structures need further investigation and in identifying the location where the stimulation should be applied and the level at which the recording is to be done. Fig. **6-7 a** gives a schematic representation of the brachial as well as the lumbar plexus. Likewise, knowledge of the peripheral nerve anatomy that is distal to the injury allows the identification of the first muscle destined to be reinnervated. Figure **6-7 b** shows the branching and the muscles innervated by the most important peripheral nerves from the upper and lower limbs. Furthermore, as previously mentioned, Table **6-1** gives detailed information about the myotomal and peripheral nerve innervation of the muscles most commonly used as recording sites in peripheral nervous system surgery.

As with other types of surgical procedures, intraoperative neurophysiology serves two main roles: mapping and monitoring.

### A. Mapping with Triggered EMG

*A1. Identification of the anatomy.* One of the important roles of electrophysiologic monitoring during peripheral nerve tumor surgery is the identification of normal anatomy, via triggered EMG recordings. Brachial plexus tumors particularly can distort the anatomy, rendering the usual landmarks that help identify the elements of the plexus unreliable. Surgical approaches to the brachial plexus can be divided into supraclavicular and infraclavicular methods. The supraclavicular approach allows access to the proximal plexus, including the nerve roots, the upper middle and proximal lower trunk, and the divisions of the upper and middle trunk (Fig. **6-8**). More distal plexus elements require an infraclavicular exposure. At surgery, intraoperative confirmation of the identity of a nerve with triggered EMG is helpful in the exposures of peripheral nerve, particularly with large tumors.

*A2. Assessing the integrity of the nervous structures postresection.* This is also achieved by triggered CMAPs, by stimulating nervous structures that have been identified before resection and assessing their integrity by comparing the amplitudes, morphologies, and stimulation thresholds of the triggered CMAPs with their preresection values. In that sense, this technique is comparable to the use of triggered CMAPs in assessing the success of decompression.

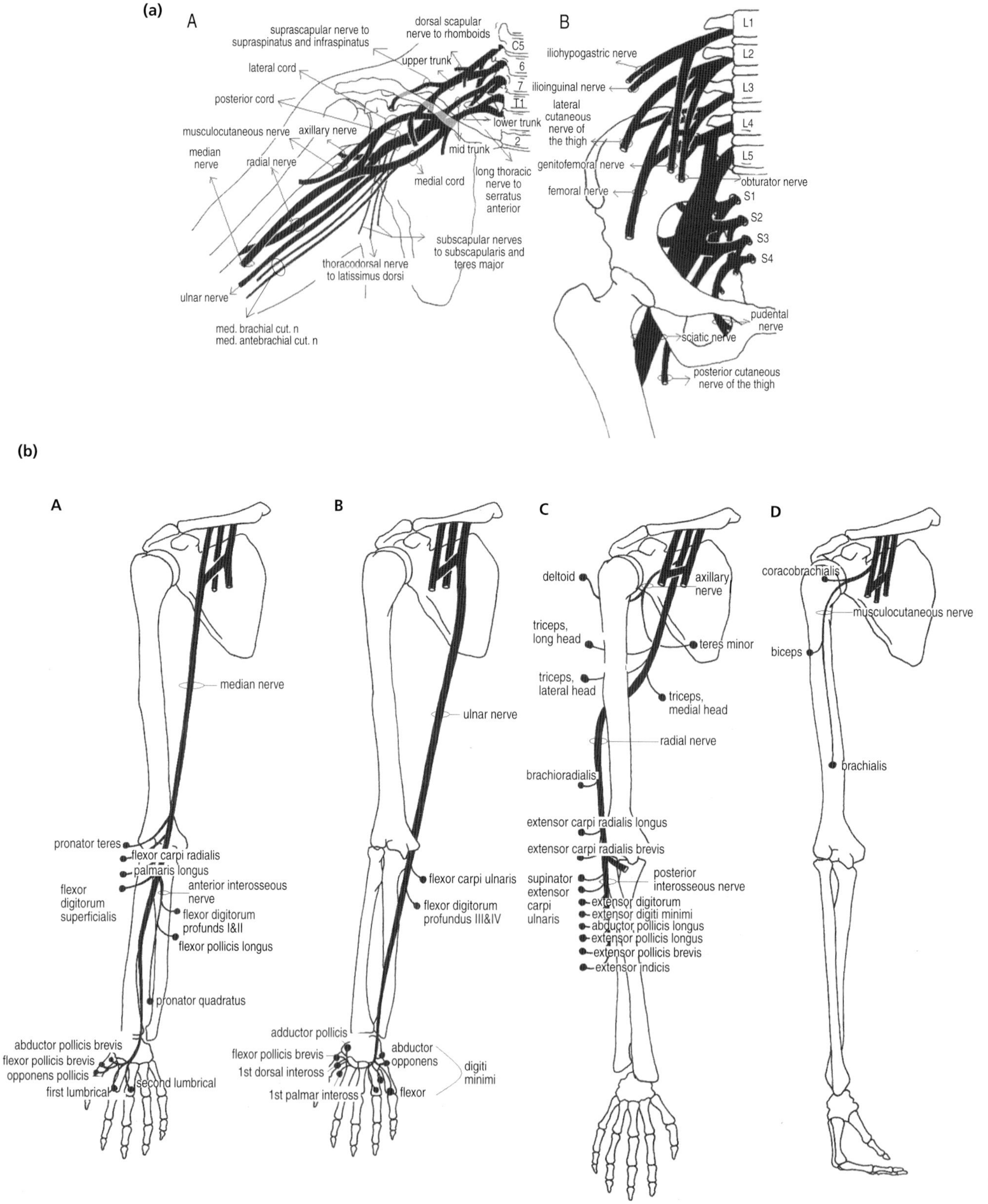


**FIGURE 6-7** Anatomy of the most commonly injured peripheral nervous structures. **a:** Nervous plexuses: **A:** brachial plexus; **B:** lumbosacral plexus. **b:** Peripheral nerves: **A:** median nerve; **B:** ulnar nerve; **C:** radial nerve; **D:** musculocutaneous nerve;

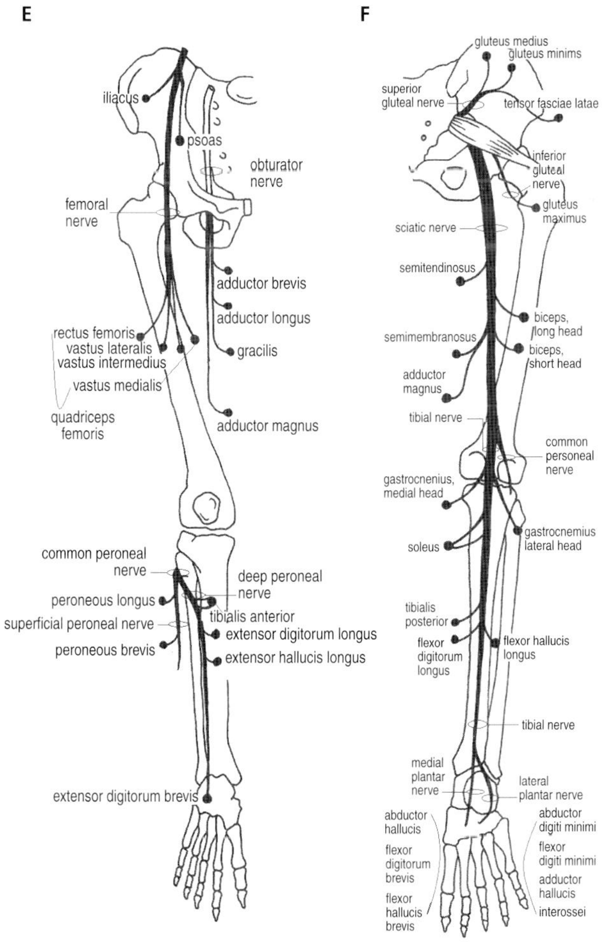


**FIGURE 6-7** *(Continued)* **E:** peripheral nerves of the lower limb, anterior aspect; **F:** peripheral nerves of the lower limb, posterior aspect.

## *B. Monitoring*

TcmMEPs and free-run EMG recordings allow continuous surveillance of the electrical transmission through the nervous structures during dissection and resection of the tumor, by continuously monitoring triggered mMEPs or "spontaneous" EMG activity, which result from mechanical irritation of the muscles innervated by the nerves situated close to the tumor.

## 2. Applying Intraoperative Neurophysiology for Safe Resection of the Tumor

As with traumatic injury of the peripheral nerves (pages 284–97), tumor resection requires a generous exposure. A wide exposure allows the safe dissection of the tumor from uninvolved structures and the testing of the parent nerves. In Fig. **6-8**, for example, a supraclavicular dissection was performed to identify the C5, C6, and C7 roots. The tumor can be seen originating from the C6 root, just proximal to the upper trunk. A principle of peripheral nerve tumor resection is the identification of nerve fascicles, which are thinly stretched on the surface of the tumor yet remain functionally important. These fascicles have been referred to as "basketed," because they resemble woven fibers on the surface of a basket (Fig. **6-8 a, c**). With the tumor in sight, the epineurium is opened with a sharp blade, at the junction of the tumor and parent nerve. The tumor may have eroded the epineurium, and this layer may be incomplete. The surface of the tumor itself is inspected, and note is taken of visible fascicles that need to be preserved. Occasionally, part of the tumor capsule triggers a vigorous EMG response, in the absence of a visible nerve. In such rare instances, the conducting portion of the capsule is left in place and the resection is subtotal. Typically, basketed fascicles are identifiable. The dissection begins over an electrically silent area of the tumor, and the fascicles are dissected sharply and then swept aside with blunt dissection, until the tumor is removed. Occasionally, the presence of multiple basketed fascicles under tension does not allow for the en bloc resection of the tumor, and the tumor is removed piecemeal rather than to impart additional stretch to the thinned nerve fascicles. During the dissection, exposure, and tumor resection, continuous tcmMEPs and free-run EMG recordings can also be done. Following resection of the tumor, the stimulation of the fascicles is done again, and the response can be compared to the initial stimulation.

Even though schwannomas are generally believed to arise eccentrically in the nerve and to displace fascicles aside, it is not uncommon for one or more fascicles to enter the substance of the tumor. Usually testing of these fascicles reveals them to be nonfunctional, and the intraoperative testing of the last remaining fascicle allows the surgeon to resect the last part of the tumor confidently (Fig. **6-8 g**). Conversely, neurofibromas are not all necessarily associated with multiple intervening fascicles, as demonstrated in Fig. **6-9**. This patient with neurofibromatosis type 1 presented with a foot drop and was found to have a large neurofibroma in the region of

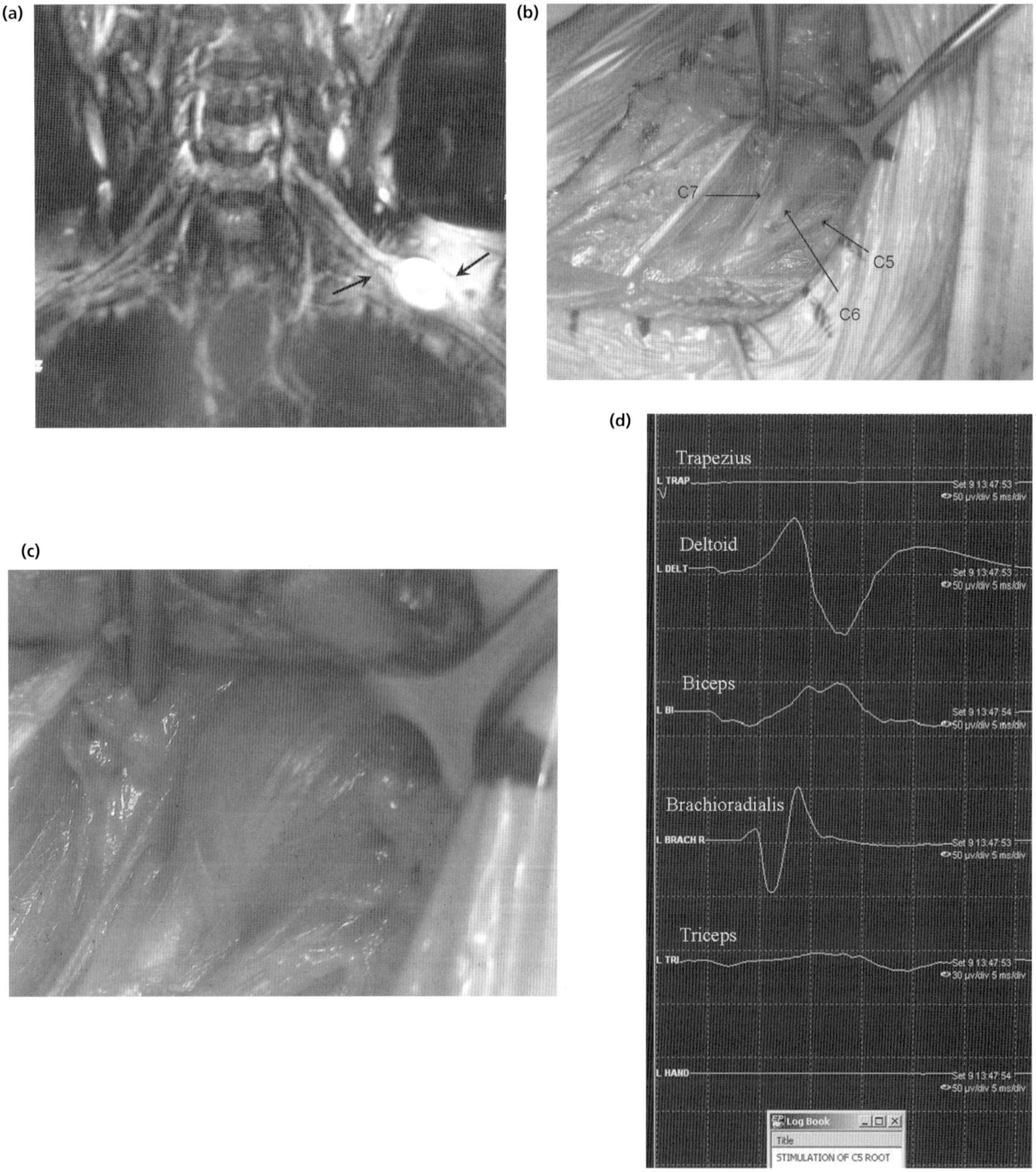


**FIGURE 6-8** Brachial plexus tumor resection under neurophysiologic guidance. **a:** Preoperative coronal T2-weighed image in a 42-year-old female with neurofibromatosis type 2, showing a left C6 schwannoma. The black arrows denote nerves that are basketed on the surface of the tumor. **b:** Intraoperative photograph shows C5, C6, and C7 nerve roots. The tumor arises from the C6 nerve, just proximal to the upper trunk. Note the presence of a fascicle on the surface of the lesion. **c:** Higher magnification view of the lesion prior to resection, showing basketed fascicle. Direct stimulation of the fascicle with a bipolar electrode resulted in activation of arm muscles, demonstrating the importance of preserving this thin fascicle. **d:** Screen shot demonstrating motor response in various upper extremity muscles following stimulation of the presumptive C5 nerve root. In this instance, triggered EMG aids in the identification of the anatomy.

(e)

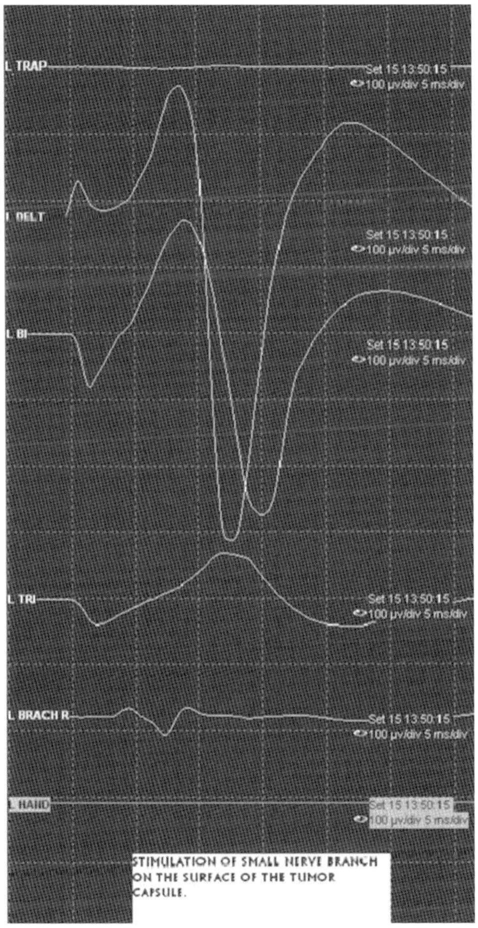


(f)

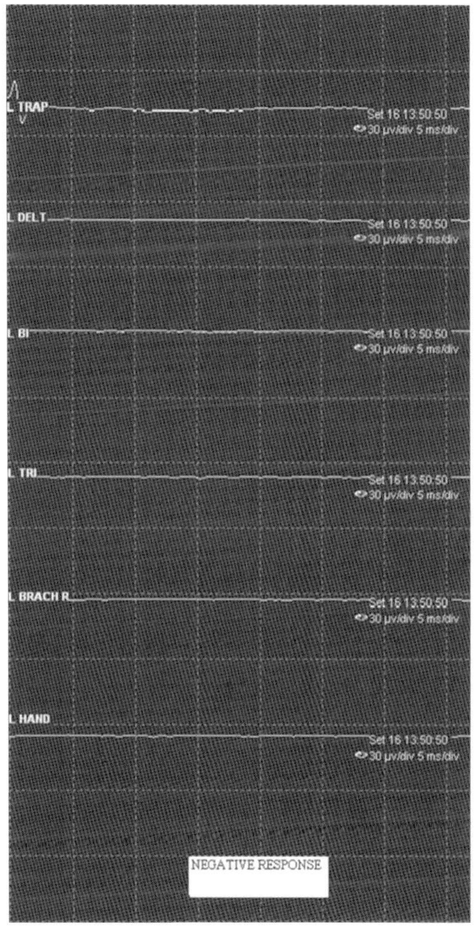


(g)

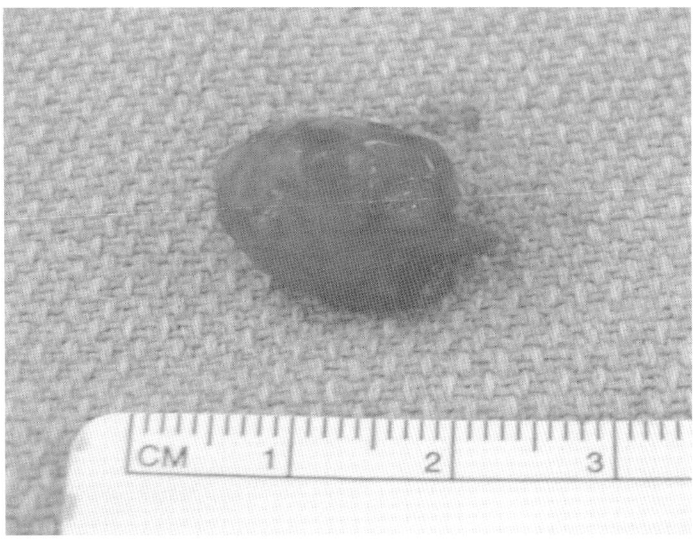


**FIGURE 6-8** *(Continued)* **e:** Screen shot following stimulation of the small fascicle demonstrating motor response in various upper extremity muscles following stimulation of the small fascicle on the tumor surface. **f:** Screen shot showing absence of motor response following stimulation of the tumor capsule above the basketed fascicle. The tumor capsule was stimulated in its entirety to identify surface nerve fascicles. **g:** Resected tumor. A small, nonconducting fascicle can be seen entering the tumor at right.

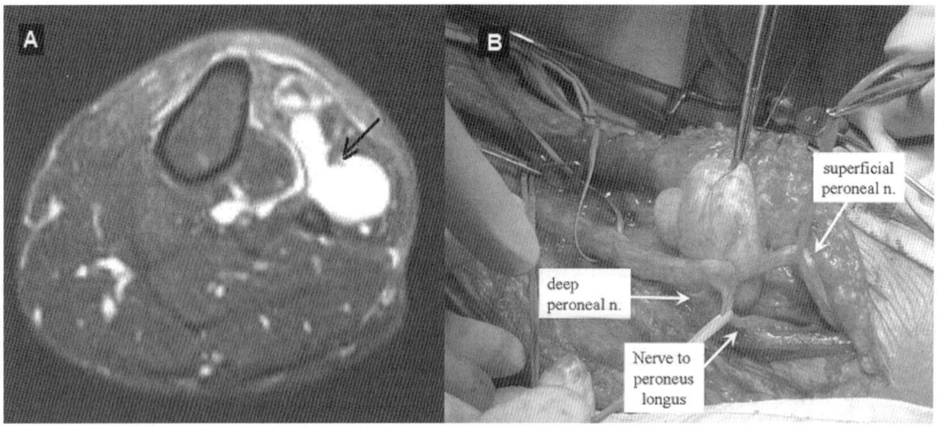


**FIGURE 6-9** Resection of a peroneal nerve tumor with the aid of intraoperative neurophysiology. Panel **A:** Axial T2-weighed MRI scan of the leg demonstrating peroneal nerve region tumor (arrow) in a patient with neurofibromatosis-type 1. The patient presented with weakness of ankle dorsiflexion and eversion. Panel **B:** Intraoperative photograph showing the tumor. The common peroneal nerve is seen at left, and the proximal deep peroneal nerve is just visible. The tumor primarily arose from the nerve to the peroneus longus. Intraoperative EMG aided in the identification of the nerves during the initial phase of the dissection.

the bifurcation of the common peroneal nerve. Triggered EMG was useful in establishing the identity of the dissected nerves, but the large tumor was easily separated from its parent nerve, without resection of a single fascicle. Hence, in spite of adequate preoperative evaluation, the intraoperative findings vary at exploration, and intraoperative monitoring is integral to the safe resection of peripheral nerve tumors.

## IV. NERVE INJURY SURGERY

### 1. Initial Evaluation and Management

Patients presenting with motor and sensory function loss that is secondary to traumatic peripheral nerve injuries present a challenge even to the experienced clinician. This is especially true in patients with brachial plexus injuries because of the enormous complexity of the anatomy and the large potential zone of injury in these patients, which spans the distance from the spinal cord to the mid-arm. A thorough history and physical examination, with an eye toward the mechanism of the trauma, is critical to making the diagnosis and determining how best to manage the nerve injury. For open injuries, particularly sharp penetrating trauma, immediate exploration and repair are usually indicated. The role of intraoperative electrodiagnostic is limited in this setting.

Closed injuries following a blunt trauma producing neurologic deficits usually are caused by stretching, compression, or avulsion of peripheral nerves. In the case of stretch injuries, they are often worsened by early exploration and frequently improve without surgical intervention, thereby mandating a period of observation. In the case of penetrating injuries from gunshots, the injury to the nerve is often indirect, in the form of a cavitation from the energy of a passing projectile, and it should be treated similarly to a closed injury, with observation (Mackinnon and Dellon, 1988; Winograd and Mackinnon, 2006).

Additional considerations in the evaluation of a nerve injury include the existence of premorbid neurologic conditions. A history of neuropathy resulting from diabetes, of previous neurologic trauma, or of a demyelinating disease may alter the patient's prognosis for neurological recovery.

### 2. Follow-up of Electrophysiologic Nerve Recovery: Indications for Surgical Exploration and Intraoperative Electrophysiologic Testing

Nerve conduction and EMG evaluation should be obtained after a sufficient period has elapsed for Wallerian degeneration to occur and for the typical electrophysiologic stigmata of injury to be detectable. These studies are used to document nerve recovery. Electrodiagnostic testing is also used to determine the anatomic distribution and degree of nerve injury, whether it is partial or complete.

Clinically, this is particularly useful in proximal injuries, such as the brachial plexus. For brachial plexus injuries, the typical time of the first clinical and electrophysiologic examination is between 4 and 6 weeks after the injury (Mackinnon and Dellon, 1988; Winograd and Mackinnon, 2006). Serial nerve conduction and EMG studies, performed every 4 to 6 weeks, document the earliest detectable recovery and typically precede any changes in clinical exam. In the case of brachial plexus injuries, full recovery may take 2 years or longer, and the surgeon evaluating these patients does not have the luxury of waiting for this period to pass. The limiting factor in these cases is the window of opportunity for motor recovery. Once deinnervated, a muscle is only receptive to re-innervation for a limited period, typically 12 to 18 months. After this time, no recovery is possible, and the muscle tissue irreversibly atrophies and becomes fibrotic. Thus, the operation must take place without excessive delay, before the corresponding myotome undergoes irreversible transformations and the opportunity for repair is lost (Belzberg et al., 2004). If nerve reconstruction is indicated, the best results are obtained if the patient is explored by 4 to 6 months, at the latest, from the time of injury, thus allowing maximal time for motor recovery after reconstruction. The latter may also be reasonably be performed for up to 1 year. Sensory nerves can be grafted after an indefinite period of deinnervation, although there is some data from animal studies that suggests that the first 6 months is the period during which the distal nerve stump is most receptive to axonal regrowth (de Medinacilli and Seaber, 1989; Fu and Gordon, 1995; Fu and Gordon, 1997; Mackinnon and Dellon, 1988; Winograd and Mackinnon, 2006).

The strategy described previously accepts that there will be some explorations in which spontaneous recovery is demonstrated intraoperatively and reconstruction is neither needed nor performed. On the other hand, surgical exploration under neurophysiologic guidance is unparalleled in its utility to the reconstructive surgeon, particularly in cases of complex injury of the brachial plexus with multiple elements injured in a deep anatomic location, for two main reasons. First, pinpointing the exact location of regenerating nerve fibers within an injured nerve can be challenging, if not impossible, with transcutaneous monitoring. Without the ability to visualize the exact anatomic area of a nerve injury, the surgeon may fail to grade nerve injuries accurately or predict the likelihood of recovery. Intraoperative electrodiagnostic testing allows the surgeon to visualize the injured nerve anatomically and stimulate it. This enables very precise testing of injured nerves, particularly in closed nerve injures and when transcutaneous testing has yielded indeterminate results.

Second, the method has an essential role in determining the degree of the nerve injury, which further signals the potential for functional recovery. Different classifications based on the type and severity have been devised, but two systems are widely used: the Seddon (Seddon, 1975) and the Sunderland (Sunderland, 1978) classifications. The Seddon classification divides injuries into three types, neurapraxia, axonotmesis, and neuronotmesis. Neurapraxia represents a transient loss of function without anatomical interruption of the nerve fibers or supporting structures. It requires no intervention. Axonotmesis represents a loss of the axon, without a loss of the supporting structures. Such injuries are followed by Wallerian degeneration of the distal segment with 2 to 4 days. Spontaneous recovery is also seen with axonal regrowth of 1–2 mm per day. The likelihood of axonal regrowth and functional recovery is governed by several factors (Burnett and Zager, 2004) . Neuronotmesis is a loss of the axon and the supporting structures as well. Scar formation is extensive, and recovery is poor without surgical intervention. The Sunderland classification divides the injuries into grades 1 through 5. A grade 1 injury is equivalent to neurapraxia according to the Seddon classification. A grade 2 injury involves an injury to the axon, without loss of the supporting structures, and is equivalent to axonotmesis. In addition to axonal loss, grade 3 injuries are associated with a loss of endoneurium and grade 4 injuries with a loss of perineurium, with only some connective tissue and epineurium remaining. Finally grade 5 injuries represent a complete transections of the nerve. Grade 4 and 5 injuries involve extensive loss of the supporting structures, resulting in scar tissue formations. They are associated with lack of spontaneous regeneration. It is these injuries that require surgical exploration, neurolysis, and possible nerve grafting.

The accuracy of this technique, employing the use of intraoperative neurophysiology at the time of surgical exploration, far exceeds what the skilled electrophysiologist can accomplish without surgery. Intraoperative testing at the time of exploration, therefore, is essential for determining which patients require reconstruction and in determining which nerves, in the case of brachial plexus injuries, require reconstruction versus further observation. Conversely, injuries which present very late, i.e. after 1 year, rarely require nerve conduction studies, because the patient's spontaneous recovery has been monitored for an extended period and a good diagnostic picture has emerged, even if spontaneous recovery is not yet complete.

## 3. The Role of Intraoperative Neurophysiologic Studies in Diagnosis, Management, and Prognosis of Peripheral Nerve Injury

In brachial plexus injuries, intraoperative neurophysiology helps in establishing the location of the injury, whether it is preganglionic, postganglionic, or both (Fig. **6-10**). This is essential for choosing the appropriate surgical strategy. Preganglionic injuries require neurotization because the proximal root has been avulsed from the spinal cord (see pages 292–93). Establishing the presence of a preganglionic injury is done with the help of the upper limb SSEPs. The modified SSEPs technique used in establishing the presence or absence of root avulsion, as well as troubleshooting methods, are given on pages 286–89.

Unlike preganglionic injuries, postganglionic injuries are amenable to direct repair, with or without nerve grafting, because the proximal nerve is in continuity with the spinal cord. In postganglionic injuries, preoperative EMG is initially used to monitor recovery (see pages 284–86). Surgical exploration with intraoperative neurophysiology carried out at an appropriate time helps establish the severity of injury and the need for nerve grafting. It is assumed that, if regeneration is successful, the electrical transmission is present and nerve grafting is not necessary.

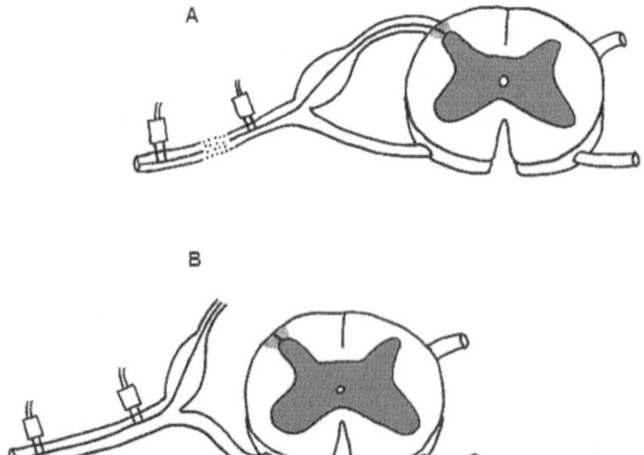


**FIGURE 6-10** Establishing the location of injury in brachial plexus surgery: postganglionic (**A**) versus preganglionic (**B**) showing root avulsion.

Thus, the necessity for nerve grafting is indicated by the absence of electrical transmission through the injured part of the peripheral nerve. Two neurophysiologic methods are used in establishing the functional continuity of the nerve: triggered EMG (CMAPs recordings) and nerve-to-nerve recordings (nerve action potentials or NAP recordings). This chapter lays out the specifics of their use in peripheral nerve repair surgery, including a systematic, step-by-step technique for stimulation and recording, as well as troubleshooting details (see pages 289–92).

Lastly, in peripheral nerve recordings, knowledge of the following are important before surgery: type and duration of the injury, extent of clinical impairment, results of serial outpatient EMG and nerve conduction studies, and scope of the surgery. It is also very important to plan the strategy in advance with the surgeon and anesthesiologist. Planning relies on a good understanding of the anatomy of brachial and lumbosacral plexus, as well as of the peripheral nerves, as mentioned previously (see Fig. **6-7** and Table **6-1**). During surgery, it is essential to maintain a good communication between the surgeon, intraoperative monitoring team, and anesthesiologist. Particularly, information such as distance between the stimulating and recording electrodes, location of the stimulation, location of the recording, and anesthetic regimen is very important.

Even though most of the references in this section relate to upper extremity or shoulder injuries, surgery for peripheral nerve injuries in the lower limbs require similar techniques.

### *A. Determining Presence or Absence of Root Avulsion*

Nerve root avulsion (preganglionic injury) can be strongly suspected and suggested preoperatively through transcutaneous electrophysiologic testing by presence of paraspinal muscle deinnervation absence of recruitable motor activity, and the presence, shortly after the injury and before Wallerian degeneration has occurred, of "normal nerve conduction" in affected nerves, which are clinically demonstrating insensate sensory territories and motor paralysis. This last finding is pathognemonic for nerve root avulsion in a preganglionic location.

In this case, the most useful test may simply be SSEPs, because it helps confirm, along with an exploration of the plexus, the presence or absence of root avulsions. General description of SSEPs technique can be found on pages 5–15. In the following, we detail a modified technique used for checking for root avulsion, including methodology, interpretation, and troubleshooting.

*A1. Methodology and interpretation of the recorded potentials.* The surgeon uses a handheld bipolar probe to stimulate the cervical roots that need evaluation. We use repetitive square pulses, with a pulse width of 0.3 msec and a frequency of 3.17 Hz. The stimulation is done with an intensity that is gradually increased from 0.5 to 5 mA. Averaged responses are recorded from the subcortical (neck) and cortical (scalp) channels. If there is transmission of the electrical volley through the dorsal nerve root, a subcortical potential is recorded in Cs5–Fz channel and, hopefully, a thalamocortical potential in the scalp channels (Cpc–Cpi, Cpc–A1; where Cpc is the contralateral parietal region and Cpi is the ipsilateral parietal region). The subcortical potential is an electronegative response, represented in our recordings by an upward deflection. This is the equivalent of P/N13 when recording is done in the Cs5–Fz channel and the stimulation of the ipsilateral median or ulnar nerves is done at the wrist. However, when stimulating at the level of brachial plexus, the latency of such response is in the order of several milliseconds, depending on the distance between the stimulation and recording sites. We usually start our recording with a baseline median or ulnar SSEPs, stimulating the peripheral nerves at the wrist. We also like to have a positive control of a robust subcortical response obtained in Cs5–Fz channel, when the surgeon stimulates a healthy part of the brachial plexus, at a similar distance from the recording site as the diseased nervous structures. This way, the positive control also indicates the latency at which future responses should be seen. The recording parameters are time base of 5 msec/div and sensitivity of 0.5 μV/div. We rarely have to decrease the time base, because the latency of this potential is, in general, around 6–8 msec. An example of preserved electrical transmission, and thus absence of root avulsion, is given in Fig. **6-11**.

*A2. Troubleshooting stimulation*

***a. Absence of stimulus artifact.*** This usually indicates defaulted stimulation. The handheld stimulator should be checked; we have encountered many nonfunctional stimulators. The distance between the tips of the stimulators should also be checked. Prongs that are placed too close together may not stimulate, particularly if blood and body fluids are present in the stimulated region, because a wet surgical field facilitates current shunting. Furthermore, all of the connections at the stimulating pad and at the stimulator headbox should be checked. A loose connection of the cable connecting the stimulating pad to the headbox can cause a lack of stimulation. A mismatch between the plug-in socket where this is plugged and the setup program in the evoked potentials EP software can be another cause (e.g. in the program setup software, the group of the recording channels is connected to the stimulator labeled "1a," but the cable is connected to the socket for the stimulator labeled "1b").

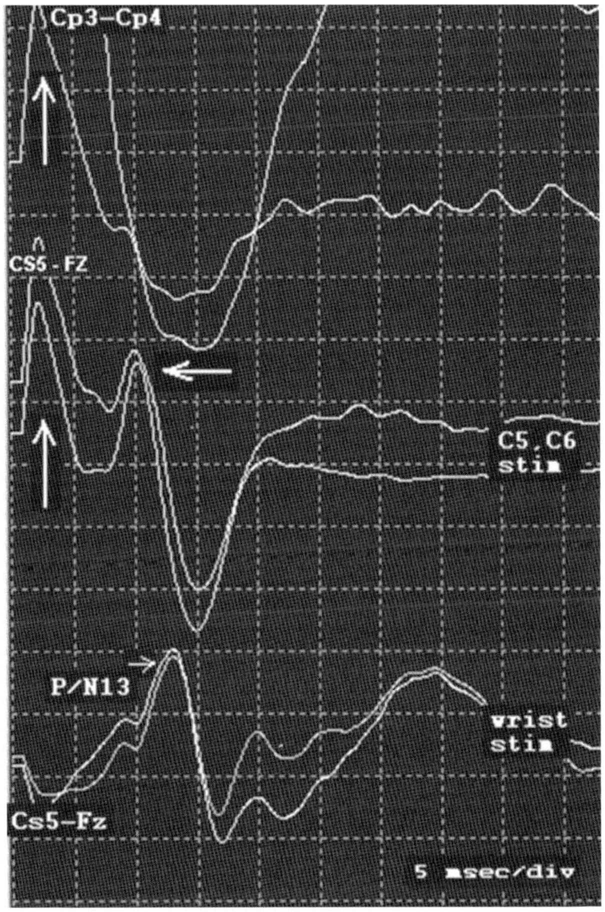


**FIGURE 6-11** SSEPs recordings in ruling out root avulsion. The first Cs5-Fz channel from the bottom shows a P/N13 (small horizontal arrow) subcortical potential obtained after transcutaneous stimulation of the right median nerve at the wrist. The second Cs5-Fz channel also shows a subcortical potential (larger horizontal arrow), recorded after stimulation of C5 and C6 roots. Notice the smaller latency of this potential compared to the P/N13 (10 msec versus 13 msec, respectively). Also notice the stimulus artifact (vertical arrows). Cp3-Cp4 channel shows a significant stimulus artifact and no obvious thalamocortical responses after stimulation of the right C5 and C6 nerve roots. However, because of the presence of a robust subcortical response, it was concluded that the right C5 and C6 ventral roots were not avulsed.

of both triggered EMG (pages 27–33) and nerve-to-nerve recordings (pages 33–36), including principles, methodology of stimulation and recording, as well as a detailed description and examples of the recorded potentials. Thus, we recommend approaching this chapter after getting familiar with both neurophysiologic techniques described on pages 27–36.

*B1. Nerve-to-nerve recordings (NAPs).* This method can be very challenging technically and in terms of interpretation because of the significant number of factors, such as variable distance between stimulating and recording electrodes, presence of a wet field leading to current shunting or spread, misappreciation of the location of the lesion, variability of the degree of severity of the lesions, and the necessity to modify the stimulation and possibly recording parameters according to the particularities of each case.

***a. Troubleshooting stimulation.*** All of the points emphasized on pages 287–88 also apply for this technique. This method is especially prone to significant stimulus artifact, mainly because of the small distance between the stimulation and recording sites that the surgeon is sometimes constrained to use. We recommend at least 5 cm between the two sites. Increasing the surgical incision can help. Recording transcutaneously is another option for certain nerves, when their trajectory is well known.

In order to decrease the stimulus artifact, it is also important that the nerve or fascicles to be stimulated are well exposed and isolated from the surrounding tissues. This will result in a decrease in the stimulus amplitude necessary to depolarize the nervous fibers. Details regarding successful stimulation for the nerve-to-nerve recording techniques (e.g. the importance of the direction of stimulation or the importance of modifying the stimulus parameters) can be found on pages 17–18.

***b. Troubleshooting recording***

1. *No NAPs recorded.* As with the SSEPs method, one has to ensure that appropriate stimulation is applied to the nerve. All of the points made on page 289 also apply here. NAPs are small-amplitude potentials, with relatively short latencies. The latter depends upon the distance between the stimulating and recording electrodes, as well as on the conduction velocity of the nerve segment. Knowledge of the distance between the stimulating and recording electrodes and the conduction velocity of the respective nerve could help approximate the expected latency of the response (Fig. **6-13**).

   It is important to try to eliminate any stimulus artifact that may obscure the presence of such an early and small potential. Appropriate recording parameters should be used, including a short time base (e.g. 2 msec/div) and a high sensitivity. This information is detailed on pages 18–21.

   As mentioned on page 18, the distance between the prongs of the recording electrode has to be at least several millimeters; this distance should be increased for larger fiber sizes, as well as for longer distances between the stimulation and recording electrodes—otherwise no NAP are recorded. On the other hand, an increased distance between the recording prongs introduces more artifact in the recording. No 60-Hz notch should be used in peripheral nerve recordings, because this may produce an effect called "ringing." In this case, with stimulation, dampened oscillation becomes part of the stimulus artifact and may look similar to a NAP (Happel and Kline, 2002).

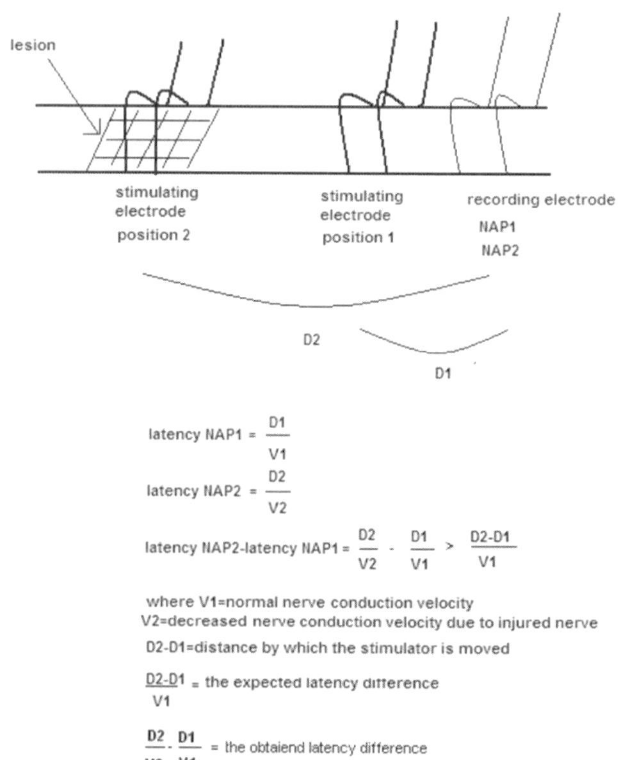


**FIGURE 6-13** Localizing the lesion on the nerve: a step-by-step stimulation and recording approach, the "inching" technique.

During surgical exploration of the peripheral nerve in one of the limbs, a tourniquet is often used for local hemostasis, producing a controlled limb ischemia. This, as well as the use of local anesthetics, can significantly affect the ability of obtaining a NAP.

If the absence of a recognizable NAP is not due to technical error or external factors, the logarithm described below, should be followed.

2. *Distinguishing between NAPs and CMAPs*. In case the stimulation is applied proximal to the lesion and the recording is done distally, CMAPs together with NAPs can be obtained. Pages 35–36 give details about the different morphologies and characteristics of the two types of evoked potentials that help in distinguishing the two.

***c. Algorithm for determining the location and the severity of the nerve lesion.*** In the following, we have provided a step-by-step approach to estimating the severity of injury and its location.

1. *Establishing a stimulation threshold for the healthy nervous tissue*. We use an intact nerve, preferably the same size as the injured nerve, for this purpose. We ask the surgeon to stimulate the healthy nerve, while maintaining the same distance between the stimulating and the recording site, as intended to be used later for the injured nerve. The chosen distance is usually approximated by the surgeon, given the presumed location of the lesion on the nerve and the fact that the lesion has to be situated in the segment of the nerve between the stimulation and recording points. This way, one establishes the depolarizing threshold as the smallest stimulus amplitude at which a response is obtained, including the smallest intensity as well as the smallest pulse width. Using the same distance between the stimulation and recording sites allows us to have a good approximation of the latency of the response, in the case of a normal conduction velocity.
2. *Exploring the injured nerve and interpreting the results*. Using the same distance, the same direction of stimulation (e.g. proximal to distal) and the stimulus parameters as described previously (as determined by the healthy nerve threshold), the surgeon stimulates the injured nerve, proximal to the lesion, and records distal to it.

   a. If the location of the lesion is known with certainty (i.e. the lesion is included between stimulation and recording sites):

**NAP response is obtained at normal stimulation threshold.** In this case, we have proven the continuity of the nerve. According to Kline and colleagues (1969), presence of a NAP proves the existence of several thousands medium-sized myelinated axons. However, the characteristics of the response can give information about the type of injury.

If there is partial but rather diffuse conduction block (as in the case of a neuropraxic lesion), the latency of the NAP is increased. Because of the temporal dispersion, its duration is increased at the cost of its amplitude, which is smaller than normal. Determining the location of such neuropraxic lesion requires a step-by-step stimulation, starting by stimulating and recording from a healthy portion of the nerve (recording NAP1) and then by slowly increasing the distance between the recording and stimulating electrodes, until the stimulating electrodes fall on the lesion (recording NAP2). In this situation, the obtained NAP2 have a longer latency than NAP1, translating in a difference in latency between NAP2 and NAP1 that is greater than that initially calculated (see Fig. **6-13**). This method is known in the literature as "inching" (Campbell, 1998).

On the other hand, latency increase is not a good measure of the extent of nerve injury. If very few myelinated axons are left intact, the latency as well as the stimulation threshold may be normal; the amplitude, however, is decreased. Thus, for more qualitative studies, one can compare the amplitude of the NAP with a baseline amplitude of a healthy nerve of similar size. This also gives a good estimate of the number of fibers affected.

If both the NAP latency as well as its amplitude are increased, presence of a low temperature should be questioned.

**NAP response is not obtained.** In this situation, the stimulus amplitude should be gradually increased. The more severe the nerve injury and smaller the number of regenerating fibers, the higher the intensity and the longer the pulse duration (width) needed to reach the stimulating threshold. For more details regarding this principle, see pages 34–35. For range of stimulus parameters used, see page 28. If no responses are obtained despite a maximal stimulation, it means that the lesion situated between the stimulating and the recording electrodes is not conducting. Furthermore, the distance between the stimulating and the recording site should be gradually decreased, until a NAP response is obtained (similar to "inching"). This helps in defining the extent of the injured segment that is nonfunctional and thus requires grafting. In special circumstances, NAPs transmission is absent, not because of a break in nerve continuity, but from post-traumatic scar compression of the nerve fascicles, impeding the axonal regeneration (Kline and Happel, 1993; Oberle et al., 1997). Recognizing this situation is difficult, and, unfortunately, neurophysiology is unhelpful, leaving the decision entirely to the surgeon, who has to rely only on the visual inspection of the nerve. In general, however, the absence of an elicited NAP is taken as a sign of lack of functional continuity.

b. If the location of the lesion is not known with certainty:

**NAP response is obtained.** If the nerve action potential is obtained at the stimulation threshold of a healthy nerve and at the same latency, it is possible that the lesion is, in fact, located either distal to the recording site or proximal to the stimulation site. Thus, the distance between the stimulating and the recording electrodes should be gradually increased. This should be done systematically, either by moving the recording electrodes distally or the stimulating electrodes proximally, until no normal NAP is obtained at the stimulation threshold. This likely signals that either the recording electrodes have reached to the lesion or that the stimulating electrodes are now on or proximal to the lesion. From this point, we use the strategy previously described, by progressively increasing the stimulus amplitude.

**NAP response is not obtained**. In such cases, the injured segment is likely included between the stimulation and recording sites. One should continue by trying to find the higher new stimulation threshold, as described above.

*B2. Triggered EMG recordings (CMAPs).* When technically feasible, EMG of the muscle(s) innervated by the injured nerve or part of the plexus may be used to look for nascent potentials that herald recovery of motor function. However, only a few axons are needed to trigger a CMAP. Thus, obtaining a CMAP cannot guarantee future recovery of voluntary contraction. On the other hand, CMAPs can be detected more easily, compared to NAPs, because of their much higher amplitudes. Even though triggering a CMAP by stimulating proximal to the lesion is easier to recognize and is reassuring, this information has to be interpreted within the clinical context, the appearance of the muscle during surgical exploration, as well as prior outpatient EMG results. In general, for establishing the continuity of the nerve, we prefer the use of NAPs instead of CMAPs recordings.

***a. Troubleshooting stimulation and recording.*** All of the points described on pages 290–91 also apply to this method.

1. *Absent CMAPs.* As already mentioned, in the case of peripheral nerve surgery, we recommend that recording is done via subdermal needles rather than intramuscular electrodes. These offer a nice compromise between surface and intramuscular recordings; they introduce less artifact than the surface stick on electrodes and "look at" a broader part of the muscle than a single pair of intramuscular electrodes. This reduces the risk of false-negative results.
   Again, if the absence of triggered CMAPs is not due to a technical error, a similar algorithm to the one described on page 291 should be applied. However, failure to trigger a CMAP does not entirely rule out the presence of a lesion in continuity, when muscle re-innervation hasn't occurred yet. This is another reason behind our preference for NAPs to CMAPs recordings.
2. *Present CMAPs.* Because CMAPs can also be triggered by direct stimulation of the muscle, one has to make sure this is not occurring as a result of current spread. We recommend introducing a reference muscle in the recording montage, situated in close proximity to the muscle of interest, yet innervated by a different nerve or root than the one investigated.

*B3. Visual inspection of the muscle contraction.* Similar to the triggered EMG recordings, visual inspection of the muscle instead of recorded CMAPs can be used to appreciate the electrical transmission through the injured nerve.

## C. Examples of common clinical scenarios

*C1. Total brachial plexus palsy.* In a patient with typical brachial plexus reconstruction, there are two common scenarios: total and partial brachial plexus palsy. In total brachial plexus palsy, nerve avulsion may be suggested by the early results of the outpatient nerve conduction and EMG/NCS studies. These will show presence of paraspinal muscles deinnervation with absence of recruitable muscle activity, and the presence of "normal nerve conduction" in the affected nerves which are clinically demonstrating insensate sensory territories and motor paralysis. This last finding is pathognemonic for nerve root avulsion in a preganglionic location. However, the most useful test for confirming root avulsion remains intraoperative SSEPs recordings, as described on pages 286–89. If the roots have been avulsed, reconstruction of the normal anatomy with excision of neuromatous segments and grafting is not possible, and nerve transfers and neurotization are employed, using intact motor nerves, such as the intercostals, spinal accessory and phrenic nerves, or unaffected regions of the brachial plexus (Allieu et al., 1984; Kawai et al., 1988; Samardzic et al., 1990). If the roots are not avulsed, a further surgical exposure of the proximal portions can be performed, and grafting of the involved segments is still possible for upper plexus injuries, particularly for the suprascapular nerve, the axillary nerve, and even the musculocutaneous nerve, by reconnecting the C5 and C6 roots with grafting to their distal uninjured portions. Intraoperative neurophysiologic recordings can also be employed in a total brachial plexus palsy if root avulsions have been ruled out, to help assess individual nerves for signs of electrical recovery in regions that are not accessible transcutaneously, allowing for greater accuracy in determining the prognosis of each affected nerve. The description of this technique was made on pages 289–92.

*C2. Partial brachial plexus palsy.* The second common scenario in brachial plexus surgery is a partial plexus palsy, typically upper plexus, with injury to the C5 and C6 roots or more distal elements of the upper plexus, possibly with involvement of C7. If these patients are explored by the fourth month, nerve conduction and EMG cannot distinguish preoperatively between patients with Sunderland grade II, grade III, or grade IV injuries, all different levels of axonotmetic injuries. At this point, Sunderland grade I injuries, neuropraxias, have recovered. Although grade II injuries recover completely without further intervention, by definition, grade III injuries only recover partially, and grade IV injuries do not recover at all. Perhaps most humbling of all in this scenario, evidence of a grade II or a grade III injury mandates further observation without reconstruction, because surgery will not likely provide better than a grade III level of functional recovery. The role of the nerve conduction studies in this case is to look for evidence of ongoing spontaneous recovery within the affected nerves that has not yet reached muscle tissue and to look for previously undetected evidence of recovery within muscle. This can only be accomplished with the use of intraoperative nerve-to-nerve and triggered EMG recordings. In addition to its utility in determining subclinical recovery within nerves, these studies can also elucidate the segments of nerve that are nonfunctional and need to be resected, although this is often determined through direct gross examination following neurolysis. Again, SSEPs recordings can be used for detection of the upper plexus roots avulsions, which mandates nerve transfers from intact motor nerve donors (Kline and Happel, 1993; Mackinnon and Dellon, 1988; Winograd and Mackinnon, 2006).

A case example is provided that demonstrates some of the concepts previously outlined. A cyclist suffered a motor vehicle accident in which he sustained, in addition to a spinal fracture, a closed stretch injury to his brachial plexus and scapulothoracic dissociation. He was followed initially by orthopedics and had a persistent axillary nerve palsy. He originally had a suprascapular palsy as well, with electrical evidence of partial function, although there was no evaluation of the suprascapular nerve at the initial EMG, making interpretation of this finding difficult. He presented for evaluation 12 months after injury. He underwent exploration and intraoperative nerve conduction studies for evaluation of his axillary and suprascapular nerves. With direct exposure, direct electrical stimulation of his suprascapular nerve and his axillary nerves triggered CMAPs in the deltoid and supraspinatus and infraspinatus muscles. Thus, electrical evidence of neural recovery was demonstrated (Fig. **6-14**). In fact, after neurolysis was complete, direct stimulation led to visibly contraction of both the supraspinatus and deltoid muscles. Further neurolysis of the suprascapular notch was performed, and no other intervention was required. As a rule, if neural regeneration can be demonstrated, nerve grafting or nerve transfer is not indicated, because there is no assurance that it will improve the quantity or quality of regeneration, and there is a significant risk of downgrading function. The patient went on to recover spontaneously additional strength in his shoulder flexion, external rotation, and abduction after neurolysis.

*C3. Difficult clinical diagnoses.* Occasionally, patients present with clinical histories and examinations that do not match up well with their electrophysiologic studies. In these cases, electrophysiologic testing can be critically important in making a diagnosis of the problem, if surgical exploration is planned. We encountered a patient, a carpenter by profession, who, while installing a kitchen island, felt a pop in his neck and developed severe pain and shoulder dysfunction after that time. He was unable

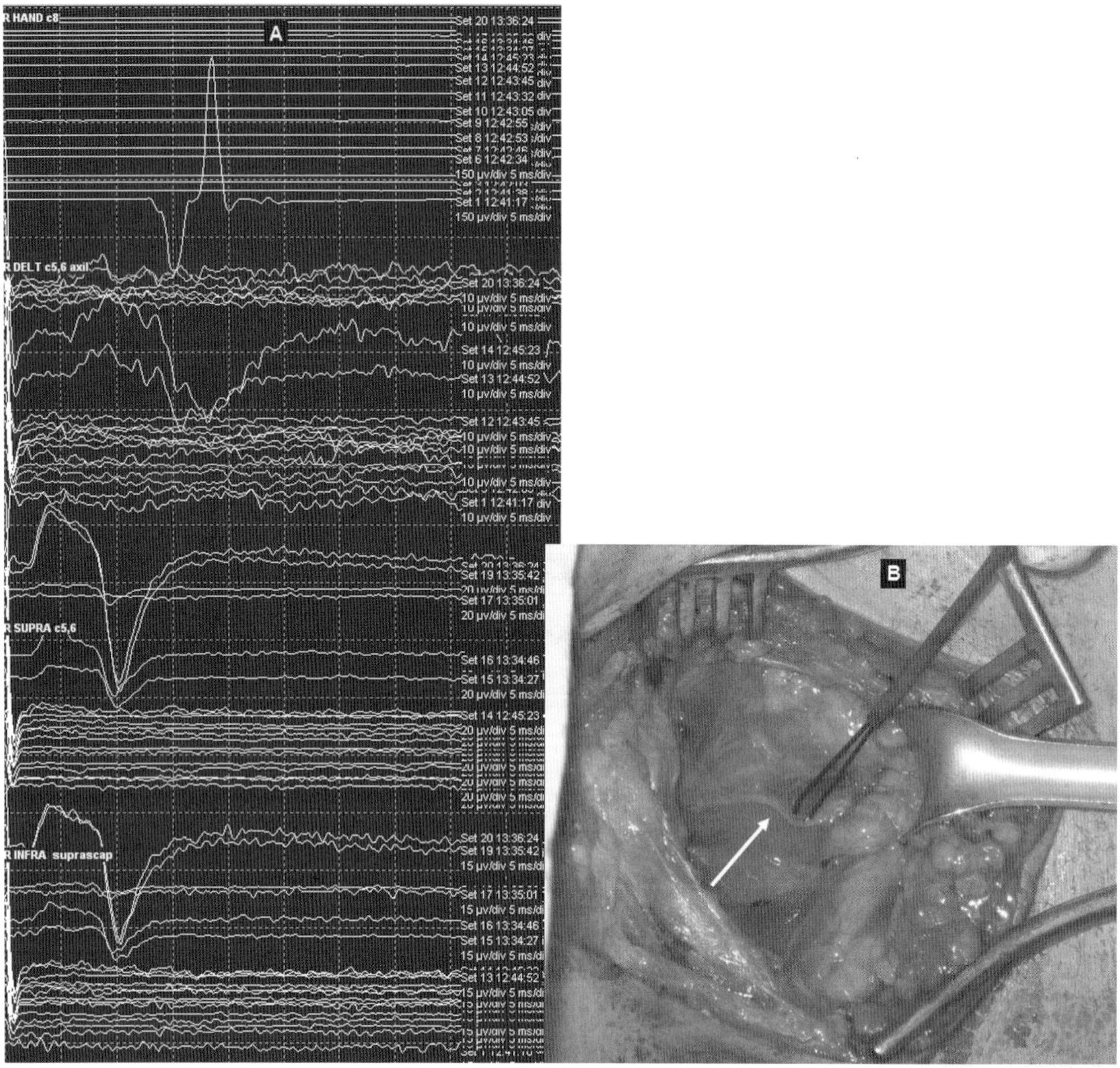


**FIGURE 6-14** Intraoperative triggered EMG recording, confirming the functional continuity of the axillary and suprascapular nerves. Panel **A** shows triggered CMAPs in the deltoid muscle, obtained after direct stimulation of the axillary nerve at 5 mA (sets 13 and 14). After stimulation of the suprascapular nerve at 0.5 and 1 mA, CMAPs were obtained in the supraspinatus and infraspinatus muscles (sets 15/16 and 19/20 respectively). Set 1 shows high-amplitude right APB CMAPs (hand channel) obtained at 0.5 mA, after stimulation of the right median nerve as a reference. Panel **B** shows a picture of intraoperative direct nerve stimulation of the suprascapular nerve. The bipolar electrode is seen on the suprascapular nerve (white arrow).

to lift his arm to a horizontal position in either a flexed or abducted posture. At 2 and 6 months after the injury, he had severe winging of his scapula and significant trapezius atrophy. He underwent transcutaneous nerve conduction and EMG evaluation of his extremity, which showed increased insertional activity and fibrillation potentials, as well as absent recruitable motor units in the trapezius muscle on both occasions. On the other hand, by EMG report, serratus anterior muscle exhibited full activation and normal recruitment. Thus, he was incorrectly diagnosed with a spinal accessory nerve injury. However, at the time of his presentation to the author, 1 year post-injury, he had demonstrable trapezial function with an intact shoulder shrug. It was also noted that his scapular winging was far worse during shoulder flexion than abduction, a typical sign of a long thoracic nerve palsy. The patient underwent exploration of his neck and axilla. He had a normal-looking trapezius muscle, which demonstrated normal triggered EMG responses with direct stimulation of the spinal accessory nerve. The serratus anterior appeared very atrophic, and this was confirmed by the failure to trigger CMAPs via stimulation. Thus, the suspected injury of the long thoracic nerve was confirmed. Because of the lack of distal triggered EMG responses, it was concluded that there was no electrical transmission through the injured portion of the nerve. A transfer of one branch of the thoracodorsal nerve to the long thoracic nerve was performed in the axilla (see Fig. **6-15**). Even though the repair was done at 1 year post-injury, the patient was eventually able to lift his arm past horizontal in both flexion and abduction of the shoulder. He was also able to lock his arm in a fully flexed position, which he had been unable to do before surgery. The winging significantly improved.

*C4. Complex peripheral nerve reconstruction.* Perhaps the quintessential application for intraoperative electrophysiology is the mapping of a neuroma in continuity. This entity, a Sunderland-Mackinnon grade VI injury, in which a nerve has suffered a closed injury and has developed a neuroma but retains partial function, is an extremely difficult treatment dilemma for the reconstructive neurosurgeon. In order to better characterize the lesion, the nerve has to be appropriately exposed, by dissecting it in a zone proximal to the neuroma, with intraneural neurolysis of the fascicles. A similar dissection distally, which often relies on the undisturbed anatomic separation of various elements and fascicles, is also performed. Adequate separation and isolation are also essential in order to decrease the stimulus amplitude that is necessary to depolarize the nerve and thus to decrease the stimulus artifact significantly. Consequently, a generous exposure is usually required. With the neuroma defined, nerve-to-nerve recording (NAPs recording) is used to determine whether conduction is present across the lesion (Tiel et al., 1996). The technique follows the same algorithm as the one described on pages 291–92.

(a)

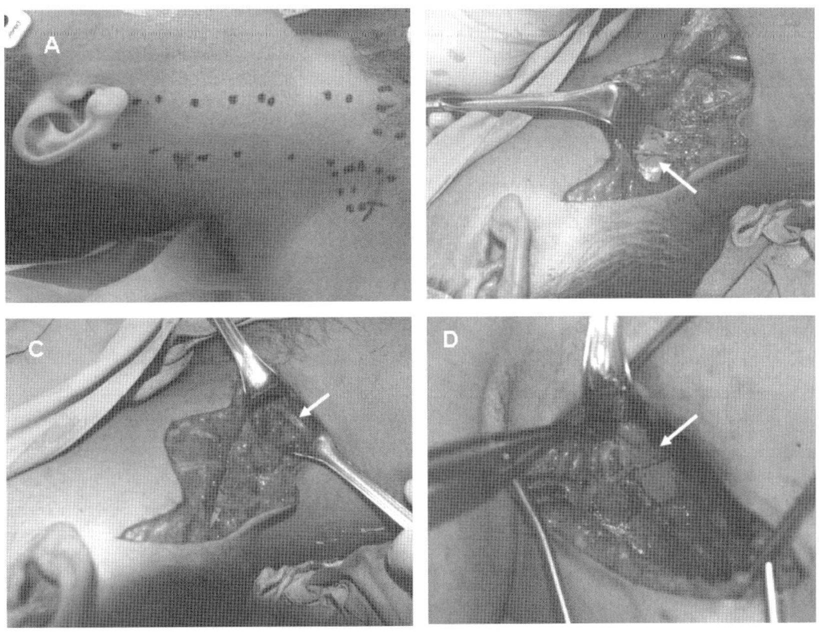


**FIGURE 6-15** Brachial plexus exploration under intraoperative neurophysiology guidance. Triggered EMG studies demonstrate functional continuity of the spinal accessory nerve, as well as inability to trigger CMAPs in the serratus anterior muscle, pointing toward a diagnosis of nonfunctional long thoracic nerve. **a**: Panel **A**: Neck marked for exposure of the spinal accessory nerve and the proximal brachial plexus, sternocleidomastoid outlined; Panel **B**: Spinal accessory nerve (white arrow) marked for testing. Panel **C**: Proximal plexus exposed at the base of the neck (white arrow). Panel **D**: Thoracodorsal nerve transfer to long thoracic nerve (white arrow) shown on microsurgical background.

(b)

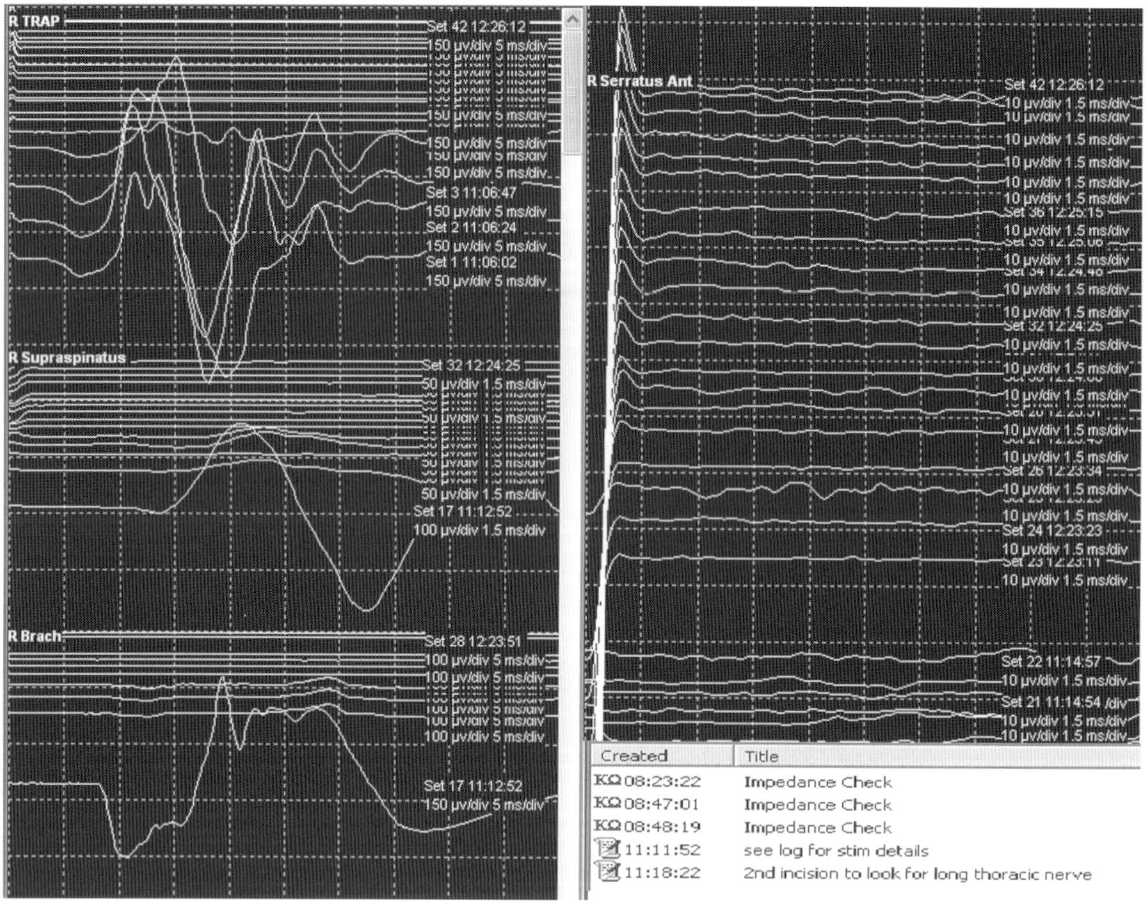


**FIGURE 6-15** *(Continued)* **b:** Triggered EMG activity. Sets 1 through 4 show high amplitude CMAPs in the right trapezius, after stimulation of the spinal accessory nerve at 0.5 mA. Set 17 shows triggered CMAPs in the right supraspinatus and brachioradialis muscles, after stimulation of the upper trunk at 1 mA. After more distal exposure, the right long thoracic nerve was stimulated at 1 through 12 mA (starting at 12:23 pm): no CMAPs were triggered in the right serratus anterior muscle.

***a. If the nerve is symmetrically involved.*** In this case, the method applies to the entire neuroma. As a result, two situations can appear.

1. *The lesion fails to conduct.* The surgical strategy is one of resection of the neuroma and nerve repair. If it is possible to re-approximate the two nerve endings primarily, by mobilizing the proximal and distal portions of the nerves, this is the preferred option. A primary repair must not be done, however, at the cost of creating tension across the repair. Among the most common causes of failure of recovery postoperatively are failure to resect the scar tissue in its entirety and failure to achieve a tension-free repair. If it is not possible to re-approximate the two nerves in a tension-free manner, interposition grafting or neurotization is necessary.
2. *The lesion conducts.* In this situation, the strategy is one of external neurolysis and resection of the scar tissue.

***b. If the nerve appears to have been asymmetrically injured.*** One option is to record along individual nerve fascicles by using the same technique as previously described and to resect and repair only the nonconducting fascicles, by incorporating them into nerve grafting (Kline et al., 2001; Mackinnon et al., 1992). For this, individual fascicles are identified and isolated, and an effort should be made to preserve the functioning fascicles. Exploration in the substance of the neuroma should thus be avoided, in order to prevent further injury through either devascularization, stretch, or incidental division of the functioning fibers during dissection. This technique is appealing, because it avoids resection of an entire neuroma containing some conducting fascicles. However, it is technically demanding and can lead to scar tissue formation.

### 4. Pitfalls

There are pitfalls that should be acknowledged when devising a reconstructive plan based on electrodiagnostic testing. Both transcutaneous nerve stimulation and intraoperative conduction studies can identify evidence of muscle re-innervation and contraction prior to voluntary muscle contraction. There is no direct correlation between intraoperative muscle contraction, generated through repeated direct neural stimulation, and clinical recovery or return to normal function. Additionally, muscle re-innervation with normal amplitude of EMG potentials may never result in functional recovery of a particular motion, if the portion tested is not representative of the larger neuromuscular anatomic unit. Extrapolations from external electromyography should be guarded at best. Without actually visualizing the entire muscle that is being tested, it is impossible to correctly identify where the muscle is being tested. It is possible for the EMG electrode to be placed within an island of healthy muscle surrounded completely by deinnervated muscle. Thus, there are severe limitations to external monitoring.

### 5. Conclusion

Intraoperative electrophysiologic testing is an invaluable tool in the armamentarium of the reconstructive peripheral nerve surgeon. It can be used to confirm diagnoses and prognoses of injured nerves in a very precise manner. It can prevent unnecessary nerve grafting, and it can add necessary precision to technically demanding procedures.

## V. ANESTHESIA CONSIDERATIONS

In general, nerve-to-nerve and nerve-to-muscle recordings (NAPs and free-run and triggered EMG recordings) are resistant to the effects of anesthetics. One definite influence is that of the muscle relaxants on the free-run and triggered EMG (fEMG and CMAPs). More so, other factors such as local ischemia (due to tourniquet use), systemic hypotension or hypothermia could affect the peripheral nerve and muscle recordings. For further details regarding the effects of anesthetics, see pages 332–34.

## REFERENCES


Agency for Healthcare Research and Quality. Data and Surveys. http://www.AHRQ.gov/data. Accessed 20 July 2009.

Allieu Y, Privat JM, Bonnel F. Paralysis in root avulsion of the brachial plexus: Neurotization by the spinal accessory nerve. Clin Plast Surg 1984;11(1):133–136.

Beatty RM, McGuire P, Moroney JM, Holladay FP. Continuous intraoperative electromyographic recording during spinal surgery. J Neurosurg 1995;82:401–405.

Belzberg AJ, Dorsi MJ, Storm PB, Moriarty JL. Surgical repair of brachial plexus injury: A multinational survey of experienced peripheral nerve surgeons. Neurosurg Focus 2004;101(3) 365–376.

Bose B, Sestokas AK, Schwartz DM. Neurophysiological detection of iatrogenic C-5 nerve deficit during anterior cervical spinal surgery. J Neurosurg Spine 2007;6(5):381–385.

Burnett M, Zager EL. Pathophysiology of peripheral nerve injury: A brief review. Neurosurg Focus 2004;16(5):1–7.

Campbell WW. The value of inching techniques in the diagnosis of focal nerve lesions. Muscle Nerve 1998;21:1554–1556.

Chappuis JL, Johnson G. Using intraoperative electrophysiologic monitoring as a diagnostic tool for determining levels to decompress in the cervical spine: A case report. J Spinal Disord Tech 2007;20(5):403–407.

Crum BA, et al. Peripheral nerve surgery. In: A Practical Approach to Neurophysiologic Intraoperative Monitoring (Husain AM, Ed.) 2008; pp 181–193. New York: Demos Medical Publishing.

de Medinacilli L, Seaber AV. Experimental nerve reconnection: Importance of initial repair. Microsurgery 1989;10:56–70.

Di Silvestre M, Parisini P, Lolli F, Bakaloudis G. Complications of thoracic pedicle screws in scoliosis treatment. Spine 2007; 32(15):1655–1661.

Fan D, Schwartz DM, Vaccaro AR, et al. Intraoperative neurophysiologic detection of iatrogenic C5 nerve root injury during laminectomy for cervical compression myelopathy. Spine 2002;27(22):2499–2502.

Fu SY, Gordon T. The cellular and molecular basis of peripheral nerve regeneration. Mol Neurobiol 1997; 14(1–2):67–116.

Fu SY, Gordon T. Contributing factors to poor functional recovery after delayed nerve repair: Prolonged axotomy. J Neurosci 1995;15(5 pt 2):3876–3885.

Giddens JL, Radomski SB, Hirshberg ED, et al. Urodynamic findings in adults with the tethered cord syndrome. J Urol 1999;161(4):1249–1254.

Gunnarsson T, Krassioukov AV, Sarjeant R, Fehlings MG. Real-time continuous intraoperative electromyographic and somatosensory evoked potential recordings in spinal surgery: Correlation of clinical and electrophysiologic findings in a prospective, consecutive series of 213 cases. Spine 2004; 29(6):677–684.

Happel L, Kline D. Intraoperative neurophysiology of the peripheral nervous system. In: Neurophysiology in Neurosurgery. A Modern Intraoperative Approach (Deletis V and Shils JL, Eds.) 2002; pp. 169–195. San Diego: Academic Press.

Holland NR, Kostuik JP. Continuous electromyographic monitoring is required to detect nerve root injury during thoracolumbar scoliosis surgery. Spine 1997;22:2547–2550.

James HE, Mulcahy JJ, Walsh JW, Kaplan GW. Use of anal sphincter electromyography during operations on the conus medullaris and sacral nerve roots. Neurosurgery 1979; 4:521–523.

Jimenez JC, Sani S, Braverman B, et al. Palsies of the fifth cervical nerve root after cervical decompression: prevention using continuous intraoperative electromyographic monitoring. J Neurosurg Spine 2005;3(2):92–97.

Kawai H, Kawabata H, Masada K, et al. Nerve repairs for traumatic brachial plexus palsy with root avulsion. Clin Orthop Relat Res 1988;(237):75–86.

Kline DG, Hackett ER, May PR. Evaluation of nerve injuries by evoked potentials and electromyography. J Neurosurg 1969; 31:128–136.

Kline, DG, Happel, LT. Denfield Lecture. A quarter century's experience with intraoperative nerve action potential recording. Can J Neurol Sci 1993;20:3–10.

Kline DG, Hudson AR, Kim D. Split nerve repair. In: Atlas of Peripheral Nerve Surgery (Kline DG, Happel LT, and Kim D, Eds.) 2001; pp. 180–181. Philadelphia: WB Saunders.

Kombos T, Suess O, Kern BC, et al. Can continuous intraoperative facial nerve electromyography predict facial nerve function following cerebellopontine angle surgery? Neurol Med Chir (Tokyo) 2000; 40:501–505.

Kosmopoulos V, Schizas C. Pedicle screw placement accuracy: S meta-analysis. Spine 2007;32(3):E111–120.

Kothbauer KF, Novak K. Intraoperative monitoring for tethered cord surgery: An update. Neurosurg Focus 2004; 16(2):E8.

Kothbauer K, Schmid UD, Seiler RW, Eisner W. Intraoperative motor and sensory monitoring of the cauda equina. Neurosurgery 1994;34:702–707.

Krassioukov AV, Sarjeant R, Arkia H, Fehlings MG. Multimodality intraoperative monitoring during complex lumbosacral procedures: Indications, techniques, and long-term follow-up review of 61 consecutive cases. J Neurosurg Spine 2005; 2(3):399–400.

Mackinnon SE, Dellon AL. Surgery of the Peripheral Nerve. 1988; New York: Thieme Medical Publisher, Inc.

Mackinnon SE, Glickman LT, Dagum A. A technique for the treatment of neuroma-in-continuity. J Reconstr Microsurg 1992; 8:379.

Matsui H, Kanamori M, Kawaguchi Y, et al. Clinical and electrophysiologic characteristics of compressed lumbar nerve roots. Spine 1997;22(18):2100–2105.

Moed BR, Ahmad BK, Craig JG, et al. Intraoperative monitoring with stimulus-evoked electromyography during placement of iliosacral screws: An initial clinical study. J Bone Joint Surg 1998;80A(4):537–546.

Mok JM, Lyon R, Lieberman JA, et al. Monitoring of nerve root injury using transcranial motor-evoked potentials in a pig model. Spine 2008;33(14):E465–E473.

Oberle J, Antoniadis G, Rath SA, Richter HP. Value of nerve action potentials in the surgical management of traumatic nerve lesions. Neurosurgery 1997;41:1337–1344.

Owen JH, Kostuik JP, Gornet M, et al. The use of mechanically elicited electromyograms to protect nerve roots during surgery for spinal degeneration. Spine 1994;19:1704–1710.

Pang D, Wilberger JE. Tethered cord syndrome in adults. J Neurosurg 1982;57:32–47.

Paradiso G, Lee GY, Sarjeant R, et al. Multimodality intraoperative neurophysiologic monitoring findings during surgery for adult tethered cord syndrome: Analysis of a series of 44 patients with long-term follow-up. Spine 2006;31(18):2095–2102.

Prass RI, Luders H. Acoustic (loudspeaker) facial electromyographic monitoring. Part I: Evoked electromyographic activity during acoustic neuroma resection. Neurosurgery 1986; 19:392–400.

Rajasekaram S, Vidyadhara S, Ramesh P, Shetty AP. Randomized clinical study to compare the accuracy of navigated and non-navigated thoracic pedicle screws in deformity correction surgeries. Spine 2007;32(2):E56–64.

Raynor BL, Lenke LG, Bridwell KH, et al. Correlation between low triggered electromyographic thresholds and lumbar pedicle screw malposition: Analysis of 4857 screws. Spine 2007; 32(24):2673–2678.

Romstock J, Strauss C, Fohebusch R. Continuous electromyographic monitoring of motor cranial nerves during cerebellopontine angle surgery. J. Neurosurg. 2000;93(4):586–593.

Sakaura H, Hosono N, Mukai Y, et al. C5 palsy after decompression surgery for cervical myelopathy: Review of the literature. Spine 2003;28(21):2447–2451.

Samardzic M, Grujicic D, Antunovic V, Joksimovic M. Reinnervation of avulsed brachial plexus using the spinal accessory nerve. Surg Neurol 1990;33(1):7–11.

Seddon HJ. Surgical Disorders of the Peripheral Nerves, 2nd ed. 1975; London: Churchill Livingstone.

Shi YB, Binette M, Martin WH, et al. Electrical stimulation for intraoperative evaluation of thoracic pedicle screw placement. Spine 2003;28(6):595–601.

Skinner SA, Transfeldt EE, Savik K. Surface electrodes are not sufficient to detect neurotonic discharges: Observations in a porcine model and clinical review of deltoid electromyographic monitoring using multiple electrodes. J Clin Monit Comput 2008;22:131–139.

Sunderland S. Nerves and Nerve Injuries, 2nd ed. 1978; London: Churchill Livingston.

Tanaka N, Nakanishi K, Fujiwara Y, et al. Postoperative segmental C5 palsy after cervical laminoplasty may occur without intraoperative nerve injury: A prospective study with transcranial electric motor-evoked potentials. Spine 2006; 31(26):3013–3017.

Tiel RL, Happel LT Jr, Kline DG. Nerve action potential recording method and equipment. Neurosurgery 1996; 39:103–109.

Togawa D, Kayanja MM, Reinhardt MK, et al. Bone-mounted miniature robotic guidance for pedicle screw and translaminar facet screw placement. Part 2: Evaluation of system accuracy. Neurosurgery 2007;60(2 Suppl 1):ONS129–139.

Viby-Mogensen J. Clinical assessment of neuromuscular transmission. Br J Anaesth 1982;54:209–223.

Von Koch CS, Quinines-Hinojosa A, Gulati M, et al. Clinical outcome in children undergoing tethered cord release utilizing intraoperative neurophysiologic monitoring. Pediatr Neurosurg 2002;37(2):81–86.

Winograd JM, Mackinnon SE. Peripheral nerve injuries: repair and reconstruction. In: Plastic Surgery, 2nd ed., Volume VII, Hand and Upper Limb, Part I (Mathes SJ and Hentz VR, Eds.), Chapter 11. 2006; Philadelphia: Saunders Elsevier.

# 7

# INTRAOPERATIVE NEUROPHYSIOLOGIC MONITORING OF THE CRANIAL NERVES AND BRAINSTEM

WAEL F. ASAAD, MIRELA V. SIMON, AND EMAD N. ESKANDAR



## I. INTRODUCTION

Brainstem and cranial nerve monitoring is called on whenever an operation to approach a brainstem or skull-base lesion puts nuclei, tracts, or nerves at risk of injury. A wide variety of surgical approaches are used to access these lesions. The particular approach is determined not simply by the location of the lesion, but largely by the path that allows access to the lesion while minimally disturbing critical structures. Therefore, the decision to monitor certain structures is derived from an understanding of this approach trajectory, as well as from knowledge of the lesion's location.

In order to avoid excessive manipulation and ensuing damage to bystander structures, the surgeon selects a cranial approach that provides line-of-sight access to the lesion while minimizing the need for retraction. Yet the high density of critical structures in the brainstem allows for few unobstructed trajectories, and even seemingly slight maneuvers can result in the injury of delicate nuclei, tracts, or nerves (Fig. **7-1 a, b**). Therefore, neurophysiologic monitoring is an indispensable component of almost any operation in or around the brainstem, and it may involve monitoring of several cranial nerves. Fig. **7-1 c** gives a schematic representation of the setup for different cranial nerves monitoring.

Somatic sensory pathways can be monitored using standard somatosensory evoked potentials (SSEPs), whereas monitoring motor structures usually involves recording the electromyogram (EMG). For many brainstem motor structures, the spontaneous EMG activity in innervated muscles has been found to be a reliable marker of mechanical, thermal, or metabolic irritation of the relevant nuclei or nerves. However, this is not always the case, and so neurophysiologic mapping of these structures, rather than continuous monitoring, may be possible in those situations.

Free and triggered EMG pertains to monitoring and mapping cranial nerve motor function distal to the cranial nerves nuclei. However, during brainstem surgery, postoperative cranial nerve dysfunction can arise not only from direct injury of the cranial nerves and nuclei but also from injury to the corticobulbar tracts innervating these nuclei. Morota et al. (1995) reported

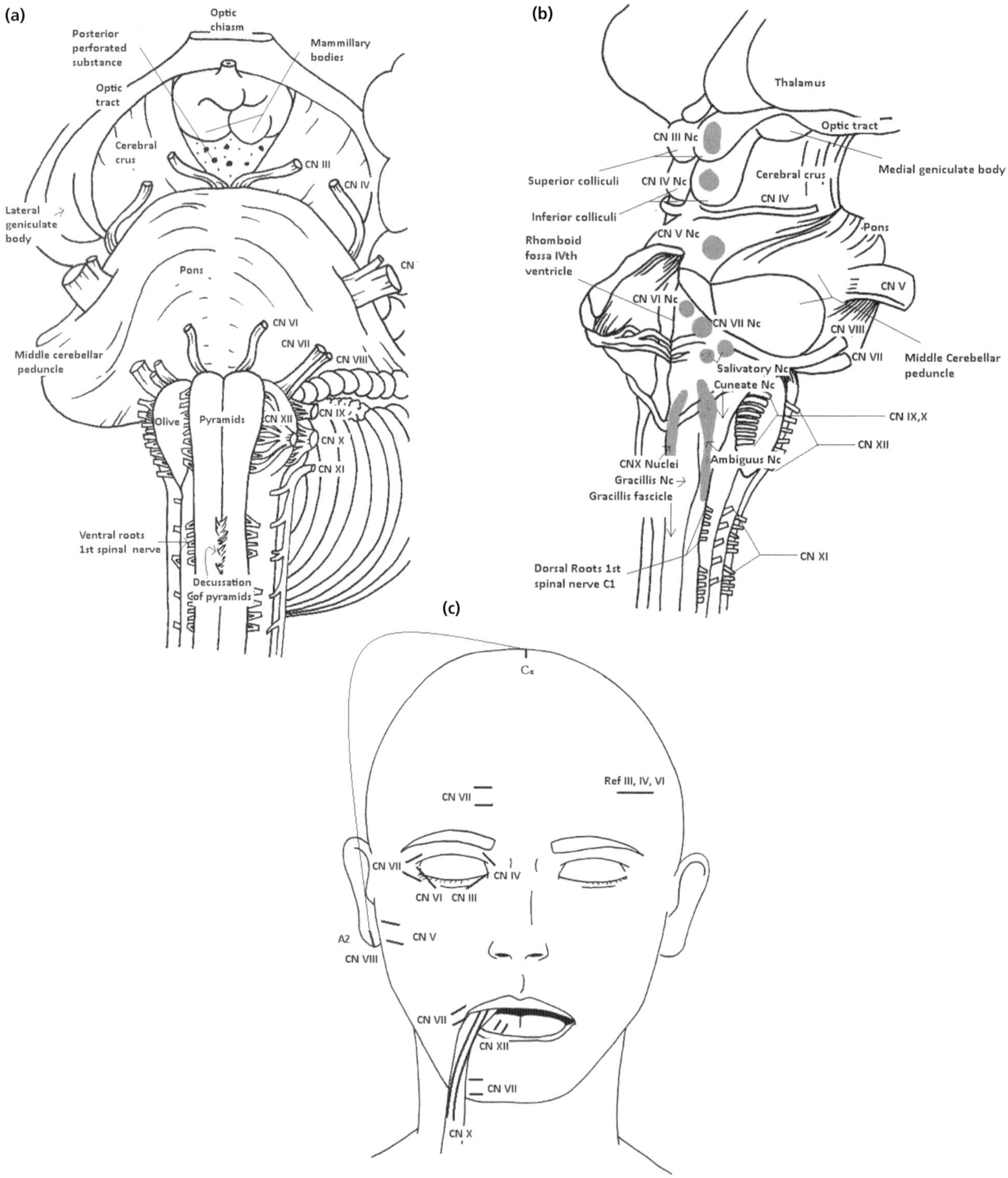


**FIGURE 7-1** Brainstem anatomy, including cranial nerves nuclei and root entry zones. Its high density of such critical structures frequently requires concomitant monitoring of several cranial nerves. **a:** Brainstem: ventral part. **b:** Brainstem: dorsolateral part. **c:** Setup for multiple right cranial nerves monitoring: CN III (medial rectus); CN IV (superior oblique); CN V (masseter); CN VI (lateral rectus); CN VII (orbicularis oculi, oris, mentalis, frontalis); CN VIII (A2–Cz); CN X (laryngeal mucles via ETT tube) and CN XII (tongue). Notice that two recording electrodes per muscles are generally used, with the exception of CN III, IV, and VI innervated muscles. These small muscles are difficult to reach; thus only one electrode per muscle is inserted, with the reference electrode placed on the contralateral forehead.

cases of postoperative deficits after resection with preserved triggered compound muscle action potential (CMAPs) in the innervated muscles of the cranial nerves, when stimulation was done distal to the cranial nerves nuclei. He thus emphasized the importance of monitoring of the corticobulbar tracts. Several authors used transcranial motor-evoked potentials (tcMEPs) in preserving the function of the cranial nerves. Deletis et al. (2009) described the methodology for successful continuous tcMEPs monitoring from the vocal cord muscles. Fukuda et al. (2008) also found that a decrease of more than 50% in the tcMEPs recorded in the facial innervated muscles predicted the facial nerve function after skull-base surgery. The advantages of MEPs recording (via transcranial or direct cortical stimulation) over that of triggered CMAPs are evident. First, MEPs offer an assessment of the entire motor pathway at risk, not only distal to the cranial nerves nuclei but also proximal to it. Second, MEPs recording technique allows direct monitoring rather than intermittent mapping of the function of the motor cranial nerves. In the following, the functional structures to be monitored or mapped are placed in the context of the surgical approach and the local neuroanatomy. The surgeon must have a clear concept of the operative trajectory and make arrangements for monitoring in advance of the procedure, and the anesthesiologist must refrain from administering agents that might interfere with the electrophysiology. The neurologist or technician performing the monitoring must have a strong understanding of normal anatomy and must attend to the progress of the operation in order to provide the best possible feedback regarding electrophysiologic changes and their interpretation.

Pages 1–46 give details about the methodology and interpretation of basic tests used in monitoring brainstem and cranial nerves: free-run and peripherally triggered EMG activity, tcMEPs, brainstem auditory evoked potentials (BAEPs), SSEPs, and nerve recordings (nerve action potentials [NAPs]). In this chapter, we expand on: (1) anatomy (see Fig. **7-1**) and operative approaches for different parts of the brainstem; (2) the particularities of monitoring cranial nerves, including troubleshooting process; the particularities of interpretation of spontaneous and triggered EMG activity in cranial nerve monitoring and its impact on and predictive value of the postoperative outcome; the association between type and timing of changes seen in the intraoperative neurophysiology recordings and different mechanisms of injury; and examples of neurophysiologic recordings in cranial nerve monitoring. Table **7-1** gives the surgical applications of cranial nerve monitoring.

For a better understanding of the brainstem anatomy, surgical approaches, and nervous structures at potential risk for surgery, see Fig. **7-1 a, b**.

## II. MIDBRAIN

### 1. Anatomy of the Midbrain and Operative Approaches

Intrinsic lesions of the midbrain can be divided into those situated primarily in the tectum, dorsally, and those located mostly in the tegmentum, ventrally. Although both areas contain a large number of functional units, a few are considered particularly worthy of preservation. Dorsal approaches endanger the decussation of the fourth cranial nerve and the collicular connections to the thalamus. Meanwhile, ventral entry into the midbrain endangers the third nerve and the cerebral peduncles.

Lesions of the dorsal midbrain in the vicinity of the colliculi can be accessed via a supratentorial (Smith et al., 2003), infratentorial, or combined approach. The patient can be positioned prone or in a semi-sitting position, allowing gravity to aide retraction of the cerebellum inferiorly. If the latter position is selected, precordial Doppler ultrasound and central venous access are recommended to detect and evacuate air emboli. The direction of the incision should be parallel to the nearest underlying fibers. For example, the fourth nerve, visualized caudal to the inferior colliculi, decussates just beneath the pial surface within the superior medullary velum; an anterior-posterior midline incision here would transect both nerves, resulting in bilateral trochlear palsies.

Meanwhile, more rostrally and laterally, fibers of the brachium of the superior and inferior colliculi also travel horizontally, carrying afferents from the geniculate nuclei. It is preferable to stay within the tectum and out of these fibers.

Medially, one must avoid entrance into the cerebral aqueduct and certainly should not violate the floor of this structure, which contains the third and fourth cranial nerve nuclei. More laterally, lying at the border of the colliculi and the peri-aqueductal gray matter, one encounters the mesencephalic trigeminal tract, which carries proprioceptive information from the face. This tract is involved in the masseter reflex arc; accordingly, damage to it does not necessarily lead to loss of facial sensation or motor function (Thomke et al., 2001).

**TABLE 7-1** Applications of cranial nerves, nuclei and brainstem monitoring/mapping procedures in surgery

| Structure | Monitoring and mapping |
|---|---|
| CN III, IV | 1. Surgeries with midbrain at risk (e.g. midbrain tumor resection)<br>2. Skull-base surgery (e.g. superior orbital fissure) |
| CN V | 1. Surgeries with pons at risk (e.g. pontine tumor resection)<br>2. Microvascular decompression of CN V and VII (trigeminal neuralgia, hemifacial spasm)<br>3. CPA tumors resection (e.g. acoustic schwanommas, facial nerve schwanommas, meningiomas, lipomas, epidermoids, arachnoid cysts)<br>4. Skull-base surgery |
| CNVI | 1. Surgeries with pons at risk<br>2. Skull-base surgery (e.g. superior orbital fissure) |
| CNVII | 1. Surgeries with pons at risk<br>2. CPA tumor resection<br>3. Intracanalicular tumor resection (e.g. acoustic schwannomas)<br>4. Microvascular decompression of CN V and VII<br>5. Skull-base surgery |
| CN VIII | 1. Surgeries with pons and lower midbrain (inferior colliculus) at risk<br>2. CPA tumor resection<br>3. Intracanalicular tumor resection<br>4. Microvascular decompression of CN V and VII<br>5. Skull-base surgery |
| CN IX | 1. Surgeries with medulla at risk<br>2. Foramen magnum pathology<br>3. Jugular foramen pathology<br>4. Skull-base surgery<br>5. Clivus surgery (bilateral monitoring recommended) |
| CN X | 1. Surgeries with medulla at risk<br>2. Foramen magnum pathology<br>3. Jugular foramen pathology<br>4. Skull-base surgery<br>5. Clivus surgery (bilateral monitoring recommended)<br>6. Neck surgery<br>7. Thyroid surgery<br>8. ACDF |
| CN XI | 1. Surgeries with medulla at risk<br>2. Foramen magnum pathology<br>3. Jugular foramen pathology<br>4. Skull-base surgery<br>5. Clivus surgery (bilateral monitoring recommended) |
| CN XII | 1. Surgeries with medulla at risk<br>2. Hypoglossal schwannoma<br>3. Foramen magnum pathology<br>4. Fourth ventricle tumors<br>5. Skull base (e.g. hypoglossal canal)<br>6. Clivus surgery (bilateral monitoring recommended)<br>7. Tongue pathology (e.g. christoma of the tongue) |
| Brainstem (involving a combination of cranial nerve monitoring, SSEPs, and tcMEPs) | 1. Tumor resection (brainstem and CPA )<br>2. Vascular surgeries involving posterior circulation<br>3. Suboccipital decompression (Chiari malformation, C1 fracture)<br>4. Skull-base surgery |

The ventral midbrain contains the cerebral peduncles laterally, the substantia nigra, just dorsal to them but still superficially accessible medially, and the red nuclei, forming much of the intra-axial mass. The third and fourth nerve nuclei are located just ventral to the cerebral aqueduct. Lesions of the ventral midbrain are generally accessed via a pterional craniotomy or some modification thereof. In these approaches, the midbrain is exposed either by dissecting down the Sylvian fissure or by raising the frontal lobe off the skull base. The third nerves are seen exiting the brainstem anteriorly (ventrally), and the contralateral fourth nerve is found looping around the midbrain laterally. The ipsilateral cerebral peduncle is oriented toward the surgeon.

## 2. Particularities of Monitoring/ Mapping of Midbrain Structures

### *A. Oculomotor Structures (CN III, CN IV, and Their Nuclei)*

For more deeply seated lesions, or those near the midline, electrophysiologic monitoring can be an important guide to the local anatomy of third- and fourth-nerve structures (Sekiya et al., 2000). There is uncertainty regarding the ability of mechanically evoked (i.e. resulting from direct physical contact between a surgical implement and a nerve or nucleus) EMG activity in the extraocular muscles to signify irritation of the brainstem oculomotor system (Schlake, Milewski, et al., 2001). Therefore, the oculomotor system is more reliably mapped than monitored. Accurate mapping requires an electrode to be placed into one or more extraocular muscles (medial rectus, for CN III; superior oblique, for CN IV), usually by an ophthalmologist after induction of anesthesia, to monitor contraction when the relevant midbrain oculomotor nuclei or tracts are electrically stimulated by the surgeon (see Fig. **1-11 a**, page 30). The reference electrode can be placed on the contralateral forehead. Stimulation technique is similar to that of the other cranial nerves and is described on page 35. In this way, the oculomotor structures can be mapped and avoided so that postoperative deficits (most often diplopia) can be minimized.

### *B. Cerebral Peduncles*

The motor homunculus is oriented medially to laterally, face-to-leg, along the surface of the peduncles; the appropriate muscles can be monitored, depending on the location of the lesion and the site of pial entry. If the underlying lesion is sufficiently superficial, the overlying corticospinal tracts may have functional gaps, created by displacement of fibers or by direct damage resulting from compression or chemical irritation by a lesion. Furthermore, based on knowledge of the homunculus, regions corresponding to a patient's established deficit can be mapped. The myelotomy can then be made in an anterior-to-posterior direction at a medial-to-lateral position that has been chosen so as to minimize injury to remaining functional tissue. At least one group has reported the successful resection of a lesion in this region utilizing intraoperative monitoring of MEPs generated by direct bipolar stimulation of the cerebral peduncles, as well as by tcMEPs (Quinones-Hinojosa et al., 2005).

In general, even when a lesion appears to present directly to the pial surface in the region of a known motor pathway, preoperative arrangements to perform electrophysiologic monitoring are useful, because standard anatomical magnetic resonance imaging (MRI) alone is not a sufficient guide to the extent and nature of the overlying tissue, especially when that tissue consists of axonal tracts that may be very thin yet nevertheless functionally significant.

# III. PONS

## 1. Anatomy of the Pons and Operative Approaches

Lesions located in the pons can be accessed either through the floor of the fourth ventricle, dorsally, or via the basis pontis, ventrally. The former is achieved through a midline suboccipital approach, traversing the obex toward the rhomboid fossa. The latter is approached most commonly through a lateral suboccipital craniotomy.

Lesions within the floor of the fourth ventricle generally are considered resectable if they present to or through the surface. However, careful planning with respect to the relevant anatomy and appropriate intraoperative monitoring can minimize the risk of new neurologic injury when this approach is necessary to access even deeper lesions. The critical structures to avoid are clustered into two groups, represented on the surface of the floor of the fourth ventricle by the facial and hypoglossal colliculi. The facial colliculi, formed by the sixth nerve nuclei and the overlying traversing fibers of the seventh nerve, are located in the base of the rostral triangle of the rhomboid fossa. These structures are within 2–4 millimeters of the surface. More caudally, the hypoglossal colliculi are less prominent than the facial colliculi; they represent the rostral portion of the hypoglossal nuclei, which are long structures (~11–12 mm) in the open portion of the medulla. Running along the midline sulcus is the medial longitudinal fasciculus (MLF), relaying fibers anteriorly from the sixth nerve nucleus and vestibular nuclei, which mediate voluntary eye movements, as well as the vestibulo-ocular reflex, and carrying posteriorly fibers, which likely mediate head movements associated with gaze shifts.

In addition, there are many important structures that are not immediately evident externally, but whose position can be inferred from external landmarks. In close association with the hypoglossal nucleus is the dorsal parasympathetic motor nucleus of the vagus nerve. This nucleus is laterally situated relative to the rostral head of the hypoglossal nucleus, but it swings dorsomedially to lie more superficially in the medulla. Lateral to these is found the Solitary Nucleus. Most laterally, at the ventrolateral angle of the rhomboid fossa and extending caudally into the medulla, are found the vestibular nuclei. They are situated between the fourth ventricle and the cerebellar peduncles more laterally. The motor and principle sensory trigeminal nuclei are also near the lateral angles of the fourth ventricle, just ventral to the superior vestibular nuclei.

Basis pontis lesions can be approached ventrolaterally. Either a far lateral suboccipital craniotomy or a larger craniotomy for a combined supra/infratentorial approach can be utilized to gain access to the ventrolateral surface of the lower brainstem (Baldwin et al., 1995). In addition, a subtemporal approach is possible for anteriorly situated lesions that may not be accessible from a ventrolateral angle (Sarma et al., 2002); however, this involves drilling into the petrous bone to expose and then mobilize the petrous carotid artery.

Within the pons, the descending motor fibers of the corticospinal and corticobulbar tracts lie clustered within the interspersed and surrounding pontine nuclei. These fibers do not rise to the surface at the level of the pons; rather, ponto-cerebellar fibers arise ventrally and curve dorsolaterally along the ventrolateral surface. A paramedian incision into the ventral pons, therefore, is generally considered optimal (Cantore et al., 1999) unless the lesion rises to the pial surface elsewhere. Indeed, the occurrence of significant or lasting new postoperative deficits has been found by some to be lowest when the location of a lesion allowed for this particular anterolateral approach (Ferroli et al., 2005).

## 2. Particularities of Monitoring/ Mapping Pons Structures

Electrophysiologic monitoring and direct stimulation can be applied to identify and preserve many of the important neural structures and pathways of the pons (Chang et al., 1999; Glasker et al., 2006 Morota et al., 1995; Morota et al., 2006).

### A. Trigeminal System

The trigeminal motor nucleus or nerve can be monitored with recording electrodes in the masseter, as shown in Fig. **1-11 b** (page 30). Monitoring of the motor part of the CN V is usually done together with CN VII and VIII monitoring in cerebello-pontine angle (CPA) tumor removal, skull-base surgery, or microvascular decompression, for trigeminal neuralgia or hemifacial spasm. During these procedures, CN V monitoring can be part of the active monitoring if directly involved in the process and thus at direct risk of injury, or it can be used as a reference during the active monitoring of the CN VII (see the following). As with the other cranial nerves (CN VII, IX, X/recurrent laryngeal nerve [RLN], XI, and XII), the monitoring involves free-run EMG and triggered EMG. The nerve irritation is depicted by the spontaneous EMG patterns described by Romstock and colleagues (2000) (see Fig. **7-2 a, b** and pages 306–9); the nerve mapping, and predissection and postdissection assessment of its integrity, is done by direct electrical stimulation of the nerve and by recording CMAPs from the masseter muscle (Fig. **7-3 a**).

The trigeminal somatosensory system, which provides sensation to the face and the anterior two-thirds of the tongue, can be monitored using standard SSEPs. Minimally, subdermal recording electrodes are placed in the region of the facial homunculus of the postcentral gyrus and over the mastoid. Sensory nerves are then tested by stimulating electrodes placed into the supra-orbital, infraorbital, or mental foramina to trigger the respective nerves of the same names, corresponding to the three divisions of the trigeminal system (Stechison et al., 1996). In addition, it is possible to map the tongue representation by applying a slight electrical stimulus to the anterior two-thirds of the tongue (Altenmuller et al., 1990).

### B. Abducens Nerve (CN VI), Nucleus, and Facial Colliculus

When working within the fourth ventricle, the abducens nucleus and the overlying facial tract fibers are superficially

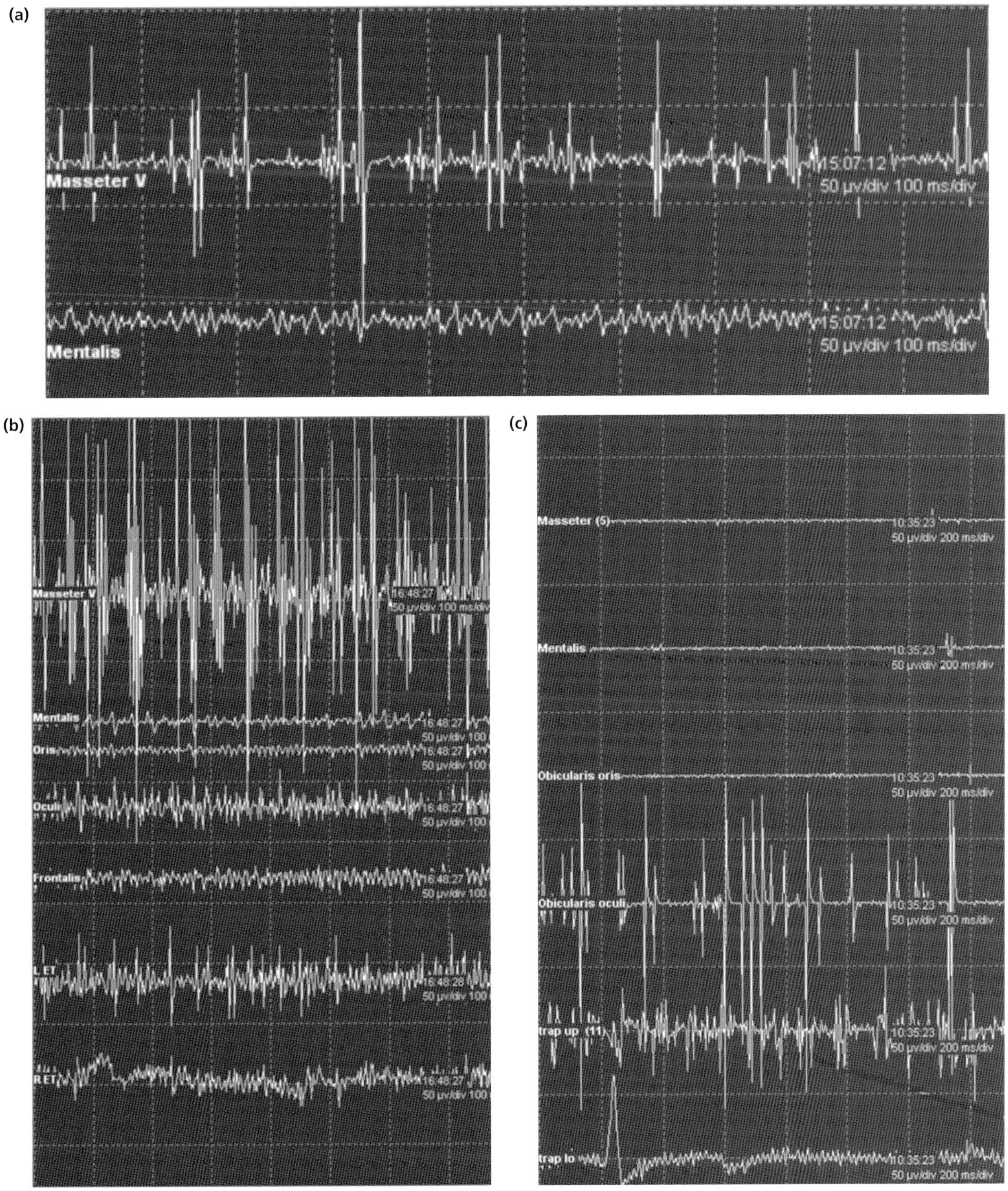


**FIGURE 7-2** Examples of B and C trains in free-run EMG activity during resection of metastatic melanoma of the brainstem. **a:** B train in masseter muscle. **b:** C train in masseter muscle. **c:** B train in orbicularis oculi and transition from a B to a C train in trapezius muscle.

(d)
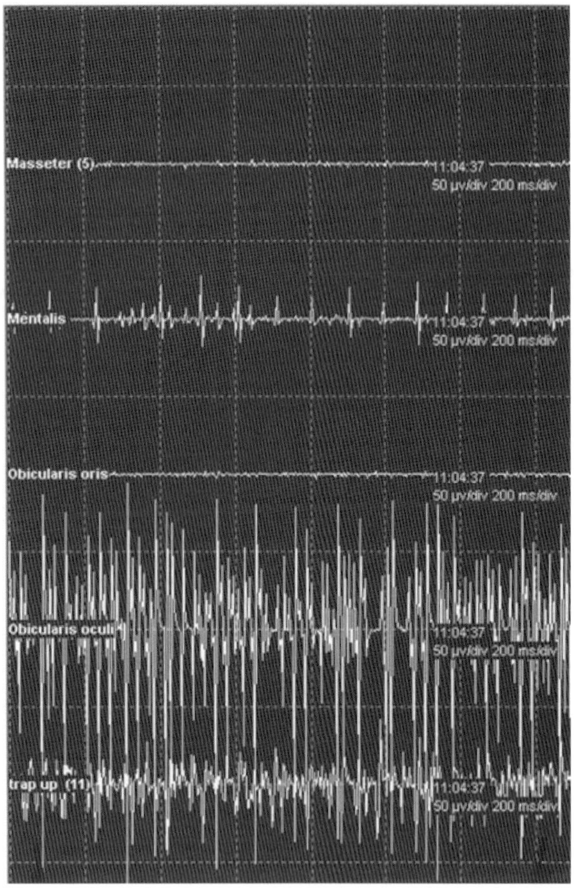


**FIGURE 7-2** *(Continued)* **d:** Increased irritation with development of C train in orbicularis oculi and appearance of a B train in the mentalis muscle.

situated and so are at risk of damage. As in the prior discussion, oculomotor mapping and facial monitoring can be used to identify the location and extent of the relevant structures. For example, stimulating in the region of the abducens nucleus can provide information regarding the integrity of the ascending MLF in the midline of the fourth ventricle. Electrically evoked activity of the abducens nucleus is assessed using recording electrodes placed in the lateral rectus muscles (see Fig. **1-11 a**, page 30). Testing the overlying facial nerve fibers can be achieved by recording CMAPs from the orbicularis oris and the orbicularis oculi. Importantly, as appears to be true for the other oculomotor cranial nerves, there is no clear predictive relationship between spontaneous EMG activity in the lateral rectus muscle and preserved CN VI function (Grabb et al., 1997); however, activity of the lateral rectus has been observed to correlate with damage to the abducens nerve.

## C. Facial Nerve (CN VII)

The facial nerve is most commonly monitored in the setting of surgery for CPA tumors, such as schwannomas and meningiomas. It is also monitored in microvascular decompressions of the facial nerve (in hemifacial spasm) and of trigeminal nerve (in trigeminal neuralgia). Recording electrodes are placed into the orbicularis oculi and orbicularis oris muscles and often into the frontalis and platysma or mentalis as well, as shown in Fig. **1-11 b**, page 30). Of note, the optimal location of the orbicularis oculi electrodes appears to be near the orbital rim (Guo et al., 2008). Thus, both spontaneous activity (free-run EMG) and the CMAP, evoked by directly stimulating the facial nerve (triggered EMG) can be recorded. The former may serve as a warning sign of impending iatrogenic injury (particularly when certain frequent or prolonged types of EMG patterns are observed), whereas

(a)
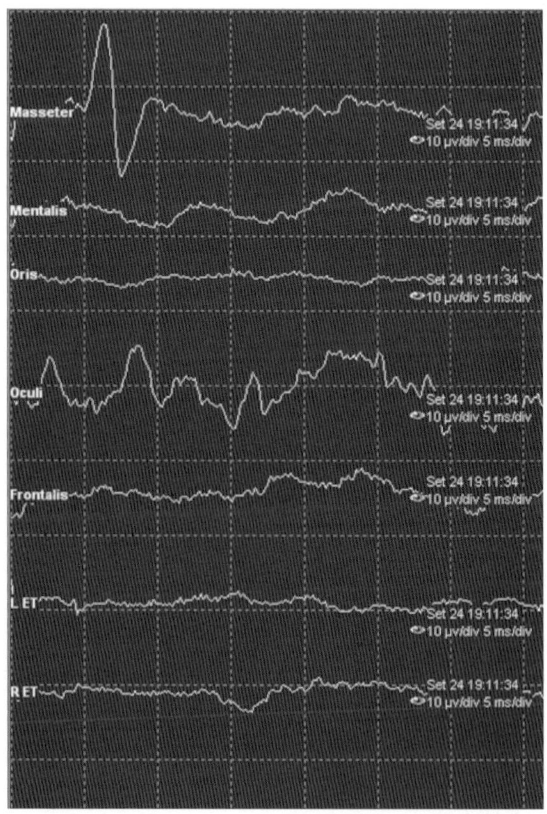


**FIGURE 7-3** Examples of triggered CMAPs: **a:** Masseter (CN V).

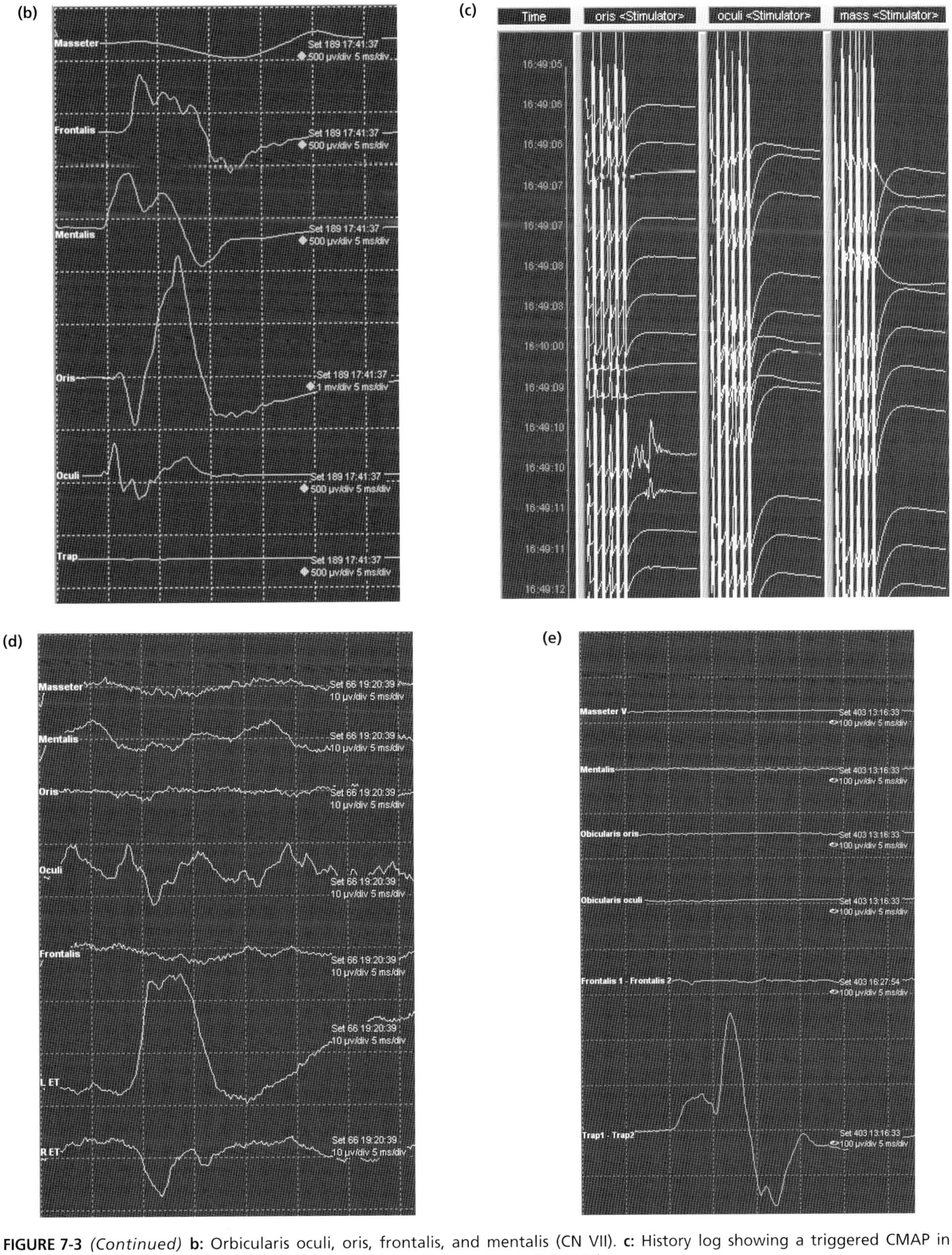


**FIGURE 7-3** *(Continued)* **b:** Orbicularis oculi, oris, frontalis, and mentalis (CN VII). **c:** History log showing a triggered CMAP in orbicularis oculi muscle. **d:** Vocal cords muscles (CN X). **e:** Trapezius muscle (CN XI).

the latter can be used to map the nerve when the anatomy has been distorted by pathology (Hone et al., 1997; Silverstein et al., 1988a; Silverstein et al., 1988b). The characteristics of both types of EMG activity are detailed as follows.

*C1. Free-run EMG activity.* Romstock et al. (2000) describe different patterns of free-run EMG activity and their value in predicting the surgical outcome. Type A are trains of EMG activity, of 100–200 $\mu$V up to 400 $\mu$V amplitude and a frequency of 60–210 Hz, lasting between a few milliseconds to seconds. It is usually observed with compression, traction, and ischemia of the nerve. Romstock et al. (2000) showed that the presence of these trains specifically correlates with a poor prognosis of the facial nerve. Spikes and bursts are usually seen when instrumenting in the close proximity of the cranial nerves. For example, large bursts (>500 $\mu$V) are typically seen in the laryngeal and mimetic muscles when arachnoidal planes between the tumor and the corresponding nerve are separated, as well as during drilling and decompressing of the internal auditory canal (IAC). B trains consist of runs of spikes or bursts at relatively regular intervals; the amplitude varies from 20 $\mu$V to 5000 $\mu$V. C trains consist of irregular, high-frequency muscle activity, with a wide range of amplitudes, from 20 to 5000 $\mu$V. Unlike A trains, B and C trains are "benign" and do not correlate with poor outcome; they usually occur at nerve contact, irrigation of the surgical field, and are consistently seen during pain-triggering surgical steps. They are overall more specific to CN IX and CN X innervated pharyngeal muscles (Prell et al., 2008; Romstock et al., 2000). Prell and colleagues (2007) studied the predictive value of the length of A trains on the surgical outcome in CPA tumor removal (vestibular schwannomas) and concluded that, in patients with previously healthy facial nerves, a mean length (per patient) of less than 0.5 seconds did not correlate with worsening of the facial nerve function. A mean length of more than 0.5 seconds, but less than 10 seconds, did result in worsening of the facial nerve function of 1 House Brackmann (HB) degree, whereas a length of more than 10 seconds resulted in a worsening of 2 or more HB degrees. Thus a "safety zone" for A trains was defined in CPA tumor resection cases, but only in patients with previously healthy nerves. In the group with previously injured nerves, all patients developed A trains, and all of them had a worsening of their deficits, including those with trains less than 0.5 seconds in length. Additionally, the prevalence of A trains was significantly higher in the patients with previous facial nerve injury than in those with previously healthy nerves (100% versus 77%). A year later, Prell extended his analysis to a different procedure, the microvascular decompressions for trigeminal neuralgia, and revisited the safety zone parameters for A trains. However, the threshold of 0.5 seconds for the mean length of A trains did not apply any longer, because all of the patients in whom the A trains were observed had a mean A train length above 0.5 seconds but no postoperative deficit. Additionally, the overall prevalence of the trains was significantly smaller in cases of microvascular decompression (MVD) of CN V for trigeminal neuralgia than in CPA tumor removals, and only 22% of the patients developed such trains. The A trains seen in MVD cases (extending to both trigeminal neuralgia and hemifacial spasm) had an overall mean length much smaller than A trains in the CPA angle tumors. Also, unlike the latter, they did not have an abrupt occurrence, but rather progressive developments from trains B and C. This progressive evolution was seen as a "warning sign" for imminent injury. In conclusion, it seems that in surgeries involving previously healthy nerves and with a lower risk of injury, the A train time threshold (signaling the imminence of significant nerve injury) is higher, frequency of such trains is lower, and their appearance is gradual. Conversely, previously damaged facial nerves have a lower tolerance to stress, thus a lower threshold for occurrence of A trains, which develop more frequently and more abruptly. This is somewhat the opposite of the case of triggered CMAPs, when diseased roots or nerves have a higher stimulation threshold for triggering CMAPs and the amplitudes are smaller. Figure **7-2** gives examples of B and C trains. Notice in figure **7-4** how cusa artifact does have the morphology of A trains, yet the amplitudes are significantly smaller. However, special attention should be paid during cusa use, because this is not an inoffensive step and nerve damage can easily occur during cusa removal of an infiltrative tumor. The main mechanisms remain stretching, compression, or ischemia of the nerve, followed by the appearance of A trains. Thus, we recommend intermittent use of cusa, which still allow observation of free-run EMG activity, in the absence of electrical artifact.

*C2. Triggered EMG activity.* Direct electrical stimulation of the nerve can be done either with bipolar or with monopolar stimulation handheld probe. The particularities of both types of stimulation are described in detail on page 27. The stimulation intensity threshold for the facial

(a)
(b)

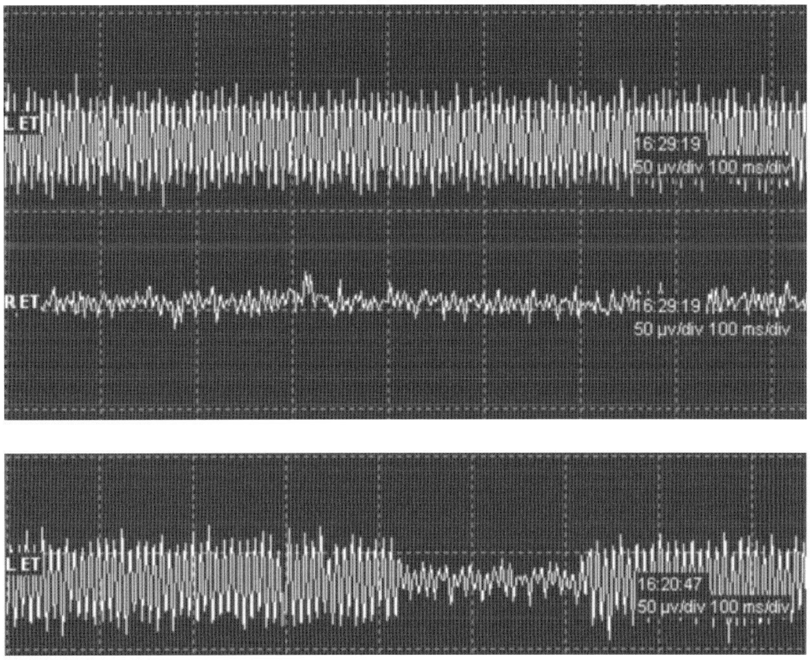


**FIGURE 7-4** Electrical artifact produced by cusa use **a:** continuous, **b:** intermittent. Notice the similar morphology and frequency (~ 110–140 Hz) with A trains. However, the amplitude of the artifact is smaller (~50 uV) than that expected with A trains.

nerve varies according to the other parameters (pulse duration), as well as to the state of the nerve (an injured nerve depolarizes at higher threshold). For a healthy nerve, we have had thresholds as low as 0.1 mA for a 0.2-msec pulse duration and up to 1 mA for 0.1-msec pulse duration.

A baseline threshold should be taken first, by stimulating the facial nerve proximal to the tumor, and before the dissection (or in general proximal to the surgery site), and then later on at the end of the surgery. Another way to do it is to use a ratio between the depolarization threshold when stimulating proximally versus stimulating distally to the surgical site, measured before and after procedure. The electrical threshold for eliciting a motor response is considered to be a good predictor of postoperative facial nerve function. Specifically, the lower the threshold is, the more likely it is that the patient will have preserved facial strength (Magliulo et al., 1997; Neff et al., 2005). Adding two more electrodes, one in the platysma and one in the frontal muscle, increases the positive predictive value of this measure (Grayeli et al., 2005). Fig. **7-3 b** and **c** shows examples of facial CMAPs. In order to increase the specificity of our triggered response and to eliminate the possibility of current spread, we always sample the masseter muscle as well. If responses are obtained in innervated muscles of both seventh and fifth nerves, we decrease the stimulus intensity until muscles innervated by one nerve only show response. There are instances when some spontaneous muscle activity is seen in the masseter, and this could be mistaken for a triggered response. In such cases, analyzing the free-run EMG at a comparable time base with the triggered EMG time base may enable recognition of EMG activity with similar morphology but random latencies. As already mentioned, facial nerve monitoring can be utilized in operations to relieve hemifacial spasm. Specifically, the disappearance of abnormal motor responses (elicited by stimulating one branch of the facial nerve and observing a response in muscles innervated by other branches) was found by some authors to predict a good postoperative outcome (Huang et al., 2009; Møller et al., 1987; Mooij et al., 2001; Yamashita et al., 2005; Zhu et al., 2008). On the other hand, Joo and colleagues (2008) found that the disappearance of the abnormal motor responses does not predict the long-term outcome.

## *D. Vestibulo-Cochlear Nerve (CN VIII)*

Surgical applications of CN VIII monitoring are given in Table **1-8** (page 41), as well as in this chapter, Table **7-1**. During these procedures, the cochlear nerve can be damaged by several mechanisms: stretch, compression, heat injury, ischemia, and vascular compromise. Additionally, rupture, occlusion, and vasospasm of the internal auditory artery can also cause ischemic damage to the cochlea. The latter can also be damaged by drilling-induced noise (see section about early drilling). Several methods exist to monitor the cochlear portion of the eighth nerve, including the cochlea. These include BAEPs, transtympanic electrocochleography (EcochG) responses, or compound action potentials (CAPs), and compound nerve action potentials of the cochlear nerve (CNAPs).

*D1. BAEPs.* The most established method is recording of BAEPs from subdermal electrodes (usually placed in the skin at the vertex and mastoid). Details about the stimulation and recording techniques, as well as about the neurophysiologic-neuroanatomic correlation, are given on pages 37–41.

In the following we describe several key steps of the intraoperative BAEPs monitoring procedures (MVD and CPA tumor resection combined) as well as the troubleshooting process.

***a. Obtaining the baselines.*** Before induction, hearing thresholds are obtained in both ears. These give an idea about the expected quality of BAEPs. Baseline BAEPs should be done in the monitored as well as the non-monitored ear. As described on page 37, because BAEPs are small-amplitude potentials, increased averaging is necessary, sometimes up to 2,000 trials. If good baselines are not obtained, the following should be considered:

1. *Is there appropriate stimulation?* First, one should make sure that the recording system is working properly. All connections should be systematically checked. A click sound should be produced at the level of the transducer: listening with the stethoscope at this level can uncover such a problem (Fig. 7-5).

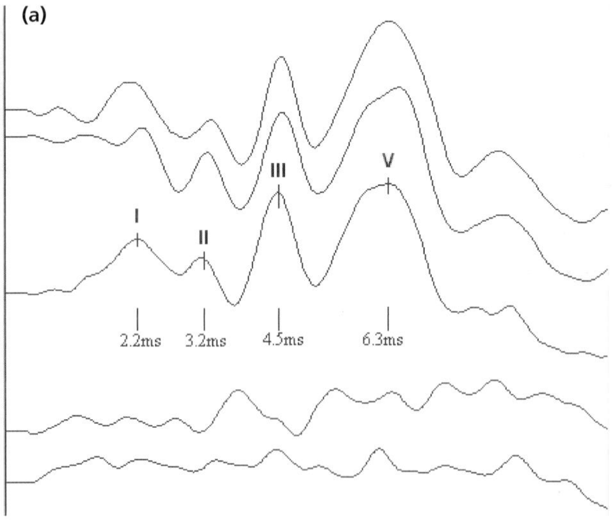


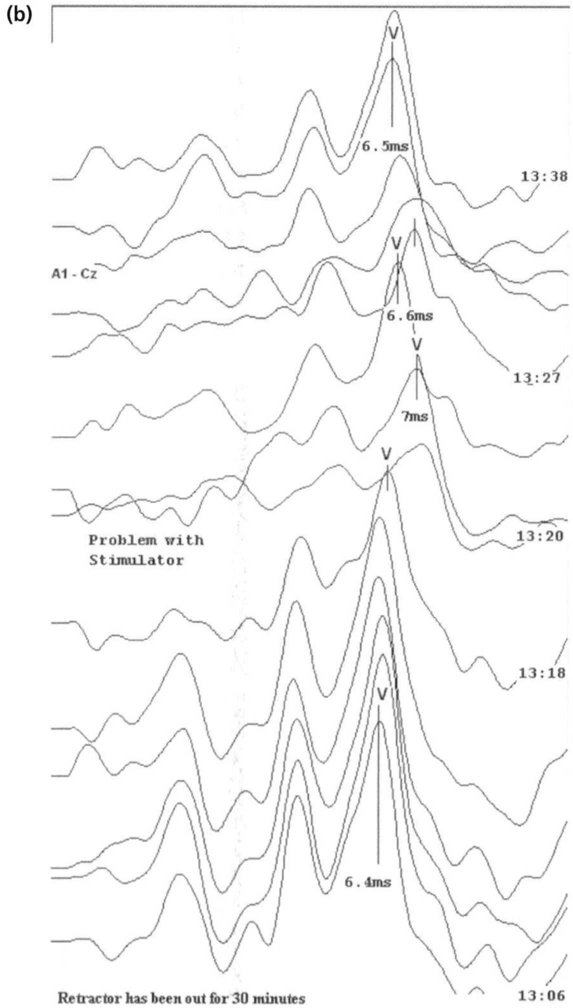


**FIGURE 7-5** Inadequate stimulation. **a:** In the beginning of the case, a nonfunctional transducer failed to deliver the stimulation click. **b:** Increased absolute latencies of wave V, with relative preservation of the interpeak latencies, are seen in trials at 13:20 and 13:27. This is due to a sudden decrease in the stimulus delivered, from 110 dB to 70 dB, which is likely due to an accidental swap between the stimulus intensity (110 dB) and white noise setup (70 dB). Once the intensity is increased to 110 dB, there is a prompt return to baseline of all absolute latencies. Sudden changes in the BAEPs, in particular when intermittent (e.g. trials showing poor BAEPs alternating with normal trials), should raise the question of intermittent, inappropriate stimulation.

2. *Is there an obstacle in the way of the stimulus?* We find that many patients do have significant amount of wax in their ears, which could interfere with the sound transmission. The length of the tube inserted in the ear is also important, because this could compromise appropriate stimulation. Extremely flexible tips can kink and block sound transmission. Cutting the tips shorter can overcome this problem. Checking the position of the ear pieces is also important.
3. *Is there too much noise and muscle artifact?* Because these potentials are very small, increased noise, particularly arising from muscle, as sometimes is encountered at the beginning of the surgery, can significantly decrease the signal-to-noise ratio. Inducing and maintaining a neuromuscular block significantly improves this ratio. However, this may become problematic later on, when EMG activity is used to monitor and map other cranial nerves (see page 321).

***b. Opening/incision time.*** During this early period of the surgery, waxing and waning of the waveforms can occur. We encounter this problem many times. A significant amount of noise, produced by bovie, can definitely worsen the morphology of the waveforms. We found it useful to pause the averaging during bovie, thus minimizing the amount of noise introduced in our recordings. We sometimes ask the surgeon to pause the bovie until we obtain reliable responses.

***c. Early drilling in the squamous part of the temporal bone, followed by IAC drilling.*** This period, which is a part of acoustic schwannomas surgery, is characterized by an increase in the ambient noise, reflected in a decrease in the signal-to-noise ratio. However, real BAEPs changes can also occur during this time, particularly affecting wave I, and can lead to its attenuation or disappearance. Thus, when BAEPs changes are seen, it is a good idea to ask the surgeon to stop the drilling until a good signal is obtained again. Real changes during this surgical step are the consequence of injury of the cochlea itself or of the distal portion of the cochlear nerve. In fact, Gouveris and colleague (2009) found that IAC drilling is one of the surgical steps most commonly involved with a significant decrease in BAEPs and EcochG potentials. The injury can be mechanical, vascular, or noise-induced. Cochlear damage can occur during drilling in the temporal bone, when the membranous labyrinth can be accidentally entered. This leads to loss of all BAEPs and deafness (Legatt, 2002). Additionally, during the drilling, noise-induced damage of the cochlea can also occur. Third, its vascular compromise, a result of injury of the labyrinthine artery, is occasionally encountered. The labyrinthine artery, also known as auditory or internal auditory artery, is a long branch of either the anterior inferior cerebellar artery, in most of the cases, or basilar (<10% cases). It accompanies the vestibulocochlear nerve through the internal acoustic meatus and is distributed to the internal ear. This artery can be easily damaged during drilling in the IAC or during the nerve dissection. Ischemic damage or infarction of the cochlea produces an abrupt loss of wave I, with subsequent loss of all waveforms. In fact, damage of the IAC was found by Nadol and colleagues (1987) to be one of the most common reasons for sudden loss of all BAEPs during CPA tumor resection. Internal auditory artery can also be damaged by disruption, compression, or coagulation during resection of the intracanalicular portion of the tumor. Thus, Ojemann (2001) has recommended trying to preserve any vessel that is entering the internal auditory meatus.

***d. Opening the dura.*** Sudden increases in latencies can be seen with dural opening, as well as changes in wave I. These are, in general, due to a decrease in temperature and probably also to a change in the environment around the acoustic nerve (AN). These changes are benign.

***e. Period of cerebellar retraction.*** Retraction of the cerebellum (Fig. 7-**6 a**) often causes stretching of the AN. The usual sign is a progressive increase in the latency of wave V. Release of the traction is usually followed by a prompt recovery (Fig. 7-**6 b**). Changes in wave I are also seen. Some have described the presence of a bifid wave I (Polo et al., 2004) or the appearance of a wave I' between waves I and IIa (Sato et al., 2009), as further described in the following. The same authors (Sato et al., 2009) have found a strong linear relationship in the changes in wave V latency and changes in I–III interpeak latency. However, wave V, believed to originate in the inferior colliculus, remains the waveform most commonly followed and measured, especially its latency: loss of this wave is associated with significant postoperative hearing loss (James et al., 2005). On the other hand, the presence of wave V at the end of the surgery is highly predictive of preserved hearing (Nadol et al., 1992). Several thresholds for increase in the wave V latency have been described as criteria for traction release. Kemick and colleagues (1990) have suggested that an optimal postoperative outcome can be achieved by stopping the tumor dissection when the changes in wave V latency are greater than 2 msec

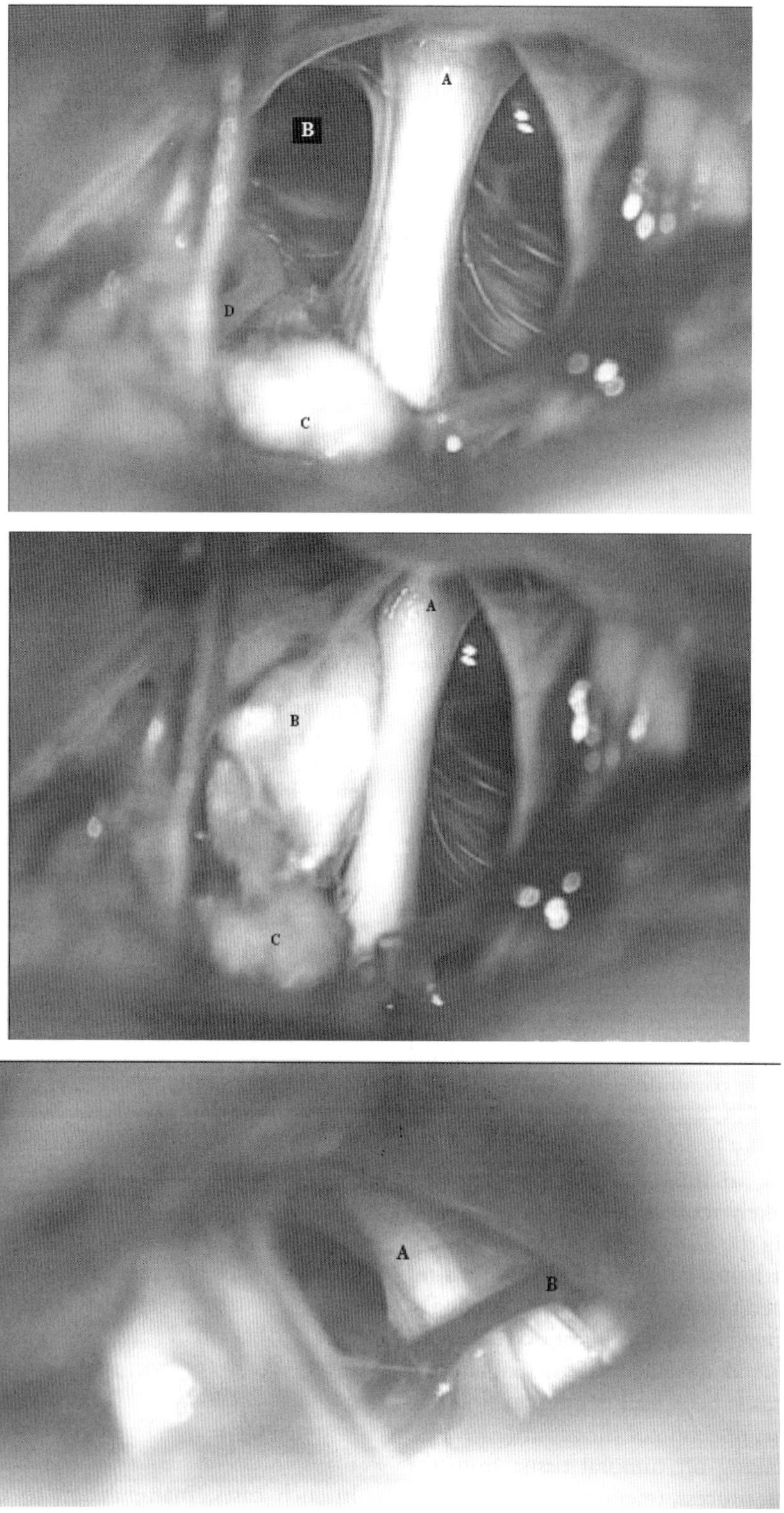


FIGURE 7-9 Microvascular decompression in trigeminal neuralgia in two different patients (**a/b** and **c**, respectively). **a:** Placement of a Teflon piece (labeled **C**) between CN V (labeled **A**) and superior cerebellar artery (labeled **D**). **B** denotes the space where a second piece of Teflon will be placed. **b:** Same patient as in **a**, this time a second Teflon piece (labeled **B**) is placed between the vessel and the nerve. **c:** A prominent branch of superior petrosal vein (labeled B) is compressing the CN V (labeled **A**).

(Fig. **7-10**). At times, the vasospasm occurs within CN VIII itself, which becomes pale. Again, irrigating the nerve with papaverine and warm saline may also lead to recovery of BAEPs (Nadol et al., 1987).

As already mentioned, because the signal-to-noise ratio of BAEPs is small, many BAEPs must be averaged to produce a reliable waveform. Therefore, awareness of any changes in this waveform depends on an accumulation of signals over time and so lags the responsible operative events. In addition, the BAEP is susceptible to a variety of systemic factors—such as changes in temperature and, to a lesser extent, depth of anesthesia—that may complicate interpretations of alterations in waveform morphology (see page 321). These factors combine to make the appreciation of gradual changes in nerve function at times difficult to ascertain. Although BAEPs monitoring has an increased sensitivity in detecting auditory damage, the specificity of the method was found to be suboptimal (Colletti et al., 1998; Yamakami et al., 2002, 2003). Thus, minimizing possible false-positive changes (such as technical-related problems) is important.

***b. Dura closure.*** During dura closure, compression of the auditory pathways can occur. This may result in postoperative hearing loss (Møller et al., 1989). As a consequence, the BAEPs monitoring should be continued during closure.

*D2. CNAPs of the cochlear nerve.* Other methods have been developed to attempt to provide a greater degree of electrophysiologic specificity and immediacy. The simplest, conceptually, is direct monitoring of the eighth nerve, in which an electrode is placed on the nerve itself to record the evoked response, a CNAP. The CNAP represents the sum of all of the NAPs arising in the nerve fibers constructing the cochlear nerve. The signal-to-noise ratio of the CNAP is much higher than that of the BAEP, with amplitudes in the 5 to 20 $\mu$V range (latencies ~ 2–4 msec). Thus, the averaging is not required, and the feedback to the surgeon is faster. Moreover, the CNAPs recordings were found by many authors to be more robust than BAEPs during the dissection procedures (Yamakami et al., 2009). It is therefore recommended that, in surgeries where BAEPs are not reliable, CNAPs recording should be tried. The same authors emphasize the importance of a steady recording and describe a newly designed electrode, consisting of a tuft of cotton attached to a flexible wire. This seems to make good contact with the cochlear nerve and allow steady recordings. The electrode is placed in the cisternal portion of the

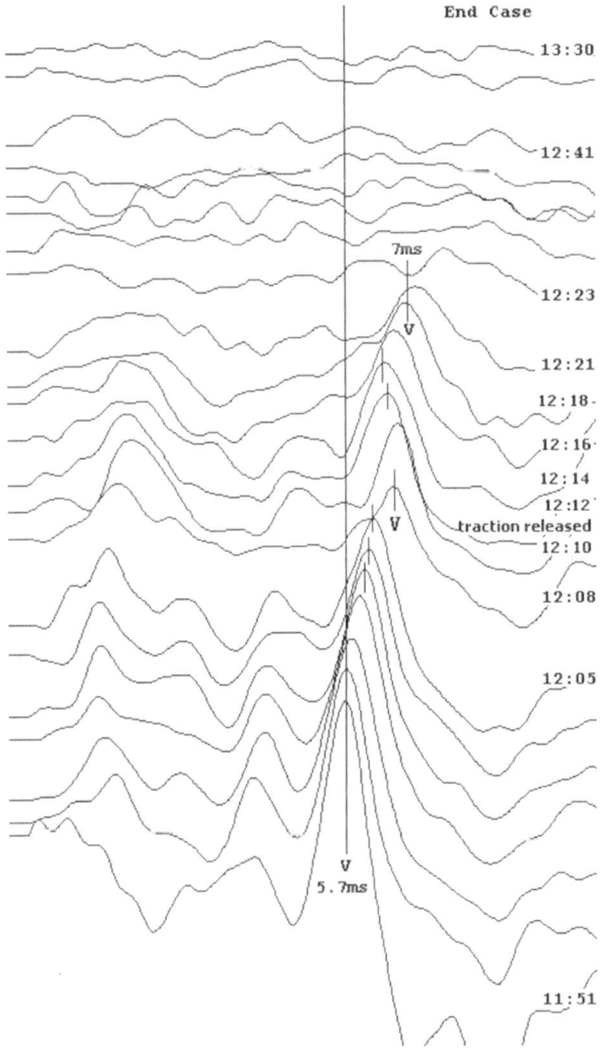


**FIGURE 7-10** Microvascular decompression for trigeminal neuralgia. Total loss of BAEPs is most likely due to vasospasm causing cochlear infarction. Notice a slow, progressive increase in the latency of wave V during the cerebellar retraction. During this time, the surgeon has been decompressing CN V by placing a Teflon between it and the irritating vessel. Until the trial at 12:05 PM, the increase in latency of wave V is considered directly related to cerebellar retraction and within acceptable limits. However, the trial at 12:08 PM requires additional averaging, and the latency of wave V increases more than predicted by the expected trend. The surgeon is promptly informed and releases the cerebellar retraction around the same time. Trials immediately following the retraction, from 12:10 to 12:16, show an improvement in the wave V latency. However, starting at 12:18, a further increase in the wave V latency is seen. The trial at 12:21 PM shows no wave I and the last well-defined wave V. No reproducible waveforms are present in the following trials. The surgeon irrigates with papaverine and warm saline the recently displaced vessel, but no improvement in BAEPs is seen. The patient will wake up with hearing deficit.

nerve, proximal to the tumor and close to the nerve root entry. Jackson et al. (2000) placed their CNAP recording electrode in the extradural portion of the nerve, with the tumor situated more proximal. However, in order to avoid false-negative results, the importance of always placing the electrode proximal to the pathological process has been emphasized by several authors (Møller, 1996; Møller et al., 1982). Thus, this type of recording is most easily achieved during the removal of the small and more distal acoustic neuromas, which may not obstruct the view in the CPA angle and allow placement of the electrode proximal to the tumor. Yamakami and colleagues (2009) showed that CNAPs preservation had a 100% positive predictive value (PPV) and negative predictive value (NPV) for serviceable hearing. Colletti et al. (1998) showed that the permanent loss of CNAPs predicts postoperative hearing loss with 100% sensitivity and specificity.

Dynamic changes in CNAPs amplitudes are seen with intracapsular tumor debulking, dissection of the tumor from the nerves, separation of VIII from VII nerve and tumor resection. It has been recommended to use sharp (rather than blunt) tumor debulking and dissection of the tumor from the nerves, using microscissors. Blind tumor dissection near the fundus of the IAC can be dangerous. In order to avoid thermal injury, it is also recommended not to use bipolar coagulation for hemostasis in the IAC.

The types of CNAPs changes have also shown correlation with certain types of injuries. Abrupt loss of CNAPs is most likely associated with a vascular injury. A gradual loss, stepwise decrease in the amplitude with increase in the latencies, is usually associated with a mechanical injury. Compression of the nerve can result in a desynchronization of the individual NAPs, causing loss or decreased amplitude in CNAP. Reversely, decompression of the nerve is sometimes followed by an improvement in the CNAPs amplitudes.

However, CNAPs recording technique suffers from at least three drawbacks. First, one must wait until the nerve is exposed to start monitoring: as already mentioned, this method works best for small and more distal tumors; if the tumor is sizable and extends more proximally, there may be very limited space for proximal placement of the electrode. In such instances, electrode placement requires sufficient nerve exposure, achieved only after some dissection is carried out; however, by that time, the nerve damage may have already occurred. Second, the cochlear nerve is very delicate, and the application of an electrode is itself a potentially traumatic event. Third, this electrode often becomes a nuisance within the operative field, and small displacement can easily occur, followed by significant changes in the recorded CNAPs.

*D3. EcochG.* Another method used for cochlear nerve monitoring is EcochG. As with CNAPs recordings, this method offers higher amplitude potentials and faster responses than BAEPs, and it should be tried when the BAEPs are nonrecordable. This involves the insertion of an electrode through the tympanic membrane into the promontory bone. Here it can measure the near-field cochlear response, corresponding to wave I of the BAEP (see page 39). Even though this cochlear potential is proximal to the site of the tumor, many manipulations that damage the nerve may also damage labyrinthine structures (e.g. ischemia). For details regarding mechanisms of injury of the cochlea, see pages 310–15.

Indeed, several studies have found reasonable correlations between the ECochG and postoperative hearing deficits (Attias et al., 2008; Battista et al., 2000; Colletti et al., 1998; Lambert et al., 1988), suggesting that such types of indirect injury may be commonly at fault. Levine and colleagues (1994) reported that, if the CAP recorded via EcochG disappeared and remained absent for more than 15 minutes, hearing was not preserved. A variation of this technique, extratympanic EcochG, does not require a myringostomy and may provide similar results (Mullatti et al., 1999). However, "pure" disconnection of the cochlea from the brainstem, observed as a preserved ECochG but absent distal BAEPs, can confound this technique; postoperative deafness is the result (Schlake et al., 1999).

*D4. Distortion-product otoacoustic emissions.* These are faint sounds thought to emanate from the movement of hair cells as a function of the difference between two presented frequencies. Their monitoring has been found to be potentially useful (Morawski et al., 2004), but a direct comparison with other types of monitoring (e.g. the ECochG) intraoperatively in humans has not yet been reported.

### E. Pons via BAEPs

In addition to detecting direct damage of the cochlear nerve or of the cochlea itself, BAEPs are a good tool to monitor the lower, middle, and upper pons. The injuries occur through similar mechanisms: compression or other mechanical trauma, thermal and vascular injury. Injury of the upper medulla and lower pons (including cochlear nucleus and superior olivary complex) results in disappearance of wave II, with equal delay in waves III and V, to total loss. Instead, wave I is unchanged and sometimes increased. This seems to be due to disinhibition of the cochlea or due to interruption of the inhibitory pathways (e.g. the olivocochlear bundle or descending brainstem auditory pathways) (Legatt, 1999).

Injury of the mid- to upper pons results in preservation of waves I, II, and III and a delay in wave V or its total loss. Injury of the auditory pathways rostral to the mesencephalon most likely will not result in BAEPs changes.

### *F. Motor Pathways*

As for the cerebral peduncles, the use of tcMEPs and direct stimulation of visible structures may help minimize risks to the motor pathways in the pons (Quinones-Hinojosa et al., 2004). Nonetheless, there are risks to the interspersed motor fibers, as well as to cortico-cerebellar pathways that traverse this region; for instance, there is a report of transient postoperative mutism in a child with a pontine cavernoma resected through this anterolateral approach (Frim et al., 1995). Note that others have reported postoperative mutism after resection of dorsal mesencephalic lesions, suggesting that this observation of this deficit may reflect one of multiple forms of damage to cerebellar input and output (Wang et al., 2002). Unfortunately, no known type of intraoperative monitoring has been shown to reduce the incidence of this deficit.

## IV. MEDULLA

### 1. Anatomy of the Medulla and Operative Approaches

Working in the vicinity of the medulla is most often necessitated by skull-base tumors in the region of the foramen magnum or along the clivus. The higher density of long tracts and lower cranial nerve nuclei within the medulla makes resection of intrinsic medullary lesions quite treacherous. This region can be approached dorsally, from a midline suboccipital craniotomy (with or without C1 laminectomy, depending on the extent of required exposure), or ventrolaterally, from a far lateral suboccipital craniotomy (craniotomy from the transverse sinus superiorly, to the sigmoid sinus laterally and to the ipsilateral occipital condyle ventrally).

In the case of intrinsic brainstem lesions, many patients develop new or further worsening neurologic symptoms in the immediate postoperative period. Fortunately, experience with brainstem glioma surgery has shown that many of these lower cranial nerve deficits improve with time (Jallo et al., 2005). There is no wide consensus regarding the safest ways to violate the surface of the medulla, and few surgeons are comfortable doing so if the lesion does not itself contact the surface. Externally salient features of the medulla include the midline sulcus, between the protuberances of the fasciculi gracili, the shallower protuberances of the cuneate fascicles more laterally, the bulges of the inferior olivary nuclei laterally, and the corticospinal tracts with their pyramidal decussation ventrally. Less clearly evident are the spinal trigeminal tract, immediately ventrolateral to the cuneate fasciculus, below the level of the inferior cerebellar peduncle, and the dorsal and ventral spinocerebellar tracts, at the lateral margins of the medulla.

The glossopharyngeal nerve (CN IX) exits as several rootlets from the more rostral portion of the medulla's posterolateral sulcus. The vagus (CN X) and accessory (CN XI) nerve rootlets exit from this same sulcus more caudally, in order. The hypoglossal (CN XII) nerve, on the other hand, exits from the anterolateral sulcus. A variety of skull-base procedures for a diverse range of tumors can place these nerves at risk as they travel within the cistern to exit the skull via the jugular foramen (IX, X, and XI) or the hypoglossal canal (XII).

Ventral lesions are more difficult, both in terms of the cranial approach and with regard to the details of the local anatomy. Access to the ventral medulla is obtained with the same approaches as previously described for the pons. Although a transoral, transclival approach to ventrally situated cavernous malformations has been employed successfully (Reisch et al., 2001), the general utility and safety of this technique is not yet clear, especially with regard to the potential incidence of meningitis. Once the ventral medulla is exposed, the choice of entry point is critical. Although it is preferable to avoid entering directly into the long tracts, especially the motor tracts, working between these tracts, particularly ventrally, is generally not feasible because of the presence of exiting nerve rootlets. Therefore, entering directly into lesions that rise to the ventral surface and either displace or have already damaged these tracts is the least likely to result in further deficits (here, motor deficits, if working ventrally, or ataxia, if working more laterally). There is at least one case report of entry into the ventrolateral medulla through the inferior olive to resect a superficially located cavernous malformation without subsequent observation of a new deficit (Oshiro et al., 2002).

## 2. Particularities of Monitoring and Mapping of Medulla Structures

Electrophysiologic monitoring of lower cranial nerves, motor systems, and sensory systems is crucial for the resection of intrinsic medullary lesions or for tumors in the vicinity of the foramen magnum. Details about motor system monitoring are found on pages 16–26, and the preceding sections. Sensory system monitoring is important in medulla protection. Detailed description of the basic stimulation and recording techniques for upper and lower limbs SSEPs, as well as their neurophysiologic-neuroanatomic correlation, are described on pages 5–15. This chapter briefly addresses the specifics of large fiber sensory monitoring in medulla surgery. Monitoring lower cranial nerve (CN IX, X, XI, and CN XII) function was believed by some authors to be a fairly reliable intraoperative guide to preserving good postoperative function (Glasker et al., 2006; Schlake, Goldbrunner, et al., 2001). Topsakal and colleagues (2008) showed that neurophysiologic monitoring and mapping of the lower cranial nerves helps decrease the postoperative deficit in high-risk group patients (e.g. tumors close to the foramen magnum and jugular foramen). Additionally, in addition to the case of skull-base and brainstem surgery, selective monitoring of certain cranial nerves may be helpful in selective procedures (e.g. RLN monitoring in thyroid surgery). A list of indications for lower cranial nerves monitoring is given in Table **7-1**. Our basic techniques of free and triggered EMG, used in cranial nerve monitoring and mapping (including types of electrodes, insertional points, and stimulation and recording parameters) has been detailed on pages 27–33. The following includes specifics for monitoring of CN IX, X, XI, and XII.

### *A. Glossopharyngeal Nerve (CN IX)*

Monitoring of the ninth nerve can be challenging because of the low-amplitude triggered CMAPs (Yingling et al., 1994) in the only CN IX-innervated muscle of the pharynx: the styplopharyngeus, which is also quite difficult to access. For this reason, misplacement of the recording electrodes in a CN X innervated pharyngeal muscle can easily occur. Thus, correct electrode insertion requires additional expertise and necessitates placement of the electrodes in the posterior wall of the pharynx, midway between the uvula and the posterior tonsillar pillar (Mishler et al., 1995). Moreover, concomitant CN IX and CN X monitoring (the latter from both pharyngeal and vocal cord muscles) has been previously emphasized (De Monte et al., 1994; Topsakal. et al., 2008), allowing distinction between CN IX or CN X stimulation. Within the same lines, Husain and colleagues (2008) have recently described an adhesive surface electrode, mounted on the laryngeal mask airway (LMA), for monitoring of the glossopharyngeal nerve. This electrode is in direct contact with the pharyngeal muscles, recording CMAPs triggered by stimulation of either CN IX or CN X nerve; a similar adhesive electrode placed on the endotracheal tube (ETT) allows recording from the CN X innervated laryngeal muscles. Moreover, the recording surface electrodes are less invasive than the needle electrodes, which could cause bleeding in the highly vascularized pharyngeal wall. In addition, at least one group has suggested that the sensory function of CN IX can be monitored by electrically stimulating in the posterior third of the tongue and recording the resulting cortical-evoked potentials (Sakuma et al., 2002).

### *B. Vagus Nerve (CN X)*

A variety of methods can be employed to monitor vagal function. These are intended to prevent damage to the nucleus ambiguous, or the vagus nerve and its branches, that could lead to dysphagia or hoarseness. Applications of CN X monitoring are given in Table **7-1**. Depending on which part of the nerve needs to be monitored, the recording electrodes can be placed at the level of the pharyngeal muscles or at the level of laryngeal muscles (see Fig. **1-11 c**, page 30), and, in order to avoid false-negative responses, stimulation of the nerve should be done proximal to the level of surgery.

As with the other cranial nerves, monitoring and mapping techniques include free and triggered EMG and tcMEPs. The advantages of tcMEPs monitoring are described on pages 299–301. Interpretation of free-run EMG patterns as part as CN X monitoring is similar to the one described on pages 306–9. Of note, B and C trains (see Fig. **7-2**) are somewhat more characteristic of pharyngeal muscles, whereas A trains are seen mostly in the facial muscles. However A trains can occur in the lower cranial nerve innervated muscles as well, especially during dissection of large tumors near the brainstem surface and intrameatal decompression. Triggered CMAPs

(Fig. **7-3 d**) are obtained with amplitudes that vary widely, from 15 to 400 $\mu$V, depending on the muscles used for recording, the type of recording electrodes, and the stimulus amplitude.

Placement of electrodes into the vocalis muscle can be achieved, through a purely endolaryngeal route or by a combined percutaneous (through the cricothyroid membrane) and endoscopic approach, to confirm placement (Leonetti et al., 1994). Or electrodes can be inserted into the false vocal cords via an endolaryngeal approach. Needle electrodes were found by some authors to offer more reliable recordings (Ulmer et al., 2008). A simpler approach uses a specially designed endotracheal tube, with electrodes along its surface, to monitor evoked movements of the vocal cords (Mikuni et al., 2004). Because hook or needle electrodes in these locations tend to become dislodged easily, many have come to prefer this last method. Alternatively, the cords can be visualized directly with a laryngoscope, observing the spontaneous movements of the cords that accompany respiration, in order to assess proximity to the vagus or its motor nucleus (Tamano et al., 2004); however, it is yet unknown whether continuous observation of these spontaneous vocal cord movements are a reliable sign of intact CN X function. Of course, irritation of vagal structures may also be evident by changes in autonomic status (e.g. bradycardia). Haerle et al. (2008) also described a new recording technique for monitoring, via a bipolar electrode, the posterior arytenoid muscle (PAM) of the larynx rather than the vocal cord muscles (VCM) CMAPs. They also gave a comparison between the two CMAPs, with the PAM CMAPs having a significantly lower amplitude than the VCM CMAPs, yet the latencies were similar (25.2 +/– 12.34 $\mu$V PAM CMAP vs. 36.5+/–18.72 $\mu$V VCM CMAP; latencies 3.01+/– 0.29 msec vs. 3.01+/– 0.24 msec). Recording from PAM thus allow bilateral monitoring of the RLNs. We use the NIM EMG ETT to record the muscle activity at the vocal cord. Height and gender seem to have a significant impact on the adequate depth of ETT insertion, in order to achieve successful neuromonitoring. Of note, misplacement of the ETT tube can occur as a result of changing head position from neutral to hyperextended with intubation. This situation has to be distinguished from malrotation of the tube, which can also cause inadequate monitoring. Thus, checking the ETT electrode impedances, as well as changes from the baseline recordings, may help (Lu et al., 2008). Alternatively, CN X monitoring can also be achieved by recording from the pharyngeal muscles, via needle or surface electrodes (Husain et al., 2008).

In addition to intermittent electrical stimulation of the vagal nerve (mapping) via a handheld monopolar stimulator, its continuous stimulation can also be achieved, and thus real-time monitoring of its function, by observing continuous generation of the CMAPs in the pharyngeal and laryngeal muscles. Ulmer and colleagues (2008) used a vagal cuff electrode, placed on the nerve, which delivered repetitive pulses at 3 Hz (pulse width 0.2 msec and intensity between 0.5 and 1.5 mA) and allowed continuous stimulation of about 60 minutes. They concluded that this allowed real-time intervention of the surgeon once the CMAPs amplitudes decreased by more than 60%.

Depending on which branch of the CN X is monitored, the stimulation of the vagal nerve needs to be done proximal to the level of the surgery; otherwise false negatives can be encountered. For example, moving away from the brainstem, recently significant steps toward the standardization of the method of stimulation of the RLN in thyroid surgery were taken (Chiang et al., 2008; Dionigi et al, 2009; Dralle et al., 2008). Indirect stimulation of the RLN by proximal stimulation of the CN X is now considered necessary. Chiang et al. (2008) have described four steps in the RLN monitoring technique: (1) V1 signal: CMAPs obtained in the vocal muscle by stimulation of the CN X, before the dissection of the RLN; (2) R1 signal: CMAPs obtained in the same muscles after stimulation of the distal RLN before dissection; (3) R2: CMAPs obtained by stimulation of the proximal RLN after dissection; (4) V2: CMAPs obtained after stimulation of CN X, after dissection and lobectomy. This technique allows not only localization of the injury site on the nerve, which otherwise looks intact, but also helps in identifying the mechanisms of injury, as well as the most dangerous steps during thyroid surgery. The same studies concluded that stretching of the nerve at the Berry ligament, either by pulling on the thyroid (tethered nerve) or its distal stretching, are among the most common mechanisms of injury. Also, bifurcation of the nerve at this level causes stretching of the anterior branch of the nerve over the ligament, which thus becomes hard to identify and easily damaged, if the posterior branch is mistaken as the entire nerve. Additionally, identification of V1 before dissection becomes important in checking the system and making sure that the stimulation is appropriate. In the absence of baseline nerve damage, a V1 should be obtained in the vocal muscles, when stimulating up to 1 mA. If no CMAPs are obtained, this is most likely due to technical failure. Moreover, using the same stimulation threshold identified for CN X for RLN stimulation should undoubtedly trigger an R1.

The impact on RLN monitoring on the surgical outcome could not be unequivocally proven in many underpowered studies because of the relatively small number of adverse events (RLN palsy) and even smaller difference of such events between the monitored and unmonitored groups (Dionigi et al., 2008; Dralle et al., 2008); yet a trend toward an additional benefit of the neuromonitoring was seen (Dralle et al., 2008). Overall, it seems that the high-risk groups are more likely to benefit from monitoring. Recently, the results of the first appropriately powered prospective randomized study (Barczynski et al., 2009) showed that the prevalence of transient and overall RLN palsy was significantly lower (by 2.9% and 3.7%, respectively) in patients with monitoring than in patients without monitoring, in the risk group. A similar trend was seen in the low-risk groups but could not be proven statistically significant. However, the role of IONM in identifying the RLN anatomy has been overall recognized. This is true in particular for identifying RLN bifurcations at the level of Berry's ligament and near the inferior thyroid artery (Barczynzki et al., 2009) as well as external branch or the superior laryngeal nerve (EBSLN). Although NPV of the IONM of RLN has been consistently reported high (close to 100%), its PPV was reported to be suboptimal (Barczynski et al., 2009; Chiang et al., 2008; Dralle et al., 2008). However, the new standardization of the IONM methodology will improve it further. Within the same lines, it is thus important to recheck the R2 and V2 responses before the wound closure, because some of the responses that disappeared during or immediately after RLN dissection may return before the wound closure. One of the explanation is that, during dissection, the nerve may be "stunned," with a return of its function later on (Chiang et al., 2008). Diamopoulos and colleagues (2009) showed the utility of free EMG monitoring in detecting the irritation and thus preventing injury of RLN associated with anterior cervical discectomy and fusion (ACDF).

### C. Accessory Nerve (CN XI)

Monitoring of the accessory nerve is relatively straightforward, requiring placement of electrodes into the trapezius muscle (see Fig. **1-11 d**, page 30). Some authors (Schlake, Goldbrunner, et al., 2001; Soo et al., 1993) further suggest placement of the electrodes in the descending part of the muscle, which receives mostly CN XI innervations (rather than in the ascending portion, which receives spinal innervation, mostly from the upper cervical roots). However, rootlets of the spinal portion of the nerve unite and ascend, entering the skull through the foramen magnum; this spinal trunk also subsequently exits the skull via jugular foramen. Thus, because surgeries at the level of foramen magnum may injure the spinal portion of the nerve as well, recording from the ascending portion of the trapezius muscle can be also helpful. Indications of CN XI monitoring are listed in Table **7-1**.

### D. Hypoglossal Nucleus and Nerve (CN XII)

One can assess the status of the hypoglossal nerve (CN XII) with electrodes placed in the tongue, ipsilaterally or bilaterally (see Fig. **1-11 e**, page 30). As with other lower cranial nerves, it is recommended that, during surgical procedures that could affect both nerves, such as clivus tumors resection, bilateral CN XII monitoring be done. This seems even more necessary because bilateral lesions result in significant difficulty of articulation and dysphagia, whereas unilateral lesions are less clinically debilitating (Topsakal et al., 2008). Two bundles make up the intracisternal segment of this nerve, and stimulation of either bundle has been observed to produce a reliable EMG response in the tongue (Mishler et al., 1995). When stimulating to map the hypoglossal nucleus itself, care must be taken with regard to possible current spread into the nearby dorsal motor nucleus of the vagus, causing bradycardia. Applications of CN XII monitoring and mapping are given in Table **7-1**.

### E. Sensory Pathways

In the region of the brainstem, damage to sensory pathways is most likely to occur when working in or near the dorsal medulla, because this is where the cuneate and gracillis fascicles are nearest the pial surface. Recording upper and lower limbs SSEPs can help identify dysfunction in the large fiber sensory system. Therefore, injury to the medullary portion of the somatosensory pathways in the medulla would most likely affect waves N18 and beyond, whereas damage to these pathways in the pons or midbrain would likely be reflected in waves N20 and P22 (believed to reflect activity in the thalamus and somatosensory cortex, respectively). With lesions at the cervico-medullary junction, affecting cuneate or gracillis nuclei, P/N13 and N30, respectively, can also be distorted, showing an increase in their duration, decreased amplitude, and poor morphology.

## V. ANESTHESIA CONSIDERATIONS

### 1. BAEPs

These short-latency evoked potentials are, in general, resistant to the action of anesthetic agents. Latency prolongations of BAEPs components and some decrease in their amplitudes have been reported with halogenated agents (desflurane, isoflurane, sevoflurane) and with propofol administration (Dubois et al., 1982; Lloyd-Thomas et al., 1990; Manninen et al., 1985; Sloan, 1998). On the other hand, nitrous oxide shows no noticeable effect on their latencies (Manninen et al., 1985). Barbiturates have been found to produce no changes in the BAEPs amplitudes, with some increases in the latencies (Drummond et al., 1987; Dubois et al., 1982; Sloan, 1998). All the previously mentioned effects are not considered to be clinically significant, and they do not affect the reliability of BAEPs recordings. On the other hand, hypothermia is known to have an effect on the BAEPs absolute and interpeak latencies. Markand and colleagues (1987) found that they increase by 7% for each 1°C drop in temperature and that at 26°C are double the value at normal body temperature. These changes are reversible with a return of the temperature to previous values, but the latencies at the same temperature may differ between the two phases of cooling and warming up. Decreases in the body temperature may also cause progressive BAEPs latency prolongations. However, changes in these components may occur more suddenly after cold saline irrigation of the local surgical field. BAEPs amplitudes can also decrease with decrease in temperature, and they usually occur first in the later components (i.e. waves III and V).

As already mentioned, muscle relaxants increase the signal-to noise ratio, thus facilitating the BAEPs recordings.

### 2. EMG Activity and Nerve Recordings

Free and triggered EMG activity used in cranial nerve monitoring is affected by the use of muscle relaxants. In fact, Shi and colleagues (2009) showed that intravenously administered vencuronium affects even more the CMAP amplitudes in laryngeal versus extremity swine muscles, with much slower recovery of the laryngeal CMAPs. On the other hand, several authors (Cengiz et al., 2008; Guo et al., 2008; Zhang et al., 2008) have found that partial neuromuscular block does not interfere with reliable facial nerve monitoring. However, in order to eliminate any such possible interference, we avoid the use of muscle relaxant when EMG activity is monitored. Additionally, tcMEPs are affected by inhalational agents. These effects, as well as the effects of muscle relaxants on both peripherally triggered (CMAPs) and centrally triggered (tcMEPs) muscle responses, are detailed on pages 327–36. CNAPs are, in general, resistant to the effects of anesthetics, both inhalational agents and total intravenous anesthesia (TIVA). Muscle relaxants do not negatively affect the recording of these potentials and, in fact, can help by decreasing the muscle artifact, thus increasing the signal-to-noise ratio.

## VI. CONCLUSIONS

Careful consideration of the local anatomy and intended surgical approach is necessary in order to obtain the most useful preoperative studies and to arrange for the proper intraoperative monitoring. Because the brainstem and cranial nerves contain a rich mixture of motor and sensory functions, working in this region requires a versatile array of neurophysiologic monitoring strategies. In the future, as the aggregate experience of neurosurgeons accrues and as the techniques for neurophysiologic brainstem and cranial nerve monitoring improve, the range of patients for whom surgery is a viable option to resect an otherwise devastating brainstem lesion will undoubtedly expand.

## REFERENCES


Altenmuller E, Cornelius CP, Buettner UW. Somatosensory evoked potentials following tongue stimulation in normal subjects and patients with lesions of the afferent trigeminal system. Electroencephalogr Clin Neurophysiol 1990;77:403–415.

Attias J, Nageris B, Ralph J, et al. Hearing preservation using combined monitoring of extra-tympanic electrocochleography and auditory brainstem responses during acoustic neuroma surgery. Int J Audiol 2008;47:178–184.

Baldwin HZ, Spetzler RF, Wascher TM, Daspit CP. The far lateral-combined supra- and infratentorial approach: Clinical experience. Acta Neurochir (Wien) 1995;134:155–158.

Barczynski M, Konturek A, Cichori S. Randomized clinical trial of visualization versus neuromonitoring of recurrent laryngeal nerves during thyroidectomy. Brit J Surg 2009;96: 240–246.

Battista RA, Wiet RJ, Paauwe L. Evaluation of three intraoperative auditory monitoring techniques in acoustic neuroma surgery. Am J Otol 2000;21:244–248.

Cantore G, Missori P, Santoro A. Cavernous angiomas of the brain stem. Intra-axial anatomical pitfalls and surgical strategies. Surg Neurol 1999;52:84–93; discussion 93–84.

Cengiz M, Ganidagli S, Alatas N, et al. Partial neuromuscular blockage levels with mivacurium during mastoidectomy allows intraoperative facial nerve monitoring. ORL J Otorhinolaryngol Relat Spec. 2008;70(4):236–241.

Chang SD, Lopez JR, Steinberg GK. Intraoperative electrical stimulation for identification of cranial nerve nuclei. Muscle Nerve 1999;22:1538–1543.

Chiang FY, Lu IC, Kuo WR, et al. The mechanism of recurrent laryngeal nerve injury during thyroid surgery—the application of intraoperative monitoring. Surgery 2008;143(6):743–749.

Colletti V, Fiorino FG. Advances in monitoring of seventh and eighth cranial nerve function during posterior fossa surgery. Am J Otol 1998;19:503–512.

Colletti V, Florino FG, Mocella S, et al. ECochG, CNAP and ABR monitoring during vestibular Schwannoma surgery. Audiology 1998;37:27–37.

Deletis V, Fernandez-Conejero I, Ulkatan S, et al.. Methodology for intraoperatively eliciting motor evoked potentials in the vocal muscles by electrical stimulation of the corticobulbar tract. Clin Neurophysiol. 2009;120(2):336–341.

De Monte F, Warf P, Al-Mefty O. Intraoperative monitoring of the cranial nerves during surgery of the jugular foramen and lower clivus. In: Intraoperative monitoring techniques in neurosurgery. (Loftus CM, Traynelis VC, Eds.) 1994;pp 205–212. New York: McGraw-Hill.

Dimopoulos VG, Chung I, Lee GP, et al. Quantitative estimation of the recurrent laryngeal nerve irritation by employing spontaneous intraoperative electromyographic monitoring during anterior cervical discectomy and fusion. J Spinal Disord Tech 2009;22(1):1–7.

Dionigi G. Use of a single bipolar electrode in the posterior arytenoids muscles for bilateral monitoring of the recurrent laryngeal nerves in thyroid surgery: Standardization of the technique. Eur Arch Otorhinolaryngol 2008;265: 1431–1432.

Dionigi G, Boni L, Rovera F, et al. Neuromonitoring and video-assisted thyroidectomy: A prospective, randomized case-control evaluation. Surg Endosc. 2009;23(5):996–1003.

Dralle H, Sekulla C, Lorenz K, et al. Intraoperative monitoring of the recurrent laryngeal nerve in thyroid surgery. World J Surg 2008;32:1358–1366.

Drummond JC, Todd MM, Schubert A, Sang H. Effect of the acute administration of high dose pentobarbital on human brain stem auditory and median nerve somatosensory evoked responses. Neurosurgery 1987;20(6):830–835.

Dubois MY, Sato S, Chassy J, Macnamara TE. Effects of enflurane on brainstem auditory evoked responses in humans. Anesth Analg 1982;61(11):898–902.

Ferroli P, Sinisi M, Franzini A, et al. Brainstem cavernomas: Long-term results of microsurgical resection in 52 patients. Neurosurgery 2005;56:1203–1212;discussion 1212–1204.

Frim DM, Ogilvy CS. Mutism and cerebellar dysarthria after brain stem surgery: Case report. Neurosurgery 1995;36:854–857.

Fukuda M, Oishi M, Saito A, et al. Facial nerve motor evoked potentials elicited by transcranial electrical stimulation for intraoperative monitoring. No Shinkel Geka 2008;36(4): 315–321.

Glasker S, Pechstein U, Vougioukas VI, Van Velthoven V. Monitoring motor function during resection of tumours in the lower brain stem and fourth ventricle. Childs Nerv Syst 2006;22(10):1288–1295.

Gouveris H, Mann W. Association between surgical steps and intraoperative auditory brainstem response and electrocochleography waveforms during hearing preservation vestibular schwannoma surgery. Eur Arch Otorhinolaryngol 2009;266:225–229.

Grabb PA, Albright AL, Sclabassi RJ, Pollack IF. Continuous intraoperative electromyographic monitoring of cranial nerves during resection of fourth ventricular tumors in children. J Neurosurg 1997;86:1–4.

Grayeli AB, Guindi S, Kalamarides M, et al. Four channel electromyography of the facial nerve in vestibular schwannoma surgery: Sensitivity and prognostic value for short-term facial function outcome. Otol Neurotol 2005;26:114–120.

Guo L, Jasiukaitis P, Pitts LH, Cheung SW. Optimal placement of recording electrodes for quantifying facial nerve compound muscle action potential. Otol Neurotol 2008;29(5):710–713.

Haerle S, Sidler D, Linder TH, Mueller W. Use of a single bipolar electrode in the posterior arytenoid muscles for bilateral monitoring of the recurrent laryngeal nerves in thyroid surgery. Eur Arch Otorhinolaryngol 2008;265:1549–1552.

Hone SW, Commins DJ, Rames P, et al. Prognostic factors in intraoperative facial nerve monitoring for acoustic neuroma. J Otolaryngol 1997;26:374–378.

Huang BR, Chang CN, Hsu JC. Intraoperative electrophysiologic monitoring in microvascular decompression for hemifacial spasm. J Clin Neurosci 2009;16(2):209–213.

Husain AM, Wright DR, Stolp BW, et al. Neurophysiological intraoperative monitoring of the glossopharyngeal nerve: Technical case report. Neurosurgery 2008;63(4):277–278.

Jackson LE, Roberson JB. Acoustic neuroma surgery: Use of cochlear nerve action potential monitoring for hearing preservation, Am J Otol 2000;21:249–259.

Jallo GI, Shiminski-Maher T, Velazquez L, et al. Recovery of lower cranial nerve function after surgery for medullary brainstem tumors. Neurosurgery 2005;56:74–77; discussion 78.

James ML, Husain AM. Brainstem auditory evoked potential monitoring: When is change in wave V significant? Neurology 2005;65:1551–1555.

Joo WI, Lee KJ, Park HK, et al. Prognostic value of intraoperative lateral spread response monitoring during microvascular decompression in patients with hemifacial spasm. J Clin Neurosci 2008;15(12):1335–1339.

Kemick JL, LaRouere ML, Kileny PR, et al. Hearing preservation following suboccipital removal of acoustic neuroma. Laryngoscope 1990;100:597–602.

Lambert PR, Ruth RA. Simultaneous recording of noninvasive ECoG and ABR for use in intraoperative monitoring. Otolaryngol Head Neck Surg 1988;98:575–580.

Legatt AD. Brainstem auditory evoked potentials: Methodology, interpretation, and clinical application. In: Electrodiagnosis in Clinical Neurology, 4th ed. (Aminoff MJ, Ed.) 1999; pp 451–484. New York: Churchill Livingstone.

Legatt AD. Mechanisms of intraoperative brainstem auditory evoked potential changes. J Clin Neurophysiol. 2002;19(5): 396–408.

Leonetti JP, Mokarry VP, Fan Z, et al. Vagal nerve monitoring during parapharyngeal space tumor removal. Skull Base Surg 1994;4:213–218.

Levine RA, Gardner JC, Fullerton BC, et al. Multiple sclerosis lesions of the auditory pons are not silent. Brain 1994 Oct; 117(Pt 5):1127–1141.

Lloyd-Thomas AR, Cole PV, Prior PF. Quantitative EEG and brainstem auditory evoked potentials: Comparison of isoflurane with halothane using the cerebral function analysing monitor. Br J Anaesth 1990;65(3):306–312.

Lu IC, Chu KS, Tsai CJ, et al. Optimal depth of NIM EMG endotracheal tube for intraoperative neuromonitoring of the recurrent laryngeal nerve during thyroidectomy. World J Surg 2008;32:1935–1939.

Magliulo G, Zardo F. Intra-operative facial nerve monitoring: Its predictive value after skull base surgery. J Laryngol Otol 1997;111:715–718.

Manninen PH, Lam AM, Nicholas JF. The effects of isoflurane and isoflurane-nitrous oxide anesthesia on brainstem auditory evoked potentials in humans. Anesth Analg 1985;64(1): 43–47.

Markand ON, Lee BI, Warren C, et al. Effects of hypothermia on brainstem auditory evoked potentials in humans. Ann Neurol. 1987;22(4):507–513.

Matthies C, Samii M. Management of vestibular schwannomas (acoustic neuromas): The value of neurophysiology for evaluation and prediction of auditory function in 420 cases. Neurosurgery 1997;40:919–929.

Mikuni N, Satow T, Taki J, et al. Endotracheal tube electrodes to map and monitor activities of the vagus nerve intraoperatively. Technical note. J Neurosurg 2004;101:536–540.

Miller MC, Spiegel JR. Identification and monitoring of the recurrent laryngeal nerve during thyroidectomy. Surg Oncol Clin N Am 2008;17(1):121–144.

Mishler ET, Smith PG. Technical aspects of intraoperative monitoring of lower cranial nerve function. Skull Base Surg 1995;5:245–250.

Møller AR. Monitoring auditory function during operations to remove acoustic tumors. Am J Otol 1996;17:452–460.

Møller AR, Jannetta PJ. Monitoring facial EMG responses during microvascular decompression operations for hemifacial spasm. J Neurosurg 1987;66:681–685.

Møller AR, Jannetta P, Møller MB. Intracranially recorded auditory nerve response in man. New interpretations of BSER. Arch Otolaryngol 1982;108:77–82.

Møller AR, Møller MB. Does intraoperative monitoring of auditory evoked potentials reduce incidence of hearing loss as a complication of microvascular decompression of cranial nerves? Neurosurgery 1989;24(2):257–263.

Mooij JJ, Mustafa MK, van Weerden TW. Hemifacial spasm: Intraoperative electromyographic monitoring as a guide for microvascular decompression. Neurosurgery 2001;49: 1365–1370; discussion 1370–1361.

Morawski K, Namyslowski G, Lisowska G, et al. Intraoperative monitoring of cochlear function using distortion product otoacoustic emissions (DPOAEs) in patients with cerebellopontine angle tumors. Otol Neurotol 2004;25:818–825.

Morota N, Deletis V. The importance of brainstem mapping in brainstem surgical anatomy before the fourth ventricle and implication for intraoperative neurophysiological mapping. Acta Neurochir (Wien) 2006;148:499–509; discussion 509.

Morota N, Deletis V, Epstein FJ, et al. Brain stem mapping: Neurophysiological localization of motor nuclei on the floor of the fourth ventricle. Neurosurg 1995;37:922–930.

Mullatti N, Coakham HB, Maw AR, et al. Intraoperative monitoring during surgery for acoustic neuroma: Benefits of an extratympanic intrameatal electrode. J Neurol Neurosurg Psychiatry 1999;66:591–599.

Mustain WD, al-Mefty O, Anand VK. Inconsistencies in the correlation between loss of brain stem auditory evoked response waves and postoperative deafness. J Clin Monit 1992;8: 231–235.

Nadol JB Jr, Chiong CM, Ojemann RG, et al. Preservation of hearing and facial nerve function in resection of acoustic neuroma. Laryngoscope 1992;102:1153–1158.

Nadol JB Jr, Levine R, Ojemann RG, et al. Preservation of hearing in surgical removal of acoustic neuromas of the internal auditory canal and cerebellar pontine angle. Laryngoscope 1987 Nov; 97(11):1287–1294.

Neff BA, Ting J, Dickinson SL, Welling DB. Facial nerve monitoring parameters as a predictor of postoperative facial nerve outcomes after vestibular schwannoma resection. Otol Neurotol 2005;26:728–732.

Ojemann RG. Retrosigmoid approach to acoustic neuroma (vestibular schwannoma). Neurosurgery 2001;48(3):553–558.

Oshiro S, Yamamoto M, Fukushima T. Direct approach to the ventrolateral medulla for cavernous malformation: Case report. Neurol Med Chir (Tokyo) 2002;42:431–434.

Polo G, Fischer C, Sindou MP, et al. Brainstem auditory evoked potential monitoring during microvascular decompression for hemifacial spasm: Intraoperative brainstem auditory evoked potential changes and warning values to prevent hearing loss—prospective study in a consecutive series of 84 patients. Neurosurgery 2004;54:97–104; discussion 104–106.

Prell J, Rampp S, Rachinger J, et al. Spontaneous electromyographic activity during microvascular decompression in trigeminal neuralgia. J Clin Neurophysiol 2008;25(4):225–232.

Prell J, Rampp S, Romstock J, et al. Train time as a quantitative electromyographic parameter for facial nerve function in patients undergoing surgery for vestibular schwannoma. J Neurosurg 2007;106:826–832.

Quinones-Hinojosa A, Alam M, Lyon R, et al. Transcranial motor evoked potentials during basilar artery aneurysm surgery: Technique application for 30 consecutive patients. Neurosurgery 2004;54:916–924; discussion 924.

Quinones-Hinojosa A, Lyon R, Du R, Lawton MT. Intraoperative motor mapping of the cerebral peduncle during resection of a midbrain cavernous malformation: Technical case report. Neurosurgery 2005;56:E439; discussion E439.

Reisch R, Bettag M, Perneczky A. Transoral transclival removal of anteriorly placed cavernous malformations of the brainstem. Surg Neurol 2001;56:106–115; discussion 115–106.

Romstock J, Strauss C, Fahlbusch R. Continuous electromyography monitoring of motor cranial nerves during cerebellopontine angle surgery. J Neurosurg 2000;93:586–593.

Sakuma J, Matsumoto M, Ohta M, et al. Glossopharyngeal nerve evoked potentials after stimulation of the posterior part of the tongue in dogs. Neurosurgery 2002;51:1026–1032; discussion 1032–1023.

Sarma S, Sekhar LN. Brain stem cavernoma excised by subtemporal-infratemporal approach. Br J Neurosurg 2002; 16:172–177.

Sato S, Yamada M, Koizumi H, et al. Neurophysiological mechanisms of conduction impairment of the auditory nerve during cerebellopontine angle surgery. Clin Neurophysiol 2009;120:329–335.

Schlake HP, Goldbrunner RH, Milewski C, et al. Intra-operative electromyographic monitoring of the lower cranial motor nerves (LCN IX–XII) in skull base surgery. Clin Neurol Neurosurg 2001;103:72–82.

Schlake HP, Goldbrunner R, Milewski C, et al. Technical developments in intra-operative monitoring for the preservation of cranial motor nerves and hearing in skull base surgery. Neurol Res 1999;21:11–24.

Schlake HP, Milewski C, Goldbrunner RH, et al. Combined intraoperative monitoring of hearing by means of auditory brainstem responses (ABR) and transtympanic electrocochleography (ECochG) during surgery of intra- and extrameatal acoustic neurinomas. Acta Neurochir (Wien) 2001;143:985–995; discussion 995–986.

Sekiya T, Hatayama T, Shimamura N, Suzuki S. Intraoperative electrophysiological monitoring of oculomotor nuclei and their intramedullary tracts during midbrain tumor surgery. Neurosurgery 2000;47:1170–1176; discussion 1176–1177.

Sekiya T, Møller AR. Avulsion rupture of the internal auditory artery during operations in the cerebellopontine angle: A study in monkeys. Neurosurgery 1987;21(5):631–637.

Shi Y, Hou V, Tucker A, Cohen J. Changes of extremity and laryngeal muscle electromyographic amplitudes after intravenous administration of vecuronium. Laryngoscope 2009;118(12):2156–2160.

Silverstein H, Smouha EE, Jones R. Routine intraoperative facial nerve monitoring during otologic surgery. Am J Otol 1988a;9:269–275.

Silverstein H, Smouha E, Jones R. Routine identification of the facial nerve using electrical stimulation during otological and neurotological surgery. Laryngoscope 1988b;98:726–730.

Sloan TB. Anesthetic effects on electrophysiologic recordings. J Clin Neurophysiol 1998;15(3):217–226.

Smith ER, Chapman PH, Ogilvy CS. Far posterior subtemporal approach to the dorsolateral brainstem and tentorial ring: Technique and clinical experience. Neurosurgery 2003;52:364–368; discussion 368–369.

Soo KC, Strong EW, Spiro RH, et al. Innervation of the trapezius muscle by the intra-operative measurement of motor action potentials. Head Neck 1993;15:216–221.

Stechison MT, Moller A, Lovely TJ. Intraoperative mapping of the trigeminal nerve root: Technique and application in the surgical management of facial pain. Neurosurgery 1996;38:76–81; discussion 81–72.

Tamano Y, Ujiie H, Kawamata T, Hori T. Continuous laryngoscopic vocal cord monitoring for vascular malformation surgery in the medulla oblongata: Technical note. Neurosurgery 2004;54:232–235; discussion 235.

Thomke F, Hopf HC. Abduction paresis with rostral pontine and/or mesencephalic lesions: Pseudoabducens palsy and its relation to the so-called posterior internuclear ophthalmoplegia of Lutz. BMC Neurol 2001;1:4.

Topsakal C, Al-Mefty O, Bulsara KR, Williford VS. Intraoperative monitoring of lower cranial nerves in skull base surgery: Technical report and review of 123 monitored cases. Neurosurg Rev 2008;31:45–53.

Ulmer C, Koch, KP, Seimer A, et al. Real-time monitoring of the recurrent laryngeal nerve: An observational clinical trial. Surgery 2008;143:359–365.

Wang MC, Winston KR, Breeze RE. Cerebellar mutism associated with a midbrain cavernous malformation: Case report and review of the literature. J Neurosurg 2002;96: 607–610.

Yamakami I, Oka N, Yamaura A. Intraoperative monitoring of cochlear nerve compound action potential in cerebellopontine angle tumour removal. J Clin Neurosci 2003;10: 567–570.

Yamakami I, Ushikubo O, Uchino Y, et al. Intraoperative monitoring of hearing function in the removal of cerebellopontine angle tumor: Auditory brainstem response and cochlear nerve compound action potential. No Shinkei Geka 2002;30:275–282.

Yamakami I, Yoshinori H, Saeki N, et al. Hearing preservation and intraoperative auditory brainstem response and cochlear nerve compound action potential monitoring in the removal of small acoustic neurinoma via the retrosigmoid approach. JNNP 2009;80:218–227.

Yamashita S, Kawaguchi T, Fukuda M, et al. Abnormal muscle response monitoring during microvascular decompression for hemifacial spasm. Acta Neurochir (Wien) 2005;147:933–937; discussion 937–938.

Yingling CD. Intraoperative monitoring of cranial nerves in skull base surgery. In: Neurotology (Jackler RK, Brackmann D, Eds.) 1994; pp 967–1002. St Louis: Mosby Year Book.

Zhang J, Zhang J, Gu YH, Liang WM. Influence of nondepolarizing relaxant on facial nerve monitoring during total intravenous anesthesia. Zonghua Yi Xue Za Zhi 2008;88(9): 591–593.

Zhu HW, Li YJ, Zhuang P, Ma K. Abnormal muscle response monitoring as a guide during microvascular decompression for hemifacial spasm. Zhongua Yi Xue Za Zhi 2008;88(39): 2767–2770.

# 8

# THE EFFECTS OF ANESTHETICS ON INTRAOPERATIVE NEUROPHYSIOLOGY STUDIES

MIRELA V. SIMON



## I. INTRODUCTION

Various types of anesthetics were found to affect neurophysiology studies used in intraoperative monitoring and mapping procedures. This is due to their different levels of mechanisms of action and influence on the excitability of neuraxis exerted at the level of cortex but also beyond. The most significant impact of the majority of anesthetics is on the synaptic transmissions, targeting NMDA (n-methyl-d-aspartic acid) and GABA (GABA aminobutyric acid) receptors, which are directly involved in the function of Na, Cl, and Ca channels. (Sloan, 2002). In the following, we have listed the most commonly used anesthetics and their particular effects on somatosensory-evoked potentials (SSEPs), motor-evoked potentials (MEPs), and electroencephalogram (EEG). However, we did not include examples of the different impacts the anesthetics have on neurophysiologic mapping and monitoring procedures. We consider that the anesthetic-induced changes seen in IOM constitute a significant part of the troubleshooting processes and thus should be further detailed within the specific context of each technique.

## II. BENZODIAZEPINES

### 1. Mechanism of Action

Benzodiazepines enhance GABA-ergic cortical inhibition by interacting with GABAR (GABA receptor), a macromolecular protein that contains binding sites for benzodiazepines, barbiturates neurosteroids, GABA, and a chloride ion-selective channel. Binding of benzodiazepines to this protein increases its Cl channel conductance and thus the inhibitory GABAR current. There are studies suggesting that the reduced number of neurons available for generation of descending pyramidal tract activity during sedation with short-acting benzodiazepines (such as midazolam) is due to an enhancement of the inhibitory action GABA-ergic cortical interneurons, which in turn decreases the excitability of the pyramidal cells (Schonle et al., 1989).

### 2. Effects on SSEPs and MEPs

The effects of midazolam on SSEPs are less dramatic than those of diazepam. Some authors (Koht et al., 1988) report an increase in their latencies with no effect on their amplitudes, whereas other authors (Sloan et al., 1990) found a 40% amplitude depression with the use of midazolam. The use of benzodiazepines results in a significant reduction of the amplitudes of muscle MEPs (mMEPs) triggered by cortical stimulation, while the cortical mapping threshold will increase.

### 3. Effects on EEG

Benzodiazepines increase the amount of EEG theta and delta slowing, with superimposed fast-frequency activities (Bond et al., 1983) in the unpremedicated patients. In doses used for premedication, diazepam increases fast-frequency activities and decreases alpha activity, corresponding to the blood levels (Fink et al., 1976).

### 4. Use in Anesthesia During SSEPs, MEPs, and EEG Procedures

Because of their excellent sedative and amnestic effect, short-acting benzodiazepines are sometimes added to total intravenous anesthesia (TIVA) or used as premedication prior to induction.

Although their effect on SSEPs does not seem dramatic enough to preclude their use in SSEPs monitoring, they do have significant effect on motor responses. Thus, their use as an adjunct for TIVA during functional mapping or surgeries requiring monitoring of the motor pathways should be avoided.

However, benzodiazepines do have a suppressive cortical effect, translating in alteration of the EEG activities, with a decrease in the overall frequencies, as well as a substantial change in the morphology, amplitudes, and frequencies of epileptiform activity—thus, we avoid the use of these drugs in electrocorticography (ECoG) recordings for seizure focus delineation in epilepsy surgery.

## III. BARBITURATES

### 1. Mechanism of Action

The mechanism of action of barbiturates is similar to that of benzodiazepines. It increases the GABA-ergic activity by binding to the barbiturates site on the GABA receptor.

### 2. Effects on SSEPs and MEPs

Barbiturates dampen the amplitudes of SSEPs and prolong their latencies, although recordings of evoked potentials can typically be obtained at high barbiturate dosages. MEPs triggered by cortical stimulation are very sensitive to barbiturates to total disappearance (Glassman et al., 1993).

### 3. Effects on EEG

The effects of barbiturates on EEG consist of an overall increase of the fast frequencies at low doses and a depressant effect, translating in burst suppression and electrocortical silence at higher doses (Kiersey et al., 1950).

### 4. Use in Anesthesia During SSEPs, MEPs, and EEG Procedures

Because of their effect in decreasing intracranial pressure (Greenbaum et al., 1975; Shapiro et al., 1973), cerebral metabolism, and cerebral blood flow (Pierce et al., 1962), some anesthesiologists use short-acting barbiturates, such as thiopental, in anesthesia during surgeries with potential risk for neurologic injury, in particular for induction. Because of their potentially significant effect on SSEPs, and especially on MEPs, we try to avoid them during functional cortical mapping or surgeries requiring monitoring of SSEPs or MEPs. Because of the significant EEG suppression they cause, we prefer not to use them during ECoG recordings, when delineation of epileptiform and epileptic foci is necessary.

## IV. PROPOFOL

### 1. Mechanism of Action

This drug was found to cause depression of cortical activity as a result of its effect on the neuronal excitability produced by selectively suppressing of Na (+) currents and L-type HVA Ca (2+) conductances in cortical neurons, as well as opening GABA (A) gated Cl (–) channels (Martella et al., 2005). Vandesteene et al. (1988) and Werner et al. (1992) described a definite effect of propofol on cerebral blood flow, cerebral oxygen consumption, and lactate and glucose metabolism. Within these lines, Van-Hemelrijck et al. (1990) showed that anesthetic doses of propofol decrease the cerebral blood flow by 30 to 40%, with a decrease in the cerebral oxygen consumption; Pinaud et al. (1990) found that, in brain-injured patients, propofol bolus of 2 mg/kg, followed by 0.15 mg/kg/mm infusion, decreased the intracranial pressure by 2 mmHg, the perfusion pressure by 25%, and the cerebral blood flow by 35%.

### 2. Effects on SSEPs and MEPs

Propofol affects the amplitudes of the motor responses and contributes to a higher cortical mapping threshold. The MEPs changes may be asymmetrical, seen more prominently in the already-affected myotomes (see Fig. **5A-4 b, c**, page 188). It also causes a progressive decrease in the amplitudes and an increase in the absolute latencies of the thalamocortical components of the SSEPs, as well as in the central conduction time (P/N13-N19 interpeak latency) (Sloan et al., 2002; Scheepstra et al., 1989). We consider that its effect on latencies is seen earlier and may be more evident than the effect on amplitudes, whereas the morphologies of the waveforms remain practically unchanged (see Fig. **5A-4 a**, page 187).

### 3. Effects on EEG

At lower doses, propofol causes a mild activation, with an increase in the fast-frequency activities, and, at high doses, a burst-suppression pattern (see Figs. **2-25, 2-26**).

### 4. Use in anesthesia during SSEPs, MEPs, and EEG procedures

Its effects on both MEPs and SSEPs are generally well tolerated in cases that require monitoring of SSEPs or MEPs, as well as in cases of functional cortical mapping. Thus, propofol is a central piece in TIVA use for anesthesia in these cases. However, there are special circumstances in which propofol-induced changes in SSEPs or MEPs could play an important role in the troubleshooting process during spine monitoring. Likewise, its influence on functional mapping results, directly related to a decrease in the cortical excitability, is extensively discussed on pages 155–56. Because of its significant depressant effect on EEG, propofol, as with any drug with potential in inducing a burst-suppression pattern, is avoided in cases of epilepsy surgery when ECoG is used to map epileptiform and epileptic activity (see pages 124–27). We also restrict its use during EEG monitoring procedures, because the presence of a burst-suppression pattern would make a reliable EEG interpretation impossible, in particular, in key moments of the surgery when an accurate analysis of the EEG is necessary (e.g. appreciation of interhemispheric asymmetries during clamping of the carotid artery in CEA procedures; see pages 75–80).

## V. NARCOTICS

### 1. Mechanism of Action

Opioids such as fentanyl, sufentanyl, alfentanil, and remifentanyl are mu receptor agonists and are thus used in anesthesia primarily for their analgesic effect.

### 2. Effects on SSEPs and MEPs

Fentanyl produces a mild increase in the latencies of SSEPs, without a significant or consistent change in amplitudes (Pathak et al., 1984). Similarly, the effect of opioids on mMEPs, triggered by cortical stimulation, is minimal, as concluded by different authors (Glassman et al., 1993; Kalkman et al., 1992; Ghaly et al., 1991). We prefer the use of remifentanyl instead of fentanyl because the former is metabolized faster than the latter, thus allowing rapid titration during its infusion. As for other drugs, we prefer opioid infusion rather than boluses because of more steady, predictable effect in the former and hard-to-predict peaks and troughs in the latter.

### 3. Effects on EEG

There are many reports in the literature suggesting that opioids may have epileptogenic effects. Smith et al. (2003) and Akcaboy et al. (2005) found that sremifentanyl increases the seizure duration during eletroconvulsive therapy (ECT), whereas others reported cases of spontaneous seizure activity following boluses of remifentanyl (Nielsen, 2004). Sufentanyl was also implicated as the cause of seizures in several cases (Brian et al., 1987; Molbegott et al., 1987). High doses of fentanyl were shown to produce seizures in rats (Carlsson et al., 1982), whereas low doses of fentanyl were thought to have induced seizures in humans (Fujimoto et al., 2003; Tempelhoff et al., 1992). Alfentanyl has consistent reports regarding its pro-epileptogenic effect, and it is frequently used to induce seizures in ECoG-guided epilepsy surgery (Cascino et al., 1993; Keene et al., 1997; Ragazzo et al., 2000; Ross et al., 2001; see also pages 121–24, Fig. **3-2 d**, page 101 and Fig. **3-7**, and page 107). Other authors (Smith et al., 1984), using two different computer analysis techniques, found that high doses of sufentanyl,

fentanyl, and morphine cause a significant increase in the low-frequency activities on the EEG.

### 4. Use in anesthesia during SSEPs, MEPs, and EEG procedures

Narcotics such as fentanyl and remifentanyl are an important component of TIVA, because of both their analgesic properties (low doses) and their anesthetic properties (high doses; Sebel et al., 1981). They are the most frequently chosen analgesics when inhalational agents have to be avoided or used in small concentrations. As with the use of propofol, the mild cortical suppressive effect of narcotics is well tolerated during monitoring procedures for SSEPs or MEPs or during functional mapping. Sloan (2002) has suggested that fentanyl can be used in reducing the spontaneous muscle activity, thus improving the recordings of mMEPs. Despite the several reports of potential pro-epileptogenic effect, we consider the use of narcotics relatively safe during cortical stimulation in functional mapping. One exception is alfentanyl, which should be avoided during cortical stimulation in functional cortical mapping because of its significant effect of lowering the seizure threshold. On the other hand, as a result of this action, alfentanyl is administered in boluses as an inducing agent for epileptic activity; thus it has potential use in demarcation of a seizure focus (see pages 121–24; and Figs. **3-2, d**, page 101, and **3-7**, page 107).

## VI. INHALATIONAL AGENTS

### 1. Mechanism of Action

Several authors have suggested that the sedative and hypnotic properties of the inhalational agents are due to an enhancement of cortical suppression by increasing the neocortical GABA (A) receptor-mediated inhibition, rather than to a decrease in cortical excitability (Hentschke et al., 2005; Osawa et al., 1994). Haghighi (1998) studied the influence of isoflurane on the motor pathways in rats and concluded that much of this drug effect is due to its effect on cortical synapses. Volatile anesthetics such as halogenated agents have an effect on both pyramidal neurons and cortical interneurons (Hicks et al., 1992). More so, Burke et al. (2000) concluded that halogenated agents significantly decrease the Na currents at the nodes of Ranvier, thus decreasing the excitability of the corticospinal axons. Other authors (Loughnan et al., 1989; Mavroudakis et al., 1994; Zentner et al., 1992) have suggested that halogenated agents have an additional suppressive effect at the spinal level, interfering with the synaptic transmission at the alpha motor neuron.

### 2. Effects on SSEPs and MEPs

Because of all of the mechanisms of actions previously described, volatile anesthetics, including nitrous oxide and halogenated agents, demonstrate a dose-dependent depression of both SSEPs and MEPs, both mMEPs and I waves (and, to a lesser extent, of D waves), with prolonged latencies and a substantially higher mapping threshold (see pages 155–56). Moreover, it is known that concomitant use of nitrous oxide ($N_2O$) and halogenated agents exerts a synergistic effect on SSEPs and MEPs. However, halogenated agents have different pontencies. Along these lines, Pathak and colleagues (1987) showed that during $N_2O$-based anesthesia, isoflurane and enflurane resulted in less SSEPs suppression than halothane (see also Fig. **5A-6**, pages 192–93).

### 3. Effects on EEG

At concentrations that produce unconsciousness, nitrous oxide produces fast oscillatory activity (Yamamura et al., 1981), with a peak frequency of about 35 Hz, more prominent over the frontal regions and directly proportional to the $N_2O$ concentration. Because the use of $N_2O$ alone is not adequate for general anesthesia, the effect of $N_2O$ on EEG is always mixed with the effects of other agents, such as halogenated agents.

The initial effect of isoflurane, at subanesthetic concentrations (<1.2%), is an overall attenuated fast-frequency (15-35 Hz) EEG background. At higher concentrations producing loss of consciousness, fast beta-frequency activity is superimposed on delta-frequency high-amplitude activity (Clark et al., 1973). At anesthetic concentrations, isoflurane produces a burst-suppression pattern, and, at end-inspired concentrations of 2.5%, complete electrical silence can be expected (Eger et al., 1971). Enflurane causes a dose-dependent loss of fast-frequency activities (Persson et al., 1978), followed by EEG burst suppression at high concentrations.

According to Porkkala et al. (1997), when oxygen is replaced with $N_2O$ during isoflurane anesthesia, the proportion of EEG suppression decreases by one-third, suggesting that a combination of $N_2O$ and isoflurane has a less suppressive EEG effect. On the other hand, it is known that halogenated agents and $N_2O$ have a synergistic effect on SSEPs and MEPs.

There have been contradictory reports about the epileptogenic effects of halogenated agents. Some authors believe that enflurane (Joas et al., 1971; Lebowitz et al., 1972; Niejadlik and Galindo, 1975; Oshima et al., 1981) and isoflurane (Poulton and Ellingson, 1984) encourage seizure activity; others (Gallager et al., 1978, for enflurane; Kofke et al., 1985, for isoflurane) sustain the opposite.

### 4. Use in Anesthesia During SSEPs, MEPs, and EEG Procedures

Because of their significant depressive effect on SSEPs and MEPs, we recommend that anesthesia protocol during cortical sensorimotor mapping and monitoring of SSEPs or MEPs excludes inhalational agents.

In general, we accept the use of inhalational agents during ECoG studies for mapping epileptiform and epileptic foci (see also pages 124–27). Inhalational agents are also used during procedures requiring EEG monitoring (e.g. CEA procedures). They are usually added to TIVA (i.e. propofol and remifentanyl), in order to decrease the concentration of the propofol and avoid a deep burst-suppression pattern, thus permitting interpretation of the EEG recordings (see pages 75–80).

## VII. ETOMIDATE

### 1. Mechanism of Action

At low doses, this drug enhances synaptic activity, which translates in an excitatory cortical effect. At high doses, it has a depressive cortical effect.

### 2. Effects on SSEPs and MEPs

At low doses, it can increase the amplitudes of SSEPs (Kochs et al., 1986; McPherson et al., 1986; Sloan et al., 1988), sometimes by more than half. Its cortical enhancing activity can also have a beneficial impact on the MEPs (Lumenta, 1991; Sloan et al., 1993) and occurs at doses that cause the desired sedation and amnesia.

### 3. Effects on EEG

The enhancing effect of etomidate seen at low doses can cause myoclonus immediately following injection at induction (Ghoneim and Yamada, 1977) and may also produce seizures in patients with epilepsy. (Rampil, 1997). However, with increasing doses it decreases alpha-frequency activity (Doenicke et al., 1982) and causes the appearance of high-amplitude theta activity, as well as burst-suppression pattern (Wauquier, 1983).

### 4. Use In Anesthesia During SSEPs, MEPs, and EEG Procedures

Because of all of these properties, etomidate is a good choice for induction and maintenance of TIVA in surgeries requiring intraoperative neurophysiologic monitoring, particularly in cases when an increase in the amplitudes of the evoked potentials is important for a reliable monitoring (e.g. symptomatic patients with small SSEPs or MEPs at baseline). Etomidate can be used instead of barbiturates such as thiopental at the induction of anesthesia to rapidly produce unconsciousness with fewer side effects (e.g. cardiovascular depression). It can also help decrease intracranial pressure without a drop in mean arterial blood pressure (Dearden and McDowall, 1985). The rapid metabolism of etomidate contributes to an easy adjustment of its concentrations.

Etomidate can decrease the afterdischarges threshold, making it less desirable during direct electrical cortical stimulation. However, at increased doses, with the appearance of the cortical suppressive effect, etomidate can increase the functional mapping threshold.

## VIII. KETAMINE

### 1. Mechanism of Action

The cortical excitatory effects of ketamine are similar to those of etomidate: it enhances the synaptic transmission, probably by acting on NMDA receptor.

### 2. Effects on SSEPs and MEPs

This effect translates in an increase in the amplitudes of SSEPs (Schubert et al., 1990). There are contradictory results regarding the effects on MEPs caused by cortical or spinal stimulation. Some authors have found that ketamine enhances these responses (Glassmann et al., 1993; Kano and Shimoji, 1974); other findings suggest that the drug produces a mild decrease in the amplitude of the MEPs and an increase in their latencies, but, overall, these effects are negligible as far as the reliability of monitoring (Yang et al., 1994). We think this variability seen in the effects on motor responses is dose dependent. The higher the doses are, the more likely is the suppression of these responses. Iida and colleagues (1997) found that depression of spinal-evoked responses at high doses of ketamine, administered intrathecally in dogs, is due to spinal axonal conduction block.

### 3. Effects on EEG

Because of its excitatory effect, ketamine produces high-amplitude theta activity as well as an increase in beta activity. Moreover, it seems to decrease the seizure threshold in patients with epilepsy.

### 4. Use in Anesthesia During SSEPs, MEPs, and EEG Procedures

Overall, as with etomidate, ketamine is a valuable adjunct in TIVA anesthesia for surgical cases that need neuromonitoring of SSEPs and MEPs, but extra caution needs to be taken regarding its hallucinatory potential and its increase in the intracranial pressure with intracranial pathology, as well as in cases of functional cortical mapping, when it can lower the threshold for stimulation-triggered epileptic activity.

## IX. DEXMEDETOMIDINE

### 1. Mechanism of Action

Dexmedetomidine is a highly selective alpha 2-adrenergic agonist. It is FDA approved to provide short-term sedation (less than 24 hours) in adult patients on mechanical ventilation. Its effects include: analgesia, sedation, and sympatholysis. It has also been recently used as an adjunct in TIVA, in an attempt to reduce the propofol concentration and thus to decrease the risk of "propofol infusion syndrome," in particular in pediatric population (Tobias, 2007).

### 2. Effects on SSEPs and MEPs

In a small series of 18 patients, Anschel and colleagues (2008) found no evidence of alteration of SSEPs, MEPs, and triggered EMG with the use of this drug. Tobias and colleagues (2008) found that an infusion rate of 0.5 μV/kg/hr of dexmedetomidine does not adversely affect the SSEPs and MEPs monitoring, provided that the depth of anesthesia is kept constant by decreasing the propofol infusion. However, the authors did not study the effects of dexmedetomidine on neurophysiologic monitoring under special conditions, such as blood loss, hypovolemia, or spine manipulation. Also, Mahmoud et al. (2007) have reported loss of MEPs associated with dexmedetomidine use in pediatric spine surgery.

### 3. Effects on EEG

Mirski and colleagues (1994) showed that dexmedetomidine increased cortical excitability and decreased the seizure threshold in a rat model. However, in a series of 6 patients, Souter et al. (2007) showed that its use in sedative doses during awake craniotomy for seizure resection did not interfere with either ECoG results for seizure foci localization or with functional mapping results. Similalry, Oda and coleagues (2007) showed that dexmedetomidine use during ECoG in patients with temporal lobe epilepsy under sevoflurane anesthesia did not change the frequency of epileptiform discharges at either 0.5 ng/mL or 1.6 ng/mL plasma concentrations. However, at both plasma concentrations, it did decrease the median frequency of the recordings.

### 4. Use in Anesthesia During SSEPs, MEPs and EEG Procedures

Due to the limited studies performed so far, with regard to the effect of dexmedetomidine on IOM procedures, we recommend caution when using this drug during SSEPs and in particular during MEPs monitoring, especially in the presence of other factors that could affect these potentials. Similarly, we recommend avoidance of its use during ECoG or functional mapping, in particular when used in combination with other anesthetics.

## X. MUSCLE RELAXANTS

### 1. Mechanism of Action

The muscle relaxants used in anesthesia are neuromuscular junction blockers. Cisatracurium is the most commonly used at our institution; however, other agents, such as succinylcholine and vecuronium, are also used.

### 2. Effect on MEPs

In order to reliably monitor mMEPs and compound muscle action potentials (CMAPs), evaluation of the degree of neuromuscular block becomes essential. There are two accepted methods used for such evaluation.

1. Measurement of CMAP (M1) produced by supramaximal stimulation of the peripheral motor nerve and the ratio between such motor response during the neuromuscular blockage (T1) and at baseline. Usually successful motor mapping or monitoring is still

considered reliable when T1 is between 5 and 50% of the baseline. (Sloan, 2002).

2. The second technique is the train of four technique (TOF). It involves supramaximal stimulation of a peripheral motor nerve with four consecutive pulses at 2 Hz. With consecutive stimulations, decremental responses in the amplitudes of CMAPs are observed, directly proportional to the degree of neuromuscular blockade. In practice, the anesthesiologist observes the total number of visible twitches obtained after this repetitive stimulation. It is accepted in general that obtaining two or more of four twitches is compatible with reliable mMEPs or CMAPs responses.

When comparing the two techniques, one of four twitches (method 2) is equivalent with a T1 of less than 10% from baseline (method 1); two of four twitches with 10–20% T1 of baseline; three of four with 20–25% T1 of baseline.

A more sensitive analysis derives from method 2 and involves comparing the amplitude of the CMAP obtained after the fourth pulse (T4) with the amplitude of the CMAP obtained after the first pulse (T1). It has been reported (Holland et al., 1998; Minahan et al., 2000) that an adequate EMG monitoring can be achieved with as much as 75% neuromuscular blockade, which would translate in an amplitude decrement of T4 of less than 100% from T1 (method 2).

It is important to assess the neuromuscular blockade in the muscles being monitored (and not in other muscles). It is known that different muscle groups are affected variably by muscle relaxants (Caffrey et al., 1986), based on their density of the neuromuscular junctions and also on their baseline strength. For example, deinnervation of the muscle, such as in cases of chronic radiculopathy, is followed by an increased susceptibility of that muscle to the action of neuromuscular blocking agents (Blair et al., 1994; Holland et al., 1998). For examples of effects of muscle relaxants on mMEPs, see Figs. **5A-3** (page 186) and **5A-5** (page 190).

Of course, these agents have no harmful effect on neurogenic MEPs such as D waves generation. In fact, they can help differentiate these triggered MEPs from muscle artifact, which may contaminate the epidural recordings.

## 3. Effects on SSEPs

The use of muscle relaxants can be beneficial during monitoring of SSEPs. They increase the signal-to-noise ratio by decreasing the muscle artifact. The benefit is seen mostly in the subcortical channel, where the electrode applied at C5 level can introduce a significant amount of muscle artifact. However, similar beneficial effect can be seen in scalp channels as well.

## 4. Effects on EEG

As with SSEPs recordings, muscle relaxants are beneficial for the EEG recording in that they increase the signal-to-noise ratio by decreasing the muscle artifact.

## 5. Use in Anesthesia During SSEPs, MEPs, and EEG Procedures

Because of their potent effects on motor responses, we strongly recommend not using muscle relaxants during motor mapping. We also prefer to avoid these agents in neuromonitoring procedures that rely heavily on monitoring of mMEPs (i.e. thoracoabdominal aneurysm repairs). We recommend using them cautiously in cases of preexisting (baseline) muscle weakness or in the presence of other variables that can have a substantial impact on the motor responses (other anesthetics with significant effect on MEPs, significant blood loss, etc.).

Another important clinical consideration is that reliable monitoring of muscle responses in the setting of an iatrogenic partial neuromuscular block necessitates avoidance of boluses of muscle relaxants and the use of a rather constant infusion rate. It can be cumbersome to use boluses of muscle relaxant at induction or intubation, especially when this is done before the final positioning of the patient. In such cases, the iatrogenic muscle relaxation interferes with obtaining reliable postposition mMEPs. For example, a bolus of 10 mg of cisatracurium may require a 30-minute to one-hour wait for its effect to wear off and before reliable muscle responses can again be elicited. The time needed for the effect of muscle relaxant to wear off depends on the baseline strength of the muscles, as well as on the other factors that may contribute to the loss of MEPs, such as other anesthetic agents (see also pages 185–87). Comparison between pre- and postposition MEPs becomes important in cases of unstable spine, when loss of MEPs can occur with inappropriate positioning (see pages 210–11). In such instances, a good coordination between the time when the baselines are taken and the time of administration of the muscle relaxant and other potentially potent anesthetics becomes important.

Whereas their effect on motor responses is not desirable, muscle relaxants can be beneficial in other neurophysiologic recordings, such as D wave recordings, nerve-to-nerve, spine-to-nerve, spine-to-spine, as well as scalp SSEPs recordings, by decreasing the muscle artifact and so increasing the signal-to-noise ratio. Thus, in circumstances when the SSEPs are of small amplitude and difficult to obtain, inducing a partial neuron-muscular blockage can be very helpful.

## XI. EFFECTS ON PERIPHERAL NERVOUS SYSTEM AND CRANIAL NERVE MONITORING

The use of all of the previously mentioned anesthetics is relatively safe in monitoring of the peripheral nervous system during surgical procedures requiring nerve-to-nerve or nerve-to-muscle recordings (nerve action potentials [NAPs] or CMAPs) such as nerve repair (nerve tumors, traumatic nerve damage) or lumbar and cervical spine cases involving root stimulation and recording from the innervated muscles.

However, **muscle relaxants** affect the CMAPs that are triggered by peripheral nerve stimulation (distal to the alpha motor neuron), as well as by stimulation of the peripheral part of the cranial nerves, and thus should be avoided in procedures requiring such neurophysiologic studies, even when only partial neuromuscular block is obtained. We make this recommendation keeping in mind that, in many cases, the motor nerve roots or the motor nerves of interest are damaged, thus requiring unexpected thresholds for stimulation (e.g. chronic radiculopathies, nerve tumors, traumatic damage with reinnervation). Thus, the presence of a partial neuromuscular block would only add another variable, precluding an objective assessment of the electrical transmission through the damaged nervous tissue. Appreciation of the intraoperative irritation of a nerve, by assessing the spontaneous activity in the innervated muscle (free EMG) can also become altered. On the other hand, when NAPs need to be obtained, muscle relaxants can be beneficial, by reducing the muscle artifact in the recordings.

Another potential exception is represented by **halogenated agents**, which were found to affect the axonal transmission through myelinated neurons (Burke et al., 2000). However, our experience shows that such peripheral recordings remain reliable during administration of an inhalational agent.

At high doses, intrathecally administered **ketamine** may decrease the triggered amplitudes of CMAPs as a result of axonal conduction block.

Spontaneous muscle activity can be reduced by spinal administration of the **fentanyl or morphine**, thus improving the triggered CMAPs recordings.

Brainstem auditory-evoked potentials, representing subcortical potentials, remain reliable during the use of the previously discussed anesthetics. **Muscle relaxants** can be beneficial in reducing the noise, in particular in cases of abnormal waveforms at baseline (e.g. acoustic neuromas); see page 321.

## REFERENCES


Akcaboy ZN, Akcaboy EY, Yigitbasł B, Bayam G, Dikmen B, Gogus N, Dilbaz N. Effects of remifentanil and alfentanil on seizure duration, stimulus amplitudes and recovery parameters during ECT. Acta Anaesthesiol Scand 2005;49(8):1068–1071.

Anschel DJ, Aherne A, Soto RG, et al. Successful intraoperative spinal cord monitoring during scoliosis surgery using a total intravenous anesthetic regimen including dexmedetomidine. J Clin Neurophysiol 2008;25(1):56–61.

Blair EA, Teeple E Jr, Sutherland RM, et al. Effect of neuromuscular blockade on facial nerve monitoring. Am J Otol 1994;15(2):161–167.

Bond A, Lader M, Shrotriya R. Comparative effects of a repeated dose regime of diazepam and buspirone on subjective ratings, psychological tests and the EEG. Eur J Clin Pharmacol 1983;24(4):463–467.

Brian JE Jr, Seifen AB. Tonic-clonic activity after sufentanyl. Anesth Analg 1987;66(5):481.

Burke D, Bartley K, Woodforth IJ, et al. The effects of volatile anesthetics on the excitability of human corticospinal axons. Brain 2000;(pt5):992–1000.

Caffrey RR, Warren ML, Becker KE Jr. Neuromuscular blockade monitoring comparing the orbicularis oculi and adductor pollicis muscles. Anesthesiology 1986 Jul;65(1):95–97.

Carlsson C, Smith DS, Keykhah MM, et al. The effects of high-dose fentanyl on cerebral circulation and metabolism in rats. Anaesth 1982 Nov;57(5):375–380.

Cascino GD, So EL, Sharbrough FW, et al. Alfentanil-induced epileptiform activity in patients with partial epilepsy. J Clin Neurophysiol 1993;10(4):520–525.

Clark DL, Hosick EC, Adam N, et al. Neural effects of isoflurane (forane) in man. Anesthesiology 1973;39(3):261–270.

Dearden NM, McDowall DG. Comparison of etomidate and althesin in the reduction of increased intracranial pressure after head injury Br J Anaesth 1985;57(4):361–368.

Doenicke A, Löffler B, Kugler J, et al. Plasma concentration and EEG after various regimens of etomidate. Br J Anaesth 1982;54(4):393–400.

Eger EI 2nd, Stevens WC, Cromwell TH. The electroencephalogram in man anesthetized with forane. Anesthesiology 1971;35(5):504–508.

Fink M, Weinfeld RE, Schwartz MA, Conney AH. Blood levels and electroencephalographic effects of diazepam and bromazepam. Clin Pharmacol Ther 1976;20(2):184–191.

Fujimoto T, Nishiyama T, Hanaoka K. Seizure induced by a small dose of fentanyl. J Anesthesiol 2003;17:55–56.

Gallagher TJ, Galindo A, Richey ET Inhibition of seizure activity during enflurance anesthesia. Anesth Analg 1978;57(1):130–132.

Ghaly RF, Stone JL, Levy WJ, et al. The effect of neuroleptanalgesia (droperiodol-fentanyl) on motor potentials evoked by transcranial magnetic stimulation in the monkey. J Neurosurg Anesthesiol 1991 Jun;3(2):117–123.

Ghoneim MM, Yamada T. Etomidate: A clinical and electroencephalographic comparison with thiopental. Anesth Analg 1977 Jul–Aug; 56(4):479–485.

Glassman SD, Shields CB, Linden RD, et al. Anesthetic effects on motor evoked potentials in dogs. Spine 1993;18(8):1083–1089.

Greenbaum R, Cooper R, Hulme A, MacKintosh IP. The effects of induction of anesthesia on intracranial pressure. In: Recent Progress in Anaesthesiology and Resuscitation (Arias A, Ed.) 1975; Amsterdam: Excerpta Medica.

Haghighi SS. Influence of isoflurane anesthesia on motor evoked potentials elicited by transcortical, brainstem and spinal root stimulation. Neurol Res 1998;20(6):555–558.

Hentschke H, Schwarz C, Antkowiak B. Neocortex is the major target of sedative concentrations of volatile anesthetics: Strong depression of firing rates and increase of GABAA receptor-mediated inhibition. Eur J Neurosci 2005;21(1):93–102.

Hicks R, Burke D, Stephen J, et al. Corticospinal volleys evoked by electrical stimulation of human motor cortex after withdrawal of volatile anaesthetics. J Physiol 1992;456:393–404.

Holland NR, Lukaczyk TA, Riley LH 3rd, Kostuik JP. Higher electrical stimulus intensities are required to activate chronically compressed nerve roots. Implications for intraoperative electromyographic pedicle screw testing. Spine 1998 Jan 15; 23(2):224–227.

Iida H, Dohi S, Tanahashi T, et al. Spinal conduction block by intrathecal ketamine in dogs. Anesth Analg 1997;85(1): 106–110.

Joas TA, Stevens WC, Eger EI 2nd. Electroencephalographic seizure activity in dogs during anaesthesia. Br J Anaesth 1971;43(8):739–745.

Kalkman CJ, Drummond JC, Ribberink AA, et al. Effects of propofol, etomidate, midazolam, and fentanyl on motor evoked responses to transcranial electrical or magnetic stimulation in humans. Anesthesiology 1992 Apr; 76(4):502–509.

Kano T, Shimoji K. The effects of ketamine and neuroleptanalgesia on the evoked electrospinogram and electromyogram in man. Anesthesiology 1974 Mar; 40(3):241–246.

Keene DL, Roberts D, Splinter WM, et al. Alfentanil-mediated activation of epileptiform activity in the electrocorticogram during resection of epileptogenic foci. Can J Neurol Sci 1997;24(1):37–39.

Kiersey DK, Bickford RG, Faulconer A. Electroencephalographic patterns produced by thiopental sodium during surgical operations: Description and classification. Br J Anesth 1950;22:141–152.

Kochs E, Treede RD, Schulte am Esche J. Increase of somatosensorially evoked potentials during induction of anaesthesia with etomidate. Anaesthesist 1986;35:359–364.

Kofke WA, Snider MT, Young RS, Ramer JC. Prolonged low flow isoflurane anesthesia for status epilepticus. Anesthesiology 1985;62(5):653–656.

Koht A, Schütz W, Schmidt G, et al. Effects of etomidate, midazolam, and thiopental on median nerve somatosensory evoked potentials and the additive effects of fentanyl and nitrous oxide. Anesth Analg 1988;67(5):435–441

Lebowitz MH, Blitt CD, Dillon JB. Enflurane-induced central nervous system excitation and its relation to carbon dioxide tension. Anesth Analg 1972;51(3):355–363.

Loughnan BA, Anderson SK, Hetreed MA, et al. Effects of halothane on motor evoked potentials recorded in the extradural space. Anesthesiology 1989;63:561–564.

Lumenta CB. Effect of etomidate on motor evoked potentials in monkeys. Neurosurgery 1991;29(3):480–482.

Mahmoud M, Sadhasivam S, Sestokas AK, et al. Loss of transcranial electric motor evoked potentials during pediatric spine surgery with dexmedetomidine. Anesthesiology. 2007;106(2):393–396.

Martella G, De Persis C, Bonsi P, et al. Inhibition of persistent sodium current fraction and voltage-gated L-type calcium by propofol in cortical neurons: Implications for its antiepileptic activity. Epilepsia 2005;46(5):624–635.

Mavroudakis N, Vandesteene A, Brunko E, et al. Spinal and brainstem SEPs and H reflex during enflurane anesthesia. Electroencephalogr Clin Neurophysiol 1994;92:82–85.

McPherson RW, Sell B, Traystman RJ. Effects of thiopental, fentanyl, and etomidate on upper extremity somatosensory evoked potentials in humans. Anesthesiology 1986; 65(6):584–589.

Minahan RE, Riley LH 3rd, Lukaczyk T, et al. The effect of neuromuscular blockade on pedicle screw stimulation thresholds. Spine 2000;25(19):2526–2530.

Mirski MA, Rossell LA, McPherson RW, Traystman RJ. Dexmedetomidine decreases seizure threshold in a rat model of experimental generalized epilepsy.

Molbegott LP, Flashburg MH, Karasic HL, Karlin BL. Probable seizures after sufentanyl. Anesth Analg 1987;66(1):91–93.

Niejadlik K, Galindo A. Electroencephaolographic seizure activity during enflurane anesthesia. Anesth Analg 1975; 54:722–725.

Nielsen J, Krøigaard M. Seizures in a 77-year-old-woman after a bolus dose of remifentanil. Acta Anaesthesiol Scand 2004;48(2):253–254.

Oda Y, Toriyama S, Tanaka K, et al. The effect of dexmedetomidine on electrocorticography in patients with temporal lobe epilepsy under sevoflurane anesthesia. Anesth Analg. 2007;105(5):1272–1277.

Osawa M, Shingu K, Murakawa M, et al. Effects of sevoflurane on central nervous system electrical activity in cats. Anesth Analg 1994;79(1):52–57.

Oshima E, Shingu K, Mori K. EEG activity during halothane anaesthesia in man. Br J Anaesth 1981;53(1):65–72.

Pathak KS, Brown RH, Cascorbi HF, et al. Effects of fentanyl and morphine on intraoperative somatosensory cortical-evoked potentials.Anesth Analg 1984;63(9):833–837.

Pathak KS, Ammadio M, Kalamchi A, et al. Effects of halothane, enflurane, and isoflurane on somatosensory evoked potentials during nitrous oxide anesthesia. Anesthesiology 1987; 66(6):753–757.

Persson A, Peterson E, Wåhlin A. EEG-changes during general anaesthesia with enflurane (Efrane) in comparison with ether. Acta Anaesthesiol Scand 1978;22(4):339–348.

Pierce EC Jr, Lambertsen CJ, Deutsch S, et al. Cerebral circulation and metabolism during thiopental anesthesia and hyperventilation in man. J Clin Invest 1962;41:1664–1671.

Pinaud M, Lelausque JN, Chetanneau A, et al. Effects of propofol on cerebral hemodynamics and metabolism in patients with brain trauma. Anesthesiology 1990;73(3):404–409.

Porkkala T, Jäntti V, Kaukinen S, Häkkinen V. Nitrous oxide has different effects on the EEG and somatosensory evoked potentials during isoflurane anaesthesia in patients. Acta Anaesthesiol Scand 1997;41(4):497–501.

Poulton TJ, Ellingson RJ. Seizure associated with induction of anesthesia with isoflurane. Anesthesiology 1984; 61(4):471–476.

Ragazzo PC, Galanopoulou AS. Alfentanil-induced activation: A promising tool in the presurgical evaluation of temporal lobe epilepsy patients. Brain Res Rev 2000;32(1):316–327.

Rampil IJ. Electroencephalogram. In Textbook of Neuroanesthesia with Neurosurgical and Neuroscience Perspectives 1997; Chapter 6, pp 193–220. New York: McGraw-Hill.

Ross J, Kearse LA Jr, Barlow MK, et al. Alfentanil-induced epileptiform activity: A simultaneous surface and depth electroencephalographic study in complex partial epilepsy. Epilepsia 2001;42(2):220–225.

Scheepstra GL, de Lange JJ, Booij LH, Ros HH. Median nerve evoked potentials during propofol anaesthesia. Br J Anaesth 1989 Jan;62(1):92–94.

Schönle PW, Isenberg C, Crozier TA, et al. Changes of transcranially evoked motor responses in man by midazolam, a short-acting benzodiazepine. Neurosci Lett 1989 Jul 3; 101(3):321–324.

Schubert A, Licina MG, Lineberry PJ. The effect of ketamine on human somatosensory evoked potentials and its modification by nitrous oxide. Anesthesiology 1990;72(1):33–39.

Sebel PS, Bovill JG, Wauquier A, Rog P. Effects of high-dose fentanyl anesthesia on the electroencephalogram. Anesthesiology 1981;55(3):203–211.

Shapiro HM, Galindo A, Wyte SR, Harris AB. Rapid intraoperative reduction of intracranial pressure with thiopentone. Br J Anaesth 1973;81(5):798–803; discussion 797.

Sloan T, Levin D. Etomidate amplifies and depresses transcranial motor evoked potentials in the monkey. J Neurosurg Anesth 1993;5:299.

Sloan TB. Intraoperative neurophysiology and anesthesia management. In: Neurophysiology in Neurosurgery (Deletis V, Shils J, Eds.) 2002; part VII, pp. 451–474. San Diego, CA: Academic Press.

Sloan TB, Fugina ML, Toleikis JR. Effects of midazolam on median nerve somatosensory evoked potentials. Br J Anaesth 1990 May;64(5):590–593.

Sloan TB, Ronai AK, Toleikis JR, Koht A. Improvement of intraoperative somatosensory evoked potentials by etomidate. Anesth Analg 1988;67(6):582–585.

Smith DL, Angst MS, Brock-Utne JG, DeBattista C. Seizure duration with remifentanil/methohexital vs. methohexital alone in middle-aged patients undergoing electroconvulsive therapy. Acta Anaesthesiol Scand 2003;47(9):1064–1066.

Smith NT, Dec-Silver H, Sanford TJ Jr, et al. EEGs during high-dose fentanyl-, sufentanyl-, or morphine-oxygen anesthesia. Anesth Analg 1984;63(4):386–393.

Souter MJ, Rozet I, Ojemann JG, et al. Dexmedetomidine sedation during awake craniotomy for seizure resection: effects on electrocorticography. J Neurosurg Anesthesiol 2007; 19(1):38–44.

Tempelhoff R, Modica PA, Bernardo KL, Edwards I. Fentanyl-induced electrocorticographic seizures in patients with complex partial epilepsy. J Neurosurg 1992;77(2):201–208.

Tobias JD. Dexmedetomidine: applications in pediatric critical care and pediatric anesthesiology. Pediatr Crit Care Med 2007;8(2):115–131. Review.

Tobias JD, Goble TJ, Bates G, et al. Effects of dexmedetomidine on intraoperative motor and somatosensory evoked potential monitoring during spinal surgery in adolescents. Paediatr Anaesth 2008;18(11):1082–1088.

Vandesteene A, Trempont V, Engelman E, et al. Effect of propofol on cerebral blood flow and metabolism on man. Anaesthesia 1988;43 Suppl: 42–43.

Van Hemelrijck J, Fitch W, Mattheussen M, et al. Effect of propofol on cerebral circulation and autoregulation in the baboon. Anesth Analg 1990 Jul;71(1):49–54.

Wauquier A. Profile of etomidate: A hypnotic, anticonvulsant and brain protective compound. Anaesthesia 1983; 38 Suppl: 26–33. Review.

Werner C, Hoffman WE, Kochs E, et al. The effects of propofol on cerebral blood flow in correlation to cerebral blood flow velocity in dogs. J Neurosurg Anesthesiol 1992;4(1):41–46.

Yamamura T, Fukuda M, Takeya H, et al. Fast oscillatory EEG activity induced by analgesic concentrations of nitrous oxide in man. Anesth Analg 1981;60(5):283–288.

Yang LH, Lin SM, Lee WY, Liu CC. Intraoperative transcranial electrical motor evoked potential monitoring during spinal surgery under intravenous ketamine or etomidate anaesthesia. Acta Neurochir (Wien) 1994;127(3–4):191–198.

Zentner J, Albrecht T, Heuser D. Influence of halothane, enflurane, and isoflurane on motor evoked potentials. Neurosurgery 1992;32:298–305.

# INDEX

Note: A page number followed by f indicates a figure; t indicates a table.